AF344321

Determining mycotoxins and mycotoxigenic fungi in food and feed

Related titles:

Mycotoxins in food: detection and control
(ISBN 978-85573-733-4)
Mycotoxins are toxic compounds produced by fungi. They are produced in foods of plant origin and pose a significant contamination risk in cereal and other foods. With its distinguished editors and international team of contributors, *Mycotoxins in food* summarises the wealth of recent research on how to assess the risks from mycotoxins, detect particular mycotoxins and control them at differing stages in the supply chain.

Detecting allergens in food
(ISBN 978-1-85573-728-0)
Allergens pose a serious risk to consumers, making effective detection methods a priority for the food industry. Bringing together key experts in the field, this important collection both reviews the range of analytical techniques available and their use to detect specific allergens such as nuts, dairy and wheat products. The first part of the book discusses methods of detection such as the use of antibodies and ELISA techniques. The second part reviews techniques for detecting particular allergens, whilst the final part explores how detection methods can be most effectively applied.

Foodborne pathogens: hazards, risk analysis and control (2nd edition)
(ISBN 978-1-84569-362-6)
Effective control of pathogens continues to be of great importance to the food industry. The first edition of *Foodborne pathogens* quickly established itself as an essential guide for all those involved in the management of microbiological hazards at any stage in the food production chain. This major new edition strengthens that reputation, with extensively revised and expanded coverage, including more than ten new chapters. Part I focuses on risk assessment and management in the food chain. Chapters in this section cover pathogen detection, microbial modelling the risk assessment procedure, pathogen control in primary production, hygienic design and sanitation, among other topics. Parts II and III then review the management of key bacterial and non-bacterial foodborne pathogens.

Details of these books and a complete list of Woodhead's titles can be obtained by:

- visiting our web site at www.woodheadpublishing.com
- contacting Customer Services (e-mail: sales@woodheadpublishing.com; fax: +44 (0) 1223 832819; tel.: +44 (0) 1223 499140 ext. 130; address: Woodhead Publishing Limited, 80 High Street, Sawston, Cambridge CB22 3HJ, UK)

If you would like to receive information on forthcoming titles, please send your address details to: Francis Dodds (address, tel. and fax as above; e-mail: francis.dodds@woodhead publishing.com). Please confirm which subject areas you are interested in.

Woodhead Publishing Series in Food Science, Technology and Nutrition: Number 203

Determining mycotoxins and mycotoxigenic fungi in food and feed

Edited by
Sarah De Saeger

WOODHEAD
PUBLISHING

Oxford   Cambridge   Philadelphia   New Delhi

Published by Woodhead Publishing Limited,
80 High Street, Sawston, Cambridge CB22 3HJ, UK
www.woodheadpublishing.com

Woodhead Publishing, 1518 Walnut Street, Suite 1100, Philadelphia,
PA 19102-3406, USA

Woodhead Publishing India Private Limited, G-2, Vardaan House,
7/28 Ansari Road, Daryaganj, New Delhi – 110002, India
www.woodheadpublishingindia.com

First published 2011, Woodhead Publishing Limited
© Woodhead Publishing Limited, 2011
The authors have asserted their moral rights.

British Library Cataloguing in Publication Data
A catalogue record for this book is available from the British Library.

ISBN 978-1-84569-674-0 (print)
ISBN 978-0-85709-097-3 (online)
ISSN 2042-8049 Woodhead Publishing Series in Food Science, Technology and Nutrition (print)
ISSN 2042-8057 Woodhead Publishing Series in Food Science, Technology and Nutrition (online)

The publisher's policy is to use permanent paper from mills that operate a sustainable forestry policy, and which has been manufactured from pulp which is processed using acid-free and elemental chlorine-free practices. Furthermore, the publisher ensures that the text paper and cover board used have met acceptable environmental accreditation standards.

Cover image courtesy of I E Tothill, © Cranfield University, UK

Typeset by Ann Buchan (Typesetters), Middlesex, UK

Contents

Part I Determining mycotoxins in food and feed

*B. Maestroni and A. Cannavan, Joint FAO/IAEA Division of
Nuclear Techniques in Food and Agriculture, International Atomic
Energy Agency (IAEA), Austria*

Part III   Development and analysis of biomarkers for mycotoxins

8   Developing biomarkers of human exposure to mycotoxins **225**
M. N. Routledge and Y. Y. Gong, University of Leeds, UK

**9   Developing mechanism-based and exposure biomarkers for mycotoxins
      in animals** **245**
*R. T. Riley and K. A. Voss, United States Department of Agriculture,
Agricultural Research Service, USA, R. A. Coulombe, Department of
Veterinary Sciences, Utah State University, USA, J. J. Pestka,
Michigan State University, USA and D. E. Williams, Oregon State
University, USA*

Part IV   Determining mycotoxigenic fungi in food and feed

**10   Rationale for a polyphasic approach in the identification of
      mycotoxigenic fungi** **279**
J. C. Frisvad, Technical University of Denmark, Denmark

Part V   Emerging methods for mycotoxin analysis in food and feed

Contributor contact details

(* = main contact)

Editor

S. De Saeger
Laboratory of Food Analysis
Faculty of Pharmaceutical Sciences
Ghent University
Harelbekestraat 72
9000 Ghent
Belgium

Email: sarah.desaeger@ugent.be

Chapter 1

B. Maestroni* and A. Cannavan
Food and Environmental Protection
    Laboratory
FAO/IAEA Agriculture and
    Biotechnology Laboratories
Joint FAO/IAEA Division of Nuclear
    Techniques in Food and
    Agriculture
Wagramerstrasse 5, P. O. Box 100
A-1400 Vienna
Austria

Email: b.m.maestroni@iaea.org

Chapter 2

E. Razzazi-Fazeli* and E. V. Reiter

Institute of Animal Nutrition
University of Veterinary Medicine
Vienna
Austria

Email:
    ebrahim.razzazi@vetmeduni.ac.at

Chapter 3

G. S. Shephard
PROMEC Unit
Medical Research Council
P.O. Box 19070
Tygerberg 7505
South Africa

Email: gordon.shephard@mrc.ac.za

Chapter 4

M. C. Spanjer
Food and Consumer Product Safety
    Authority
Croesestraat 77a
3522 AD Utrecht
The Netherlands

Email: martien.spanjer@vwa.nl

Chapter 5

I. Goryacheva*
Saratov State University
Astrakhanskaya 83
410012 Saratov ul.
Russia

Email: goryachevaiy@mail.ru

S. De Saeger
Laboratory of Food Analysis
Faculty of Pharmaceutical Sciences
Ghent University
Harelbekestraat 72
9000 Ghent
Belgium

Email: sarah.desaeger@ugent.be

Chapter 6

H. Z. Senyuva* and J. Gilbert
FoodLife International Ltd
ODTU Teknokent Ikizler Binasi
Zemin Kat No: Ara-1 Çankaya 06531
Ankara
Turkey

Email:
    hamide.senyuva@foodlifeint.com
    john.gilbert@foodlifeint.com

Chapter 7

J. O. De Beer*
Scientific Institute of Public Health
Juliette Wytmanstreet 14
B1050 Brussels
Belgium

Email: jacques.debeer@wiv-isp.be

C. Van Poucke
Laboratory of Food Analysis
Faculty of Pharmaceutical Sciences
Ghent University
Harelbekestraat 72
9000 Ghent
Belgium

Email: christof.vanpoucke@ugent.be

Chapter 8

M. N. Routledge and Y. Y. Gong*
Molecular Epidemiology Unit
Leeds Institute of Genetics, Health
    and Therapeutics
University of Leeds
Leeds LS2 9JT

Email: medyg@leeds.ac.uk
        y.gong@leeds.ac.uk

Chapter 9

R. T. Riley* and K. A. Voss
Toxicology and Mycotoxin Research
    Unit
United States Department of Agricul-
    ture, Agricultural Research Service
    (USDA-ARS)
Athens, GA
USA

Email: Ron.riley@ars.usda.gov

R. A. Coulombe, Jr.
Department of Veterinary Sciences
Utah State University
Logan, UT
USA

J. J. Pestka
Department of Food Science and
    Human Nutrition
Michigan State University
East Lansing, MI
USA

D. E. Williams
Department of Environment and
    Molecular Toxicology
Oregon State University
Corvallis, OR
USA

Chapter 10

J. C. Frisvad
Department of Systems Biology
Center for Microbial Biotechnology
Technical University of Denmark
Søltofts Plads
Building 221, room 204
2800 Kongens Lyngby
Denmark

Email: jcf@bio.dtu.dk

Chapter 11

A. Moretti*
Institute of Sciences of Food Produc-
    tion (ISPA)
National Research Council (CNR)
Via Amendola 122/D
70126 Bari
Italy

Email: antonio.moretti@ispa.cnr.it

F. Munaut and F. Van Hove
Mycothèque de l'Université catho-
    lique de Louvain (BCCM™/MUCL)
Applied Microbiology, Earth and Life
    Institute
Université catholique de Louvain
Croix du Sud 3 box 6
1348 Louvain-la-Neuve
Belgium

Email:
    francoise.munaut@uclouvain.be
    francois.vanhove@uclouvain.be

Chapter 12

D. W. Brown,* R. A. E. Butchko and
    R. H. Proctor
Bacterial Foodborne Pathogens and
    Mycology Research
United States Department of Agricul-
    ture, Agricultural Research Service
    (USDA ARS)
1815 N. University St.
Peoria
Illinois 61604
USA

Email: daren.brown@ars.usda.gov

Chapter 13

M. Santamaria
Institute for Biomedical Technologies
National Research Council (CNR)
Via Amendola 122/D
70126 Bari
Italy

Email:
    monica.santamaria@ba.itb.cnr.it

Chapter 14

I. E. Tothill
Advanced Diagnostics and Sensors
    Group
Cranfield Health
Cranfield University
Cranfield
Bedfordshire
MK43 0AL

Email: i.tothill@cranfield.ac.uk

Chapter 15

J. Diana Di Mavungu* and S. De
    Saeger
Laboratory of Food Analysis
Faculty of Pharmaceutical Sciences
Ghent University
Harelbekestraat 72
9000 Ghent
Belgium

Email:
    jose.dianadimavungu@ugent.be
    sarah.desaeger@ugent.be

Chapter 16

C. B. Singh and D. S. Jayas*
Biosystems Engineering
University of Manitoba
Winnipeg, MB
R3T 5V6
Canada

Email: digvir_jayas@umanitoba.ca

Woodhead Publishing Series in Food Science, Technology and Nutrition

Preface

Over the last few decades it has become increasingly clear that mycotoxins play a significant role in food and feed safety. Indeed, mycotoxins have been shown to be the principal threat regarding chronic toxicity. Legislative limits for a range of mycotoxins continue to develop worldwide resulting in an increased number of official controls deriving from national food safety plans and for food trade purposes. This book therefore focuses on recent developments in the determination of mycotoxins and mycotoxigenic fungi in food and feed.

A mycotoxin test procedure is a multi-stage process generally consisting of three steps: sampling, sample preparation and analytical determination. The sampling phase is the largest source of variability of the test procedure. The official sampling protocols are still complicated and very challenging in practical terms. Further extensive research on sampling plans is mandatory, taking into account the real risk to human health together with the economic perspective. New developments in sample preparation focus on faster, environmentally friendly, cost effective and fit-for-purpose extraction and clean-up methods in food, feed, biological tissue and bodily fluids. Screening immunochemical and confirmatory chromatographic analytical methods are widely used; a clear trend towards multi-mycotoxin analysis and more precisely towards LC–MS/MS has been noticed.

Quality assurance in mycotoxin analysis is of the utmost importance. Notwithstanding the general acceptance of the benefits of adopting a performance criteria-based approach, some countries have a regulatory framework which requires the publication of 'official methods' in their own regulations. Food control laboratories should continuously follow actual progress in analysis development and statistical method validation within an accredited quality environment such as prescribed in the ISO 17025 norm. Further attention towards a harmonized method validation procedure is necessary.

In order to understand possible links between mycotoxins and human disease or animal disease outbreaks, it is necessary to measure the exposure to the toxin in question. Advances in analytical techniques have resulted in the development and use of various biological markers (biomarkers) which allow more accurate and objective assessment of exposure at the individual level. The development and determination of validated exposure as well as mechanism-based biomarkers is critical to reduce the existing uncertainty in the risk assessment of most mycotoxins.

Fungal isolates involved in mycotoxicoses are preferably identified by a polyphasic approach in order to avoid mistakes, starting at genus level and further to species level using a combination of morphological, physiological, nutritional and chemical data. The identification is validated by PCR-based molecular methods which can be considered under two main complementary approaches: by targeting conserved functional genes or regions of taxonomical interest, or by focusing on the mycotoxigenic genes. The possibility of using a highly standardized, rapid and practical DNA barcoding protocol that can be easily used both by researchers involved in species definition studies and by non-experts for practical uses is currently investigated. However, in order to assess the risks related to the presence of mycotoxigenic fungi in food and feedstuffs reliably, one should also investigate whether or not the mycotoxin genes are expressed. Further progress in transcriptomics, proteomics and metabolomics will continue to advance the understanding of fungal secondary metabolism, providing insight into how to reduce mycotoxin contamination of crop plants and the food/feed derived therefrom.

Fungal secondary metabolites, mycotoxins and food safety will continue to be of critical interest to a variety of researchers for years to come. Innovations take place at a rapid pace, for example through new nanotechnology-based biosensing techniques and non-destructive spectroscopic techniques. Furthermore, the discovery of masked mycotoxins and the inherent analytical challenges will be the subject of future research.

Sarah De Saeger

Part I

Determining mycotoxins in food and feed

1

Sampling strategies to control mycotoxins

B. Maestroni and A. Cannavan, Joint FAO/IAEA Division of
Nuclear Techniques in Food and Agriculture, International Atomic
Energy Agency (IAEA), Austria

Abstract: Mycotoxins can have adverse effects on human and animal health, productivity, economics and trade. Efficient and cost-effective sampling protocols and analytical tools and methods are needed for the detection and control of mycotoxins worldwide. Effective testing schemes depend on sound analytical methods and on sampling plans that generate results that reflect the actual concentrations present in consignments or lots of produce. Test results can be used to implement regulatory decisions on the suitability of lots of food for consumption or trade. Several studies have been conducted to gain knowledge on the variability of mycotoxins, and enabling the establishment of sampling plans for the control of mycotoxins in several commodities. Some official sampling protocols for the control of mycotoxins in food and feed are discussed, including those formulated by the European Commission for several mycotoxins and by Codex Alimentarius for aflatoxins in peanuts, corn and treenuts. Even when using accepted methods or protocols, there are uncertainties associated with the mycotoxin test procedure. This chapter describes options to reduce the total variability associated with a mycotoxin test procedure, and discusses the design and the performance of sampling plans. Producing safe and good quality food is a prerequisite to ensuring consumer health and successful domestic and international trade, and a key to the sustainable development of national agricultural resources. Therefore, a holistic approach for the control of mycotoxins, which includes the adoption of the best agricultural practices in the field and throughout the whole farm-to-fork chain, the best sampling practices, the use of validated and fit-for-purpose methods, trained professionals, and participation in integrated food control systems is important.

Key words: food control systems, operating characteristics curves, sampling plans, variability, uncertainty.

Objectives
The aim of this chapter is to introduce the reader to the basic issues related to sampling food for mycotoxins and the need for sound sampling plans. The reader is strongly advised to consult the reading guide at the end of this chapter to view the relevant available literature and selected web resources. The authors hope that the information provided in this chapter will stimulate further research into appropriate sampling methods and approaches.

1.1   Food safety and the requirements for international food trade

Agriculture is a dominant component of the global economy and pressure to produce abundant, available and safe high quality foods for the world's ever growing population has had a worldwide impact on agricultural practices. The ability both to detect contaminated food products and to trace their origin is of major concern to regulatory authorities, trading partners and the food industry owing to the rapid increase in cross-border trading of food commodities. The occurrence of mycotoxins in foods can have profound implications especially for developing countries, including health and economic impacts due to losses in grain and other staple foods and diminished animal production. The Food and Agriculture Organization of the United Nations (FAO) has estimated that 25% of crops worldwide are contaminated with mycotoxins and that the food losses amount to one billion tonnes per year. As a result, food security is challenged and trade is hampered. To ensure the safety of food many countries throughout the world have established effective food control systems (FAO, 2007). The problem of aflatoxin residues in food is mainly an issue for developing countries and can best be addressed by targeted international assistance, as advocated by Wagacha and Muthomi (2008) for mycotoxin problems in Africa. The socio-economic status of the majority of inhabitants of sub-Saharan Africa predisposes them to consume mycotoxin-contaminated products either directly, or in processed food at various points of the food chain. Mycotoxins in food affect human and animal health, productivity, the economy and trade. Clearly there is a need for efficient and cost-effective sampling protocols and analytical tools and methods that can be used for the detection and control of mycotoxins worldwide.

It is recognized that international trade in food plays an increasingly important role in achieving food security for many countries. One of the benefits is the availability of a broader choice of nutritious foods for the consumer. Food trade provides exporting countries with foreign exchange, contributing to economic development and an improvement in standards of living (WHO, 2002). FAO has worked for many years on capacity building of food control systems (FAO, 2006) as an effective way to demonstrate the equivalence of food safety systems and food quality, thereby promoting trade. Since much of the food for the developed world is produced by developing countries, it is important that the developed world shares the responsibility to ensure that effective farm to fork food safety systems are in place.

To prevent mycotoxins from entering the food production chain, controls should preferably be applied at the raw materials stage. It is of vital importance that consignments of food are tested to establish that they are 'fit for purpose'. Effective testing schemes depend both on sound analytical methods and on sampling plans designed in such a way that the results generated from analysis of the test samples reflect the actual concentrations present in consignments or lots of produce. The test results can then be used to implement regulatory decisions on the suitability of lots of food for consumption or trade (Whitaker *et al.*, 2007b). The underpinning requirement is that the sampling plan adopted to acquire information about the mycotoxin contamination is truly reliable and representative.

1.2   Principles of food and feed sampling for mycotoxin analysis

1.2.1   Sampling food and feed for mycotoxin analysis

The impacts of mycotoxin contamination of agricultural commodities on human and animal health as well as on domestic and international trade are increasingly recognized in both developed and developing countries.

In general, developed countries have enacted regulatory limits to protect consumers from exposure to mycotoxins. In many developing countries, however, regulation is insufficient and certain agricultural commodities, including dietary staple foods, can contain unacceptably high levels of mycotoxins (Whitaker *et al.*, 2009). In these countries, maximum limits should be set at a level which is reasonably achievable by following good agricultural and manufacturing practices and is consistent with consumer protection, considering the risk related to the consumption of the food (Cheli *et al.*, 2009). Van Egmond provided an excellent compilation of the regulations worldwide relating to mycotoxins (2002, 2007).

The analysis of mycotoxins requires not only validated, reliable analytical methods, and regulatory limits against which the analytical result is compared, but also validated sampling methods that are representative and practicable (easy to apply, quick and cost effective). This is especially important for trade goods or goods that are moved in large quantities (Stroka *et al.*, 2004).

In regulatory control, it is important to be able to estimate as accurately as possible the true levels of a mycotoxin in a commodity so that correct decisions can be made about its suitability for consumption. This can only be achieved through the collection of truly representative samples, which requires carefully designed sampling plans. The consequences of using a poorly or inappropriately designed or implemented sampling plan can include health issues, trade rejections, false information for risk assessors and managers and litigation problems.

The adoption of well designed sampling plans and the early detection of mycotoxin-contaminated lots is essential in the food industry to ensure that mycotoxins are excluded from further processing/manufacturing stages. For most commodities, the production and marketing system acts as a mixer where many different lots are blended together during handling, storage and processing (Pittet, 1995). Issues such as lot traceability, raw material specifications, quality control and quality assurance programmes and training are important in detecting and segregating mycotoxin-contaminated materials at the farm or at the first point of marketing.

The overall objective of sampling is to provide representative samples for analysis, the results of which can be used as a basis for 'fitness for purpose' investigations (Miraglia *et al.*, 2005). It is important to understand that sampling plans may have different objectives. For example, an acceptable sampling plan for quality control purposes may be very different from a sampling plan for commodities at harvest. In general, sampling plans may be prepared for monitoring, which means conducting a planned sequence of observations or measurements with a view to obtaining an overview of the state of compliance with food law (EC 882/2004), for surveillance, which means a careful observation of one or more food businesses operators or their activities (EC 882/2004) and for targeted sampling. Monitoring is both a preliminary and a routinely performed activity and should be undertaken to protect the health of the population and to support trade. The number of samples to be collected for monitoring should be proportional to the food consumption rate and take into account the amount of domestic production and the amount of imports. Surveillance is undertaken whenever data from monitoring reveals that standard/legal values have been exceeded and it aims to provide a basis for centralized and qualified feedback (FAO, 2005). Targeted sampling is undertaken when there is a concrete suspicion that mycotoxins are present in excessive amounts, based on previous detection or a history of trade rejections. Targeted sampling focuses on specific sample populations which are either likely to be non-compliant, for example, goods produced or stored under bad conditions or food derived from animals showing clinical signs of intoxication, or are intended for more sensitive consumers such as babies or immunocompromized patients.

The Joint FAO/WHO Expert Committee on Food Additives has considered sampling requirements for the surveillance of mycotoxins (WHO, 2002). The committee noted that very little work has been done to address the need for specific sampling plans for surveys, which are the key to obtaining quality data for risk assessment studies. It was recommended that data for risk characterization should be obtained using effective and validated sampling protocols. The protocols should reflect the selection of the sampling sites within the food chain and geographically, also taking into consideration differences in agro-climatic conditions.

Even with well designed sampling plans, accurate estimation of the mycotoxin concentration in large quantities of bulk commodities is very difficult, owing to the large variability associated with the mycotoxin test procedure, which includes sampling, sample preparation, and analytical steps. Because of the inevitable errors associated with each step of the testing procedure, the mycotoxin concentration in a lot cannot be measured with absolute certainty and individual analytical results, as well as the estimated results for a lot, should always be reported with an estimate of the uncertainty.

1.2.2   Understanding variation

Fungal development and mycotoxin production are 'spot processes' and are significantly correlated to, for example, the type of crop, crop variety, agronomic practices, the weather conditions during growth and harvest, storage and processing conditions and the toxigenic potential of the different mould species.

The UK Home Grown Cereal Authority (HGCA, 2004) showed that grain quality can be extremely variable, not just between fields, but even within individual ears of grain. Grain quality will differ because of various factors, such as soil variations, geographical orientation of the field, sowing date, crop rotation, weather conditions at harvest, machinery used, moisture variation during the day and between trailer loads and variations between dried and undried grain.

Mycotoxins in cereals can originate either in the field during plant growth, or during storage. *Aspergillus* and *Penicillium* are mostly responsible for the production of aflatoxins (AF) and ochratoxin A (OTA), respectively, during storage. Mycotoxins produced under storage are often concentrated in 'hot spots' (Whitaker *et al.*, 1974; Whitaker, 2003; Whitaker and Johansson, 2005) as a result of a sudden fungal attack and can occur when the grain is stored for some time under optimal conditions for both the growth of the fungi and mycotoxin formation. For example, this could happen when moist grain is left for some time before it is passed through a hot-air dryer, with the highest risk of mycotoxin production being in the middle of the bulk lot, or when the grain is being dried from the bottom up; the highest mycotoxin risk in this case is in the undried grain on the top. This generally results in a heterogeneous distribution throughout a lot.

Fusarium species are mainly associated with mycotoxin production during plant growth in wet and cold weather conditions and can produce fumonisins and trichothecenes, for example, deoxynivalenol (DON) and/or nivalenol, and zearalenone. The distribution of *Fusarium* toxins is generally more homogeneous than the toxins produced during storage, being more likely to be attributable to mixing during handling and manipulation at harvest and further stratification during storage and transport. This is supported by the studies by Hart and Schabenberger (1998) and Biselli *et al.* (2008), which showed that DON was spread less heterogeneously than OTA in truckloads of wheat.

Knowledge of the variability of mycotoxins is essential for the design of effective sampling plans. A number of papers have been published on the variability of aflatoxins and other mycotoxins in various commodities, including peanuts (Whitaker and Wiser, 1969; Whitaker *et al.*,1994, 1999), raw shelled peanuts (Whitaker *et al.*, 1970, 1972, 1974, 1979, 1994, 1995, 1996a; Vandeven *et al.*, 2002), corn (Whitaker and Dickens, 1983; Whitaker *et al.*, 1979 1998, 2001, 2007b; Shotwell *et al.*, 1974; Johansson *et al.*, 2000a), green coffee (Vargas *et al.*, 2004, 2005; Whitaker *et al.*, 2004; Whitaker and Johansson, 2005a), pistachios (Shatzki, 1995a, 1995b), almonds (Whitaker *et al.*, 2006, 2007a), figs (Sharman *et al.*, 1994), hazelnuts (Ozay *et al.*, 2006) and ginger (Trucksess *et al.*, 2009).

It was shown that there was a large variability among the aflatoxin results for ten replicate samples from each of six lots of shelled peanuts, with the maximum result being, for some lots, four to five times the average lot concentration. The variability tended to decrease as the average lot concentration increased. The distribution of the ten laboratory sample results for each lot was positively skewed, meaning that more than half of the results were below the 'true' (mean) lot concentration. These observations are also generally true for other mycotoxins and other commodities (Whitaker and Park, 1993; Whitaker *et al.*, 1998, 2000; Hart and Schabenberger, 1998; Johansson *et al.*, 2000a; Cucullu *et al.*, 1986).

1.2.3   Theoretical distributions

In general, increasing the number of sample results improves the characterization of mycotoxin variability and facilitates more accurate mathematical modelling of contaminated food and feed. Several different mathematical models have been evaluated to describe the experimentally observed distributions of mycotoxins in contaminated corn and peanut lots (FAO, 1993). Four different theoretical distributions are generally considered; the lognormal, negative binomial, normal and compound gamma distributions. Except for the normal distribution, these theoretical distributions are positively skewed and have characteristics similar to the observed distributions of mycotoxins (Whitaker *et al.*, 1996b). The suitability of a theoretical distribution to accurately fit an observed distribution of sample test results is measured by a statistical goodness of fit (GOF) test, for example, the Kolmogorov–Smirnov (KS) test, the Chi-squared test or the power divergence (PD) test.

1.2.4   Sampling plan

A mycotoxin sampling plan may be defined as a mycotoxin test procedure that generates a test result, coupled to a defined acceptance/rejection limit, usually a regulatory limit, to which the test result is compared to check whether or not the lot meets the sanitary quality control criteria (Johansson *et al.*, 2000a).

A mycotoxin test procedure is a multi-stage process generally consisting of three steps: sampling from the target population, sample preparation and analysis (quantification). Sampling consists of all operations which, applied to a lot of an agricultural product, lead to an aggregate sample/laboratory sample. The sampling step specifies how the sample will be selected or taken from the lot and the size of the sample. Sample preparation is the process of grinding, homogenizing and sub-sampling in order to obtain an analytical portion (test portion), which is solvent-extracted and analysed using an approved and validated analytical procedure to quantify the mycotoxin concentration.

The mycotoxin concentration of a lot is usually estimated by measuring the mycotoxin concentration in a small representative sample taken from the lot (the laboratory sample). Based on the measured mycotoxin concentration in the laboratory sample, a decision is made about the quality of the lot. For example, in a regulatory environment, decisions will be made to classify the lot as acceptable or unacceptable based upon a comparison of the measured sample concentration to an accepted limit (for example, a legal limit, or a quality control level). If the sample concentration does not accurately reflect the lot concentration, the lot may be misclassified and there may be undesirable economic and/or health consequences. Sampling plans should be designed to minimize this possibility.

1.2.5 Sample selection

To try to overcome the problem caused by the heterogeneous distribution of mycotoxins in food and feed, the sample selected for analysis should be an accumulation of many small portions, called incremental samples, taken randomly from many different locations throughout the lot (Whitaker *et al.*, 1970; Whitaker and Dickens, 1983).

The incremental samples together form an aggregate sample. If the aggregate sample is larger than required for the laboratory sample, it should be blended and sub-divided until the desired laboratory sample size is achieved. This process of sub-sampling is critical and should be carefully conducted in order to ensure that the sample remains representative of the lot. Two frequent errors that can compromise the representativeness of the sample are taking too few incremental samples and taking incremental samples of inadequate mass (samples are too small).

Static sampling

When drawing an aggregate sample from a static container, a probing pattern should be developed so that product can be collected from different locations in the lot. An example of several probing patterns used by the United States Department of Agriculture (USDA) to collect aggregate samples from peanut lots is shown in Fig. 1.1. The ISO 24333:2009 standard (ISO, 2009) requires, in particular cases, eight incremental samples to be taken according to the pattern in Fig. 1.2.

According to Codex Alimentarius (FAO, 1993) the probes should be carefully selected on the basis of the type of container, since all the units should have the same chance of being selected. The ISO 24333:2009 standard gives examples, in Annex B, of the devices that can be used to sample static lots. Examples include manual concentric tapered sampling probes such as open shafts with several apertures, gravity type sampling probes with extension rods and T shaped handles, mechanical sampling devices such as suction or vacuum sampling devices and instruments used to take samples from sacks or bags, including Archimedes' screw sampling probes.

Dynamic sampling

Random sampling can be more nearly achieved when taking increments from a moving stream as the product is transferred from one location to another using, for example, a conveyor belt. The increments should be collected along the entire length and across the entire cross-section of the moving stream (see Fig. 1.3). At regular intervals the flowing stream of product can be diverted, for example from a hopper, into collection vessels. For very large lots this methodology is rather time consuming, since the procedure implies drawing samples at regular intervals of time which may be all night and day, with possible interruptions to the procedure.

1   Front

x	o	
		x
	x	o
o		x
x		

2   Front

	o	x
x		
o	x	
x		o
		x

3   Front

	x	
x	o	
		x
x		o
o	x	

4   Front

	x	
	o	x
x		
o		x
	x	o

5   Front

		x
o	x	
x		o
	x	
	o	x

6   Front

x		
	x	o
o		x
	x	
x	o	

7   Front

x		o
	x	
	o	x
x		
o	x	

8   Front

o		x
	x	
x	o	
		x
	x	o

9   Front

o		x
x		o
	x	
	o	x
x		

10   Front

x		o
o		x
	x	
x	o	
		x

11   Front

	x	
		x
x		
	x	
		x

12   Front

	x	
x		
		x
	x	
x		

x = 5 Probe patterns
x + 0 = 8 Probe patterns

Fig. 1.1   Example of several five- and eight-sampling probe patterns used by the United States Department of Agriculture to sample peanuts for grade.

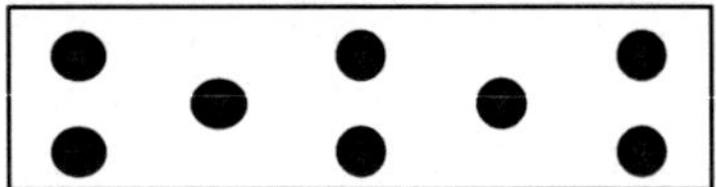

Fig. 1.2   Example of the eight-point probing pattern used for lots from 15t up to 30t according to ISO 24333:2009.

The deployment of automated sampling equipment, such as cross-cut samplers, can greatly assist the process (Codex Alimentarius, 2004).

Whether using automatic or manual methods, small increments of product should be collected and composited at frequent and uniform intervals throughout the entire time that the product flows past the sampling point. According to Pierre Gy (Pitard, 1993) it is important when sampling from a conveyor belt that the sides of the cutting device are strictly parallel and that the cutter traverses the entire stream at uniform speed, resulting in an equal representation of the entire width of the belt in the final sample. When sampling from a conveyor belt it is also important to respect the centre of gravity rule; that is, any particle having its centre of gravity inside the delimited incremental sample should be included in the increment. When using a cross-stream sampler, the top edges must be such that a

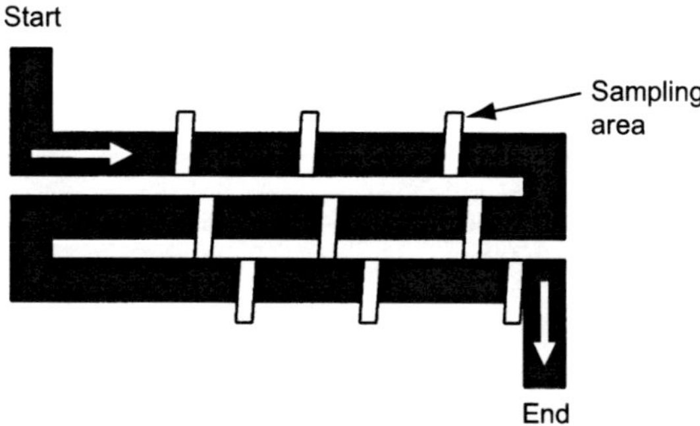

Fig. 1.3 Example of dynamic sampling from a conveyor belt.

particle hitting the device will either fall inside or outside depending on their centre of gravity (Petersen *et al.*, 2005).

When food lots are stored in bins/containers/silos where the access is limited, dynamic sampling as the product is either being put into or removed from the container is the recommended sampling method (FAO, 2001). The control of ambient conditions, in this case, is of utmost importance to avoid possible fungal attack and the production of mycotoxins.

1.2.6   Examples of established sampling plans

Examples of sampling protocols for the control of mycotoxins in food and feed are those formulated by the EU for several mycotoxins (EC 401/2006 and EC 152/ 2009), by Codex Alimentarius for aflatoxins in peanuts, corn and treenuts (Codex Alimentarius Commision,1995, revision, 2009), by ISO 24333:2009, and by the USDA for aflatoxins in several commodities (FDA, 2009). Some of these protocols are discussed below.

For mycotoxins other than aflatoxins, general principles, schemes and sampling plans adopted so far are mainly based on those for aflatoxins. More work is needed in this area, especially in the refinement of sampling plans according to the variability of certain mycotoxins in food commodities under specific agro-ecological production conditions and taking into account the real risk to human health and the economic perspective.

1.3   International guidance on sampling food and feed for mycotoxin analysis

1.3.1   European legislation

European Community regulations and decisions are directly applicable in the member states of the European Union. These legislative texts set limits for certain contaminants in foodstuffs, as well as detailing the official methods of control and

Table 1.1   Example of the sampling plan for cereals and cereal products

Lot weight T (tons)	Weight or number of sublots	Number of incremental samples	Aggregate sample weight (kg)
$T \geq 1500$	500 tons	100	10
$300 < T < 1500$	3 sublots	100	10
$50 < T \leq 300$	100 tons	100	10
$20 < T \leq 50$		100	10
$10 < T \leq 20$		60	6
$3 < T \leq 10$		40	4
$1 < T \leq 3$		20	2
$0.5 < T \leq 1$		10	1
$0.05 < T \leq 0.5$		5	1
$T \leq 0.05$		3	1

requirements for complying with food laws. The text with direct relevance to sampling is European Commission Regulation (EC) No 401/2006 (2006), laying down the methods of sampling and analysis for the official control of the levels of mycotoxins in foodstuffs. In this regulation, the European Commission has brought together, the sampling protocols and the performance criteria for the methods of analysis to be used for the official control of mycotoxins in foodstuffs, mainly for official local authorities and port health authorities. The official sampling protocols are very challenging in practical terms. The regulation states that an alternative sampling plan may be used, provided that it can be shown to be 'as representative as possible' and equivalent to the official plan. However, the directive lacks guidance about what is considered to be 'as representative as possible' (Van Egmond *et al.*, 2007).

Since there are significant differences in the distributions of mycotoxins in different commodities, the Regulation provides for different sampling plans according to the type of food product. An example of a sampling plan for cereals and cereal products is shown in Table 1.1. According to the Regulation, 100 incremental samples of 100 g should be taken from any lot of cereals exceeding 20 tonnes. In the case of pistachios, the aggregate sample weight is 30 kg, which represents a challenge because it is neither easy nor economical to transport aggregate samples of this size to the laboratory. Significant financial investments in terms of large-volume grinders or 'slurry' processing equipment are also required at the laboratory in order to handle such samples.

EC 401/2006 provides the first published guidance for sampling at retail, in terms of the number of incremental samples and the weight of the aggregate sample. The European sampling plan regulations have been shown to be impractical in some instances. Spanjer (2006, 2007) summarized some of the practical difficulties encountered; for example in the import control of treenuts packed in sacks, two food inspectors would need half a working day to sample only one container on just one ship.

1.3.2   The ISO 24333:2009 standard: cereals and cereal products: sampling

This standard specifies requirements for the dynamic or static sampling, by manual or mechanical means, of cereals and cereal products, to assess their quality and condition. The standard is applicable to sampling for the determination of heterogeneously distributed contaminants, for example, mycotoxins.

Annex A of the standard describes the general types of mechanical sampling devices used for dynamic lots. Annex B gives examples of the devices used to sample static lots and of sample dividers. It also contains a guide for the selection of the proper sampling tools for cereals and cereal products. The standard states that sampling should be carried out under dynamic conditions using mechanical devices. Only when this is not possible should a manual sampling plan be implemented. Dynamic sampling methods should be adapted to the speed at which the products are flowing. The standard also gives rules concerning the patterns to be followed when sampling from static lots that have a depth less than 9 m (rail or road wagons, lorries, bulk tankers or ships). An example of the distribution of the sampling points for eight points is shown in Fig. 1.2. The standard provides sampling procedures for obtaining the minimum mass of laboratory sample for dynamic and static lots.

1.3.3   Codex general standard for contaminants and toxins in food and feed (Codex Stan 193-1995, revision 2009)

Codex Stan 193-1995, revision 2009, contains the main principles recommended for dealing with contaminants and toxins and lists the maximum levels for contaminants and associated sampling plans for commodities moving in international trade, such as peanuts and treenuts. According to this standard, contaminant levels in food and feed shall be as low as reasonably achievable through best practices such as good agricultural practice (GAP) and good manufacturing practice (GMP). National measures regarding food and feed contamination should avoid the creation of unnecessary barriers to international trade.

Annex 1 of Codex Stan 193-1995 contains the sampling plan for total aflatoxins in peanuts intended for further processing. The sampling plan calls for a single 20 kg laboratory sample of shelled peanuts (27 kg of unshelled peanuts) to be taken from a peanut lot and tested against a maximum level of total aflatoxins in $\mu g\ kg^{-1}$. The number of incremental samples to be taken from different locations throughout the lot, depends on the weight of the lot, with a minimum of 10 and a maximum of 100, as shown in Table 1.2. The weight of the incremental samples should be a minimum of 200 g. The standard also provides criteria for the sample preparation and the performance of analytical methods used in the analysis of the aflatoxins.

Annex 2 of the standard contains the sampling plan for aflatoxin contamination in ready-to-eat treenuts (almonds, hazelnuts and pistachios) and treenuts destined for further processing. The sampling plan also specifies a 20 kg aggregate sample, to be tested against maximum levels for total aflatoxins in 'ready-to-eat' and 'destined for further processing' treenuts of 10 and 15 $\mu g\ kg^{-1}$, respectively.

Table 1.2   Number of incremental samples to be taken depending on the weight of the lot according to the sampling plan for total aflatoxins in peanuts intended for further processing (Codex Standard 193-1995)

Lot weight T (tons)	Number of incremental samples
$T \geq 15$	100
$10 < T < 15$	80
$5 < T \leq 10$	60
$1 < T \leq 5$	40
$T \leq 1$	10

Table 1.3   Maximum levels, required number and laboratory sample size for total aflatoxins in treenuts (almonds, hazelnuts, and pistachios) 'ready-to-eat' and 'destined for further processing' (Codex Standard 193-1995)

	Ready-to-eat	Destined for further processing
Maximum level ($\mu g\ kg^{-1}$)	10	15
Number of laboratory samples	2	1
Laboratory sample size (kg)	10	20

Table 1.4   Minimum number and size of incremental samples as a function of the lot weight for total aflatoxins in treenuts (almonds, hazelnuts, and pistachios) (Codex Standard 193-1995)

Lot weight tons – (T)	Minimum number of incremental samples	Minimum size of incremental sample (kg)	Minimum size of aggregate sample (kg)
$T \geq 15$	100	0.2	20
$10 < T < 15$	75	0.267	20
$5 < T \leq 10$	50	0.4	20
$1 < T \leq 5$	25	0.8	20
$T \leq 1$	10	2	20

The two sampling plans are illustrated in Table 1.3. The number and size of the incremental samples will vary with lot (sublot) size. Table 1.4 is used to determine the number of incremental samples (between 10 and 100) to be taken from lots or sub-lots of various sizes below 25 tonnes. Criteria are provided for the sample preparation and the performance of analytical methods used in the analysis of the aflatoxins in treenuts.

1.3.4   USDA sampling plans for aflatoxins

The United States Food and Drug Administration (USDA) has well defined sampling procedures for aflatoxins (Park and Pohland, 1989). These take account of the commodity type, whether samples are to be taken at retail or from bulk

Table 1.5 Incremental sample size and aggregate sample size required for the aflatoxins control programmes in peanuts and tree nuts by the US Food and Drug Administration (FDA, 2009)

Commodity	Control programme	Minimum number of incremental samples	Incremental sample size (kg)	Minimum aggregate sample size (kg)
Peanuts roasted in shell (only for domestic runner variety)	Monitoring Surveillance	15 75	0.454 0.454	6.8 34
Tree nuts (except in-shell Brazil nuts and all pistachio nuts in import status) shelled, in-shell slices, pieces, or flour	Monitoring Surveillance	10 50	0.454 0.454	4.5 22.7

commodities and the lot size. For each commodity the minimum number of sub-samples to be taken and the minimum unit size are specified as shown in Table 1.5 (FDA, 2009).

1.4 Uncertainty estimation and designing sound sampling plans for mycotoxin analysis in food and feed

1.4.1 Uncertainty of the test procedure

Table 1.6 shows the mycotoxin measurement process, starting with sampling of the lot and ending with analytical determination. Each step contributes to the uncertainty of measurement (Ramsey and Ellison, 2007). Even when using accepted sample selection, sample preparation and analytical methods (Campbell *et al.*, 1986; Whitaker, 2006), or applying official protocols, there are uncertainties associated with each of the steps in the mycotoxin test procedure. The mycotoxin

Table 1.6 Mycotoxin measurement process

	Sampling		Sample preparation			Analysis	
Lot sampling	Primary sampling	Sub-sampling	Preparation of the laboratory sample	Analtical sample	Test portion	Test aliquot	
Collection of several increments to form the aggregate sample	Comminution and/or splitting	Further comminution and/or splitting	Milling, wet milling, splitting, homogenization	Selection of the sample for chemical analysis	Chemical analysis	Quantification of mycotoxin concentration	Mycotoxin test results

concentration in the lot cannot be measured with absolute certainty and the result should always be reported with an estimate of the uncertainty.

Each of the three steps of the mycotoxin test procedure has an associated variance that contributes to the total variance of the testing scheme. In order to evaluate the sampling strategy, each of these components must be measured and/ or modelled and their relationship understood. Amongst the statistical measures of variability only the variance is additive. Therefore, it is assumed that the total variance, V_T associated with a mycotoxin test procedure is the sum of the sampling variance (V_S), sample preparation variance (V_{SP}), and analytical variance (V_A) (equation [1.1]).

$$V_T = V_S + V_{SP} + V_A \qquad [1.1]$$

Cheli *et al.* (2009) summarized the variability associated with each of these steps (Table 1.7). The data indicate that for small sample sizes the sampling phase is the largest source of variability of the mycotoxin test procedure for maize and peanuts contaminated with fumonisin and aflatoxin.

1.4.2   Random and systematic aspects of uncertainty

The bias is the difference between the test result (mean) and the true or reference value. Bias accounts for the systematic component of the uncertainty. Accuracy is the closeness of agreement between the result of a measurement and the true/ reference value of the quantity being measured. Precision, measured as variability, is defined as the closeness of agreement between independent test results obtained under stipulated conditions. Precision accounts for the random component of the uncertainty.

1.4.3   Sampling variance

The explanation of the uncertainty of sampling is very well described by Ramsey and Thompson (2007): 'We extract a small amount of material (the sample) to determine the composition of a much larger body (the target). This sample should ideally have exactly the same composition as the target, but never does. The discrepancy gives rise to uncertainty from sampling'.

The contribution of the sampling variance to the total variance has been evaluated and quantified in several products. Researchers have developed equations to describe the sampling variance for several commodities and mycotoxins (Whitaker *et al.*, 1972, 1974, 1976, 1979b, 1993, 1998, 2000; Schatzky *et al.*, 1995a, 1995b; Hart and Schabenberger, 1998; Johansson *et al.*, 2000c). These equations show that, especially for small laboratory sample sizes, the sampling step is usually the largest source of variability associated with the mycotoxin test procedure. The sampling variance decreases with increasing lot concentration and laboratory sample size.

The sampling error is inevitably large because of the uneven distribution of mycotoxins amongst contaminated particles within a lot. Studies on aflatoxins in

Table 1.7 Variability associated with each step of the test procedure, sampling (V_S), sample preparation (V_{SP}) and analysis (V_A) expressed as a percentage of the total variance (V_T). Adapted from Cheli *et al.* (2009), with permission

Matrix	Mycotoxin	Limit (μ kg^{-1})	Aggregate sample size (kg)	Sample homogenizer	Subsample (g)	Aliquots for analysis	V_S/V_T (%)	V_{SP}/V_T (%)	V_A/V_T (%)	Reference
Shelled corn	Aflatoxin	20	0.91	Romer mill	50	1	75.6	15.9	8.5	Whitaker (2006)
Shelled corn	Aflatoxin	20	4.54	Romer mill	100	2	55.2	29.1	15.7	Whitaker (2006)
Shelled corn	Aflatoxin	20	1.13	Romer mill	50	1	77.8	20.5	1.7	Johansson *et al.* (2000c)
Cottonseed	Aflatoxin	20	4.54	Romer mill	100		96.8	2.6	0.7	Whitaker *et al.* (1976)
Wheat	Deoxynivalenol	5000	0.454	Romer mill	25	1	22	56	22	Whitaker (2000)
Shelled corn	Aflatoxin	20	5	Romer mill	100	1	59.8	34.5	5.7	Johansson *et al.* (2000a)
Peanut	Aflatoxin	100	2.27		100		92.7	7.2	0.1	Whitaker *et al.* (1994)
Shelled corn	Fumonisin	2000	1.1		25	1	61	18.2	20.8	Whitaker *et al.* (1998)
Shelled corn	Fumonisin	2000	2	Romer mill	25	1	69			Whitaker *et al.* (2007b)
Almond	Aflatoxin	15	10	Romer mill	100	1	96			Whitaker *et al.* (2006)
Green coffee	Ochratoxin A	5	1	Romer RAS mill	25	1	72.6	26.4	1.0	Vargas *et al.* (2004)
Powdered ginger	Aflatoxin		0.14	NA	5	1	87		13	Trucksess *et al.* (2009)
Powdered ginger	Ochratoxin A		0.14	NA	5	1	97.1		2.9	Trucksess *et al.* (2009)

corn and peanuts suggest that about 0.1% of the kernels in a lot are contaminated and the concentration in a single kernel may be extremely high. Cucullu *et al.* (1986) reported that the contamination of aflatoxins could reach 1×10^6 µg kg^{-1} and 5×10^6 µg kg^{-1} for one single peanut kernel and cottonseed, respectively. Shotwell *et al.* (1974) reported findings over 4×10^5 µg kg^{-1} for aflatoxins in a maize kernel.

In order to perform reliable sampling, in theory each individual unit within a lot must have the same probability of being selected. In most cases this is impossible or impractical; consider, for example, sampling a ship load consisting of 500 tonnes of wheat. Two main aspects of the sampling step are critical for the reduction of the uncertainty, the selection techniques and the number and the size of the incremental samples selected from the lot. It is important to note that sampling variance also depends on the mycotoxin/matrix combination. In corn, for example, the curves describing the relationship between the sampling coefficient of variation (CV) and the mycotoxin concentration show the same trend, but the values of CV are higher for aflatoxin than for fumonisin and DON (Johansson *et al.*, 2000b; Whitaker *et al.*, 1998, 2000).

Systematic effects in sampling are caused by the heterogeneity of the lot, combined with the inability of the sampling method to reflect this heterogeneity properly (Ramsey and Thompson, 2007; Whitaker, 2006). Static lot sampling bias can be caused by, for example, a sampling probe that does not allow larger particles into the probe, a probe that does not reach every location in the shipment and use of a single probing point in a poorly mixed lot. Systematic effects may be difficult to quantify, but they can be reduced, for example by selecting proper sampling devices or by increasing the size of the sample. If the sample for analysis comprises the entire lot, the systematic effects or bias arising from sampling will be negated (although not the bias arising from the analytical stage). In almost all cases this is impossible and/or impractical, but increasing the sample size will give a better representation of the whole lot. Other measures such as grinding solid materials to reduce the particle size, either of the whole lot or of a relatively large sample, and efficient mixing can also reduce bias.

Random effects in sampling are mainly caused by variations in the composition of the sample in space or in time, by the use of different sampling methods, by the sampling procedure or the handling of the sample (e.g. by different samplers) and by variability in the performance of the sampling equipment. The most obvious approach to reducing the random effects is to increase the number of samples taken, which will result in a smaller standard deviation of the mean result. An equivalent approach is to increase the number of sub-samples or increments taken to produce one aggregate sample for analysis. A careful investigation of the variations in time and space, carried out as part of the validation of the sampling procedure, might be needed to select the proper sampling frequency or spatial distribution for the given quality requirement. Collecting too many samples will be more expensive, but will not necessarily give more or better information and thus should be avoided. The methodologies and equipment employed in collecting incremental samples are crucial in reducing the errors associated with the myco-toxin test procedure.

Table 1.8   Methodology for estimation of precision and bias of sampling and analysis in the empirical approach according to the Eurachem/Citac guide (Ramsey and Thompson, 2007)

	Sampling	Analysis
Precision	Perform duplicate sampling	Perform replicated analysis
Bias	Use a reference sampling lot (target), participate in inter-organizational sampling trial	Use certified reference materials (CRM)

Measurement uncertainty arising from sampling: the Eurachem/Citac guide
Eurolab, Nordtest and the UK RSC Analytical Methods Committee have jointly produced a Eurachem/Citac guide on measurement uncertainty arising from sampling (Ramsey and Ellison, 2007). Two main approaches to quantify uncertainty are described in the guide, the empirical approach and the modelling approach.

The empirical (also defined as 'experimental', 'retrospective' or 'top-down') approach uses repeated sampling and analysis, under various conditions, to quantify the effects caused by factors such as heterogeneity of the analyte in the sampling target and variations in the application of one or more sampling protocols. This approach relies on overall reproducibility estimates from either in-house or inter-organizational measurement trials, without necessarily trying to quantify any of the sources of uncertainty individually. The Eurachem/Citac document gives guidance on how to quantify the systematic and random errors in sampling and analysis to provide an estimation of the overall uncertainty (Table 1.8).

The guide describes four types of methods to estimate the combined uncertainty empirically. The 'duplicate method' often gives a reasonably reliable estimate of uncertainty. This method is the simplest and probably the most cost-effective of the four methods described by the guide for the empirical approach. It is based upon a single sampler carrying out the same sampling protocol and taking duplicate samples from at least eight sampling lots, selected at random. If only one lot exists, then all eight duplicates can be taken from it, but the uncertainty estimate will only be applicable to that one lot. Both of the duplicate samples are sub-sampled, resulting in two separate test samples. Duplicate test portions are drawn from both test samples and each is then analysed in duplicate (i.e. duplicate chemical analysis). This system of duplicated sampling and chemical analysis is known as a 'balanced design' (Fig. 1.4). To calculate uncertainty, the random component of the uncertainty can be estimated by applying analysis of variance (ANOVA) to the measurements of concentration on the duplicated samples. A 'samplers guide' is available on the internet which offers excel spreadsheets for ANOVA calculation (Groen, C).

The modelling (also defined as 'theoretical', 'predictive' or 'bottom-up') approach uses a predefined model that identifies separately each of the components of the uncertainty and sums them in order to make an overall estimate. Models from Gy's sampling theory (see next section) can sometimes be used in this approach to estimate some of the components. Further examples are given in the Eurachem/Citac guide.

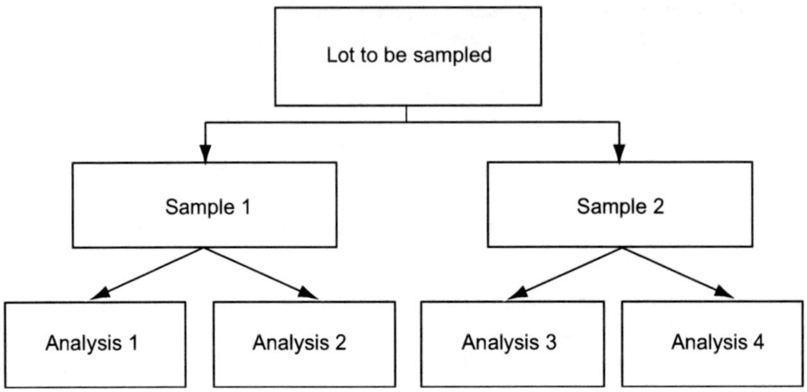

Fig. 1.4   A balanced design.

Theory of sampling (TOS)

The theory of sampling (TOS) was introduced by Pierre Gy in 1950. The TOS provides a description of all errors involved in the sampling of heterogeneous materials as well as the tools necessary for their evaluation, elimination and/or minimization. A comprehensive summary of the TOS has been prepared by Petersen *et al.* (2005), and provides many practical examples. The total measurement error, which Gy called the global estimation error (GEE), is the sum of the total sampling error (TSE) and total analytical error (TAE).

The components of TSE can be divided into two major groups:

1. Errors of incorrect sampling. Examples are gross errors and as such they are excluded from estimates of uncertainty. Incorrect sampling errors are unpredictable and arise from sampling equipment and procedures that do not follow the rules of sampling correctness defined in the sampling theory. These errors can be minimized or eliminated by carefully checking the performance of the equipment and procedures, by replacing inappropriate devices and procedures with those that comply with sampling rules and by sufficient training of sampling personnel.
2. Errors of correct sampling. These are a consequence of the material structure and are therefore inevitable, leading Gy to describe sampling as 'a science that falls in the province of the calculus of probability'. The material heterogeneity can be divided into two classes, constitution heterogeneity (CH) and distribution heterogeneity (DH).

CH is inherently dependent on the physical properties (composition, shape, size, density, etc.) of the particles making up the lot. Mixing and blending does not change the CH. The only way to alter the CH is by crushing/comminuting, for example by milling grain to reduce particle size. DH is dependent on the spatial distribution of the particles in the lot and shows the local stratification/segregation of particles in the lot. For example, particles with large differences in size and/or density tend to segregate or stratify heavily, with the smallest and/or densest particles at the bottom of the lot. Thus, all particles do not have the same probability of being selected for sampling.

Gy's fundamental sampling error (FSE) is the minimum error of an ideal sampling protocol. It depends on the number of critical particles in the samples. The Eurachem/Citac Guide provides Gy's formula and accompanying explanation and examples on the estimation of the relative variance of the FSE.

Gy's formula is:

$$S^2_{(FSE)} = Cd^3 \left(\frac{1}{M_S} - \frac{1}{M_L} \right) \approx \frac{Cd^3}{M_S} \qquad [1.2]$$

where M_S is the mass of the sample, M_L is the mass of the lot, C is the sampling constant and d is a measure of the coarsest fragment size (top 95% retaining sieve mesh size).

According to the Eurachem/Citac Guide, the empirical approach to the quantification of uncertainty tends to be more generally applicable and does not depend on prior knowledge of all of the sources of uncertainty. This approach is less time consuming and therefore less costly. The modelling approach gives more information about the individual sources of uncertainty and is perhaps more appropriate when elaborating a sampling plan that should be long term and with a very specific purpose.

1.4.4 Sample preparation variance

Once a representative primary aggregate sample has been selected from the lot, a laboratory sample must be taken from the aggregate sample. To save money and time, the mass of the primary sample should be reduced before transportation, storage or analysis. This is a critical phase; it is important that sub-sampling retains the representativeness of the lot.

The laboratory sample must be processed/comminuted for mycotoxin quantification. It is highly recommended that the entire laboratory sample is comminuted before a test portion is taken for analysis. The particle size must be sufficiently small to allow effective mixing, giving a distribution of the mycotoxins that is as homogeneous as possible, thereby reducing variability in the sample preparation step (Spanjer *et al.*, 2006). This also applies to the preparation of the aggregate sample to obtain the laboratory sample. The most efficient grinders are those that can reduce the particle size of the laboratory sample to the smallest size possible. Several studies have been conducted on sample reduction techniques. The water slurry technique is the preparation of a homogeneous paste by blending an already milled sample with an appropriate amount of water at high speed in a slurry mixer. Velasco and Morris (1976) and Schatzki and Toyofuku (2003) demonstrated that the application of the slurry technique reduces the clogging of samples that have a high oil content and, in addition, produces a smaller particle size and more homogeneous samples. Spanjer *et al.* (2006) also showed that for pistachios, coffee beans and spices the slurry technique gives a lower variability than dry milling.

Petersen *et al.* (2004) presented a comprehensive survey of 17 types of field/ laboratory mass reduction equipment and compared them with respect to accuracy (bias), reproducibility (precision), representativeness, material loss, operation time as well as ease of cleaning. It was shown that many devices have inherent design faults. All of the tested sample splitters and related dividers were orders of magnitude superior to various spoon and shovel methods. It was concluded that grab sampling and shovelling methods should be totally avoided. Although grab sampling is fast, easy and cheap, it also leads to heavily biased samples.

To find a robust, quick and efficient method for soil splitting in the field, Gerlach *et al.* (2002) evaluated five sampling techniques: riffle splitting, paper cone splitting, fractional shovelling, coning and quartering and grab sampling. Riffle splitting performed best, while grab sampling performed by far the worst, with 15–20% bias.

Despite the use of correct sample reduction procedures, a certain degree of variation among different laboratory samples is unavoidable and represents the main source of sample preparation variability. Small particle size, large sample size and low contamination levels are associated with a reduction of the sample preparation variability.

1.4.5   Analytical variance

Recent advances in analytical methodology have been applied to improve significantly the capabilities for the efficient detection and quantification of mycotoxins in agricultural commodities. As discussed above, the variance associated with the analytical step is usually lower than for the sampling or sample preparation steps. The analytical variance is a function of the mycotoxin concentration and the number of aliquots analysed.

Whitaker and co-workers (1996a, 2003) showed that an increase in the mycotoxin concentration generated a higher analytical variance, whilst an increase in the number of aliquots analysed reduced the variance. Whitaker also discussed the influence of the analytical method on the variance. He demonstrated that an HPLC method for the analysis of aflatoxins in corn produced less variability than TLC and enzyme-linked immunosorbant assay (ELISA) methods. European Commission Regulation EC 401/2006 (2006) specifies minimum performance criteria at different levels of contamination for methods of analysis for mycotoxins and this is an integral part of the sampling protocol.

1.4.6   Reducing variability of a mycotoxin test procedure

As discussed above, biases have the potential to occur in the sample selection process and sample preparation process, and in the quantification steps of the mycotoxin test procedure. Biases should be the easiest component of uncertainty to control and reduce to acceptable levels, but methods to reduce bias are difficult to evaluate because of the difficulty in knowing the 'true' mycotoxin concentration of the lot (Whitaker *et al.*, 2009). However, biases can be minimized by ensuring that sample selection and sample preparation equipment are continuously checked for performance. Analytical methods must undergo a validation process to show that they are 'fit for purpose' and biases can be minimized and corrected by the use of certified reference materials if available, or by the use of recovery tests.

In summary, the only way to obtain a precise estimate of the 'true' lot concentration is to reduce the total variability of the mycotoxin test procedure by reducing the variability associated with each step, sampling, sample preparation and analysis.

- Sampling variability can be reduced by increasing the size of the aggregate sample.
- Sample preparation variability can be reduced by increasing the analytical portion size and/or increasing the degree of comminution (number of particles per unit mass).
- Analytical variance can be reduced by increasing the number of aliquots analysed and/or by using a better performing analytical method (less uncertainty).

1.4.7 Uncertainty estimation and 'fit for purpose' concept

The Eurachem/Citac Guide emphasizes that uncertainty of sampling must be embedded in the concept of fitness for purpose. The uncertainty level tolerated by the user of the results should be carefully considered before designing a testing plan. If the performance level is set too high (too stringent), the investigation will be unnecessarily expensive. Ramsey and Thompson (2007) analysed the best division of resources between sampling and analysis. In general, measurements should be performed in such a way that the uncertainty is the lowest that can be achieved. However, reducing the uncertainty of a measurement result involves rapidly escalating costs. The true cost of a decision is the sum of the measurement costs and the costs deriving from incorrect decisions. This sum has a minimum value at some particular level of uncertainty and this uncertainty level defines fitness for purpose (Ramsey *et al.*, 2001). As a rule of thumb, an inverse-square relationship can be applied between cost and variability (measured as standard deviation): if the total standard deviation is cut to half, the cost will increase four times (Minkkinen, 2004).

To summarize, the contribution of the sampling to the overall uncertainty is occasionally small, but is often dominant (larger than 90% of the total measurement variance). This suggests the need for an increased proportion of the total expenditure involved to be invested in sampling, rather than chemical analysis, in order to reduce total uncertainty and achieve fitness for purpose.

1.4.8 Operating characteristics curves

Whitaker *et al.* (1970) developed operating characteristic (OC) curves for several commodities. An OC curve is a plot that has a unique shape for a particular

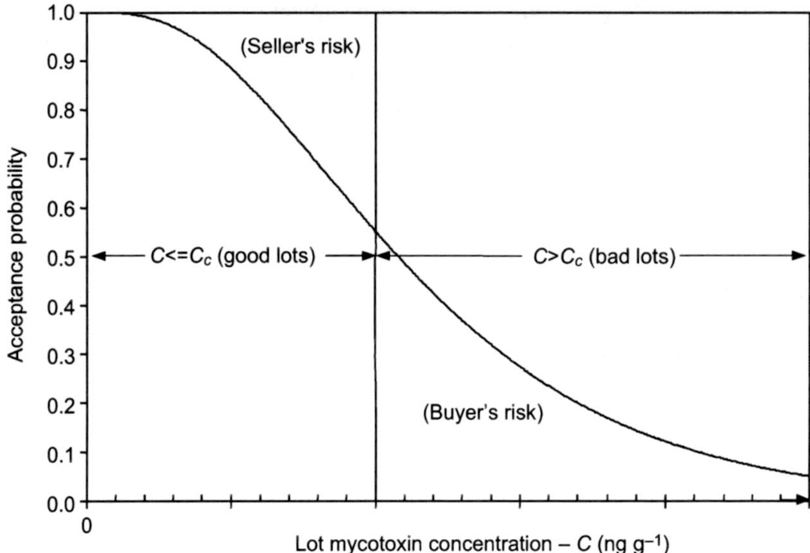

Fig. 1.5   Operating characteristics curves.

sampling plan, showing the relationship between the probability of accepting a lot with a particular mycotoxin concentration and the lot mycotoxin concentration (Fig. 1.5). The OC curve can be considered a footprint of a specific sampling plan and is defined for designated values of sample size, degree of comminution, sub-sample size, analytical method, number of analyses and an accept/reject limit.

For a given sampling plan design, lots with a mycotoxin concentration M will be accepted with a certain probability $P(M)$ (the acceptance probability) which is the probability that a sample test result, M, arising from the sampling plan is less than or equal to the sample accept/reject level, M_c.

$$P(M) = \text{prob}(M < M_c) \tag{1.3}$$

For a given sampling plan, it is possible to calculate the probability of rejecting a good lot (type I error, false positive sample) or accepting a bad lot (type II error, false negative sample) as a function of the mycotoxin concentration and the risk associated with a specific sampling plan. The accept/reject level (the 'sample acceptance limit') may be a legal limit or a product quality limit. The areas delimited by the curve and the accept/reject level (see Fig. 1.5) describe an economic risk (area above the curve) or a consumer risk (area below the curve).

1.4.9   Design of sound sampling plans

Since the slope of the OC curve has high economic and health relevance, it is crucial, when designing a sampling plan, to maximize the slope of the OC in order to reduce consumer and producer risks and minimize the risk of lot misclassification (Johansson *et al.*, 2000b; Whitaker, 2006). To evaluate the performance of a specific mycotoxin sampling plan, information must be available on the variability associated with the mycotoxin test procedure and the distribution of the mycotoxin test results.

Johansson *et al.* (2000b) showed that, in the case of sampling plans for aflatoxins in shelled corn, both false positive and false negative risks can be reduced by increasing the size of the aggregate sample, the degree of sample comminution, the subsample size, or the number of aliquots that are analysed. MacArthur *et al.* (2006) examined some retail-sampling strategies for the measurement of mycotoxins in dried vine fruit and nuts. A simulation was used to examine how the uncertainty associated with measurement results could be expected to change with the number of increments used to form the aggregate sample. The simulation accounted for one, 10 and 30 1-kg increments. For simulated measurements based on 10 increments, the probability of a false negative result was less than 5% for lots containing 24 μg kg^{-1} of OTA. If less than 10 increments were taken, the uncertainty was very high. On the other hand, increasing the number of increments beyond 50, given the analytical uncertainty, resulted in little improvement in measurement uncertainty (Spanjer, 2007).

The important conclusion of this study is that a sampling design that has not been adequately researched cannot be said to offer either effective protection to consumers or value for money. For example, a design that is effective for the measurement of OTA in dried fruit is likely to leave consumers largely unprotected if it is applied to the measurement of aflatoxins in pistachio nuts; whereas a design that is effective for the measurement of aflatoxins in pistachio nuts would represent a waste of resources if applied to the measurement of OTA in dried fruits (Spanjer, 2007). The extensive data available on the variability associated with sampling, sample preparation and analysis, has enabled the establishment of a series of sampling plans for the control of aflatoxins, DON, fumonisins and OTA in several commodities, as summarized in Table 1.9.

1.5 Quality assurance and quality control procedures in sampling and arrival of the samples at the analytical laboratory

1.5.1 Quality assurance and quality control procedures in sampling

The Eurachem/Citac guide (Ramsey and Thompson, 2007) mentions the importance of implementing validation and regular quality control procedures for the sampling step of a sampling plan. Quality assurance/quality control procedures applicable to the sampling step include the following:

- Ensure the presence of trained samplers.
- Ensure the availability of written instructions regarding sampling and transport of the samples to the laboratory.
- Ensure the availability of a sheltered unloading area, proper unloading equipment and appropriate sampling devices.

Table 1.9   Sampling plans for mycotoxins in different commodities

	Aflatoxin	Fumonisin	Deoxynivalenol	Ochratoxin A
Peanuts	Whitaker *et al.* (1974, 1995, 1999), Knutti and Schlatter (1982)			
Peanut butter	Waltking (1980)			
Maize	Johansson *et al.* (2000a)	Whitaker (2000); Whitaker *et al.* (1998, 2007b)		
Cottonseed	Velasco *et al.* (1975), Whitaker *et al.* (1976), Park *et al.* (2000)			
Wheat			Whitaker *et al.* (2002)	
Coffee				Vargas *et al.* (2006)
Ginger roots	Trucksess *et al.* (2009)			Trucksess *et al.* (2009)
Pistachios	Schatzki (1995a,b, 2004)			
Almonds	Whitaker *et al.* (2007a)			
Hazelnuts	Ozay *et al.* (2007)			
Figs	Sharman *et al.* (1994)			
Feedstuff	Coker *et al.* (2000)			
Grapes				Battilani *et al.* (2006)

- Use clean sampling equipment and sample bags and containers free of contamination to avoid any cross-contamination.
- Take increments from the same lot.
- Avoid any changes which would affect the mycotoxin content, i.e. do not open packaging in adverse weather conditions or expose samples to excessive moisture or sunlight.
- Take the correct number of incremental samples of the appropriate weight at various places distributed throughout the lot.
- Ensure that access lanes in the storage facility are accessible in all directions.
- Place samples collected in a clean, dry, opaque, leak-proof container that can be securely sealed.
- Record as much information as possible about the lots from which samples have been taken to provide as much traceability data as possible.
- Store all samples in a cool dark place and segregate different lots. Apply good storage practices.
- Dispatch samples to the laboratory as soon as possible after sampling.
- Ensure samplers are wearing the correct personal protective equipment and apply good sampling practices.

1.5.2 Arrival at the laboratory and sample preparation

After the actual sampling, the aggregate sample may go through a number of further steps before reaching the laboratory. In this process all possible precautions should be taken to avoid any alteration of the sample by human error, spillage, contamination, packaging, adhesion of critical components to the sides of containers/bags, loss of fine particles during handling or crushing, moisture uptake/loss, biodegradation, and so on.

Upon arrival at the laboratory, the aggregate sample container should be checked to ensure that it is sealed. The accompanying documentation should give details of whether the consignment is intended for direct human consumption or will be subjected to sorting and/or other physical treatment before human consumption. The fine grinding, slurrying and mixing of the laboratory sample should be carried out using a process that has been previously validated as fit for purpose. The sample preparation procedures must be validated at the laboratory and available as standard operating procedures (SOPs).

According to European legislation, if the case applies, formal official aggregate samples are mixed and then split at the laboratory into three sub-samples and only then can each sub-sample be homogenized. In the case of products intended for direct human consumption, one analytical sample, one defence sample and one reference sample are taken from each sub-sample (laboratory sample). For every official aggregate sample taken from a batch of pistachios, for example, nine samples in total are obtained from the homogenized subsamples: three analytical samples, three defence samples and three reference samples.

1.5.3   Analysis

For official control purposes the analysis should be carried out by an official laboratory accredited according to ISO/IEC 17025 (ISO, 2005). It is becoming more common for laboratories to accredit their sampling procedures (Minkkinen, 2004). The basic requirements for accreditation are that the sampling equipment is correct, the uncertainties of the methods have been estimated, procedures are regularly audited and personnel have been adequately trained for their jobs.

1.6   Strengthening national food control systems

1.6.1   Training sampling inspectors

To implement sound sampling plans, it is essential that samplers are well trained and competent in carrying out their jobs. Without training on at least the basics of sampling theory, it is difficult to obtain representative samples, especially when dealing with heterogeneous contaminants. However, very few universities or academic institutions provide courses on sampling. At the national level, the responsibility for the education and training of samplers lies with national inspection bodies, which are the official institutions that share responsibility for food and feed quality and safety.

Capacity building and technical training requires, inter alia, basic infrastructure and investments, information technologies, knowledge of the national food control strategy, food legislation, food inspection services, food control laboratories, collaboration and cooperation among control agencies, sound scientific and technical expertise, and financial resources. In order to be effective, trained food inspection officials should have well planned food inspection programmes, should understand their duties and responsibilities, and should maintain close collaboration with other food control services. This requires adequate management, training and equipment.

1.6.2   Sampling responsibility

The Eurachem/Citac guide (Ramsey and Ellison, 2007) suggests that the responsibility for sampling should be clearly assigned to one organization/individual, while defining responsibilities for the other steps of the measurement process. If possible, the analyst who performs the analysis should also be involved in the planning of the experimental design before analysis. If this is not possible, it is important that the responsibility for sampling, chemical analysis and data analysis is shared between the individuals/organizations involved. Information about the contributions arising from the different steps in the measurement procedure to the overall uncertainty should also be shared among all participants/organizations.

1.6.3   Holistic approach to control of mycotoxins

Producing safe and good quality food is a prerequisite for successful domestic and international trade and key to the sustainable development of national agricultural resources. Despite many years of research and the introduction of good practices in the food production, storage and distribution chain, mycotoxins continue to be a problem, especially in developing countries. In this context it is of utmost importance that a holistic approach is adopted for the control of mycotoxins.

Mycotoxins may originate in the field and therefore mycotoxin control must start during crop production. Unquestionably, prevention is the best method for controlling mycotoxin contamination. FAO has published codes of practice for the prevention and control of mycotoxins; for example, the code of practice for the prevention and reduction of aflatoxin contamination in peanuts (Codex Alimentarius Commission, 2004; CAC/RCP 55-2004), and the code of practice for the prevention and reduction of ochratoxin A contamination in wine (Codex Alimentarius Commission, 2007; CAC/RCP 63-2007).

Pre-harvest mycotoxin formation can be partially controlled through good agricultural practices (GAP) which include insect management, good irrigation and mineral nutrition, crop rotation and the use of mould resistant crop varieties. Rapid field detection methods can be of great help in this respect. Post-harvest prevention procedures include good storage procedures, minimization of moisture exposure, insect infestation prevention and cleaning and disinfection of storage containers and transportation equipment (Boutrif and Canet, 1998). In addition, in storage, the initial grain condition is critical. Good quality, clean, sound grain is easier to maintain in storage than physically damaged grain.

The development of internationally harmonized regulatory measures for mycotoxins is paramount in the global strategy for minimizing mycotoxin contamination while maximizing the availability of food. To this end, national participation in the process of the elaboration of guidelines by bodies such as the Codex Alimentarius Commission is important. This allows discussion of the problems encountered in many countries and the development of consensus guidelines to address such issues. Interaction with the Codex Committees is especially important for developing countries, who can both input into and benefit from the guidelines developed (Boutrif, 1995).

1.6.4   Conclusion

The adoption of the best agricultural practices in the field and throughout the whole farm–fork chain, coupled to the best sampling practices and the use of validated and fit-for-purpose methods, together with accreditation and participation in proficiency testing are the recommended means of ensuring the recognition of mycotoxins test results worldwide. Assistance by international organizations such as FAO may be necessary, particularly in developing countries, to stimulate and implement the necessary food control systems (FAO, 2001).

Disclaimer

The views expressed in this text do not necessarily reflect those of the IAEA or FAO, or the governments of their Member States.

The mention of names of specific companies or products (whether or not indicated as registered) does not imply any intention to infringe proprietary rights, nor should it be construed as an endorsement or recommendation on the part of IAEA or FAO.

1.7   Sources of further information and advice

1.7.1   Books on sampling for mycotoxins

The authors are not aware of any single volume dedicated specifically to sampling for mycotoxins. Individual chapters can be found in the following books:

Adams J. and Whitaker T. B. (2004). 'Peanuts, Aflatoxin and the U.S. Origin Certification Program', in *Meeting the Mycotoxin Menace*, Barug D., van Egmond H., Lopez-Garcia R., van Osenbruggen T. and Visconti A. (eds), Wageningen Academic Publishers, The Netherlands, 183–96.

Anon (2004). 'Sampling', in *Grain Fungal Diseases and Mycotoxin Reference*, USDA, Grain Inspection, Packers and Stockyards Administration (GIPSA), 24–9.

Miraglia M., De Santis B., Pannunzi E., Debegnac F. and Brera C. (2008). 'Mycotoxin concentration data quality: the role of sampling', in *Mycotoxins Detection Methods, Management, Public Health and Agricultural Trade*, Leslie J., Bandyopadhyay R. and Visconti A. (eds), CAB International, Oxfordshire, UK, 185–94.

Whitaker T. B. (2001). 'Sampling techniques', in *Mycotoxin Protocols*, Trucksess M. W. and Pohland A. E. (eds), Series: Methods in Molecular Biology, Vol. 157, Humana Press, Totowa, New Jersay, 11–24.

Whitaker T. B. (2004). 'Sampling for mycotoxins', in *Mycotoxins in Food: Detection and Control*, Magan N. and Olsen M. (eds), National Food Administration, Sweden. Woodhead Food Series No. 103, Woodhead Publishing, Cambridge, UK, 69–87.

Whitaker T. B. (2005). 'Sampling feeds for mycotoxin analysis', in *Mycotoxin Blue Book*, Diaz D. (ed.), Nottingham University Press, Bath, UK, 1–23.

Whitaker T. B., Dickens J. W. and Giesbrecht F. J. (1991). 'Testing animal feedstuffs for mycotoxins: sampling, subsampling, and analysis', in *Mycotoxins and Animal Foods*, Smith D. and Henderson R. (eds), CRC Press, Boca Raton, Florida, 153–64.

Whitaker T. B., Hagler W. M. Jr., Johansson A. A., Giesbrecht F. G. and Trucksess M. W. (2001). 'Sampling shelled corn for fumonisin', in *Mycotoxins and Phycotoxins in Perspective at the Turn of the Millennium*, deKoe W. J., Samson R. A., Van Egmond H. P., Gilbert J. and Sabino M. (eds), Wageningen Academic, The Netherlands, 97–107.

1.7.2   Gy's theory of sampling

Gy's sampling theory is well covered in the following books:

Gy P. M. (1998). *Sampling for Analytical Purposes*, John Wiley and Sons, Chichester, Sussex, UK.

Pitard F. F. (1993). *Pierre Gy's Sampling Theory and Sampling Practice*, 2nd edn, CRC Press, Boca Raton, Florida.

1.7.3   Web resources

European Mycotoxin Awareness Network (EMAN)
http://www.mycotoxins.org/
http://services.leatherheadfood.com/foodline/guides.aspx
http://www.food.gov.uk/foodindustry/guidancenotes/foodguid/mycotoxinssampling

1.7.4   Codex Alimentarius standards

The Codex Alimentarius web page can be accessed at: http://www.codexalimentarius.net/web/index_en.jsp#
A list of Codex standards can be accessed at: http://www.codexalimentarius.net/web/standard_list.do?lang=en

1.7.5   European legislation

The European legislation can be accessed at: http://eur-lex.europa.eu/en/index.htm

1.7.6   Uncertainty resources

Groen C., Bjerre Hansen J., Magnusson B., Nordbotten A., Krysell M., Jebjerg Andersen K. and Lund U. (2007). *Nordic Innovation Centre NT technical report TR 604: Uncertainty from Sampling – A Nordtest handbook for sampling planners on sampling quality assurance and uncertainty estimation*, based upon the Eurachem International Guide estimation of measurement uncertainty arising from sampling http://www.nordicinnovation.net/nordtestfiler/tr604.pdf
http://www.eurachem.org/guides/pdf/UfS_2007.pdf

1.7.7   Whitaker's research work

A complete list of Whitaker's publications can be found at: http://www.bae.ncsu.edu/usda/www/researchpaperstbw.htm

1.8   References

Battilani P., Barbano C., Rossi V., Bertuzzi T. and Pietri A. (2006). 'Spatial distribution of OTA in vineyard and sampling design to assess must contamination'. *J Food Protection*, **69**(4), pp. 884–90.

Biselli S., Persin C. and Syben M. (2008). 'Investigation of the distribution of deoxynivalenol and ochratoxin A contamination within a 26 t truckload of wheat kernels'. *Mycotoxin Res*, **24**(2), 98–104.

Boutrif E. and Canet C. (1998). 'Mycotoxin prevention and control: FAO programmes'. *Revue Méd. Vét.*, **149**(6), 681–94.

Boutrif E. (1995). 'FAO programmes for prevention, regulation and control of mycotoxins in food'. *Natural Toxins*, **3**, 322–6.

Campbell A. D., Whitaker T. B., Pohland A. E., Dickens J. W. and Park D. L. (1986). 'Sampling, sampling preparation, and sampling plans for foodstuffs for mycotoxin analysis'. *Pure Appl Chem*, **58**, 305–14.

Cheli F., Campagnoli A., Pinotti L., Fusi E. and Dell'Orto V (2009). 'Sampling feed for mycotoxins: acquiring knowledge from food'. *Ital J Anim Sci*, **8**, pp.5–22.

Codex Alimentarius Commission (1995). *Codex Stan 193-1995: Codex general standard for contaminants and toxins in food and feed.* FAO, Rome.

Codex Alimentarius Commission (2004). *CAC/RCP 55-2004: Code of practice for the prevention and reduction of Aflatoxins contamination in peanuts*. FAO, Rome.

Codex Alimentarius Commission (2007). *CAC/RCP 63-2007: Code of practice for the prevention and reduction of Ochratoxin A contamination in wine*. FAO, Rome.

Coker R. D., Nagler M. J., Defize P. R., Derksen G. B., Buchholz H., Putzka H. A., Hoogland H. P., Roos A. H., Boenke A. (2000). 'Sampling plans for the determination of aflatoxin B_1 in large shipments of animal feedstuffs'. *J AOAC Int*, **83**, 1252–8.

Cucullu A. F., Lee L. S., Mayne R.Y. and Goldblatt L. A. (1986). 'Determination of aflatoxin in individual peanuts and peanuts sections'. *J Am Oil Chem Soc*, **43**, 89.

European Commission (2006). Commission Regulation (EC) 401/2006 of, 23 February, 2006 laying down the methods of sampling and analysis for the official control of the levels of mycotoxins in foodstuffs. *Official Journal of the European Union*, **L** 70, 12–34; Annex I, part B.4.

Food and Agriculture Organization (FAO) (1993). *Sampling Plans for Aflatoxin Analysis in Peanuts and Corn*. FAO Food and Nutrition Paper 55, Rome, Italy.

Food and Agriculture Organization (FAO) (2001). *FAO/WHO Safety Evaluation of Certain Mycotoxins in Food*. WHO Food Additives Series, No. 47. Food and Nutrition Paper 74.

Food and Agriculture Organization (FAO) (2005). *Second FAO/WHO Global Forum of Food Safety Regulators*, 12–14 October, 2004, Bangkok, Thailand. Building effective food safety systems, proceedings of the forum, http://www.fao.org/docrep/meeting/008/y5871e/y5871e00.htm#Contents (accessed 31 January, 2010)

Food and Agriculture Organization (FAO) (2006). *Strengthening National Food Control Systems. Guidelines to assess capacity building needs*. ftp://ftp.fao.org/docrep/fao/0009/a0601e/a0601e00.pdf

Food and Agriculture Organization (FAO) (2007). *Strengthening National Food Control Systems. A quick guide to assess capacity building needs*. ftp://ftp.fao.org/docrep/fao/0010/a1142e/a1142e00.pdf

Food and Drug Administration (FDA) (2009). *Investigations Operations Manual*. http://www.fda.gov/ICECI/Inspections/IOM/default.htm (accessed 31 January, 2010)

Gerlach R.W., Dobb D.E., Raab G.A. and Nocerino J.M. (2002). 'Gy sampling theory in environmental studies. 1. Assessing soil splitting protocols'. *J Chemometrics*, **16**, 321–8.

Groen C., http://www.samplersguide.com, DHI Water, Denmark

Hart L. P. and Schabenberger O. (1998). 'Variability of vomitoxin in truckloads of wheat scab epidemic year'. *Plant Dis*, **82**, 625–30.

HGCA (2004). *Grain Sampling from Field to Buyer: Understanding Variation*. Guides and Guidelines/G18/Published: 9 June (2004 (accessed 31 January, 2010) http://www.hgca.com/cms_publications.output/2/2/Publications/Publication/Grain%20sampling%20from%20field%20to%20buyer%20-%20understanding%20variation.mspx?fn=showandpubcon=1259

ISO (2005). ISO 17025:2005. *General Requirements for the Competence of Testing and Calibration Laboratories*. ISO International Organization for Standardization, Genève, Switzerland.

ISO (2009). ISO 24333:2009. *Cereals and Cereal Products – Sampling*. ISO International Organization for Standardization, Genève, Switzerland.

Johansson A. S., Whitaker T. B., Giesbrecht F. G., Hagler Jr W. M. and Young J. H. (2000a). 'Testing shelled corn for aflatoxin, Part III: evaluating the performance of aflatoxin sampling plans'. *J AOAC Int*, **83**, 1279–84.

Johansson A.S., Whitaker T. B., Giesbrecht F. G., Hagler Jr. W. M. and Young, J. H. (2000b). 'Testing shelled corn for aflatoxin, Part II: modelling the observed distribution of aflatoxin test results'. *J AOAC Int*, 83:1270–1278.

Johansson A. S., Whitaker T. B., Hagler Jr W. M., Giesbrecht F. G., Young J. H. and Bowman D. T. (2000c). 'Testing shelled corn for aflatoxin, Part I: estimation of variance components'. *J AOAC Int*, **83**, 1264–9.

Knutti R. and Schlatter C. (1982). 'Distribution of aflatoxin in whole peanut kernels, sampling plans for small samples'. *Z Lebensm Unters Forsch*, **174**, 122–128.

MacArthur R., MacDonald S., Brereton P. and Murray A. (2006). 'Statistical modelling as an aid to the design of retail sampling plans for mycotoxins in food'. *Food Additives Contaminants*, **23**, 84–92.

Minkkinen P. (2004). 'Practical applications of sampling theory'. *Chemometrics and Intelligent Laboratory Systems*, **74**, 85–94.

Miraglia M., De Santis B., Minardi V., Debegnach F. and Brera C. (2005). 'The role of sampling in mycotoxin contamination: An holistic view'. *Food Additives Contaminants*, **1**, 31–6.

Ozay G., Seyhan F., Yilmaz A., Whitaker T. B., Slate A. B. and Giesbrecht F. (2006). 'Sampling hazelnuts for aflatoxin: uncertainty associated with sampling, sample preparation, and analysis'. *J AOAC Int*, **89**, 1004–11.

Ozay G., Seyhan F., Yilmaz A., Whitaker T. B., Slate A. B. and Giesbrecht F. (2007). 'Sampling hazelnuts for aflatoxin: effect of sample size and accept/reject limit on reducing the risk of misclassifying lots'. *J AOAC Int*, **90**, 1028–35.

Park D. L. and Pohland (1989). 'Sampling and sample preparation for detection and quantitation of natural toxicants in food and feed'. *J Assoc Off Anal Chem*, **72**(3), 399–404.

Park D. L., Whitaker T. B., Giesbrecht F. G. and Njapau H. (2000). 'Performance of three pneumatic probe samplers and four analytical methods used to estimate aflatoxins in bulk cottonseed'. *J AOAC Int*, **83**, 1247–51.

Petersen L., Dahl C. K. and Esbensen K. H. (2004). 'Representative mass reduction in sampling; a critical survey of techniques and hardware'. *Chemometrics and Intelligent Laboratory Systems*, **74**, 95–114.

Petersen L., Minkkinen P., Esbensen K. H. (2005). 'Representative sampling for reliable data analysis: theory of sampling'. *Chemometrics and Intelligent Laboratory Systems*, **77**, 261–77.

Pitard F. F. (1993). *Pierre Gy's Sampling Theory and Sampling Practice*, 2nd edn, CRC Press.

Pittet A. (1995). 'Keeping the mycotoxins out: experience gathered by an international food company'. *Natural Toxins*, **3**(4), 281–7.

Ramsey M. H. and Thompson M. (2007). 'Uncertainty from sampling, in the context of fitness for purpose'. *Accreditation Quality Assurance*, **12**, 503–13.

Ramsey M. H. and Ellison S. L .R. (eds) (2007). Eurachem/EUROLAB/CITAC/Nordtest/ AMC Guide: Measurement uncertainty arising from sampling: a guide to methods and approaches. Eurachem. Available from http://www.eurachem.org/guides/pdf/ UfS_2007.pdf

Ramsey M. H., Lyn J. and Wood R. (2001). 'Optimised uncertainty at minimum overall cost to achieve fitness-for-purpose in food analysis', *Analyst*, **126**, 1777–83.

Schatzki T. F. (1995a). Distribution of aflatoxin in pistachios. 1. Lot distributions. *J Agric Food Chem*, **43**, 1561–5.

Schatzki T. F. (1995b). Distribution of aflatoxin in pistachios. 2. Distribution in freshly harvested pistachios. J Agric Food Chem, **43**, 1566–9.

Schatzki T. F. and Toyofuku N. (2003). 'Sample preparation and presampling of pistachios'. *J Agric Food Chem*, **51**, 6068–72.

Schatzki, T.F., Toyofuku, N. 2004). 'Sampling and sample preparation of pistachios', in *Proceedings Meeting: The Mycotoxin Menace*. Barug D., Van Egmond H. P., Lopez-Garcia R., Van Osenbruggen W. A. and Visconti A. (eds), Wageningen Academic Publishers, The Netherlands, 221–35.

Sharman M., Macdonald S., Sharkey A. and Gilbert J. (1994). 'Sampling bulk consignments of dried figs for aflatoxins analysis'. *Food Additives Contaminants*, **11**(1), 17–23.

Shotwell O. L., Goulden M. L. and Hessletine C. W. 1974). 'Aflatoxin: distribution in contaminated corn'. *Cereal Chem*, **51**, 492–6.

Spanjer M. C. (2006). 'Theory and criteria for practical sampling' conference materials presented at *The World Mycotoxin Forum, the IV Conference*, Cincinnati, Ohio, USA, November 6–8, 2006.

Spanjer M.C. (2007). 'Sampling for grain quality'. *Stewart Postharvest Review*, **3**(6), 1–6, Stewart Postharvest Solutions Ltd, UK, Published on line.

Spanjer M. C., Scholten J. M., Kastrup S., Jörissen U., Schatzki T. F. and Toyofuku N. (2006). 'Sample comminution for mycotoxin analysis: Dry milling or slurry mixing?' *Food Additives Contaminants*, **23**, 73–83.

Stroka J., Spanjer M., Buechler S., Barel S., Kose G. and Anklama E. (2004). 'Novel sampling methods for the analysis of mycotoxins and the combination with spectroscopic methods for the rapid evaluation of deoxynivalenol contamination'. *Toxicol Lett*, **153**, 99–107.

Trucksess M. W., Whitaker T. B., Weaver C. M., Slate A. B., Giesbrecht F. G., Rader J. I. and Betz J. M. (2009). 'Sampling and analytical variability associated with the determination of total aflatoxins and ochratoxin A in powdered ginger sold as a dietary supplement in capsules'. *J Agric Food Chem*, **57**, 321–5.

Vandeven M., Whitaker T. B. and Slate A. B. (2002). 'Statistical approach for risk assessment of aflatoxin sampling plan used by manufacturers for raw shelled peanuts'. *J AOAC Int*, **85**, 925–32.

Van Egmond H. P. (2002). 'Worlwide regulations for mycotoxins', in *Mycotoxins and Food Safety*, Series Advances in Experimental Medicine and Biology, Vol. 504, DeVries J. W., Trucksess M. W. and Jackson L. S. (eds), Proceedings of the American Academic Society Symposium *Mycotoxins and Food Safety* held in Washington DC, 21–23 August, 2000. Kluwer Academic/Plenum Publishers, New York.

Van Egmond H. P., Schothorst R. C. and Jonker M. A. (2007). 'Regulations relating to mycotoxins in food: perspectives in a global and European context', *Anal Bioanal Chem*, **389**(1), 147–57.

Vargas E. A., Whitaker T. B., Santos E. A., Slate A. B., Lima F. B. and Franca R. C. A. (2004). 'Testing green coffee for ochratoxin A, Part I: estimation of variance components'. *J AOAC Int*, **87**(4), 884.

Vargas E. A., Whitaker T. B., Santos E. A., Slate A. B., Lima F. B. and Franca R. C. A. (2005). 'Testing green coffee for ochratoxin A, Part II: observed distribution of Ochratoxin A test results'. *J AOAC Int*, **88**(3), 780.

Vargas E. A., Whitaker T. B., Santos E. A., Slate A. B., Lima F. B. and Franca R. C. A. (2006). 'Testing green coffee for ochratoxin A, Part III: performance of ochratoxin A sampling plan. *J AOAC Int*, **89**(4), 1021–26.

Velasco J. and Morris S. L. (1976). 'Use of water slurries in aflatoxin analysis'. *J Agric Food Chem*, **24**, 320–40.

Velasco J., Whitaker T. B. and Whitten M. E. (1975). 'Sampling cottonseed lots for aflatoxin contamination'. *J Am Oil Chemists' Soc*, **52**, 191–5.

Wagacha J. M. and Muthomi J. W. (2008). 'Mycotoxin problem in Africa: current status, implications to food safety and health and possible management strategies' *Int J Food Microbiol*, **124**(1), 1–12.

Waltking A. (1980). "Sampling and preparation of samples of peanut butter for aflatoxins analysis', *J Assoc Off Anal Chem*, **63**(1), 103–6.

Whitaker T. B. (2000). 'Sampling techniques', in *Methods in Molecular Biology, Vol. 157, Mycotoxin Protocols*. Trucksess M. W. and Pohland A. E. (eds), Humana Press, Totowa, NJ, 11–24.

Whitaker T. B. (2003). 'Standardisation of mycotoxin sampling procedures: an urgent necessity'. *Food Control*, **17**, 233–7.

Whitaker T. B. (2004). 'Sampling for mycotoxins', in *Mycotoxins in Food: Detection and Control*. Magan N. and Olsen M. (eds), Woodhead Publishing, Cambridge, UK, 69–81.

Whitaker T. B. (2006). 'Sampling food for mycotoxins'. *Food Additives Contaminants*, **23**(1), 50–61

Whitaker T. B. and Wiser E. H.. (1969). 'Theoretical investigations into the accuracy of sampling shelled peanuts for aflatoxin'. *J Am Oil Chemists' Soc*, **46**, 377–9.

Whitaker T. B. and Dickens J. W. (1979). 'Estimation of the distribution of lots of shelled peanuts according to aflatoxin concentrations'. *Peanut Sci*, **6**, 124–6.

Whitaker, T. B. and Dickens J. W. (1983). 'Evaluation of a testing program for aflatoxin in corn meal'. *J Assoc Off Anal Chem*, **66**, 1055–8.

Whitaker T. B. and Park D. L. (1993). 'Problems associated with accurately measuring aflatoxin in food and feeds: errors associated with sampling, sample preparation, and analysis'. in *The Toxicology of Aflatoxins: Human Health, Veterinary, and Agricultural Significance*. Eaton D. L. and Groopman J. D. (eds), Academic Press, San Diego, CA, 433–50.

Whitaker T. B. and Johansson A. J. (2005). 'Sampling uncertainties for the detection of chemical agents in complex food matrices'. *J Food Protection*, **68**(6).

Whitaker T. B., Dickens J. W. and Wiser E. H. (1970). 'Design and analysis of sampling plans to estimate aflatoxin concentrations in shelled peanuts'. *J Am Oil Chemists' Soc*, **47**, 501–4.

Whitaker T. B., Dickens J. W., Monroe R. J. and Wiser E. H. (1972). 'Comparison of the observed distribution of aflatoxin in shelled peanuts to the negative binomial distribution'. *J Am Oil Chemists' Soc*, **49**, 590–3.

Whitaker T. B., Dickens J. W. and Monroe R. J. (1974). 'Variability of aflatoxin test results'. *J Am Oil Chemists' Soc*, **51**, 214–21.

Whitaker T. B., Whitten M. E. and Monroe R. J. (1976). 'Variability associated with testing cottonseed for aflatoxin'. *J Am Oil Chemists' Soc*, **53**, 502–5.

Whitaker T. B., Dickens J. W. and Monroe R. J. (1979). 'Variability associated with testing corn for aflatoxin'. *J Am Oil Chemists' Soc*, **56**, 789–94.

Whitaker T. B., Dowell F. E., Hagler Jr. W. M., Giesbrech F. G. and Wu J. (1994). 'Variability associated with sampling, sample preparation, and chemical testing of farmers stock peanuts'. *J Assoc Off Anal Chem Int*, **77**, 107–16.

Whitaker T. B., Springer J., Defize P. R., deKoe W. J. and Coker R. (1995). 'Evaluation of sampling plans used in the United States, United Kingdom, and The Netherlands to test raw shelled peanuts for aflatoxin'. *J Assoc Off Anal Chem Int*, **78**, 1010–8.

Whitaker T. B., Horwitz W., Albert R. and Nesheim S. (1996a). 'Variability associated with methods used to measure aflatoxin in agricultural commodities'. *J Assoc Off Anal Chem Int*, **79**, 476–85.

Whitaker T. B., Giesbrecht F. G. and Wu J. (1996b). 'Suitability of several statistical models to simulate observed distributions of sample test results in inspection of aflatoxin-contaminated peanut lots'. *J Assoc Off Anal Chem Int*, **79**, 981–8.

Whitaker T. B., Trucksess M., Johansson A. S., Giesbrecht F. G, Hagler Jr. W. M. and Bowman D. T. (1998). 'Variability associated with testing corn for fumonisin'. *J Assoc Off Anal Chem Int*, **81**, 1162–8.

Whitaker T. B., Hagler Jr. W. M. and Giesbrecht F. G. (1999). 'Performance of sampling plans to detect aflatoxin in farmers' stock peanut lots by measuring aflatoxin in high-risk grade components'. *J Assoc Off Anal Chem Int*, **82**, 264–70.

Whitaker T. B., Hagler Jr. W. M., Giesbrecht F. G. and Johansson A. S. (2000). 'Sampling, sample preparation, and analytical variability associated with testing wheat for deoxynivalenol'. *J Assoc Off Anal Chem Int*, **83**, 1285–92.

Whitaker T. B., Hagler Jr. W. M., Johansson A. S., Giesbrecht F. G. and Trucksess M. W. (2001). 'Distribution among sample rest results when testing shelled corn lots for fumonisin'. *J Assoc Off Anal Chem Int*, **84**, 770–776.

Whitaker T. B., Hagler Jr. W. M., Griesbrecht, F. G. and Johansson, A. S. (2002). 'Sampling wheat for deoxynivalenol', in *Mycotoxin and Food Safety*. DeVries J. W., Trucksess M. W. and Jackson L. S. (eds), Kluwer Academic/Plenum Publisher, NY, USA, 73–83.s

Whitaker T. B., Slate A. B., Jacobs M., Hurley J. M., Adams J. and Giesbrecht F. G. (2006). 'Sampling almonds for aflatoxin, Part I: estimation of uncertainty associated with sampling, sample preparation, and analysis'. *J AOAC Int*, **89**(4), 1027–34.

Whitaker T. B., Slate A. B., J. M. Hurley and Giesbrecht F. G.. (2007a). 'Sampling almonds for aflatoxin, Part II: estimating risks associated with various sampling plan designs'. *J AOAC Int*, **90**(3), 778–85.

Whitaker T. B., Doko M. B., Maestroni B. M., Slate A. B. and Ogunbanwo B. F. (2007b). 'Evaluating the performance of sampling plans to detect Fumonisin B_1 in maize lots marketed in Nigeria'. *J AOAC Int*, **90**, 1050–9.

Whitaker T. B., Slate A. B, Doko M. B., Maestroni B. M. and Cannavan A. (2009). *Sampling Procedures to Detect Mycotoxins in Agricultural Commodities*, Springer Verlag, UK (in press).

World Health Organisation (WHO) (2002). Report series 906, *Fifty-sixth report of the Joint FAO/WHO Expert Committee on Food Additives, Evaluation of certain mycotoxins in food*, Geneva.

2

Sample preparation and clean up in mycotoxin analysis: principles, applications and recent developments

E. Razzazi-Fazeli and E. V. Reiter, University of Veterinary Medicine, Austria

Abstract: Risk assessment and monitoring of mycotoxins in agricultural commodities is associated with an increasing demand for appropriate analytical methods. Owing to the introduction of regulatory limits for various mycotoxins, the determination of mycotoxins in food and feed has become an important research issue in routine and governmental laboratories. The most prevalent methods for qualitative and quantitative determination, with the exception of enzyme linked immunosorbent assay (ELISA), are not able to analyse mycotoxins in complex matrices without sample pretreatment. In the same way as for all other analytes, sample preparation is one of the most crucial and important steps in the whole chain of analysis. There are different strategies used to determine mycotoxins in agricultural commodities as well as in biological samples, for example animal tissues, urine and so on. There are various approaches, such as solid phase extraction and liquid–liquid extraction; however, the most important and popular sample preparation technique used in mycotoxin analysis is immunoaffinity chromatography. In recent years there have been efforts to introduce selective and cost-effective strategies in sample pretreatment of mycotoxins. This chapter describes the significance of sample clean up in the analysis of major and abundant mycotoxins. Various conventional methods that are applied to sample preparation of mycotoxins in food, feed and biological samples will be discussed. Recent developments are included and different strategies for achieving more efficient extraction and separation are compared and presented.

Key words: extraction, immunoaffinity columns, mycotoxins, sample clean up, solid phase extraction.

2.1   Introduction

The elimination of matrix compounds is the major goal of sample preparation and is a key step in modern and sophisticated chemical analysis. Different conventional and modern sample pretreatment methods are currently used in different laboratories. Generally, most sample pretreatment procedures are time consuming and labour intensive and some are poorly suited to analysing large volumes of samples. While chromatographic techniques as well as immunological methods have been automated, sample preparation, especially based on immunoaffinity enrichment in mycotoxin analysis, still have to be carried out manually. Additionally, the cost of such single used immunoaffinity columns is much higher than that of conventional solid-phase extraction. However, the selection of a preparation method is dependent upon the mycotoxin(s) to be analysed, their concentration level, the sample matrix and the instrumental measurement technique. To improve the sample clean-up and increase the effectiveness of matrix elimination, additional innovative approaches have been taken in the last decades.

Determination of mycotoxins in food and feed as well as in biological samples is always challenging owing to the diversity of sample matrices and the variety of molecular species that need to be determined. A sophisticated analysis flow includes sampling the commodity (feed and food), sample extraction, sample preparation and finally chromatographic or immunochemical analysis (Fig. 2.1). Improvements and validation of the whole analysis chain from sample pretreatment to final analysis are necessary to provide reliable results. With this in mind, sample preparation is an essential stage especially in the analysis of mycotoxins in food and feed as well as in biological matrices. The significance of sample preparation in chromatographic analysis becomes obvious if the cost and time taken, as well as source of errors throughout the analysis are considered. According to R.E. Majors (no date), sample preparation takes about 61% of the time and cost of analysis (https://cp.chem.agilent.com/Library/slidepresentation/Public/Trends%20in%20Sample%20Preparation%20for%20Chromatography.pdf). In general, optimized sample pretreatment would enormously help to generate accurate and consistent analytical results. The selectivity of the applied clean up strategy influences directly the sensitivity and reliability of the whole analytical method. Particularly, in the field of mycotoxin analysis in food and feed, enormous progress in sample preparation has been achieved in recent years.

2.2   Methods used for extraction and clean up of mycotoxins from complex matrices

In most cases sample preparation itself comprises several processes. The mycotoxins have to be extracted prior to further treatment. Subsequently, sample preparation includes two steps of extraction and clean up. Both are crucial and cannot be separated from each other. They strongly influence the recovery of the specific

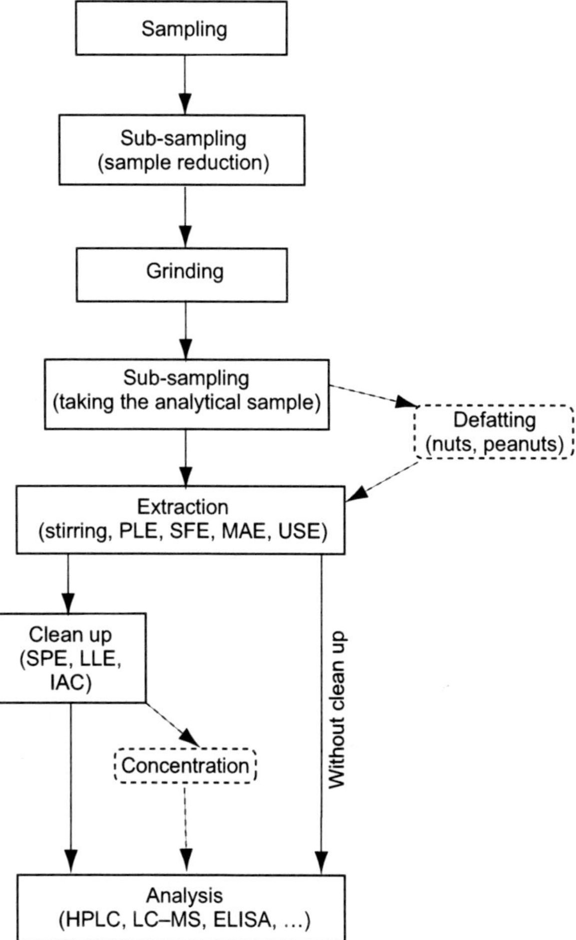

Fig. 2.1   Flow diagram of sample pretreatment in mycotoxin analysis (PLE = pressurized liquid extraction; SFE = supercritical fluid extraction; MAE = microwave assisted extraction; USE = ultrasonic extraction; SPE = solid phase extraction; LLE = liquid liquid extraction; IAC = immunoaffinity chromatography; HPLC = high performance liquid chromatography; LC–MS = liquid chromatography mass spectrometry; ELISA = enzyme linked immunosorbent assay).

compounds and therefore the accuracy of the whole chromatographic method. The extraction is often performed with organic solvents, for instance methanol, acetonitrile or acetone, depending on the physical properties of the intended analyte. While polar mycotoxins such as deoxynivalenol (DON) can be easily extracted with water, other compounds need non-polar organic solvents to be extracted. The extraction techniques should meet requirements to be fast and effective and, of course, economic. In addition, the environmental aspects such as pollution by use of chlorinated solvents should not be waived.

2.2.1   Extraction

When analysing mycotoxins in complex matrices, either by using chromatographic separation or immunoassays, solvent extraction is usually required to release the mycotoxin from the matrix. However, following clean up of the extract, it is then necessary to reduce matrix effects in chromatography. In general, mycotoxins can be extracted from feed and food matrices by shaking or blending with mixtures of water or other polar solvents such as methanol or acetonitrile. In the case of liquid samples, such as fruit juices or milk, liquid–liquid partitioning could be applied, now being replaced by solid-phase or immunoaffinity column clean up.

Sample extraction should simplify and improve sample preparation by combining the appropriate organic solvent, sorbent chemistry, device format and methodology. The most important aspects (parameters) in the extraction procedure are the physicochemical properties of the mycotoxins. An appropriate organic solvent should be chosen in order to extract the analyte with the highest efficiency, which also depends on the sample matrix and the type of clean up. In this case the polarity of the mycotoxins to be extracted is of major concern. Therefore, some extraction solvents are more suitable than others. Although either methanol or acetonitrile mixtures with water are proposed for the extraction, different recovery rates have been reported using either methanol (MeOH)–water or acetonitrile(ACN) –water for the extraction of mycotoxin contaminated samples. Possible reasons for the different recoveries are either interference with matrix compounds, or possible absorption of water into the dry matter (Stroka *et al.*, 1999). In the case of aflatoxins (AF), the most studied mycotoxins, the extraction efficiency was evaluated in some cases. A higher loss of water in the extractant using ACN–water or acetone–water than MeOH–water as extraction solvent was registered in infant formula, animal feed and paprika. The highest loss of water appeared via ACN– water contents (60:40 and 70:40), which was accompanied by very high recoveries. In contrast, extraction techniques with mixtures of MeOH–water have provided more reliable results. The authors proposed that the loss of water is a result of water absorption of the matrix (Stroka *et al.*, 1999).

Möller and Nyberg have evaluated the extraction of AF from peanut meal (certified reference material) using four different solvents, where mixtures of chloroform–water (100+10), ACN–water (60+40), acetone–water (85+15) and MeOH–water (80+20) were tested. The best results were obtained with the solvents consisting of chloroform–water and ACN–water, which have shown good analytical results near the certified value while the others obtained unsatisfactory recoveries (Möller and Nyberg, 2004). In contrast, in a third study dealing with extraction of AFB_1 from sorghum and maize, the use of acetone resulted in an increase of extracted AFB_1 (Bradburn *et al.*, 1995).

In the case of fumonisin (FUM) some investigations were taken to evaluate the extraction performance of mixtures of ACN–MeOH–water, MeOH–water and also under acidified and alkaline conditions. The results show that a mixture of ACN–MeOH–water (25–25–50) is better than conventional MeOH–water mixtures and even much better than alkaline or acidified extractants (Scott *et al.*,

1999). However, in a further evaluation of FUM extraction solvents on corn-based samples, only MeOH–water and acidified MeOH–water mixtures were used, followed by means of immunoaffinity clean up. A higher recovery for maize gluten using acidic extraction was shown (Meister, 1999). For a more polar toxin such as DON, extraction using only water is common. Extraction solvents such as MeOH–water (50+50) or ACN–water (84+16) have been evaluated in wheat and maize and ACN–water was found to have fewer matrix interferences (Trenholm *et al.*, 1985). In case of T-2 toxin, an ACN–water extraction technique was developed and tested in comparison to often suggested MeOH–water mixtures. It was shown that ACN is more effective in the extraction of complex matrices. Nevertheless, in simple matrices MeOH is as suitable as ACN (Trebstein *et al.*, 2009).

Another important aspect in the extraction procedure is the sample to solvent ratio. A rule of thumb in the extraction of mycotoxins is that about 4–5 ml are used for the extraction of 1 g of contaminated material. Some column and ELISA kit producing companies suggest the use of only 2 ml extractant per gram sample, while sometimes higher sample–solvent ratios (e.g. 10 ml.g^{-1}) are proposed. Furthermore, the extraction time plays a key role, which has been suggested to be at least about 10 min shaking with a magnetic stirrer. However, stirring times up to one hour are also recommended to guarantee a complete extraction. If a high speed mixer (Ultra-Turrax®) is used, a few minutes are suggested to be efficient enough. Experiments on extraction efficiency were mainly performed in the last two decades of the 20th century. It was reported that it is easier to extract spiked samples than naturally contaminated ones (Scott, 1995). The extraction efficiency was tested at numerous peanut to solvent ratios and different methanol concentrations. It was shown that a methanol concentration >55% did not enhance the extraction yield. On the other hand a higher solvent to sample ratio (>5) was suggested to be desirable for the extraction of AF from peanuts (Whitaker *et al.*, 1984). The optimal extraction conditions were calculated to be 60% methanol and a solvent to sample ratio of 10.8 in the case of peanuts (Whitaker *et al.*, 1986).

Extraction procedures for mycotoxins have been optimized and adapted for each mycotoxin in different matrices. However, possible interactions caused by water absorption or salting out effects have to be taken into account. Although MeOH is the dominating solvent in extraction procedures, it has often been shown that ACN–water mixtures are more suitable in some matrices. Beside the commonly used extraction techniques in the last decade, emerging extraction techniques such as pressurized liquid extraction, supercritical fluid extraction, microwave-assisted extraction or ultrasonic extraction have been evaluated and established in mycotoxin analysis. Nevertheless, after application of an effective and optimized extraction technique, additional clean up steps are necessary in most cases to eliminate matrix interferences in chromatographic techniques.

Pressurized liquid extraction/accelerated solvent extraction
Pressurized liquid extraction (PLE) has been initially applied in environmental analysis but also in the analysis of naturally occurring contaminants such as mycotoxins. PLE is synonymous with accelerated solvent extraction (ASE), which

is the trade name of Dionex. The advantage of this method is the use of low amounts of organic solvents during the extraction procedure making it known as an eco-friendly technique. In PLE, conventional solvents at high temperatures (100–180 °C) and pressures (1500–2000 psi equivalent to 10.3–13.8 MPa) are used in order to improve the extraction of analytes from the matrix. In general, fast and efficient extraction from solid samples is achieved. In the case of water as extractant this technique is called subcritical water extraction (SWE). The sample is placed in an extraction cell and the extraction is carried out under pressure at increased temperatures, where the temperature is responsible for breaking bonds within the matrix and leads to diffusion of the favoured analytes. In addition, the solvent volume and extraction time are reduced; therefore the extraction time is only about 15 min. Before extraction, the samples are either ground or sieved. Additionally, the sample needs to be dried when using non-polar solvents for extraction because redundant water affects the extraction efficiency.

PLE can be conducted in two different ways, either in a static mode, where the sample is kept in the solvent under constant pressure, or in dynamic mode where the solvent is pressed through the sample continuously and allows better contact between the extractant and sample. The most important goal of the method is the removal of co-extracting substances. In PLE, dynamic mode high extraction volumes are obtained and therefore often additional clean-up steps are performed (Mendiola *et al.*, 2007; Carabias-Martínez *et al.*, 2005; Richter *et al.*, 1996). Tools for pressurized extraction are commercially available; therefore PLE is widely used in routine analysis of food and environmental samples. The application of PLE in the analysis of mycotoxins was carried out to extract ochratoxin A (OTA) from rice samples (Juan *et al.*, 2008a,b). It was also applied in the extraction of OTA from bread samples (González-Osnaya *et al.*, 2006) as well as from breakfast and infant cereals (Zinedine *et al.*, 2010). An approach for the determination of zearalenone (ZEA) was reported for the extraction of wheat and corn samples (Pallaroni and von Holst, 2003, 2004) and the analysis of ZEA in swine feed and cereals (Urraca *et al.*, 2004). Additionally, PLE has been applied in the analysis of FUM, for the extraction of maize (D'Arco *et al.*, 2008). In AF determination PLE has been applied in the clean up of pistachios (Sheibani and Ghaziaskar, 2009). PLE was also applied in the simultaneous extraction of the *Fusarium* toxins DON, ZEA and FUM (Royer *et al.*, 2004). In comparison to previously known methods, the authors observed similar recoveries (85–104% if only one mycotoxin is of interest, 40–90% by analysing co-occurring mycotoxins), good reproducibility and the ease of automatization. A further advantage of PLE is that no additional filtration step is needed.

Supercritical fluid extraction
Supercritical fluid extraction (SFE) uses the distinctive properties of supercritical fluids to assist the extraction of analytes from the matrix. In most cases carbon dioxide under critical conditions is used. An advantage of this technique is the possibility that CO_2 can be used at low temperatures accompanying the possibility to extract temperature sensible analytes such as flavours or fragrances. However,

a disadvantage of the method is the use of the non-polar extraction solvent CO_2. As a consequence polar modifiers (co-solvents) such as methanol have to be added to increase the ability to extract polar substances.

SFE can be performed either in static or dynamic mode using CO_2 as extractant similar to PLE. Different pre-extraction procedures for liquid and solid samples are performed. While solid samples need to be dried, freeze-dried or mixed with an inert gas, the liquid samples have to be either adsorbed on an inert substrate or co-injected with the extraction solvent into the extraction cell (Mendiola *et al.*, 2007). There are only few papers reporting the application of SFE to the analysis of mycotoxins. The main problem is probably the extraction of polar substances from non-polar food products (Anklam *et al.*, 1998). SFE has been applied in the extraction of AF from medicinal plants such as *Zizyphi fructus* (Liau *et al.*, 2007) and even in paprika powder, where high amounts of AF were registered in the lipophilic residue (Ehlers *et al.*, 2006). Furthermore AFs have also been extracted from soil (Starr and Selim, 2008). Ambrosino and co-workers have used SFE for the extraction of beauvericin (BEA) from maize (Ambrosino *et al.*, 2004). In the case of trichothecenes the co-extraction of DON and T-2 toxin from feed has been reported (Huopalahti *et al.*, 1997). Moreover, methods for the determination of five B-trichothecenes in wheat (Josephs *et al.*, 1998) and ZEA in maize flour (Zougagh and Ríos, 2008) have been described. SFE has been applied to several mycotoxins where overall recoveries ranged from 53–104%. Since the introduction of modern SFE techniques in last few decades (mid-1980s), this method has not found broad application in mycotoxin analysis.

Microwave assisted extraction

In microwave assisted extraction (MAE) the samples are extracted within a vessel, where samples and extractant are in contact, when electromagnetic energy is transformed into thermal energy. The use of heated solvents is responsible for increased mass transfer, which leads to improved analyte extraction. By applying this technique, both extraction time and the amount of extraction solvent are reduced. After sample loading, the solvent is added and the vessel is closed. The extraction takes place by heating and following irradiation of the vessel, which takes about 15–30 min. Different commercially available microwave extraction systems allow parallel extraction of up to 50 samples (Sparr Eskilsson and Björklund, 2000). MAE is only applicable for thermally stable compounds. As an example this technique has been implemented in ZEA extraction of wheat and corn samples (Pallaroni *et al.*, 2002, Pallaroni and von Holst, 2003). Although MAE has been widely used in the extraction of contaminants and nutrients from foodstuffs, it has not been utilized often for extraction of mycotoxins in food and feed. MAE provided average recoveries from 79–92% in the analysis of ZEA contaminated samples.

Ultrasonic extraction

In ultrasonic extraction (USE), ultrasonic vibration is applied to ensure contact between the matrix and the solvent. The extraction efficiency is not as high as of those previously mentioned techniques. In general, samples are weighed in a vessel together with the extraction solvent and treated by sonification for a defined period of time in an ultrasonic bath. The ultrasonic extraction is a rapid extraction technique; however, a subsequent clean up step is often necessary. A disadvantage of USE is the lack of automatization (Ridgway *et al.*, 2007). USE has been compared with other extraction techniques for the extraction of some mycotoxins. In the extraction of AF from hazelnuts comparable results were obtained by three techniques: homogenization, USE and matrix solid-phase dispersion (Bacaloni *et al.*, 2008). Furthermore, the suitability of USE for the extraction of OTA from muscles of different animals was evaluated, where inappropriate results were obtained in contrast to other tested extraction procedures such as MAE and PLE (Guillamont *et al.*, 2005). USE has also been applied for the extraction of ZEA contaminated corn (Pallaroni and von Holst, 2003). By comparing MAE, USE and PLE with conventional shaking and blending, the authors showed that ASE and MAE have a higher extraction efficiency (Pallaroni and von Holst, 2003). Usually, the extraction techniques are coupled to HPLC or LC–MS. However, blending is a proven method and is still in routine use in most laboratories, since the cost and the expenses of other techniques are too high.

Liquid-liquid partitioning
The liquid-liquid partitioning or extraction (LLE) is based on the distribution of analytes in two non-miscible phases, where the analyte is transferred from one phase into the other. Conventionally, LLE is carried out in a separatory funnel. The phases are usually an aqueous solvent (hydrophilic) and a hydrophobic organic solvent. The distribution between the phases is a result of thermodynamic forces until equilibrium between both phases is achieved (Cantwell and Losier, 2002). LLE was the first extraction method used for the analysis of AF in feed and food samples from the beginning of the mycotoxin story in the 1960s but has been replaced more and more by other extraction and clean up strategies. The major drawback of this method is the high amount of applied organic and formerly chlorinated solvents. Other difficulties include the formation of emulsions. As a consequence this method is being replaced either by solid-phase extraction or immunoaffinity extraction. However, LLE is still applied in the sample preparation of some mycotoxins. LLE is still used, for instance, as a second clean up step when analysing trace mycotoxin concentrations in complex matrices for the analysis of AFs and OTA in duplicate diets (Sizoo and van Egmond, 2005) or the determination of OTA in pig tissues prior to immunoaffinity clean up (Monaci *et al.*, 2004). Further applications are the extraction of OTA-contaminated coffee using phosphoric acid and dichloromethane for extraction (Pittet and Royer, 2002) or in a modified manner for the extraction of OTA-contaminated milk using MeOH (González-Osnaya *et al.*, 2008). The ergot alkaloids were extracted by LLE prior to solid-phase extraction clean up in rye samples (Mohamed *et al.*, 2006) and *Alternaria* toxins in rape meal and sunflower meal using chloroform (Nawaz *et al.*, 1997). Cyclopiazonic acid, an AF-co-occurring substance, has been extracted

using a chloroform-free extraction technique, where acetonitrile–alkaline water was employed (Hayashi and Yoshizawa, 2005). Patulin (PAT) in apple juice was determined after LLE using ethylacetate (Murillo *et al.*, 2008, Murillo-Arbizu *et al.*, 2009). On the other hand in case of OTA-contaminated cereals Solfrizzo *et al.* reported that LLE is not suitable for the clean up of all cereals owing to impure extracts. Recoveries from 68–97% can be obtained using LLE with the exception of ergot alkaloids or cyclopiazonic acid where lower recoveries were observed.

New developments in LLE include the use of aqueous two-phase polymeric systems, aqueous two-phase micellar systems or two-phase reversed micellar systems (Mazzola *et al.*, 2008). As an example of these trends, OTA was extracted from wine, must or beer by applying these techniques. In the supramolecular extraction, reversed micelles are formed by coacervation. An amphiphile-based supramolecular solvent is formed by dispersion of decanoic acid in tetrahydrofuran–water and added to the liquid sample; the extraction is carried out by stirring and centrifugation. A supramolecular solvent strategy aims to simplify sample extraction without special equipment. The advantages of this method are that very low extraction volumes are needed and high amounts of analyte are concentrated in a small volume (Ballesteros-Gómez *et al.*, 2009). The extraction of AFM_1 from milk and derived products was carried out by means of a sodium citrate solution. The idea behind this strategy is to resuspend proteins, which are known to bind AFM_1 and allow the transportation of AFM_1 in the aqueous phase. The aqueous phase, which is obtained after heating and centrifugation, is analysed directly using chromatographic methods (Anfossi *et al.*, 2008). Furthermore, AFs were extracted from maize using aqueous sodium dodecyl sulfate (Maragos, 2008).

A modified liquid extraction procedure, liquid-phase microextraction (LPME), was proposed for the extraction of OTA from wine, where only small amounts of extraction solvent are placed in a porous fibre connected to syringe needles on each extreme. The fibre is located in a vial together with the sample (González-Peñas *et al.*, 2004). The extraction is carried out by stirring with a magnetic stirrer until the equilibrium between the sample and the fibre is reached. The advantages of the method are the low costs of the extraction equipment and the possibility of parallel extractions (González-Peñas *et al.*, 2004).

Another alternative to LLE is solid-supported LLE, which uses diatomaceous earth particles to perform the extraction. The cartridges are packed with specially cleaned and sized diatomaceous-earth. There are commercially available products such as EXtrelut® or ChemElut®. The aqueous sample is adsorbed and distributed into a thin film over the solid support. Afterwards a water-immiscible organic solvent such as dichloromethane, chloroform, methyl t-butyl ether or ethyl acetate are passed through the cartridge to percolate down and extract the analytes from the aqueous layer throughout the column while interfering substances remain unextracted. In this case phase separation occurs in the interface between the two liquid phases and on the surface of the diatomaceous earth particle. This strategy of modified LLE was used to extract DON from the urine and faeces of animals fed with DON prior to applying the extract to immunoaffinity columns (Valenta *et al.*, 2003; Dänicke *et al.*, 2004).

Application of LLE either as a single clean up or coupled with other clean up methods has been performed in wheat (Schuhmacher *et al.*, 1997), infant food (Lom-baert *et al.*, 2003; Schothorst *et al.*, 2005), an alcoholic fermentation process (Garda *et al.*, 2005), beverages, juice and soft drinks (Melchert and Pabel, 2004). Furthermore, Schuhmacher *et al.* (1997) also investigated EXtrelut® performance during the interlaboratory comparison analysis study of DON and other fusarium toxins.

2.2.2   Solid-phase extraction/multifunctional columns

Analyte enrichment is the next critical step after sample extraction. Numerous techniques are used in sample preparation. The main goals of the intended sample clean up techniques are the elimination of matrix interferences as well as the preconcentration of analytes. Formerly solvent-consuming techniques such as LLE were used, which are nowadays mainly replaced by solvent-saving methods using solid-phase extraction (SPE).

SPE gained enormous interest in the field of analytical chemistry including pharmaceutical analysis, food and environmental analysis. This was due to the fact that SPE is more efficient than LLE, easy to perform, rapid and can simply be automated. Moreover, the amount of organic solvent used and the extraction time are reduced and there is no necessity to use chlorinated solvents. In general, sorbent particles are placed in plastic tubes between porous frits and are based either on a silica matrix or polymeric phases. Owing to the interactions between the analyte and the matrix they are adsorbed to the solid phase and after a washing step the analyte is eluted, commonly with organic solvents (Fig. 2.2). Nowadays, a wide range of sorbents are available, such as silica, alumina, diatomaceous earth or Florisil® as well as modified silica, porous polymers or carbon (Poole, 2003). The most common used SPE columns and cartridges contain modified silica like C18, phenyl or aminopropyl bonded phases and so on (Scott, 1993; Poole, 2003). Furthermore, other specific SPE stationary phases based on ion-exchange, molecular recognition or immunoaffinity are often applied in mycotoxin clean up and are discussed later in detail.

SPE can be either used for analyte enrichment or matrix elimination. It can be applied as a reversed phase or a normal phase separation. Reversed phase SPE is characterized by a polar or moderately polar mobile phase, a non-polar analyte and a non-polar stationary phase, where hydrophilic silica surfaces are modified by hydrophobic groups. In normal phase SPE the analyte of interest and the stationary phase are polar, while the mobile phase is non-polar. Retention of the analyte in the column takes place owing to interactions between the functional groups of the analytes and the surface of the sorbent.

The disadvantages of SPE are the variations in reproducibility caused by sorbent properties and additionally often low capacities in retaining impurities, which interfere in subsequent determination (Poole, 2003). An important advantage of SPE is the ease of online coupling (Rodriguez-Mozaz *et al.*, 2007).

Another type of modified SPR, multifunctional columns, has only been used

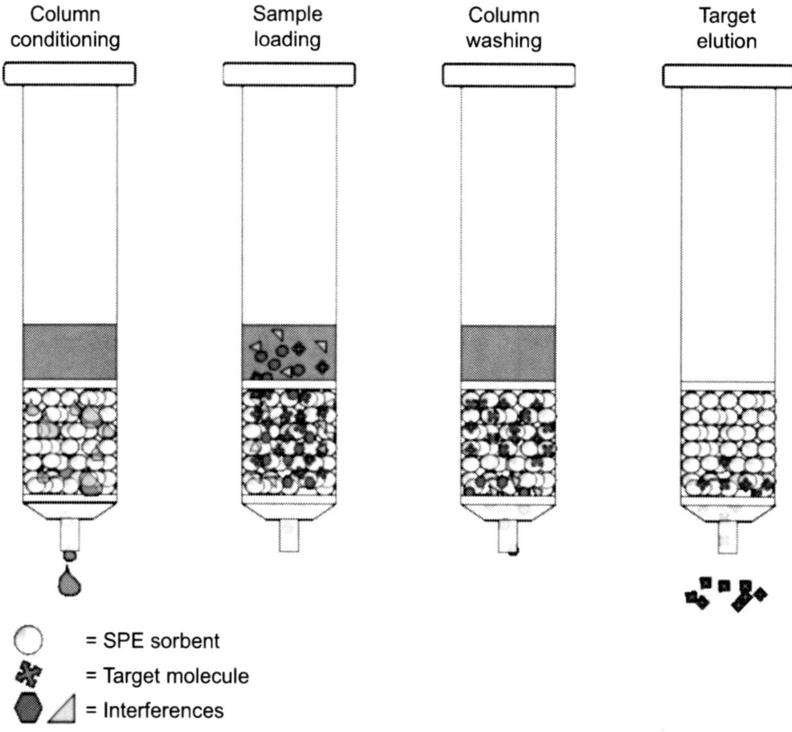

Fig. 2.2   Basic steps in clean up using solid-phase extraction (SPE).

for the analysis of mycotoxins since the 1980s (Wilson and Romer, 1991). The columns are commercially available under the trade name of Mycosep® and/or Multisep® and can be used in a one-step clean up (Fig. 2.3). The adsorbent consists of alumina, active charcoal, celite, polymers and ion-exchange resins in a specially designed column (Weingaertner *et al.*, 1997; Krska, 1998). This column has a filter disc with a rubber flange in the bottom that contains a porous frit and one-way valve, while the adsorbent materials are packed on the upper side of the filter disc. The extract is filtered and then transferred into the tube of the Mycosep column. Purification is performed by pushing the flange into the tube. The crude extract will pass through the adsorbent material and the purified extract remains in the upper side of the adsorbent packing material. The purified extract is either evaporated under nitrogen or can be injected directly into HPLC or GC after derivatization. The multifunctional clean up strategy is a quick sample purification method, which is reliable, robust and shows appropriate recovery. This fast clean up method has frequently been employed for the analysis of type A and B trichothecenes in food and feed samples. These types of adsorbent are able to remove impurities but are less selective in sample preparation.

SPE columns were often used in routine analysis during the last decade, but in the clean up of mycotoxins they have been more and more often replaced by

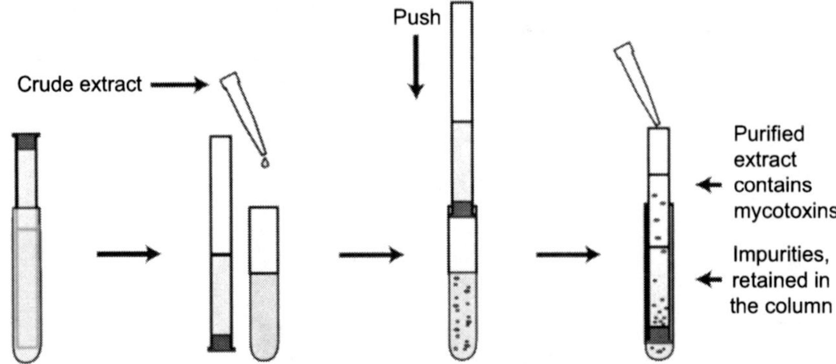

Fig. 2.3   Purification steps using multifunctional columns. Originated by Romer Labs®.

immunoaffinity chromatography. There are plenty of reports on the use of SPE in the analysis of different mycotoxins such as trichothecenes, AF, OTA or FUM. Additionally, there are some applications in the clean up of ergot alkaloids, enniatins and other seldom analysed mycotoxins. SPE has also been found to be a suitable tool in multitoxin clean-up prior to LC–MS determination. The following, recently published, examples are not intended to be exhaustive. Some examples include the analysis of *Alternaria* toxins, where SPE was successfully applied in the clean up of tomato paste (Fente, 1998) and apple juice (Delgado *et al.*, 1996) and in the analysis of alternariol and its monomethylether in fruit juices and common beverages (Lau *et al.*, 2003) as well as in carrots (Solfrizzo *et al.*, 2004). In the AF analysis mainly IAC clean up is applied. However a few applications of SPE clean up have been reported, for instance an automated SPE clean up for AF contaminated drupes and olive leaves (Alcaide-Molina *et al.*, 2009). Previous applications are the sample purification of rice (Park and Kim, 2006) or the use of aluminium oxide columns to clean up AF-contaminated commodities like almonds, Brazil nuts or pistachios (Sobolev and Dorner, 2002) and the use of a Florisil®-based column to clean up imported products such as peanuts, corn rice, cotton seed and various nuts (Sobolev, 2007). The clean up of citrinin-contaminated grain and food samples was developed using polyamide SPE columns (Meister, 2004). A Finnish group reported the clean up of enniatins together with beauvericin in a corn-based material (Kokkonen and Jestoi, 2009) and also in the eggs of laying hens prior to LC–MS determination (Jestoi *et al.*, 2009).

Recently some applications of the SPE clean up of ergot alkaloids have been reported in rye flour (Mohamed *et al.*, 2006; Storm *et al.*, 2008) and cereal-based feed (Ruhland and Tischler, 2008), as well as in cereals and derived products (Krska *et al.*, 2008). In the sample preparation of FUM anion exchange columns (e.g. SAX) or hydrophilic lipophilic balance sorbents (e.g. Oasis™ HLB) are often applied. Several SPE columns were tested in the analysis of corn and corn-based feed, where C18 was found to be the most appropriate, after ACN–water extraction (Dilkin *et al.*, 2001). Hydrolyzed fumonisin B$_1$ (HFB$_1$) was determined in alkaline treated corn using ion-exchange clean up columns (Scott and Lawrence, 1996).

Furthermore, free and bound FUM were determined in heat processed corn (Park *et al.*, 2004). FB_1 and FB_2 were purified by tandem coupling of SAX and C18 columns in corn and derived products (Miyahara *et al.*, 1996). Additionally, FB_1 and its metabolite aminopentol-1 (also known as HFB_1), which is formed by hydrolysis after alkaline treatment, was determined in swine liver (Pagliuca *et al.*, 2005). However, SPE clean up of FUM has been replaced more and more by IAC clean up.

A wide range of SPE applications is found in the clean up of OTA. SPE clean up was applied to contaminated grains, where the performance of two commercially available SPE columns was evaluated (Buttinger *et al.*, 2004). In addition SPE clean up was applied to maize (Pelegri *et al.*, 1997), cereals and feed (Biancardi and Riberzani, 1996; Vega *et al.*, 2009), wine and beer (Varelis *et al.*, 2006; Bacaloni *et al.*, 2005; Reinsch *et al.*, 2005; Jornet *et al.*, 2000), dried vine fruits (Galvis-Sánchez *et al.*, 2008), coffee, either prior to chromatographic analysis or flow through immunoassay (Sibanda *et al.*, 2002) and in biological fluids such as urine (Domijan *et al.*, 2003) or river water samples (Almeda *et al.*, 2009). Similarly PAT, which mainly occurs on fruits and derived products, was purified using SPE devices. In apple puree a C18 column was necessary to remove matrix interferences with LLE (Valle-Algarra *et al.*, in press); in addition, a two-step SPE was applied to apple juice, apple puree and apples (Boonzaaijer *et al.*, 2005) and a single C18 SPE to apple juice (Li *et al.*, 2007).

In the extraction and clean up of trichothecenes, compromises have to be undertaken owing to the different polarity of these type-A and type-B trichothecenes. Macrocyclic and other trichothecenes in wheat were determined using different commercially available SPE columns and self-made (poly)glycidyl methacrylate-divinylbenzene (GMA-DVB) columns, where GMA-DVB and Oasis™-HLB obtained the highest effectiveness especially for macrocyclic trichothecenes (Stecher *et al.*, 2007). A further approach for determination of B-trichothecenes in wheat was applied by Valle-Algarra *et al.* (2005) where MycoSep® columns, cartridges made of alumina–charcoal–silica or alumina–charcoal–C18 silica mixtures showed the best performance. The application of commercially available SPE columns showed that by rinsing the columns with ACN prior to sample application, the recovery was enhanced (Jestoi *et al.*, 2004). Commercial SPE columns (Bond Elut®) were applied to the analysis of 12 trichothecenes and ZEA after ACN–water extraction. By applying this method, higher recoveries were obtained for polar trichothecenes, where for less polar metabolites decreased recoveries were registered (Klötzel *et al.*, 2006). An often codetermined mycotoxin, ZEA, has been purified using SPE in corn and other cereals (Llorens *et al.*, 2002) and grains (Zöllner *et al.*, 1999).

Other rarely determined mycotoxins such as sterigmatocystin in grains (Versilovskis *et al.*, 2008) and moniliformin in corn (Munimbazi and Bullerman, 1998) were also cleaned up using SPE. A great advantage of SPE in contrast to IAC is observed in the multitoxin analysis, which has been forced owing to the necessity to analyse as many toxins as possible. Several benefits are found in SPE clean up particularly owing to the increasing importance of LC–MS techniques.

Because of the different polarities a range of compromises have to be undertaken to allow clean up with SPE. Additionally, a few applications propose the use of several clean up tools for one extract to recover all the mycotoxins. As an example one part of the extract was applied to a NH$_2$-SPE column followed by C18-SPE while the other part was purified using SAX in the mycotoxin analysis of sweet pepper (Monbaliu *et al.*, 2009). Furthermore, medical and aromatic herbs were extracted and the metabolites they contained (AF, OTA, ZEA, DON, T-2, CIT and FUM) were purified in particular using commercial SPE columns (Santos *et al.*, 2009). The application of commercial SPE in co-extraction of mycotoxins was further reported for the determination of six mycotoxins in corn silage (Garon *et al.*, 2006), AF and OTA in beer (Ventura *et al.*, 2006), OTA and ZEA in soil (Mortensen *et al.*, 2003) or 18 mycotoxins and their metabolites in bovine milk (Sorensen and Elbaek, 2005). In recent years, generally recoveries of >70% were reported for SPE with the exception of some ergot alkaloids or enniatins. SPE can be easily automated for high-throughput applications and can be seen as a cheap alternative to IAC. Several approaches regarding the automation of SPE clean up coupled to the HPLC analysis have been undertaken. This type of sample preparation has the advantage that a large number of samples can be prepared in parallel.

Matrix solid phase dispersion
Matrix solid phase dispersion (MSPD) is a modification of SPE, where a dispersion of a matrix and a solid phase is used to extract viscous, solid or semi-solid samples (Barker, 2007). In MSPD, the samples are treated through blending, accompanied by interaction with the solid phase. In most cases diatomaceous earth is used as solid phase. The blended mixture is subsequently transferred into a column and consequently extracted using an appropriate solvent. There are, of course, some aspects of the application of the sample in MSPD that differ from SPE. While in SPE the sample is applied on the top of the solid material, in MSPD the sample has to be disrupted completely and is distributed, together with solid support material, homogenously in the column. The major point is that a stronger interaction between the sample and the solid material is expected. A benefit of this method is the saving of solvent in comparison to conventional SPE (Barker, 2007). The applicability of this technique was evaluated for the AF extraction of hazelnuts (Bacaloni *et al.*, 2008), in olive oils (Cavaliere *et al.*, 2007), high pigment samples (Hu *et al.*, 2006) and peanuts (Blesa *et al.*, 2003), AFM$_1$ in cheese (Cavaliere *et al.*, 2006), the ZEA extraction of rainbow trout tissues (Lagana *et al.*, 2004) and PAT in apple juice and apple products (Wu *et al.*, 2008, 2009). The use of this special technique in mycotoxin analysis is limited to a certain extent. In MSPD, recoveries from 70–110% were obtained.

Solid-phase microextraction
Solid-phase microextraction (SPME) is a rather new method of sample pretreatment. In this method the sample is adsorbed to a sorbent in a first step and in the following step the analyte is desorbed into the analytical column (Pawliszyn, 2002). The SPME is based on a solventless extraction and selective adsorption

onto a coating medium. In this technique, the sorbent is packed in cartridge devices. The most common SPME system consists of a fused silica fibre coated with a thin film of sorbent, which contains a polymeric coating and is placed between two fibre holders. The intended analytes are adsorbed by the fibre and later desorbed for subsequent determination. This simple technique allows rapid extraction of volatile and semi-volatile substances. An advantage of this method is the reduced sample volume (Pawliszyn, 2002). The benefits of SPME include ease of automatization prior to GC or LC applications (Risticevic *et al.*, 2009). Only some examples of SPME application in mycotoxin analysis have been published such as clean up of OTA in beer (Aresta *et al.*, 2006) and urine samples (Vatinno *et al.*, 2007, 2008), AF in food samples (Nonaka *et al.*, 2009), PAT in fruit juice and fruit samples (Kataoka *et al.*, 2009), T-2 toxin in water samples (Lee *et al.*, 1999) and mycophenolic acid in cheese (Zambonin *et al.*, 2002). Recoveries of SPME applications were reported to be >66%.

Immunoaffinity chromatography

The combination of a selective sample preparation such as immunoaffinity chromatography and HPLC with sensitive detection will lead to a method of greater selectivity and sensitivity. Immunoaffinity chromatography uses the selectivity of the extremely specific interactions between antigens and antibodies to enrich and/ or separate antigens from complex mixtures. Nowadays, immunoaffinity chromatography is an important sample preparation strategy, which can eliminate the interfering matrix components very effectively and enrich the analyte, resulting in better detection and quantification limits.

IAC in mycotoxin analysis was originally developed as a modification of SPE and is based on the binding of the immobilized specific antibodies against the target compound. Either silica or sepharose gels are used as support material for the antibodies which are either polyclonal (pAbs) or monoclonal (mAbs). The characteristics of pAbs are ease of production accompanied by lower costs. Further, pAbs are a mixture of specific antibodies representing the whole IgG fraction. On the other hand mAbs are subjected to long-term development, but have the advantage of having the same specificity for batch-to-batch production.

IAC consists of anti-mycotoxin antibodies coupled covalently to an appropriate carrier and stored in phosphate buffered saline. The mycotoxins will bind to the specific antibodies when the sample extract is loaded into the column. The impurities will be removed without retention during loading and washing steps. Afterwards the analyte can be eluted through the denaturation of antibodies with appropriate organic solvents such as methanol or acetonitrile (Fig. 2.4).

The binding to specific antibodies and the formation of Ab–Ag complexes without binding of matrix compounds enables very clean extracts to be obtained (Delaunay-Bertoncini and Hennion, 2004). However, antibodies often show cross reactivities with structurally similar compounds so that determination of a whole class of analytes antibodies with high cross reactivities should be preferred.

The only disadvantage of commercially available IACs is the lack of reusability and therefore the higher cost compared with other sample pretreatment strategies.

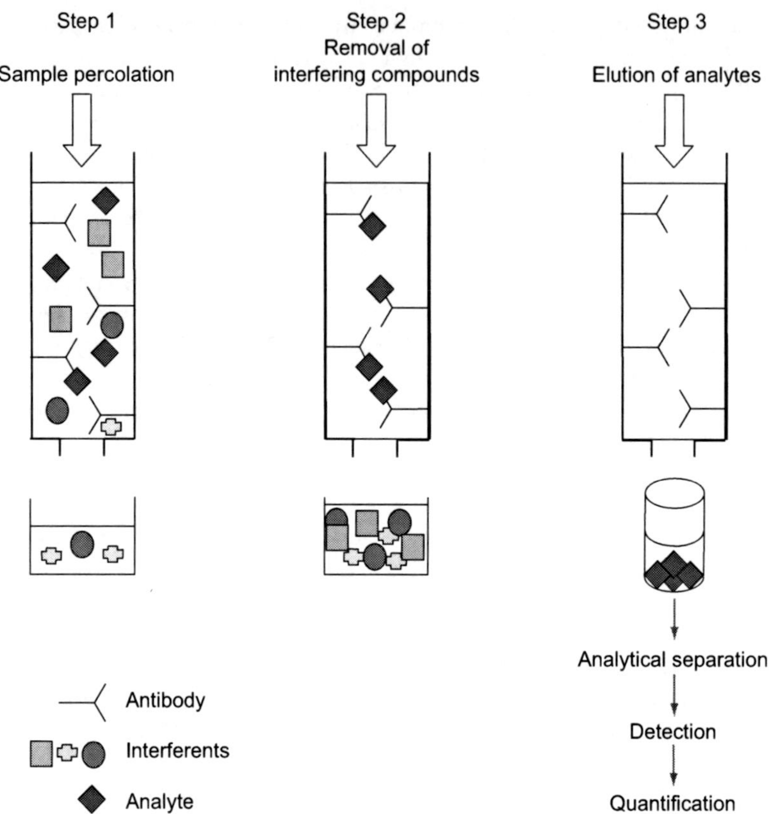

Fig. 2.4   Schematic overview of the immunoaffinity clean up. First published by
Delaunay-Bertoncini *et al.* (2001).

The stability of the used antibodies varies. Therefore, the conditions of sample
application, washing and elution have to be optimized for each antibody itself,
which is carried out by the producer. The capacity of the immunoaffinity columns
depends on the amount of immobilized antibodies. Nowadays, a wide range of
IACs are commercially available and have been successfully applied to the clean
up of the most common mycotoxins. New developments in immunoaffinity clean
up are mainly based on multitoxin extraction, which allows high enrichment and
clean chromatograms. Many papers have been published using IAC as a clean up
strategy for AF. Some newly published examples dealing with this clean up
technique are the analysis of AFM_1 in milk (Hussain *et al.*, 2010; Lee *et al.*, 2009)
or human urine samples (Romero *et al.*, 2010). AF have been further determined
in pistachios (Ariño *et al.*, 2009), spices (O'Riordan and Wilkinson, 2008), figs
(Bircan *et al.*, 2008), dried fruits (Juan *et al.*, 2008c), rice (Reiter *et al.*, 2010), or
in cereals (Tam *et al.*, 2006).

Recently published applications of IAC clean up, dealing with the extraction of
FUM from corn, reported that MeOH may not be sufficient to eluate quantitatively

the bound FUM (Oh *et al.*, 2009). FUM have also been analysed by means of IAC in tissue (Tardieu *et al.*, 2008), milk (Gazzotti *et al.*, 2009) or in corn and corn-based products (Oh *et al.*, 2009; Silva *et al.*, 2009; Senyuva *et al.*, 2008).

Examples of the applications of IAC clean up to the analysis of OTA in contaminated samples are those in serum (Ghali *et al.*, 2008), milk (Bascarán *et al.*, 2007), grapes and grape juices (Czerwiecki *et al.*, 2005), wine (Coronel *et al.*, in press), coffee (Noba *et al.*, 2009), wheat bread (Bento *et al.*, 2009), or cereals (Zaied *et al.*, 2009) and derived products (Kabak, 2009).

In the analysis of trichothecenes, DON, T-2 and HT2 toxin have been mainly analysed by IAC, since there are no IACs commercially available for other trichothecenes. DON was measured in cereals and derived products, for instance wheat (Bensassi *et al.*, 2010; Neumann *et al.*, 2009), wheat flour (Moazami and Jinap, 2009) but also in biological samples such as swine serum (He *et al.*, 2009). T-2 and HT2 toxin were analysed in cereals and oats (Trebstein *et al.*, 2008; Meister, 2008). The toxin ZEA was assessed in grains (Manova and Mladenova, 2009), animal feed and feed ingredients (Campbell and Armstrong, 2007) as well as in water samples (Gromadzka *et al.*, 2009) or biological samples (Songsermsakul *et al.*, 2006) by means of IAC. In IAC clean-up recoveries of <70% are reported.

Several multitoxin approaches were carried out using IAC columns, where antibodies against more than one mycotoxin were immobilized. In this case AF and OTA, which are often co-occurring have been determined together in paprika (Shundo *et al.*, 2009) and breakfast cereals (Villa and Markaki, 2009). Furthermore, for clean up of maize samples, an IAC able to bind six types of mycotoxins (Lattanzio *et al.*, 2007) has been applied. In another study, a multitoxin IAC has been used for the determination of AF, OTA and ZEA in air samples (Wang *et al.*, 2008). A combination of both, clean up techniques, namely SPE and IAC, has been reported (Ofitserova *et al.*, 2009). Even in IAC some attempts to automate sample clean up were performed; however, the limitations are the sensitivity of the antibodies to organic solvents and the pressure instability of the columns (Chan *et al.*, 2004; Eskola *et al.*, 2002). Another important restriction of IAC is that specific antibodies are needed and for some mycotoxins, no antibodies have been developed yet.

The predominantly applied methods, SPE and IAC have been compared in the case of OTA extraction in several papers, for instance in barley and wheat samples (Jestoi and Kokkonen, 2008) or in cereals, raisins and green coffee beans (Sugita-Konishi *et al.*, 2006) as well as in wine, must and beer (Saez *et al.*, 2004). In the analysis of DON, comparison of SPE and IAC was carried out in cereal-based food (Klötzel *et al.*, 2005) and swine serum (He *et al.*, 2009). In ZEA, sample preparation has been compared in corn and cereals (Llorens *et al.*, 2002; Zöllner *et al.*, 1999). In most comparative studies, IAC was able to detect lower concentrations and could eliminate matrix interferences in contrast to SPE. Compared with IAC, SPE columns are rapid tools, cheaper to obtain, easy to perform with the possibility of automatization, and have a higher capacity and are less specific (Llorens *et al.*, 2002; Zöllner *et al.*, 1999).

2.2.3   Dilute-and-shoot

The dilute-and-shoot approach, which is based on the analysis of crude extracts without clean up, was developed for multitoxin analysis using MS as detector. The precursor of this technique is simple dilution prior to LC–MS analysis. However, co-eluting compounds are able to suppress or even enhance the signals. The matrix effect in LC–MS is a known phenomenon and is a strong limitation of dilute-and-shoot-approaches, which can lead to false positive or negative results. However, when using an appropriate sample preparation strategy, these interferences are eliminated to a certain extent. Suppression varies from sample to sample and has to be carefully investigated for each matrix (Nelson and Dolan, 2002). To obtain reliable results it is necessary to apply the standards in blank matrices. In addition, different dilution solvents have to be investigated to find the most suitable for the intended matrix (Mol *et al.*, 2008). The problem of ion suppression can be overcome using internal standards, most preferably isotope-labelled ones. Examples of multitoxin analysis using the dilute-and-shoot technique are the analysis of seven different matrices, analysed for their content of 13 to 24 different mycotoxins (Spanjer *et al.*, 2008). According to some multitoxin approaches, it is possible to determine more than 100 toxins in parallel, currently 186 toxins and metabolites (Vishwanath *et al.*, 2009). By applying this strategy, depending on the intended analyte, low recoveries (<20%) and good recoveries (>90%) have been reported. It has to be mentioned that the success of a dilute-and-shoot method highly depends on the sensitivity of the individual instruments and can be reached only by using the most sophisticated and highly developed LC–MS instruments. Multitoxin methods can be recommended for mycotoxins, where maximum levels exist or will possibly come into force. However, the necessity of analysing hundreds of mycotoxins in one run should be questioned.

2.3   Recent developments

Sample preparation techniques are undergoing a continuous process of method improvement and development. In some cases efforts have been undertaken to focus either on the reusability or on minimizing costs in the intended clean up for their applicability in developing countries. Other new developments aim to develop new selective binding ligands. Some recent developments and sample preparation strategies, which were applied to the mycotoxin analysis, are presented.

2.3.1   Immuno-ultrafiltration

Recently a new strategy based on immunoglobulin recognition of mycotoxins has been utilized which is different from immunoaffinity chromatography. Immuno-ultrafiltration (IUF) is based on selective interactions between antigens and antibodies; however, the critical immobilization step is avoided by applying the antibodies to the extract in their free form. IUF is carried out in an ultrafiltration (UF) device using a membrane with pores that are narrow enough to retain the

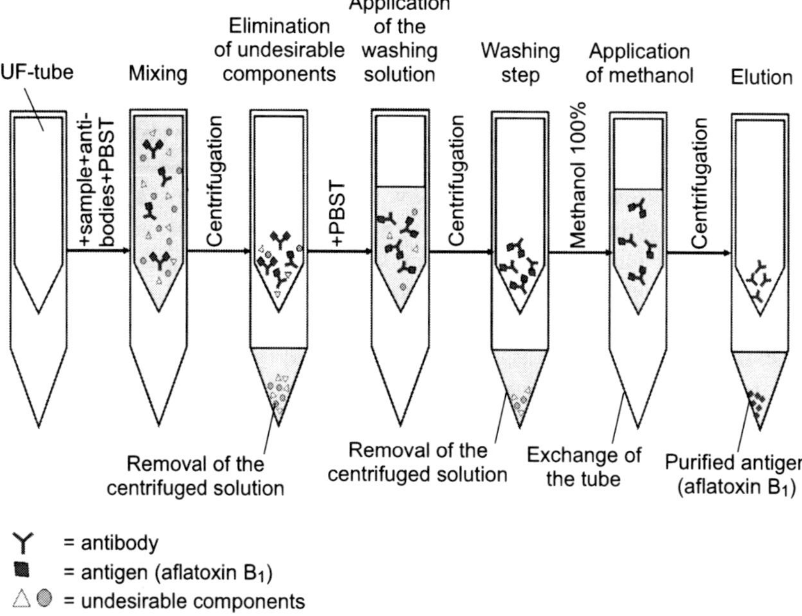

Fig. 2.5 Schematic overview of the immuno-ultrafiltration (IUF) clean up procedure. Reproduced with permission from Reiter *et al.* (2009). Copyright Wiley-VCH Verlag GmbH and Co. KGaA.

antibodies. After mixing the diluted sample extract with an excess of antibodies and, if necessary, incubating it, a centrifugation step is carried out to remove unbound matrix compounds, which are small enough to pass through the membrane. After washing the UF device to remove non-specific bound components, the antigen–antibody complex is dissociated and the analyte is eluted by centrifugation (Fig. 2.5). IUF, which formerly has been used on biological samples (Haasnoot *et al.*, 2002; Westcott *et al.*, 1997), has been successfully applied twice in the clean up of mycotoxins. Reiter *et al.* have applied this technique in the determination of AFs in contaminated rice and maize (Reiter *et al.*, 2009). In another study, IUF was used in the clean up of DON in contaminated maize (Böhm *et al.*, 2008). The clean up, including dilution of solvents, a washing step and elution, has to be optimized for each mycotoxin and antibody. Because of the differences in binding kinetics after application of extract, an incubation step was necessary in the case of AF. However, for DON, incubation with antibodies was found to have no effect. In both applications, recoveries > 76% were obtained. It was shown that matrix interferences can be removed successfully. Additionally, the method is a cheap alternative in developing countries, if there is the possibility of producing the antibodies in-house. Generally, clean up methods based on antibody–antigen interactions are more expensive since the antibody production is very cost intensive. However, another disadvantage is the lack of reusability.

2.3.2   Sol-gel-based immunoaffinity chromatography

Sol-gel immunoaffinity chromatography is based on the encapsulation of antibodies in the pores of a silica glass matrix. During the encapsulation, conformation of the antibodies will not be altered since the preparation is performed under mild conditions. The sol-gel method opened up a new perspective for preparing packing materials for immunoaffinity columns (Cichna-Markl, 2006).

In principle, the preparation is started with the hydrolyzation of a precursor such as tetramethoxysilane. In the following step the precursor is added to the biomolecules or antibodies and in the condensation a gel is formed. In the ongoing alteration the solvents are removed and the glass is formed. Finally, the glass is comminuted with a mortar and placed into columns. The steps in sample loading and extraction are the same as those in immunoaffinity clean up. However, gentle elution conditions are chosen to prevent antibody disruption. The columns are reconditioned using PBS. In comparison to other immobilization techniques, sol-gel technology is a very cheap alternative and less time-consuming. Further, sol-gel columns showed high storage stability, where storage is possible at room temperature.

The sol-gel technique is an appropriate tool to produce reusable IAC. The sol-gel process is a well-known technique for entrapping selective compounds (Cichna-Markl, 2006). In this technique the critical immobilization step is circumvented and possible confirmation changes, which occur by the binding of antibodies on solid support material, are prevented. The sol-gel technique was introduced for the first time in the clean up of mycotoxins in the analysis of DON in food and feed samples and a reusability of the columns of approximately 20 times was reported (Brenn-Struckhofova *et al.*, 2007). It was also successfully applied in the analysis of DON in conventional and organic food and feed (Klinglmayr *et al.*, 2010). In addition, the co-immobilization of anti-DON and anti-ZEA antibodies was successfully applied to the extraction of DON and ZEA from wheat and derived products (Brenn-Struckhofova *et al.*, 2009). Recovery rates ranging from 82–111% have been reported.

2.3.3   Molecular imprinted polymers (MIP)

MIP-based technology is a new type of SPE, where monomers are polymerized to a highly cross-linked polymer. During the polymerization binding sites are formed, which are able to recognize target molecules as well as structurally related compounds. In MIP technology various polymerization techniques have been performed (Haginaka, 2009). Throughout preparation, the structure of the compound is held stable by the cross-linked polymeric structure. For MIP polymerization (Fig. 2.6) different matrices are used such as acrylic and vinyl polymers (e.g. vinylpyridine, 3-acrylamidopropyl trimethylammonium chloride, styrene), organic polymers (e.g. polyphenols, polyurethanes) or those based on silica and titanium dioxide (Pichon and Haupt, 2006). After removal of the imprinted molecules, the polymer shows complementary binding sites, which are able to recognize target molecules (Pichon and Haupt, 2006). The polymer contains

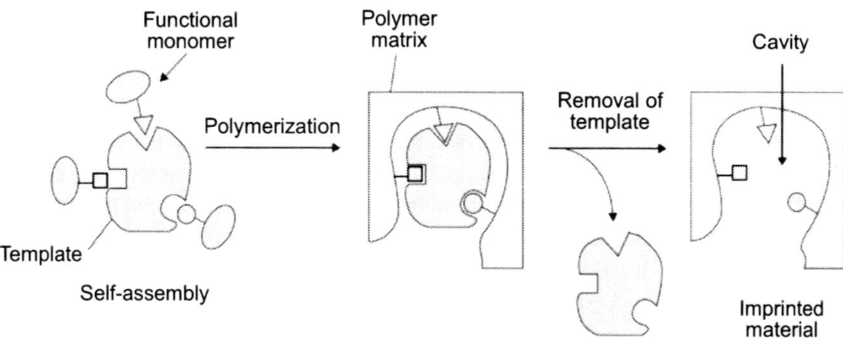

Fig. 2.6   Key steps in the molecular imprinting process. Reproduced from Baggiani *et al.* (2010).

complementary shape, size and functionalities that interact with the intended molecule (Baggiani *et al.*, 2008). The advantages of this technology are the physical and chemical stability of the polymers during extraction, the reusability and even the low costs of production, in contrast to IACs (Baggiani *et al.*, 2008). Further advantages are long storage conditions, without loss of affinity and the possibility of applying this technique in various formats (Mahony *et al.*, 2005). The limitations of this technology are leakage of the columns, which has been overcome by the use of structural analogues (Haginaka, 2009). MIPs have been applied in mycotoxin analysis, where a few applications to OTA extraction have been published (Maier *et al.*, 2004; Jodlbauer *et al.*, 2002; Zhou *et al.*, 2004; Yu and Lai, 2007). MIPs have been developed and applied for the sample clean up of ZEA (Urraca *et al.*, 2006), FUM (De Smet *et al.*, 2009) and DON (Pascale *et al.*, 2008) contaminated samples. Recoveries ranging from 62–103% were reported for these applications.

2.3.4   Aptamers

Aptamers are known as short DNA and RNA strains, which are able to detect target molecules (Hamula *et al.*, 2006). DNA aptamers are single stranded oligonucleotides, which bind specifically to a target molecule and to recognize features of the target molecules in a similar manner to antibodies. DNA aptamers are selected from oligonucleotides libraries, by the SELEX technique (systematic evolution of ligands by exponential enrichment) and then synthesized by means of PCR (Nery *et al.*, 2009). The advantages of this new technique are stability and the fast synthesis in contrast to the use of antibodies. Furthermore, they are easy to couple with fluorescent dye or another marker and are suitable for assays. Moreover aptamers are easy to pack into columns where they are used in the same way as conventional IACs (Cruz-Aguado and Penner, 2008). The first application of DNA-aptamers in mycotoxin analysis was carried out for OTA determination (Cruz-Aguado and Penner, 2008).

2.4   Conclusions

In the analytical process, in most cases, the sample extract cannot be injected directly into the instrument. Since chromatographic techniques are the most applied methods in mycotoxin analysis, novel approaches to sample preparation are needed to address problems that deal with the sensitivity and selectivity of the whole analytical system. There are many techniques available in mycotoxin analysis prior to instrumental analysis and new sample pretreatment strategies have gained some interest. An important aspect is that the measurement uncertainty of analytical procedures not only depends on mycotoxin concentrations and the matrix but also strongly on sample preparation. The future development of sample pretreatment methods will consequently aim to introduce more novel and selective strategies. More effort and research should be performed in the field of sample preparation of mycotoxins, since this is a bottleneck in the whole analytical procedure. Additionally, the introduction of ultra fast LC, known as UPLC and/or ultra-fast GC, aimed to reduce the analysis time. Consequently, in future the automation of the sample clean up is demanded. However, strategies based on selective interactions between the analyte and the stationary phase will still play important roles in mycotoxin analysis. These are methods mainly based on immunoaffinity recognition but also on molecular imprinted polymers as well as aptamers.

2.5   Acknowledgements

The work was supported by the European Union-funded Integrated Project BIOTRACER (contract 036272) under the 6th RTD Framework. BIOTRACER has 46 project partners from 24 countries (including four INCO countries).

2.6   References

Alcaide-Molina, M., Ruiz-Jiménez, J., Mata-Granados, J. M. and Luque de Castro, M. D. (2009) 'High through-put aflatoxin determination in plant material by automated solid-phase extraction on-line coupled to laser-induced fluorescence screening and determination by liquid chromatography-triple quadrupole mass spectrometry'. *Journal of Chromatography A*, **1216**, 1115–25.

Almeda, S., Arce, L., Benavente, F., Sanz-Nebot, V., Barbosa, J. and Valcãįrcel, M. (2009) 'Comparison of off- and in-line solid-phase extraction for enhancing sensitivity in capillary electrophoresis using ochratoxin as a model compound. *Analytical and Bioanalytical Chemistry*, **394**, 609–15.

Ambrosino, P., Galvano, F., Fogliano, V., Logrieco, A., Fresa, R. and Ritieni, A. (2004) 'Supercritical fluid extraction of Beauvericin from maize'. *Talanta*, **62**, 523–30.

Anfossi, L., Calderara, M., Baggiani, C., Giovannoli, C., Arletti, E. and Giraudi, G. (2008) 'Development and application of solvent-free extraction for the detection of aflatoxin M_1 in dairy products by enzyme immunoassay'. *Journal of Agricultural and Food Chemistry*, **56**, 1852–7.

Anklam, E., Berg, H., Mathiasson, L., Sharman, M. and Ulberth, F. (1998) 'Supercritical fluid extraction (SFE) in food analysis: A review'. *Food Additives and Contaminants: Part A: Chemistry, Analysis, Control, Exposure and Risk Assessment*, **15**, 729–50.

Aresta, A., Palmisano, F., Vatinno, R. and Zambonin, C. G. (2006) 'Ochratoxin A determination in beer by solid-phase microextraction coupled to liquid chromatography with fluorescence detection: A fast and sensitive method for assessment of noncompliance to legal limits'. *Journal of Agricultural and Food Chemistry*, **54**, 1594–8.

Ariño, A., Herrera, M., Estopañan, G., Rota, M. C., Carramiñana, J. J., Juan, T. and Herrera, A. (2009) 'Aflatoxins in bulk and pre-packed pistachios sold in Spain and effect of roasting'. *Food Control*, **20**, 811–4.

Bacaloni, A., Cavaliere, C., Faberi, A., Pastorini, E., Samperi, R. and Lagana, A. (2005) 'Automated on-line solid-phase extraction-liquid chromatography-electrospray tandem mass spectrometry method for the determination of ochratoxin a in wine and beer'. *Journal of Agricultural and Food Chemistry*, **53**, 5518–25.

Bacaloni, A., Cavaliere, C., Cucci, F., Foglia, P., Samperi, R. and Lagana, A. (2008) 'Determination of aflatoxins in hazelnuts by various sample preparation methods and liquid chromatography-tandem mass spectrometry'. *Journal of Chromatography A*, **1179**, 182–9.

Baggiani, C., Anfossi, L. and Giovannoli, C. (2008) 'Molecular imprinted polymers as synthetic receptors for the analysis of myco- and phyco-toxins'. *Analyst*, **133**, 719–30.

Baggiani, C., Anfossi, L. and Giovannoli, C. (2010). *Artificial Systems for Molecular Recognition of Mycotoxins, Mycotoxins in Food, Feed and Bioweapons*. M. Rai and A. Varma (eds), Springer, Berlin, 3–20.

Ballesteros-Gómez, A., Rubio, S. and Pérez-Bendito, D. (2009) 'Potential of supramolecular solvents for the extraction of contaminants in liquid foods'. *Journal of Chromatography A*, **1216**, 530–9.

Barker, S. A. (2007) 'Matrix solid phase dispersion (MSPD)'. *Journal of Biochemical and Biophysical Methods; Sample Preparation*, **70**, 151–62.

Bascarán, V., De Rojas, A. H., Chouciño, P. and Delgado, T. (2007) 'Analysis of ochratoxin A in milk after direct immunoaffinity column clean-up by high-performance liquid chromatography with fluorescence detection'. *Journal of Chromatography A*, **1167**, 95–101.

Bensassi, F., Zaied, C., Abid, S., Hajlaoui, M. R. and Bacha, H. (2010) 'Occurrence of deoxynivalenol in durum wheat in Tunisia'. *Food Control*, **21**, 281–5.

Bento, J. M. V., Pena, A., Lino, C. M. and Pereira, J. A. (2009) 'Determination of ochratoxin A content in wheat bread samples collected from the Algarve and Bragança regions, Portugal: Winter, 2007'. *Microchemical Journal*, **91**, 165–9.

Biancardi, A. and Riberzani, A. (1996) 'Determination of Ochratoxin a in Cereals and Feed by SAX-SPE Clean Up and LC Fluorimetric Detection'. *Journal of Liquid Chromatography and Related Technologies*, **19**, 2395–407.

Bircan, C., Barringer, S. A., Ulken, U. and Pehlivan, R. (2008) 'Increased aflatoxin contamination of dried figs in a drought year'. *Food Additives and Contaminants: Part A: Chemistry, Analysis, Control, Exposure and Risk Assessment*, **25**, 1400–8.

Blesa, J., Soriano, J. M., Molto, J. C., Marin, R. and Manes, J. (2003) 'Determination of aflatoxins in peanuts by matrix solid-phase dispersion and liquid chromatography'. *Journal of Chromatography A*, **1011**, 49–54.

Böhm, C., Cichna-Markl, M., Brenn-Struckhofova, Z. and Razzazi-Fazeli, E. (2008) 'Development of a selective sample clean-up method based on immuno-ultrafiltration for the determination of deoxynivalenol in maize'. *Journal of Chromatograpy A*, **1202**, 111–7.

Boonzaaijer, G., Bobeldijk, I. and Van Osenbruggen, W. A. (2005) 'Analysis of patulin in Dutch food, an evaluation of a SPE based method'. *Food Control*, **16**, 587–91.

Bradburn, N., Coker, R. D. and Blunden, G. (1995) 'A comparative study of solvent extraction efficiency and the performance of immunoaffinity and solid phase columns on the determination of aflatoxin B_1'. *Food Chemistry*, **52**, 179–85.

Brenn-Struckhofova, Z., Cichna-Markl, M., Böhm, C. and Razzazi-Fazeli, E. (2007) 'Selective sample cleanup by reusable sol-gel immunoaffinity columns for determination of deoxynivalenol in food and feed samples'. *Analytical Chemistry*, **79**, 710–7.

Brenn-Struckhofova, Z., Füreder, C., Cichna-Markl, M. and Razzazi-Fazeli, E. (2009) 'Co-isolation of deoxynivalenol and zearalenone with sol-gel immunoaffinity columns for their determination in wheat and wheat products'. *Journal of Chromatography A*, **1216**, 5828–37.

Buttinger, G., Fuchs, E., Knapp, H., Berthiller, F., Schuhmacher, R., Binder, E.-M. and Krska, R. (2004) 'Performance of new clean-up column for the determination of ochratoxin A in cereals and foodstuffs by HPLC-FLD'. *Food Additives and Contaminants*, **21**, 1107–14.

Campbell, H. M. and Armstrong, J. F. (2007) 'Determination of zearalenone in cereal grains, animal feed, and feed ingredients using immunoaffinity column chromatography and liquid chromatography: Interlaboratory study'. *Journal of AOAC International*, **90**, 1610–22.

Cantwell, F. F. and Losier, M. (2002) Chapter 11 'Liquid–liquid extraction', in *Sampling and Sample Preparation for Field and Laboratory*, Pawliszyn, J. (ed.), Comprehensive Analytical Chemistry, Volume 37, Elsevier.

Carabias-Martínez, R., Rodríguez-Gonzalo, E., Revilla-Ruiz, P. and Hernández-Méndez, J. (2005) 'Pressurized liquid extraction in the analysis of food and biological samples'. *Journal of Chromatography A*, **1089**, 1–17.

Cavaliere, C., Foglia, P., Guarino, C., Marzioni, F., Nazzari, M., Samperi, R. and Lagana, A. (2006) 'Aflatoxin M$_1$ determination in cheese by liquid chromatography–tandem mass spectrometry'. *Journal of Chromatography A*, **1135**, 135–41.

Cavaliere, C., Foglia, P., Guarino, C., Nazzari, M., Samperi, R. and Lagana, A. (2007) 'Determination of aflatoxins in olive oil by liquid chromatography–tandem mass spectrometry'. *Analytica Chimica Acta*, **596**, 141–8.

Chan, D., Macdonald, S. J., Boughtflower, V. and Brereton, P. (2004) 'Simultaneous determination of aflatoxins and ochratoxin A in food using a fully automated immunoaffinity column clean-up and liquid chromatography-fluorescence detection'. *Journal of Chromatography A*, **1059**, 13–6.

Cichna-Markl, M. (2006) 'Selective sample preparation with bioaffinity columns prepared by the sol-gel method'. *Journal of Chromatography A*, **1124**, 167–80.

Coronel, M. B., Sanchis, V., Ramos, A. J. and Marin, S. (2009) 'Assessment of the exposure to ochratoxin A in the province of Lleida, Spain'. *Food and Chemical Toxicology*, **47**, 2847–52.

Cruz-aguado, J. A. and Penner, G. (2008) 'Determination of Ochratoxin A with a DNA Aptamer'. *Journal of Agricultural and Food Chemistry*, **56**, 10456–61.

Czerwiecki, L., Wilczyska, G. and Kwiecie, A. (2005) 'Ochratoxin A: an improvement clean-up and HPLC method used to investigate wine and grape juice on the Polish market'. *Food Additives and Contaminants: Part A: Chemistry, Analysis, Control, Exposure and Risk Assessment*, **22**, 158–62.

D'arco, G., Fernández-Franzón, M., Font, G., Damiani, P. and Mañes, J. (2008) 'Analysis of fumonisins B$_1$, B$_2$ and B$_3$ in corn-based baby food by pressurized liquid extraction and liquid chromatography/tandem mass spectrometry'. *Journal of Chromatography A*, **1209**, 188–94.

Dänicke, S., Valenta, H., Döll, S., Ganter, M. and Flachowsky, G. (2004) 'On the effectiveness of a detoxifying agent in preventing fusario-toxicosis in fattening pigs'. *Animal Feed Science and Technology*, **114**, 141–57.

De Smet, D., Dubruel, P., Van Peteghem, C., Schacht, E. and De Saeger, S. (2009) 'Molecularly imprinted solid-phase extraction of fumonisin B analogues in bell pepper, rice and corn flakes'. *Food Additives and Contaminants: Part A: Chemistry, Analysis, Control, Exposure and Risk Assessment*, **26**, 874–84.

Delaunay-Bertoncini, N. and Hennion, M.-C. (2004) 'Immunoaffinity solid-phase extraction for pharmaceutical and biomedical trace-analysis – Coupling with HPLC and CE – Perspectives'. *Journal of Pharmaceutical and Biomedical Analysis*, **34**, 717–36.

Delaunay-Bertoncini, N., Pichon, V. and Hennion, M.-C. (2001) 'Immunoextraction: a highly selective method for sample preparation'. *LC-GC Europe*, **14**, 162–72.

Delgado, T., Gómez-Cordovés, C. and Scott, P. M. (1996) 'Determination of alternariol and alternariol methyl ether in apple juice using solid-phase extraction and high-performance liquid chromatography'. *Journal of Chromatography A*, **731**, 109–14.

Dilkin, P., Mallmann, C. A., De Almeida, C. A. A. and Corrêa, B. (2001) 'Robotic automated clean-up for detection of fumonisins B_1 and B_2 in corn and corn-based feed by high-performance liquid chromatography'. *Journal of Chromatography A*, **925**, 151–7.

Domijan, A.-M., Peraica, M., Miletic´-Medved, M., Lucic´, A. and Fuchs, R. (2003) 'Two different clean-up procedures for liquid chromatographic determination of ochratoxin A in urine'. *Journal of Chromatography B: Analytical Technologies in the Biomedical and Life Sciences*, **798**, 317–21.

Ehlers, D., Czech, E., Quirin, K.-W. and Weber, R. (2006) 'Distribution of aflatoxins between extract and extraction residue of paprika using supercritical carbon dioxide'. *Phytochemical Analysis*, **17**, 114–20.

Eskola, M., Kokkonen, M. and Rizzo, A. (2002) 'Application of manual and automated systems for purification of ochratoxin A and zearalenone in cereals with immunoaffinity columns'. *Journal of Agricultural and Food Chemistry*, **50**, 41–7.

Fente, C. A. (1998) 'Determination of alternariol in tomato paste using solid phase extraction and high-performance liquid chromatography with fluorescence detection'. *Analyst*, **123**, 2277–80.

Galvis-Sánchez, A. C., Barros, A. S. and Delgadillo, I. (2008) 'Method for analysis dried vine fruits contaminated with ochratoxin A'. Papers presented at the *3rd International Symposium on Recent Advances in Food Analysis – Food Analysis, 2007, Analytica Chimica Acta*, **617**, 59–63.

Garda, J., Martins Macedo, R., Faria, R., Bernd, L., Carla Dors, G. and Badiale-Furlong, E. (2005) 'Alcoholic fermentation effects on malt spiked with trichothecenes'. *Food Control*, **16**, 423–8.

Garon, D., Richard, E., Sage, L., Bouchart, V., Pottier, D. and Lebailly, P. (2006) 'Mycoflora and multimycotoxin detection in corn silage: experimental study'. *Journal of Agricultural and Food Chemistry*, **54**, 3479–84.

Gazzotti, T., Lugoboni, B., Zironi, E., Barbarossa, A., Serraino, A. and Pagliuca, G. (2009) 'Determination of fumonisin B_1 in bovine milk by LC–MS/MS'. *Food Control*, **20**, 1171–4.

Ghali, R., Hmaissia-Khlifa, K., Ghorbel, H., Mezigh, C., Maaroufi, K. and Hedili, A. (2008) 'Determination of ochratoxin A in human blood serum by high-performance liquid chromatography: method validation and comparison of extraction procedures'. *Analytical Letters*, **41**, 2452–64.

González-Osnaya, L. G., Soriano Del Castillo, J. M., Moltó Cortés, J. C. and Mañes Vinuesa, J. (2006) 'Extraction and analysis of ochratoxin A in bread using pressurised liquid extraction and liquid chromatography'. *Journal of Chromatography A*, **1113**, 32–6.

González-Osnaya, L., Soriano, J. M., Moltó, J. C. and Mañes, J. (2008) 'Simple liquid chromatography assay for analyzing ochratoxin A in bovine milk'. *Food Chemistry*, **108**, 272–6.

González-pPeñas, E., Leache, C., Viscarret, M., Pérez De Obanos, A., Araguás, C. and López De Cerain, A. (2004) 'Determination of ochratoxin A in wine using liquid-phase microextraction combined with liquid chromatography with fluorescence detection'. *Journal of Chromatography A*, **1025**, 163–8.

Gromadzka, K., Waskiewicz, A., Golinski, P. and Swietlik, J. (2009) 'Occurrence of estrogenic mycotoxin – Zearalenone in aqueous environmental samples with various NOM content'. *Water Research*, **43**, 1051–9.

Guillamont, E. M., Lino, C. M., Baeta, M. L., Pena, A. S., Silveira, M. I. N. and Vinuesa,

J. M. (2005) 'A comparative study of extraction apparatus in HPLC analysis of ochratoxin A in muscle'. *Analytical and Bioanalytical Chemistry*, **383**, 570–5.

Haasnoot, W., Kemmers-Voncken, A. and Samson, D. (2002) 'Immunofiltration as sample cleanup for the immunochemical detection of ß-agonists in urine'. *Analyst*, **127**, 87–92.

Haginaka, J. (2009) 'Molecularly imprinted polymers as affinity-based separation media for sample preparation'. *Journal of Separation Science*, **32**, 1548–65.

Hamula, C. L. A., Guthrie, J. W., Zhang, H., Li, X.-F. and Le, X. C. (2006) 'Selection and analytical applications of aptamers'. *TrAC Trends in Analytical Chemistry – On-site Instrumentation and Analysis*, **25**, 681–91.

Hayashi, Y. and Yoshizawa, T. (2005) 'Analysis of cyclopiazonic acid in corn and rice by a newly developed method'. *Food Chemistry*, **93**, 215–21.

He, J., Li, X.-z. and Zhou, T. (2009) 'Sample clean-up methods, immunoaffinity chromatography and solid phase extraction, for determination of deoxynivalenol and deepoxy deoxynivalenol in swine serum'. *Mycotoxin Research*, **25**, 89–94.

Hu, Y.-Y., Zheng, P., Zhang, Z.-X. and He, Y.-Z. (2006) 'Determination of aflatoxins in high-pigment content samples by matrix solid-phase dispersion and high-performance liquid chromatography'. *Journal of Agricultural and Food Chemistry*, **54**, 4126–30.

Huopalahti, R. P., Ebel, J. and Henion, J. D. (1997) 'Supercritical fluid extraction of mycotoxins from feeds with analysis by LC/UC and LC/MS'. *Journal of Liquid Chromatography and Related Technologies*, **20**, 537–51.

Hussain, I., Anwar, J., Asi, M. R., Munawar, M. A. and Kashif, M. (2010) 'Aflatoxin M$_1$ contamination in milk from five dairy species in Pakistan'. *Food Control*, **21**, 122–4.

Jestoi, M. and Kokkonen, M. (2008) 'Comparison of two sample preparation techniques for the determination of ochratoxin A in grains'. *Journal of Liquid Chromatography and Related Technologies*, **31**, 912–25.

Jestoi, M., Ritieni, A. and Rizzo, A. (2004) 'Analysis of the fusarium mycotoxins fusaproliferin and trichothecenes in grains using gas chromatography mass spectrometry'. *Journal of Agricultural and Food Chemistry*, **52**, 1464–9.

Jestoi, M., Rokka, M., Järvenpää, E. and Peltonen, K. (2009) 'Determination of fusarium mycotoxins beauvericin and enniatins (A, A$_1$, B, B$_1$) in eggs of laying hens using liquid chromatography-tandem mass spectrometry (LC–MS/MS)'. *Food Chemistry*, **115**, 1120–7.

Jodlbauer, J., Maier, N. M. and Lindner, W. (2002) 'Towards ochratoxin A selective molecularly imprinted polymers for solid-phase extraction'. *Journal of Chromatography A*, **945**, 45–63.

Jornet, D., Busto, O. and Guasch, J. (2000) 'Solid-phase extraction applied to the determination of ochratoxin A in wines by reversed-phase high-performance liquid chromatography'. *Journal of Chromatography A*, **882**, 29–35.

Josephs, R. D., Krska, R., Grasserbauer, M. and Broekaert, J. A. C. (1998) 'Determination of trichothecene mycotoxins in wheat by use of supercritical fluid extraction and high-performance liquid chromatography with diode array detection or gas chromatography with electron capture detection'. *Journal of Chromatography A*, **795**, 297–304.

Juan, C., Molto, J. C., Lino, C. M. and Manes, J. (2008a) 'Determination of ochratoxin A in organic and non-organic cereals and cereal products from Spain and Portugal'. *Food Chemistry*, **107**, 525–30.

Juan, C., Zinedine, A., Idrissi, L. and Mañes, J. (2008b) 'Ochratoxin A in rice on the Moroccan retail market'. *International Journal of Food Microbiology*, **126**, 83–5.

Juan, C., Zinedine, A., Moltó, J. C., Idrissi, L. and Mañes, J. (2008c) 'Aflatoxins levels in dried fruits and nuts from Rabat-Salé area, Morocco'. *Food Control*, **19**, 849–53.

Kabak, B. (2009) 'Ochratoxin A in cereal-derived products in Turkey: Occurrence and exposure assessment'. *Food and Chemical Toxicology*, **47**, 348–52.

Kataoka, H., Itano, M., Ishizaki, A. and Saito, K. (2009) 'Determination of patulin in fruit juice and dried fruit samples by in-tube solid-phase microextraction coupled with liquid chromatography-mass spectrometry'. *Journal of Chromatography A*, **1216**, 3746–50.

Klinglmayr, C., Nöbauer, K., Razzazi-Fazeli, E. and Cichna-markl, M. (2010) 'Determination of deoxynivalenol in organic and conventional food and feed by sol–gel immunoaffinity chromatography and HPLC-UV detection'. *Journal of Chromatography B*, **878**, 187–93.

Klötzel, M., Schmidt, S., Lauber, U., Thielert, G. and Humpf, H.-U. (2005) 'Comparison of different clean-up procedures for the analysis of deoxynivalenol in cereal-based food and validation of a reliable HPLC method'. *Chromatographia*, **62**, 41–8.

Klötzel, M., Lauber, U. and Humpf, H.-U. (2006) 'A new solid phase extraction clean-up method for the determination of 12 type A and B trichothecenes in cereals and cereal-based food by LC–MS/MS'. *Molecular Nutrition and Food Research*, **50**, 261–9.

Kokkonen, M. and Jestoi, M. (2009) 'A multi-compound LC–MS/MS method for the screening of mycotoxins in grains'. *Food Analytical Methods*, **2**, 128–40.

Krska, R. (1998) 'Performance of modern sample preparation techniques in the analysis of Fusarium mycotoxins in cereals'. *Journal of Chromatography A*, **815**, 49–57.

Krska, R., Stubbings, G., Macarthur, R. and Crews, C. (2008) 'Simultaneous determination of six major ergot alkaloids and their epimers in cereals and foodstuffs by LC–MS-MS'. *Analytical and Bioanalytical Chemistry*, **391**, 563–76.

Lagana , A., Faberi, A., Fago, G., Marino, A., Pastorini, E. and Samperi, R. (2004) 'Application of an innovative matrix solid-phase dispersion solid-phase extraction liquid chromatography tandem mass spectrometry analytical methodology to the study of the metabolism of the estrogenic mycotoxin zearalenone in rainbow trout liver and muscular tissue'. *International Journal of Environmental Analytical Chemistry*, **84**, 1009–1016.

Lattanzio, V. M. T., Solfrizzo, M., Powers, S. and Visconti, A. (2007) 'Simultaneous determination of aflatoxins, ochratoxin A and Fusarium toxins in maize by liquid chromatography/tandem mass spectrometry after multitoxin immunoaffinity cleanup'. *Rapid Communications in Mass Spectrometry*, **21**, 3253–61.

Lau, B. P.-Y., Scott, P. M., Lewis, D. A., Kanhere, S. R., Cléroux, C. and Roscoe, V. A. (2003) 'Liquid chromatography-mass spectrometry and liquid chromatography-tandem mass spectrometry of the Alternaria mycotoxins alternariol and alternariol monomethyl ether in fruit juices and beverages'. *Journal of Chromatography A*, **998**, 119–31.

Lee, J. E., Kwak, B.-M., Ahn, J.-H. and Jeon, T.-H. (2009) 'Occurrence of aflatoxin M_1 in raw milk in South Korea using an immunoaffinity column and liquid chromatography'. *Food Control*, **20**, 136–8.

Lee, P. K., Krystin Kee, S. Y., Ng, W. and Gopalakrishnakone, P. (1999) 'Determination of trichothecene toxin (T2 mycotoxin) in aqueous sample with solid phase microextraction technique followed by gas chromatography with flame ionization detection'. *Journal of High Resolution Chromatography*, **22**, 424–6.

Li, J.-H., Wu, R.-N., Hu, Q.-N. and Wang, J.-N. (2007) 'Solid-phase extraction and HPLC determination of patulin in apple juice concentrate'. *Food Control*, **18**, 530–4.

Liau, B.-C., Jong, T.-T., Lee, M.-R. and Chang, C.-M. J. (2007) 'Supercritical fluid extraction and quantification of aflatoxins in *Zizyphi fructus* by liquid chromatography/ atmospheric pressure chemical ionization tandem mass spectrometry'. *Rapid Communications in Mass Spectrometry*, **21**, 667–73.

Llorens, A., Mateo, R., Mateo, J. J. and Jimenez, M. (2002) 'Comparison of extraction and clean-up procedures for analysis of zearalenone in corn, rice and wheat grains by high-performance liquid chromatography with photodiode array and fluorescence detection'. *Food Additives and Contaminants: Part A: Chemistry, Analysis, Control, Exposure and Risk Assessment*, **19**, 272–81.

Lombaert, G. A., Pellaers, P., Roscoe, V., Mankotia, M., Neil, R. and Scott, P. M. (2003) 'Mycotoxins in infant cereal foods from the Canadian retail market'. *Food Additives and Contaminants*, **20**, 494–504.

Mahony, J. O., Nolan, K., Smyth, M. R. and Mizaikoff, B. (2005) 'Molecularly imprinted polymers – potential and challenges in analytical chemistry'. *Analytica Chimica Acta: Interpretation, Design and Selection of Biomolecular Interactions*, **534**, 31–9.

Maier, N. M., Buttinger, G., Welhartizki, S., Gavioli, E. and Lindner, W. (2004) 'Molecularly imprinted polymer-assisted sample clean-up of ochratoxin a from red wine: Merits and limitations'. *Journal of Chromatography B: Analytical Technologies in the Biomedical and Life Sciences*, **804**, 103–11.

Majors, R. E. (n.d) *Trends in Sample Preparation for Chromatography*. Agilent Technologies. Available from: https://cp.chem.agilent.com/Library/slidepresentation/Public/Trends%20in%20Sample%20Preparation%20for%20Chromatography.pdf (accessed 28 July 2010).

Manova, R. and Mladenova, R. (2009) 'Incidence of zearalenone and fumonisins in Bulgarian cereal production'. *Food Control*, **20**, 362–365.

Maragos, C. M. (2008) 'Extraction of anatoxins B_1 and G_1 from maize by using aqueous sodium dodecyl sulfate'. *Journal of AOAC International*, **91**, 762–7.

Mazzola, P. G., Lopes, A. M., Hasmann, F. A., Jozala, A. F., Penna, T. C., Magalhaes, P. O., Rangel-Yagui, C. O. and Jr, A. P. (2008) 'Liquid–liquid extraction of biomolecules: an overview and update of the main techniques'. *Journal of Chemical Technology and Biotechnology*, **83**, 143–57.

Meister, U. (1999) 'Effect of extraction and extract purification on the measurable fumonisin content of maize and maize products. Tests on the efficiency of acid extraction and use of immunoaffinity columns'. *Mycotoxin Research*, **15**, 13–23.

Meister, U. (2004) 'New method of citrinin determination by HPLC after polyamide column clean-up'. *European Food Research and Technology*, **218**, 394–9.

Meister, U. (2008) 'Analysis of T-2 and HT-2 toxins in oats and other cereals by means of HPLC with fluorescence detection'. *Mycotoxin Research*, **24**, 31–9.

Melchert, H. U. and Pabel, E. (2004) 'Reliable identification and quantification of trichothecenes and other mycotoxins by electron impact and chemical ionization–gas chromatography–mass spectrometry, using an ion-trap system in the multiple mass spectrometry mode: Candidate reference method for complex matrices'. *Journal of Chromatography A*, **1056**, 195–9.

Mendiola, J. A., Herrero, M., Cifuentes, A. and Ibañez, E. (2007) 'Use of compressed fluids for sample preparation: Food applications'. *Journal of Chromatography A – Advances in Sample Preparation – Part I*, **1152**, 234–46.

Miyahara, M., Akiyama, H., Toyoda, M. and Saito, Y. (1996) 'New procedure for fumonisins B_1 and B_2 in corn and corn products by ion pair chromatography with *o*-phthaldialdehyde postcolumn derivatization and fluorometric detection'. *Journal of Agricultural and Food Chemistry*, **44**, 842–47.

Moazami, E. F. and Jinap, S. (2009) 'Optimisation of the determination of deoxynivalenol in wheat flour by HPLC and a comparison of four clean-up procedures'. *Food Additives and Contaminants: Part A: Chemistry, Analysis, Control, Exposure and Risk Assessment*, **26**, 1290–7.

Mohamed, R., Gremaud, E., Richoz-Payot, J., Tabet, J.-C. and Guy, P. A. (2006) 'Quantitative determination of five ergot alkaloids in rye flour by liquid chromatography-electrospray ionisation tandem mass spectrometry'. *Journal of Chromatography A*, **1114**, 62–72.

Mol, H. G. J., Plaza-Bolanos, P., Zomer, P., De Rijk, T. C., Stolker, A. A. M. and Mulder, P. P. J. (2008) 'Toward a Generic Extraction Method for Simultaneous Determination of Pesticides, Mycotoxins, Plant Toxins, and Veterinary Drugs in Feed and Food Matrixes'. *Analytical Chemistry*, **80**, 9450–9.

Möller, T. E. and Nyberg, M. (2004) 'Efficiency of different extraction solvent mixtures used in analyses of aflatoxins from a certified peanut meal reference material. *Food Additives and Contaminants*, **21**, 781–5.

Monaci, L., Tantillo, G. and Palmisano, F. (2004) 'Determination of ochratoxin A in pig tissues by liquid-liquid extraction and clean-up and high-performance liquid chromatography'. *Analytical and Bioanalytical Chemistry*, **378**, 1777–82.

Monbaliu, S., Van Poucke, C., Van Peteghem, C., Van Poucke, K., Heungens, K. and De

Saeger, S. (2009) 'Development of a multi-mycotoxin liquid chromatography/tandem mass spectrometry method for sweet pepper analysis'. *Rapid Communications in Mass Spectrometry*, **23**, 3–11.

Mortensen, G. K., Strobel, B. W. and Hansen, H. C. B. (2003) 'Determination of zearalenone and ochratoxin A in soil'. *Analytical and Bioanalytical Chemistry*, **376**, 98–101.

Munimbazi, C. and Bullerman, L. B. (1998) 'High-performance liquid chromatographic method for the determination of moniliformin in corn'. *Journal of AOAC International*, **81**, 999–1004.

Murillo, M., González-Peñas, E. and Amézqueta, S. (2008) 'Determination of patulin in commercial apple juice by micellar electrokinetic chromatography'. *Food and Chemical Toxicology*, **46**, 57–64.

Murillo-Arbizu, M., Amézqueta, S., González-Peñas, E. and De Cerain, A. L. (2009) 'Occurrence of patulin and its dietary intake through apple juice consumption by the Spanish population'. *Food Chemistry*, **113**, 420–3.

Nawaz, S., Scudamore, K. A. and Rainbird, S. C. (1997) 'Mycotoxins in ingredients of animal feeding stuffs: I. determination of Alternaria mycotoxins in oilseed rape meal and sunflower seed meal'. *Food Additives and Contaminants: Part A: Chemistry, Analysis, Control, Exposure and Risk Assessment*, **14**, 249–62.

Nelson, M. D. and Dolan, J. W. (2002) 'Ion suppression in LC–MS–MS – a case study'. *LCGC North America*, **20**, 24–32.

Nery, A. A., Wrenger, C. and Ulrich, H. (2009) 'Recognition of biomarkers and cell-specific molecular signatures: Aptamers as capture agents'. *Journal of Separation Science*, **32**, 1523–30.

Neumann, G., Lombaert, G. A., Kotello, S. and Fedorowich, N. (2009) 'Determination of deoxynivalenol in soft wheat by immunoaffinity column cleanup and LC-UV detection: Interlaboratory study'. *Journal of AOAC International*, **92**, 15–25.

Noba, S., Uyama, A. and Mochizuki, N. (2009) 'Determination of ochratoxin A in ready-to-drink coffee by immunoaffinity cleanup and liquid chromatography tandem mass spectrometry'. *Journal of Agricultural and Food Chemistry*, **57**, 6036–40.

Nonaka, Y., Saito, K., Hanioka, N., Narimatsu, S. and Kataoka, H. (2009) 'Determination of aflatoxins in food samples by automated on-line in-tube solid-phase microextraction coupled with liquid chromatography-mass spectrometry'. *Journal of Chromatography A*, **1216**, 4416–22.

O'Riordan, M. J. and Wilkinson, M. G. (2008) 'A survey of the incidence and level of aflatoxin contamination in a range of imported spice preparations on the Irish retail market'. *Food Chemistry*, **107**, 1429–135.

Ofitserova, M., Nerkar, S., Pickering, M., Torma, L. and Thiex, N. (2009) 'Multiresidue mycotoxin analysis in corn grain by column high-performance liquid chromatography with postcolumn photochemical and chemical derivatization: Single-laboratory validation'. *Journal of AOAC International*, **92**, 15–25.

Oh, K. S., Scott, P. M. and Chung, S.-H. (2009) 'Incomplete recoveries of fumonisins present in naturally contaminated corn foods from an immunoaffinity column'. *Journal of AOAC International*, **92**, 496–501.

Pagliuca, G., Zironi, E., Ceccolini, A., Matera, R., Serrazanetti, G. P. and Piva, A. (2005) 'Simple method for the simultaneous isolation and determination of fumonisin B_1 and its metabolite aminopentol-1 in swine liver by liquid chromatography–fluorescence detection'. *Journal of Chromatography B*, **819**, 97–103.

Pallaroni, L. and Von Holst, C. (2003) 'Comparison of alternative and conventional extraction techniques for the determination of zearalenone in corn'. *Analytical and Bioanalytical Chemistry*, **376**, 908–12.

Pallaroni, L. and Von Holst, C. (2004) 'Development of an extraction method for the determination of zearalenone in corn using less organic solvents'. *Journal of Chromatography A*, **1055**, 247–9.

Pallaroni, L., Von Holst, C., Eskilsson, C. and Björklund, E. (2002) 'Microwave-assisted extraction of zearalenone from wheat and corn'. *Analytical and Bioanalytical Chemistry*, **374**, 161–6.

Park, J. W. and Kim, Y.-B. (2006) 'Effect of Pressure Cooking on Aflatoxin B_1 in Rice'. *Environmental Science and Technology*, **35**, 2401–6.

Park, J. W., Scott, P. M., Lau, B. P.-Y. and Lewis, D. A. (2004) 'Analysis of heat-processed corn foods for fumonisins and bound fumonisins'. *Food Additives and Contaminants: Part A: Chemistry, Analysis, Control, Exposure and Risk Assessment*, **21**, 1168–78.

Pascale, M., De Girolamo, A., Visconti, A., Magan, N., Chianella, I., Piletska, E. V. and Piletsky, S. A. (2008) 'Use of itaconic acid-based polymers for solid-phase extraction of deoxynivalenol and application to pasta analysis'. *Analytica Chimica Acta*, **609**, 131–8.

Pawliszyn, J. (2002) 'Chapter 13 Solid phase microextraction; Comprehensive Analytical Chemistry', in Pawliszyn, J. (ed.), *Sampling and Sample Preparation for Field and Laboratory*. Volume 37 ed., Elsevier.

Pelegri, J. M., Velazquez, C., Sanchis, V. and Canela, R. (1997) 'Solid phase extraction on sax columns as an alternative for ochratoxin A analysis in maize'. *Revista Iberoamericana de Micologia*, **14**, 194–6.

Pichon, V. and Haupt, K. (2006) 'Affinity separations on molecularly imprinted polymers with special emphasis on solid-phase extraction'. *Journal of Liquid Chromatography and Related Technologies*, **29**, 989–1023.

Pittet, A. and Royer, D. (2002) 'Rapid, low cost thin-layer chromatographic screening method for the detection of ochratoxin A in green coffee at a control level of 10 µg/kg'. *Journal of Agricultural and Food Chemistry*, **20**, 243–7.

Poole, C. F. (2003) 'New trends in solid-phase extraction'. *TrAC Trends in Analytical Chemistry*, **22**, 362–73.

Reinsch, M., Töpfer, A., Lehmann, A. and Nehls, I. (2005) 'Determination of ochratoxin A in wine by liquid chromatography tandem mass spectrometry after combined anion-exchange/reversed-phase clean-up'. *Analytical and Bioanalytical Chemistry*, **381**, 1592–5.

Reiter, E. V., Cichna-markl, M., Chung, D.-h., Zentek, J. and Razzazi-fazeli, E. (2009) 'Immuno-ultrafiltration as a new strategy in sample clean up of aflatoxins'. *Journal of Separation Science*, **32**, 1729–39.

Reiter, E. V., Vouk, F., Böhm, J. and Razzazi-Fazeli, E. (2010) 'Aflatoxins in rice – a limited survey of products marketed in Austria'. *Food Control*, **21**, 988–91.

Richter, B. E., Jones, B. A., Ezzell, J. L., Porter, N. L., Avdalovic, N. and Pohl, C. (1996) 'Accelerated Solvent Extraction: A Technique for Sample Preparation'. *Analytical Chemistry*, **68**, 1033–9.

Ridgway, K., Lalljie, S. P. D. and Smith, R. M. (2007) 'Sample preparation techniques for the determination of trace residues and contaminants in foods'. *Journal of Chromatography A – Advances in Sample Preparation – Part II*, **1153**, 36–53.

Risticevic, S., Niri, V., Vuckovic, D. and Pawliszyn, J. (2009) 'Recent developments in solid-phase microextraction'. *Analytical and Bioanalytical Chemistry*, **393**, 781–95.

Rodriguez-Mozaz, S., Lopez De Alda, M. J. and Barceló, D. (2007) 'Advantages and limitations of on-line solid phase extraction coupled to liquid chromatography-mass spectrometry technologies versus biosensors for monitoring of emerging contaminants in water'. *Journal of Chromatography A – Advances in Sample Preparation – Part I*, **1152**, 97–115.

Romero, A. D. C., Ferreira, T. R. B., Dias, C. T. D. S., Calori-domingues, M. A. and Da Gloria, E. M. (2010) 'Occurrence of AFM_1 in urine samples of a Brazilian population and association with food consumption'. *Food Control*, **21**, 554–8

Royer, D., Humpf, H.-U. and Guy, P. A. (2004) 'Quantitative analysis of Fusarium mycotoxins in maize using accelerated solvent extraction before liquid chromatography/atmospheric pressure chemical ionization tandem mass spectrometry'. *Food Additives and Contaminants: Part A: Chemistry, Analysis, Control, Exposure and Risk Assessment*, **21**, 678–92.

Ruhland, M. and Tischler, J. (2008) 'Determination of ergot alkaloids in feed by HPLC'. *Mycotoxin Research*, **24**, 73–9.

Saez, J. M., Medina, A., Gimeno-adelantado, J. V., Mateo, R. and Jimenez, M. (2004) 'Comparison of different sample treatments for the analysis of ochratoxin A in must, wine and beer by liquid chromatography'. *Journal of Chromatography A*, **1029**, 125–33.

Santos, L., Marín, S., Sanchis, V. and Ramos, A. J. (2009) 'Screening of mycotoxin multicontamination in medicinal and aromatic herbs sampled in Spain'. *Journal of the Science of Food and Agriculture*, **89**, 1802–7.

Schothorst, R. C., Jekel, A. A., Van Egmond, H. P., De Mul, A., Boon, P. E. and Van Klaveren, J. D. (2005) 'Determination of trichothecenes in duplicate diets of young children by capillary gas chromatography with mass spectrometric detection'. *Food Additives and Contaminants*, **22**, 48–55.

Schuhmacher, R., Krska, R., Weingaertner, J. and Grasserbauer, M. (1997) 'Interlaboratory comparison study for the determination of the Fusarium mycotoxins deoxynivalenol in wheat and zearalenone in maize using different methods'. *Fresenius' Journal of Analytical Chemistry*, **359**, 510–5.

Scott, P. M. (1993) 'Recent developments in methods of analysis for mycotoxins in foodstuffs'. *TrAC Trends in Analytical Chemistry*, **12**, 382–6.

Scott, P. M. (1995) 'Mycotoxin methodology'. *Food Additives and Contaminants: Part A: Chemistry, Analysis, Control, Exposure and Risk Assessment*, **12**, 395–403.

Scott, P. M. and Lawrence, G. A. (1996) 'Determination of hydrolysed fumonisin B_1 in alkali-processed corn foods'. *Food Additives and Contaminants: Part A: Chemistry, Analysis, Control, Exposure and Risk Assessment*, **13**, 823–32.

Scott, P., Lawrence, G. and Lombaert, G. (1999) 'Studies on extraction of fumonisins from rice, corn-based foods and beans'. *Mycotoxin Research*, **15**, 50–60.

Senyuva, H. Z., Ozcan, S., Cimen, D. and Gilbert, J. (2008) 'Determination of fumonisins B_1 and B_2 in corn by liquid chromatography/mass spectrometry with immunoaffinity column cleanup: Single-laboratory method validation'. *Journal of AOAC International*, **91**, 598–606.

Sheibani, A. and Ghaziaskar, H. S. (2009) 'Pressurized fluid extraction for quantitative recovery of aflatoxins B_1 and B_2 from pistachio'. *Food Control*, **20**, 124–8.

Shundo, L., De Almeida, A. P., Alaburda, J., Lamardo, L. C. A., Navas, S. A., Ruvieri, V. and Sabino, M. (2009) 'Aflatoxins and ochratoxin A in Brazilian paprika'. *Food Control*, **20**, 1099–102.

Sibanda, L., De Saeger, S. and Van Peteghem, C. (2002) 'Optimization of solid-phase cleanup prior to liquid chromatographic analysis of ochratoxin A in roasted coffee'. *Journal of Chromatography A*, **959**, 327–30.

Silva, L., Fernández-franzón, M., Font, G., Pena, A., Silveira, I., Lino, C. and Mañes, J. (2009) 'Analysis of fumonisins in corn-based food by liquid chromatography with fluorescence and mass spectrometry detectors'. *Food Chemistry*, **112**, 1031–7.

Sizoo, E. A. and Van Egmond, H. P. (2005) 'Analysis of duplicate, 24-hour diet samples for aflatoxin B_1, aflatoxin M_1 and ochratoxin A'. *Food Additives and Contaminants*, **22**, 163–72.

Sobolev, V. S. (2007) 'Simple, rapid, and inexpensive cleanup method for quantitation of aflatoxins in important agricultural products by HPLC'. *Journal of Agricultural and Food Chemistry*, **55**, 2136–41.

Sobolev, V. S. and Dorner, J. W. (2002) 'Cleanup procedure for determination of aflatoxins in major agricultural commodities by liquid chromatography'. *Journal of AOAC International*, **85**, 642–5.

Solfrizzo, M., Avantaggiato, G. and Visconti, A. (1998) 'Use of various clean-up procedures for the analysis of ochratoxin A in cereals'. *Journal of Chromatography A*, **815**, 67–73.

Solfrizzo, M., De Girolamo, A., Vitti, C., Visconti, A. and Van Den Bulk, R. (2004) 'Liquid chromatographic determination of alternaria toxins in carrots'. *Journal of AOAC International*, **87**, 101–6.

Songsermsakul, P., Sontag, G., Cichna-markl, M., Zentek, J. and Razzazi-Fazeli, E. (2006) 'Determination of zearalenone and its metabolites in urine, plasma and faeces of horses by HPLC-APCI-MS'. *Journal of Chromatography B*, **843**, 252–61.

Sorensen, L. K. and Elbaek, T. H. (2005) 'Determination of mycotoxins in bovine milk by liquid chromatography tandem mass spectrometry'. *Journal of Chromatography B*, **820**, 183–96.

Spanjer, M. C., Rensen, P. M. and Scholten, J. M. (2008) 'LC–MS/MS multi-method for mycotoxins after single extraction, with validation data for peanut, pistachio, wheat, maize, cornflakes, raisins and figs'. *Food Additives and Contaminants: Part A: Chemistry, Analysis, Control, Exposure and Risk Assessment*, **25**, 472–89.

Sparr Eskilsson, C. and Björklund, E. (2000) 'Analytical-scale microwave-assisted extraction'. *Journal of Chromatography A*, **902**, 227–50.

Starr, J. M. and Selim, M. I. (2008) 'Supercritical fluid extraction of aflatoxin B_1 from soil'. *Journal of Chromatography A*, **1209**, 37–43.

Stecher, G., Jarukamjorn, K., Zaborski, P., Bakry, R., Huck, C. W. and Bonn, G. K. (2007) 'Evaluation of extraction methods for the simultaneous analysis of simple and macrocyclic trichothecenes'. *Talanta*, **73**, 251–7.

Storm, I. D., Rasmussen, P. H., Strobel, B. W. and Hansen, H. C. B. (2008) 'Ergot alkaloids in rye flour determined by solid-phase cation-exchange and high-pressure liquid chromatography with fluorescence detection'. *Food Additives and Contaminants: Part A: Chemistry, Analysis, Control, Exposure and Risk Assessment*, **25**, 338–346.

Stroka, J., Petz, M., Joerissen, U. and Anklam, E. (1999) 'Investigation of various extractants for the analysis of aflatoxin B_1 in different food and feed matrices'. *Food Additives and Contaminants*, **16**, 331–338.

Sugita-Konishi, Y., Tanaka, T., Nakajima, M., Fujita, K., Norizuki, H., Mochizuki, N. and Takatori, K. (2006) 'The comparison of two clean-up procedures, multifunctional column and immunoaffinity column, for HPLC determination of ochratoxin A in cereals, raisins and green coffee beans'. *Talanta*, **69**, 650–5.

Tam, J., Mankotia, M., Mably, M., Pantazopoulos, P., Neil, R. J., Calway, P. and Scott, P. M. (2006) 'Survey of breakfast and infant cereals for aflatoxins B_1, B_2, G_1 and G_2'. *Food Additives and Contaminants*, **23**, 693–99.

Tardieu, D., Auby, A., Bluteau, C., Bailly, J. D. and Guerre, P. (2008) 'Determination of Fumonisin B_1 in animal tissues with immunoaffinity purification'. *Journal of Chromatography B*, **870**, 140–4.

Trebstein, A., Seefelder, W., Lauber, U. and Humpf, H.-U. (2008) 'Determination of T-2 and HT-2 toxins in cereals including oats after immunoaffinity cleanup by liquid chromatography and fluorescence detection'. *Journal of Agricultural and Food Chemistry*, **56**, 4968–75.

Trebstein, A., Marschik, S., Lauber, U. and Humpf, H.-U. (2009) 'Acetonitrile: the better extractant for the determination of T-2 and HT-2 toxin in cereals using an immunoaffinity-based cleanup?'. *European Food Research and Technology*, **228**, 519–29.

Trenholm, H. L., Warner, R. M. and Prelusky, D. B. (1985) 'Assessment of extraction procedures in the analysis of naturally contaminated grain products for deoxynivalenol (vomitoxin)'. *Journal of the Association of Official Analytical Chemists*, **68**, 645–9.

Urraca, J. L., Marazuela, M. D. and Moreno-Bondi, M. C. (2004) 'Analysis for zearalenone and [alpha]-zearalenol in cereals and swine feed using accelerated solvent extraction and liquid chromatography with fluorescence detection'. *Analytica Chimica Acta – Papers presented at the VIIIth International Symposium on Analytical Methodology in the Environmental Field and XIIIth Meeting of the Spanish Society of Analytical Chemistry, University of A Coruna, Spain – 21–24 October, 2003*, **524**, 175–83.

Urraca, J., Marazuela, M. and Moreno-Bondi, M. (2006) 'Molecularly imprinted polymers applied to the clean-up of zearalenone and alpha-zearalenol from cereal and swine feed sample extracts'. *Analytical and Bioanalytical Chemistry*, **385**, 1155–61.

Valenta, H., Dänicke, S. and Döll, S. (2003) 'Analysis of deoxynivalenol and de-epoxydeoxynivalenol in animal tissues by liquid chromatography after clean-up with an immunoaffinity column'. *Mycotoxin Research*, **19**, 51–5.

Valle-Algarra, F. M., Medina, A., Gimeno-Adelantado, J. V., Llorens, A., Jiménez, M. and Mateo, R. (2005) 'Comparative assessment of solid-phase extraction clean-up procedures, GC columns and perfluoroacylation reagents for determination of type B trichothecenes in wheat by GC-ECD'. Talanta, **66**, 194–201.

Valle-Algarra, F. M., Mateo, E. M., Gimeno-Adelantado, J. V., Mateo-Castro, R. and Jiménez, M. (2009) 'Optimization of clean-up procedure for patulin determination in apple juice and apple purees by liquid chromatography'. *Talanta*, **80**, 636–42.

Varelis, P., Leong, S.-l. L., Hocking, A. and Giannikopoulos, G. (2006) 'Quantitative analysis of ochratoxin A in wine and beer using solid phase extraction and high performance liquid chromatography fluorescence detection'. *Food Additives and Contaminants: Part A: Chemistry, Analysis, Control, Exposure and Risk Assessment*, **23**, 1308–15.

Vatinno, R., Aresta, A., Zambonin, C. G. and Palmisano, F. (2007) 'Determination of ochratoxin A in human urine by solid-phase microextraction coupled with liquid chromatography-fluorescence detection'. *Journal of Pharmaceutical and Biomedical Analysis*, **44**, 1014–8.

Vatinno, R., Vuckovic, D., Zambonin, C. G. and Pawliszyn, J. (2008) 'Automated high-throughput method using solid-phase microextraction–liquid chromatography–tandem mass spectrometry for the determination of ochratoxin A in human urine'. *Journal of Chromatography A – 10th International Symposium on Hyphenated Techniques in Chromatography and Hyphenated Chromatographic Analysers*, **1201**, 215–21.

Vega, M., Muñoz, K., Sepúlveda, C., Aranda, M., Campos, V., Villegas, R. and Villarroel, O. (2009) 'Solid-phase extraction and HPLC determination of Ochratoxin A in cereals products on Chilean market. **Food Control, 20**, 631–4.

Ventura, M., Guillén, D., Anaya, I., Broto-Puig, F., Lliberia, J. L., Agut, M. and Comellas, L. (2006) 'Ultra-performance liquid chromatography/tandem mass spectrometry for the simultaneous analysis of aflatoxins B_1, G_1, B_2, G_2 and ochratoxin A in beer'. *Rapid Communications in Mass Spectrometry*, **20**, 3199–204.

Versilovskis, A., Bartkevics, V. and Mikelsone, V. (2008) 'Sterigmatocystin presence in typical Latvian grains'. *Food Chemistry*, **109**, 243–8.

Villa, P. and Markaki, P. (2009) 'Aflatoxin B_1 and ochratoxin A in breakfast cereals from athens market: Occurrence and risk assessment'. *Food Control*, **20**, 455–61.

Vishwanath, V., Sulyok, M., Labuda, R., Bicker, W. and Krska, R. (2009) 'Simultaneous determination of 186 fungal and bacterial metabolites in indoor matrices by liquid chromatography/tandem mass spectrometry'. *Analytical and Bioanalytical Chemistry*, **395**, 1355–72.

Wang, Y., Chai, T., Lu, G., Quan, C., Duan, H., Yao, M., Zucker, B.-A. and Schlenker, G. (2008) 'Simultaneous detection of airborne aflatoxin, ochratoxin and zearalenone in a poultry house by immunoaffinity clean-up and high-performance liquid chromatography'. *Environmental Research*, **107**, 139–144.

Weingaertner, J., Krska, R., Praznik, W., Grasserbauer, M. and Lew, H. (1997) 'Use of Mycosep multifunctional clean-up columns for the determination of trichothecenes in wheat by electron-capture gas chromatography'. *Fresenius' Journal of Analytical Chemistry*, **357**, 1206–10.

Westcott, J. Y., Sloan, S. and Wenzel, S. E. (1997) 'Immunofiltration purification for urinary leukotriene E4 quantitation'. *Analytical Biochemistry*, **248**, 202–10.

Whitaker, T. B., Dickens, J. W. and Giesbrecht, F. G. (1984) 'Effects of methanol concentration and solvent: Peanut ratio on extraction of aflatoxin from raw peanuts'. *Journal of the Association of Official Analytical Chemists*, **67**, 35–6.

Whitaker, T. B., Dickens, J. W. and Giesbrecht, F. G. (1986) 'Optimum methanol concentration and solvent/peanut ratio for extraction of aflatoxin from raw peanuts by modified AOAC method II'. *Journal of the Association of Official Analytical Chemists*, **69**, 508–10.

Wilson, T. J. and Romer, T. R. (1991) 'Use of Mycosep Multfunctional Cleanup Column for Liquid Chromatographic Determination of Aflatoxins in Agricultural Products'. *Journal of AOAC International*, **74**, 951–56.

Wu, R.-N., Dang, Y.-L., Niu, L. and Hu, H. (2008) 'Application of matrix solid-phase dispersion-HPLC method to determine patulin in apple and apple juice concentrate'. *Journal of Food Composition and Analysis*, **21**, 582–6.

Wu, R.-N., Han, F.-L., Shang, J., Hu, H. and Han, L. (2009) 'Analysis of patulin in apple products by liquid-liquid extraction, solid phase extraction and matrix solid-phase dispersion methods: a comparative study'. *European Food Research and Technology*, **228**, 1009–14.

Yu, J. C. C. and Lai, E. P. C. (2007) 'Determination of ochratoxin A in red wines by multiple pulsed elutions from molecularly imprinted polypyrrole'. *Food Chemistry*, **105**, 301–10.

Zaied, C., Abid, S., Zorgui, L., Bouaziz, C., Chouchane, S., Jomaa, M. and Bacha, H. (2009) 'Natural occurrence of ochratoxin A in Tunisian cereals'. *Food Control*, **20**, 218–22.

Zambonin, C. G., Monaci, L. and Aresta, A. (2002) 'Solid-phase microextraction-high performance liquid chromatography and diode array detection for the determination of mycophenolic acid in cheese'. *Food Chemistry*, **78**, 249–54.

Zhou, S., Lai, E. C. and Miller, J. D. (2004) 'Analysis of wheat extracts for ochratoxin A by molecularly imprinted solid-phase extraction and pulsed elution'. *Analytical and Bioanalytical Chemistry*, **378**, 1903–6.

Zinedine, A., Blesa, J., Mahnine, N., El Abidi, A., Montesano, D. and Mañes, J. (2010) 'Pressurized liquid extraction coupled to liquid chromatography for the analysis of ochratoxin A in breakfast and infants cereals from Morocco'. *Food Control*, **21**, 132–5.

Zöllner, P., Jodlbauer, J. and Lindner, W. (1999) 'Determination of zearalenone in grains by high-performance liquid chromatography-tandem mass spectrometry after solid-phase extraction with RP-18 columns or immunoaffinity columns'. *Journal of Chromatography A*, **858**, 167–74.

Zougagh, M. and Ríos, Á. (2008) 'Supercritical fluid extraction of macrocyclic lactone mycotoxins in maize flour samples for rapid amperometric screening and alternative liquid chromatographic method for confirmation'. *Journal of Chromatography A*, **1177**, 50–7.

3

Chromatographic separation techniques for determination of mycotoxins in food and feed

G. S. Shephard, Medical Research Council, South Africa

Abstract: This chapter discusses the use of various chromatographic separation techniques in the determination of mycotoxins in foods and feeds. Although thin layer chromatography and gas chromatography are still in use, the separation technique that has come to predominate for mycotoxin analysis is high-performance liquid chromatography (HPLC), coupled to a range of detectors. This chapter will describe current trends in HPLC separations, the use of ultraviolet and fluorescence detectors and, where necessary, the derivatization of mycotoxins to enable sensitive detection by these spectrophotometric detectors. Confirmation of mycotoxin analysis can be provided by diode array detection, whereas a description of the use of mass spectrometry (MS) to provide multi-mycotoxin analysis and unequivocal identification is contained in a separate chapter (Chapter 4).

Key words: chromatographic separation, derivatization, electrophoretic separation, fluorescence detector, high-performance liquid chromatography, ultra performance liquid chromatography, ultraviolet detector.

3.1   Introduction

The analysis of fungal-derived toxins has been a challenge for analytical chemists dating back 50 years to the discovery of the aflatoxins in a consignment of peanut meal imported into the United Kingdom from Brazil. The significant human and animal health implications of ingestion of these toxins and the ensuing legislative requirements in many countries has required the development of ever more sensitive and accurate analytical methods. Mycotoxins are naturally occurring contaminants in a wide range of staple foods (CAST, 2003). The levels at which they typically occur are in the micrograms per kilogram (μg kg^{-1}) (ppb) range. Earlier chapters in this book have described their extraction from a variety of matrices and the various clean-up methods available to remove significant amounts of co-extractants that have the potential to interfere with the analytical determination. However, even the most sophisticated clean-up methods such as immunoaffinity columns (IACs), do not necessarily achieve pure solutions of analyte amenable to direct measurement, although in some specific instances, these purified solutions can be used with derivatization to determine the mycotoxin of interest by direct fluorometry (Duncan *et al.*, 1998).

More generally, the analytical process requires a further stage of separation of the analyte from other impurities still remaining in the purified extract. The use of anti-mycotoxin antibodies in a variety of immunological assay formats (see following chapters) can provide this discrimination. However, these assays are mostly semi-quantitative and for a fully quantitative analysis, a separation technique such as chromatography, electrophoresis or electrochromatography is required. As some of the most important mycotoxins such as the aflatoxins and fumonisins also occur as a number of different analogues, chromatographic separation is required to quantify each analogue individually.

Mycotoxins are mostly low molecular mass polar organic compounds produced by a secondary metabolism of fungi and are soluble in a range of organic and aqueous organic solvents. For this reason, they are amenable to separation by a range of chromatographic techniques, including thin layer chromatography (TLC), gas chromatography (GC) and HPLC.

A variety of detectors suitable for determination of mycotoxins in the ppb range coupled to the separation techniques mentioned above. As HPLC has become the ubiquitous method of choice, UV or fluorescence detectors have found wide application in mycotoxin determination. These systems use either the natural UV absorption or fluorescence of the mycotoxin of interest or suitable derivatization methods have been developed to allow appropriately sensitive detection. The successful coupling of HPLC to MS using atmospheric pressure ionization techniques such as electrospray ionization (ESI) and atmospheric pressure chemical ionization (APCI), has allowed this powerful technique to be used for simultaneous mycotoxin analysis and confirmation. The full description of these advances and their application to multi-mycotoxin analysis will be dealt with in Chapter 4.

3.2   Thin-layer chromatography in mycotoxin analysis of food and feed

TLC was the first chromatographic method to be applied to mycotoxin determination and is still in routine use in many laboratories, especially in developing countries, for aflatoxin analysis. After suitable extraction and clean up, aflatoxins are readily separated on silica gel TLC plates using any one of a number of solvent mixtures such as acetone–chloroform (1+9), benzene–methanol–acetic acid (90+5+5), ether–methanol–water (96+3+1) or chloroform–isopropanol (99+1) (Trucksess, 2000). The aflatoxins separated on silica gel plates may be readily visualized under long wavelength UV light. The nomenclature of the toxins as 'B' and 'G' was derived from their blue and green fluorescence colours observed under such a light. Visual inspection of the plates and comparison with a similarly chromatographed set of authentic standards can yield semi-quantitative results. This basic system of TLC has been improved in a number of ways, including the use of high-performance TLC plates, two-dimensional TLC, bi-directional TLC and improved quantification by the use of fluorescence densitometry (Sydenham and Shephard, 1996). Comparison of instrumental high-performance TLC with HPLC showed that although the former did not achieve as good a resolution, it could match the latter with respect to precision, accuracy and sensitivity (Tosch *et al.*, 1984).

Apart from the aflatoxins, TLC methods have been approved as AOAC International official methods for a number of other mycotoxins, including patulin (PAT), deoxynivalenol (DON), zearalenone (ZON) and ochratoxin A (OTA) (Trucksess, 2000). For those toxins lacking sufficient natural fluorescence, suitable spray reagents have been employed, such as $AlCl_3$ (with heating at 120 °C) for DON and 3-methyl-2-benzothiazoline hydrazone (MBTH) for PAT. Instrumental quantification of the spots formed by PAT has been achieved using either a TLC-scanner in absorbance–reflectance mode at 275 nm, which can achieve a detection limit of 3 ng PAT per chromatographic spot (Lin *et al.*, 1993) or by application of a charge couple device (CCD) camera with a detection limit of 5 ng PAT per spot (Welke *et al.*, 2009).

Apart from the shift in TLC R_f values achieved by the use of different solvent systems, specific confirmation methods have been developed for certain mycotoxins separated on TLC plates (Trucksess, 2000). Possible interferences in the analysis of cottonseed for aflatoxins can be distinguished by a spray of sulphuric acid, which turns aflatoxins yellow or yellow-blue and allows distinction between the mycotoxin and common interferences. ZON is visible as a greenish-blue spot under short wavelength UV light (256 nm), but can be confirmed with a spray of $AlCl_3$ and subsequent heating at 130 °C, where after it becomes visible as a blue spot under long wavelength UV light (365 nm). OTA is directly visible as a greenish-blue fluorescent spot, which under conditions involving an alkaline spray of ammonia, $AlCl_3$ or sodium bicarbonate, changes to a more intense blue. Fumonisins have been confirmed in maize and maize-based food products by reversed-phase high-performance TLC and visualization with a spray reagent of 0.5% vanillin in 97% sulphuric acid–ethanol (Pittet *et al.*, 1992). After heating at 120 °C, fumonisin B_1 (FB_1) was identified as a blue-purple spot.

The more recently discovered fumonisin mycotoxins were originally purified using TLC as a guide during column fractionation of the fungal extracts (Cawood *et al.*, 1991). The fumonisin analogues were separated on silica gel plates using p-anisaldehyde or ninhydrin spray reagents with heating to visualize the individual spots. Although determined almost universally by HPLC, TLC methods have also been developed for quantification of fumonisins in maize and rice. The method for rice required the developed plate to be dipped rather than sprayed with p-anisaldehyde reagent (Dawlatana *et al.*, 1995). A method involving extraction of maize with methanol–water, clean up on strong anion exchange solid phase extraction (SPE) cartridges, derivatization with fluorescamine and separation of the preformed FB_1 derivative on reversed-phase TLC plates was tested in a collaborative trial in 14 laboratories (Shephard and Sewram, 2004). Results were comparable to HPLC, which showed the method to be feasible for the analysis of fumonisins.

Most TLC methods developed have relied on clean up of extracts using silica columns or SPE cartridges. However, further improvements in the technique can be achieved by using IACs for purification. This has been applied to the analysis of FB_1 in which maize extracts cleaned up on IAC were separated on a reversed-phase TLC plate and quantified by fluorodensitometry down to a limit of 0.1 mg kg^{-1} following derivatization with fluorescamine spray reagent (Preis and Vargas, 2000). Similarly, extracts of peanut butter, paprika and pistachio nuts have been purified on IAC and the aflatoxins separated on a standard silica gel plate (Stroka *et al.*, 2000a). Quantification by TLC scanner achieved quantification limits significantly lower than current legislative requirements in the European Community.

Another variation in TLC is the use of overpressured layer chromatography (OPLC), which is a technique designed to integrate the benefits of HPLC and TLC. OPLC is a forced flow technique in which external pressure is applied to the chromatoplate which is sealed on all edges and the mobile phase is applied by pump. This system has been used to separate fumonisin analogues produced by Fusarium verticillioides and F. proliferatum on reversed-phase plates (Katay *et al.*, 2001) and to determine aflatoxins in maize, wheat, fish and red paprika on silica gel plates (Otta *et al.*, 2000; Moricz *et al.*, 2007).

3.3   Gas chromatography in mycotoxin analysis of food and feed

Although GC is a powerful separation technique, its application to mycotoxin analysis has been limited. Mycotoxins are polar compounds that mostly require derivatization prior to injection into a GC column. Analysis of fumonisins by GC has been extremely limited and involved hydrolysis of the tricarballylic esters and separation of the hydrolyzed backbones as trimethylsilyl or trifluoroacetyl derivatives on capillary GC (Shephard, 1998).

For PAT analysis, both derivatized and underivatized extracts have been separated on capillary GC columns with MS detection in selected ion monitoring mode. The use of capillary GC–MS as a confirmatory method for PAT has been demonstrated by Llovera *et al.* (1999) and Roach *et al.* (2000). The former used electron impact ionization, whereas the latter used negative ion chemical ionization. Using a hexachlorobenzene internal standard, Llovera *et al.* achieved a detection limit in apple juice of between 4 and 10 µg l^{-1}, depending on the extraction and clean-up method used. Quantitative determination of PAT was achieved using trimethylsilyl derivatives and GC–MS with electron impact source (Cunha *et al.*, 2009; Marks, 2007). Recovery losses in the extraction and clean-up steps of the analysis as well as possible matrix effects within the mass selective detector were compensated by the use of a stable isotope internal standard of ^{13}C5-7 patulin added to the sample aliquot (Cunha *et al.*, 2009). This method achieved a quantification limit of 0.4 µg kg^{-1}, well within the limits required by EU legislation. A previous criticism of the use of derivatives has been that many ions in the MS arise from fragments of the derivative rather than from the PAT portion of the molecule, making the mass spectrum less diagnostic for patulin (Roach *et al.*, 2000).

The group of mycotoxins that have been extensively determined by GC are the trichothecenes (TRICs). These mycotoxins form a large group of fungal secondary metabolites, which are conveniently divided into four types, A to D, according to their chemical structures. They are characterized by the 12,13-epoxy-trichothec-9-ene ring system. Of the four different types, type A and B occur as contaminants of staple food supplies and have been the focus of much analytical attention and method development. They differ by the presence or absence of a carbonyl function at C-8. Type B TRICs, which possess the carbonyl function, show a weak UV absorption and are best represented by DON and nivalenol (NIV), contaminants which are found on a wide range of cereals. T-2 toxin, HT-2 toxin, monoacetoxyscirpenol, diacetoxyscirpenol and scirpentriol are all examples of type A TRICs, which lack a carbonyl at C-8 and show no UV absorption band.

The weak or lack of UV absorbance shown by TRICs has resulted in the development of a range of GC methods with suitable detectors such as flame ionization (FID), electron capture (ECD) and MS. Of these three methods, FID has had limited use and most publications have reported either ECD or MS detection (Koch, 2004). MS or tandem MS (MS/MS) has the advantage of sensitivity, as well as providing confirmatory evidence in the form of characteristic fragment ions. The TRICs are oxygenated polar compounds and require adequate derivatization to increase volatility before they can be injected into a GC column. However, since they are structurally similar and possess similar chemical properties, GC offers the advantage of being capable of determining a range of TRICs simultaneously. The type B TRICs possess a conjugated carbonyl moiety, which makes them amenable to ECD detection, whereas the type A TRICs need to be suitably derivatized for ECD, usually by fluoroacylation. Hence common derivatization reactions at the hydroxyl moieties of TRICs involve the formation of trimethylsilyl (TMS) ethers (for type B TRICs by ECD or type A by MS), or trifluoroacetyl, pentafluoro-propionyl and heptafluorobutyryl ester derivatives.

For TMS derivatives, care must be exercized in optimizing the reaction so as to avoid multiple reaction products, such as mono-, di- and tri-TMS ethers of DON (Sydenham and Shephard, 1996). These problems can be overcome by using mixtures of derivatization reagents such as 1-(trimethylsilyl)imidazole, trimethylchlorosilane and N,O-bis(trimethylsilyl)acetamide (Mateo *et al.*, 2001).

The silylating reagent, N,N-dimethyl-trimethylsilyl-carbamate has been proposed as a suitable single reagent for silylation of DON, NIV and diacetoxyscirpenol (Eke and Torkos, 2004). Reaction conditions needed to be optimized so as to produce a single GC-FID peak for NIV. The analytical method was applied to the determination of DON, NIV, T-2 toxin and 4,15-diacetoxyscirpenol in maize grits and semolina (Eke *et al.*, 2004). MS detection limits of 0.05–0.35 mg kg^{-1} were slightly lower than those of 0.30 to 0.47 mg kg^{-1} achieved for GC-FID. For fluoroacylation of type B TRICs, the heptafluorobutyryl esters are preferable to trifluoroacetyl esters in terms of response, but are unsuitable for determination of 15-acetylDON owing to stereochemical hindrance during the derivatization reaction (Mateo *et al.*, 2001). Quantitative methods have used different internal standards, with Mirex being used for ECD (Koch, 2004) or n-docosane, neosolaniol or alpha-chloralose being employed in FID or MS (Eke *et al.*, 2004; Jestoi *et al.*, 2004; Schothorst and Jekel, 2001).

GC–MS has been applied to TRIC analysis and identification using electron impact ionization, negative chemical ionization and positive chemical ionization (Melchert and Pabel, 2004). These authors provide a list of key fragment ions of TMS derivatives of various TRICs for toxin identification using the above three ionization methods in an ion-trap system operating in the multiple mass spectrometry mode with the aim of identifying possible intentional criminal or natural food contamination.

Recent advances in GC techniques have been applied to type A and B TRIC analysis by comprehensive two-dimensional GC-time-of-flight MS (Jelen and Wasowicz, 2008). Type A TRICs (scirpentriol, HT-2 toxin, T-2 toxin, 15-monoacetoxyscirpenol, 4,15-diacetoxyscirpenol and triacetoxyscirpenol) were analysed as trifluoroacetyl esters, whereas the type B TRICs (DON, 3-acetylDON, fusarenone-X and NIV) were derivatized to form TMS ethers. The extracts of samples of wheat grain were analysed without clean up as the analytes could be separated from matrix constituents using the vast peak capacity of two-dimensional GC. Detection limits for the various TRICs were between 10 and 50 µg kg^{-1}. Each sample was split so as to be derivatized separately, the TMS ethers providing a better detector response for the type B TRICs, whereas the type A TRICs responded better as trifluoroacyl derivatives.

3.4   High-performance liquid chromatography in mycotoxin analysis of food and feed

Mycotoxins of interest in food and feed contamination are mostly relatively small polar compounds, ideally suited to separation by reversed-phase HPLC using a range of mobile phase compositions made up of water, methanol and acetonitrile or mixtures of these components. Common spectrophotometric HPLC detectors, such as UV and fluorescence, have found widespread application. Recent advances in the coupling of HPLC to mass spectrometry via atmospheric pressure ionization techniques has resulted in the development of analytical methods for simultaneous quantification and confirmation using appropriate fragment ions and multiple reaction monitoring. Also, the use of mass spectrometric detection has provided a basis for multi-mycotoxin analysis, using a compromise extraction solvent for all toxins of interest and a 'dilute-and-shoot' approach, which precludes the necessity of finding a clean-up technique for all the toxin analytes. This approach is fully described in Chapter 4.

3.4.1 High-performance liquid chromatography columns and mobile phases

As in the majority of reversed-phase HPLC systems, the standard column used for mycotoxin analysis contains packing of silica particles of diameter 5 μm or less with a surface modified with a hydrophobic layer, mostly with octadecyl (C18) moieties. As analytical methods using conventional spectrophotometric detectors are mainly restricted to single mycotoxins or structurally similar mycotoxin analogues (such as the determination of aflatoxins B_1 (AFB_1), B_2 (AFB_2), G_1 (AFG_1) and G_2 (AFG_2) or the determination of fumonisins B_1, B_2 and B_3), the majority of mycotoxin HPLC separations are performed isocratically. Nevertheless, in some instances, such as the analysis of PAT in raw apple juice, gradients may be necessary to remove late eluting contaminants from the HPLC column (Sydenham *et al.*, 1995). The determination of FB_1 in rodent feed using 9-fluorenylmethyl-chloroformate is another example of the use of gradient elution to solve difficult separation problems (Holcomb *et al.*, 1993).

The introduction of multi-mycotoxin analytical methods using either MS/MS or fluorescence detection has resulted in an increased use of gradient elution to achieve chromatographic separation in a reasonable run time. In this manner, aflatoxins, OTA and ZON in poultry house air samples were isolated on an IAC containing antibodies to all these mycotoxins and analysed by reversed-phase HPLC with fluorescence detection utilizing a mobile phase whose composition was changed after the elution of the aflatoxin analogues in order to elute OTA and ZON in a total run time of 30 min (Wang *et al.*, 2008a).

Mobile phase compositions are selected during method development to provide optimum chromatographic performance with regard to peak shape, peak separation and retention times. An example of this process was the optimization of the analytical method for aflatoxins in commodities and at levels of importance for EU legislation, which led to the investigation of the influence of mobile phase composition on the separation of AFB_1, AFB_2, AFG_1 and AFG_2 (Stroka *et al.*, 2000b). It was found that mixtures of water and methanol led to broad peaks and long retention times, whereas water–acetonitrile mixtures could not achieve baseline resolution of all four aflatoxin analogues. Consequently, a tertiary mixture of water, methanol and acetonitrile was chosen as the optimum.

For analysis of commodities at low contamination levels, it may also be necessary to make large injections into the HPLC column. In this case, the relative elution strengths of injection solvent and mobile phase become important and need to be matched such that a large injection volume allows the analyte to become concentrated at the head of the HPLC column prior to its separation (Stroka *et al.*, 2000b). A method specifically developed to analyse, 24-hour diet samples for AFB_1, aflatoxin M_1 (AFM_1) and OTA in the Netherlands achieved adequate sensitivity using large injection volumes of 2.5 ml of aqueous injection solution containing 15% methanol (Sizoo and Van Egmond, 2005).

For mycotoxins such as the fumonisins and OTA, which contain free carboxylic acid moieties, the pH of the aqueous mobile phase is lowered to suppress ionization of the acid groups and enhance chromatographic peak shape. For fumonisins, depending on the HPLC column packing, typical elution solvent mixtures contain approximately 80% methanol and 20% 0.1 M sodium dihydrogen phosphate adjusted to an apparent pH of 3.35 (Sydenham *et al.*, 1996), whereas mobile phases for OTA analysis typically contain 1–2% glacial acetic acid (Visconti *et al.*, 2001a). Although FB_1 is optimally determined at the above pH, its two partially hydrolyzed forms could only be fully separated from FB_1 and themselves by using a higher pH value of 6.0, at which point the peak shape of FB_1 had deteriorated (Shephard *et al.*, 1994).

Certain important mycotoxins are highly polar molecules, requiring HPLC mobile phases with very low organic content. PAT can be determined by reversed-phase HPLC using mobile phases of 3–10% acetonitrile (MacDonald *et al.*, 2000). The ultimate determinant of chromatographic performance in this analysis is the separation of PAT from impurities in the analytical sample (mostly apple or fruit juices), the main impurity being 5-hydroxymethylfurfural (HMF), a by-product produced by the heat treatment during juice processing. Chromatographic separation is achieved by adjusting mobile phase composition and flow rate. Moniliformin (MON) is a highly polar acidic mycotoxin (pKa 1.70), which occurs naturally as a sodium or potassium salt. It can be chromatographed by reversed-phase HPLC using ion-pair chromatography with a tetrabutylammonium salt as the ion pair reagent (Munimbazi and Bullerman, 2001).

Separation of mycotoxins by reversed-phase HPLC can generally be satisfactorily achieved on a wide range of commercial columns. However, mycotoxins such as tenuazonic acid and related tetramic acids require more careful selection of the packing material. Tenuazonic acid is a beta-diketone with metal chelating properties and also contains a secondary amine moiety. Both these properties are detrimental to efficient chromatographic performance on silica-based packing materials. The amine moiety is susceptible to interactions with active silanol groups on the packing material, thus causing broad, tailing chromatographic peaks. Beta-diketones are strong metal chelators and tend to bind to residual metal ions present in packing materials that have been insufficiently demineralized. In extreme cases, the analyte can be completely adsorbed by the column packing material. Efficient chromatographic separation of tenuazonic acid from its diastereomer, *allo*-tenuazonic acid, and from a related mycotoxin, 3-acetyl-5-isopropyltetramic acid, was achieved using a commercial octadecylsilica packing with a high (30%) carbon loading (ODS30) (Shephard *et al.*, 1991).

A recent development in liquid chromatography has been the introduction of columns (typically, 2.1 mm internal diameter and 10–15 cm length) packed with octadecylsilica particles of 1.7 µm diameter, which together with high pressure solvent delivery, has been commercialized under the name ultraperformance liquid chromatography® (UPLC). The advantage of UPLC is that it provides rapid sample throughput and reduced solvent consumption by reducing analysis times. To date, a limited number of applications of UPLC to mycotoxin analysis have been published. Coupled with MS/MS detection, rapid multitoxin analytical methods have been developed. The four aflatoxin analogues and OTA were separated in only 3.2 min and subsequently detected by ESI-MS/MS operating in positive mode for the aflatoxin analogues and in negative mode for OTA (Ventura *et al.*, 2006). Seventeen mycotoxins and an internal standard were determined in two separate chromatographic runs, one each for positive and negative ESI-MS/MS, on a UPLC instrument operating a 7-minute gradient with a 2-minute re-equilibration time (Ren *et al.*, 2007). Twelve mycotoxins including aflatoxins, fumonisins, DON, OTA, ZON, T-2 and HT-2 toxins have been determined in a range of commodities (maize, walnuts, biscuits and breakfast cereals) by gradient UPLC and positive ESI-MS/MS, with a time between injections of 8.5 min (Frenich *et al.*, 2009).

Although the octadecylsilica column in its various commercial formats has become the workhorse of mycotoxin determination by HPLC, analysis of mycotoxins extracted from certain complex matrices can present difficulties in separation of the analyte from co-eluting impurities. In such cases, the different selectivity offered by alternative surface chemistries, such as phenylhexyl or pentafluorophenyl, may be advantageous. ZON and OTA were determined in soil samples of various types and organic matter contents using a phenylhexyl column eluted with a mobile phase of approximately 73% methanol and 27% aqueous phosphoric acid (Mortensen *et al.*, 2003). FB_1 and its fully hydrolyzed aminopentol were extracted from swine liver and separated on a phenylhexyl column using a gradient of up to 50% acetonitrile followed by a column wash of 100% acetonitrile (Pagliuca *et al.*, 2005).

3.4.2   Detectors

Although the MS or MS/MS detector has resulted in great advances in mycotoxin analysis with regard to multitoxin determination and simultaneous confirmation, UV and fluorescence detectors remain the standards for quantification of mycotoxins in most matrices. Based on sensitivity and specificity, fluorescence detection is preferred, although UV is the normal detection system for mycotoxins with strong UV absorption bands, such as PAT (at, 276 nm) and MON (at 229 nm). Although variable wavelength UV detectors operating at a fixed wavelength are frequently used, an advantage of the diode array UV detector is that it can generate a simultaneous UV spectrum of the HPLC peak, thus providing an element of confirmation in the analysis. The application of this can be clearly demonstrated for the confirmation of PAT during the HPLC separation of PAT from co-extracted compounds in the determination of the contamination of raw apple juice (Sydenham *et al.*, 1995).

Some mycotoxins possess natural fluorescence, which can be utilized for their analytical determination. The coupling of HPLC separation with fluorescence detection has been used as the method of choice for interlaboratory validations and the development of official aflatoxin methods (Arranz *et al.*, 2006; Brera *et al.*, 2007; Senyuva and Gilbert, 2005; Stroka *et al.*, 2001, 2003; Trucksess *et al.*, 2008). However, in the aqueous mixtures used for reversed-phase chromatography, the fluorescence of AFB_1 and AFG_1 are significantly quenched (Kok, 1994). This is generally overcome by the pre- or post-column derivatization of these two analogues at the reactive 8,9-double bond of the dihydrofuran moiety. The former can be performed using trifluoroacetic acid (TFA), which causes hydration of the 8,9-bond to produce the hemiacetals AFB_{2a} and AFG_{2a}. Suitable post-column derivatives can be synthesized by reacting the 8,9-bond with a reactive halogen. The original method, which was developed in the 1980s, used post-column addition of a saturated iodine solution and subsequent heating at 60–75 °C in a reaction coil (Shepherd and Gilbert, 1984; Thiel *et al.*, 1986). Detection limits for this method were of the order 20 pg/injection for AFB_1. An added advantage of automated post-column methods is that a measure of confirmation can be achieved by observing the decrease in the height of the peaks representing AFB_1 and AFG_1 upon switching off of the derivatization mechanism. However, the iodination method also has several disadvantages, including the need for a separate pump and a heated reaction coil, which can cause peak broadening, and the possible crystallization of iodine in incorrectly operated systems. Consequently, reaction systems utilizing the more reactive halogen, bromine, were developed with the added advantage of a greater analyte response than achieved with iodine (Stroka *et al.*, 2000b; O'Riordan and Wilkinson, 2009). Post-column bromination can be cleanly accomplished in either one of two ways, the simplest being electrochemical generation in a so-called Kobra cell. For this method, potassium bromide is dissolved in an acidified mobile phase. The alternative method requires a pulseless pump for post-column addition of pyridinium bromide perbromide (PBPB) and the use of a short reaction coil at ambient temperature.

The bromination methods have found wide applicability and include the determination of aflatoxins in a diverse range of matrices, including peanut butter, pistachio paste, fig paste and paprika powder (Stroka *et al.*, 2000c), baby food (Stroka *et al.*, 2001), hazelnut paste (Senyuva and Gilbert, 2005) and maize (Brera *et al.*, 2007). Photochemical derivatization is another more economical post-column derivatization method. It involves passing the HPLC column eluate through a reaction coil wound around a UV light at ambient temperature, which causes hydration of AFB_1 and AFG_1 to their respective hemiacetals. A recent comparison of this method with the Kobra cell and iodine methods for peanuts and maize showed that the methods were analytically equivalent for peanuts, but that photochemical derivatization gave a slightly high bias for maize (Waltking and Wilson, 2006).

Enhancement of fluorescence for mycotoxin detection can be achieved by incorporation in the HPLC system of specific cyclodextrins (CDs) in the mobile phase (Galaverna *et al.*, 2008; Maragos *et al.*, 2008). CDs, which are cyclic oligosaccharides composed of multiple subunits of glucose, contain an internal cavity that can act as a host site for smaller molecules by forming an inclusion complex. Besides influencing chromatographic (and electrophoretic) properties, interaction with CDs can lead to the enhancement of the analyte's fluorescence properties, the exact mechanism of which is not clear (Maragos *et al.*, 2008). The possible application of various CDs in the analysis of aflatoxins, OTA, ZON and a fluorescent derivative of T-2 toxin has been described (Maragos *et al.*, 2008).

Mycotoxins such as the fumonisins and type A trichothecenes lack a UV absorption band and are hence 'invisible' to UV and fluorescence detectors. As a consequence, these mycotoxins require derivatization for measurement by HPLC with conventional spectrophotometric detectors. A number of fluorogenic reagents have recently been investigated for the HPLC determination of T-2 and HT-2 toxins as ester derivatives (Lippolis *et al.*, 2008) Of these, pyrene-1-carbonyl cyanide and 2-naphthoyl chloride produced better performance with respect to sensitivity and selectivity than 1-naphthoyl chloride or 1-anthroyl cyanide and can be used for simultaneous determination of these two toxins in raw cereals.

For fumonisin, although a number of fluorescence reagents have been used, including fluorescamine (Ross *et al.*, 1991), 4-fluoro-7-nitrobenzofurazan (NBDF) (Scott and Lawrence, 1992), 1-dimethylaminonaphthalene-5-sulphonyl chloride (dansyl chloride) (Scott and Lawrence, 1992), 6-aminoquinolyl N-hydroxysuccinimidylcarbamate (Velazquez *et al.*, 1995) and 9-fluorenylmethyl-chloroformate (FMOC) (Holcomb *et al.*, 1993), the most useful derivatives have been those produced using o-phthaldialdehyde (OPA) as a pre-column fluorogenic derivatizing agent (Shephard *et al.*, 1996). Coupled with strong anion exchange SPE or immunoaffinity column clean up, two methods for the determination of fumonisins in maize using OPA derivatives were validated by collaborative study and achieved AOAC International official status (Sydenham *et al.*, 1996; Visconti *et al.*, 2001b). Although OPA is a recognized reagent for amino acid analysis, it has the draw back of producing unstable derivatives, which requires careful timing of the derivatization reaction and the HPLC injection. This problem has been addressed in various ways such as the use of a different reaction partner to replace 2-mercapto-ethanol in the derivatization reaction (N-acetyl-cystein was found to be most suitable for a stable derivative with OPA; Stroka *et al.*, 2002), the automation of the procedure for pre-column derivatization (Dilkin *et al.*, 2001), the use of post-column OPA derivatization (Akiyama *et al.*, 1998) or the use of a related but more stable reagent, naphthalene dicarboxaldehyde (NDA) (Bennett and Richard, 1995).

Less conventional detectors, such as the evaporative light scattering detector (ELSD), have found limited use in mycotoxin analysis. The ELSD has been used to determine fumonisins in fungal culture material (Plattner, 1995) and as an aid in facilitating the purification of fumonisin standards (Wilkes *et al.*, 1995). It was reported that the calibration curve for all three fumonisin analogues was non-linear and followed a second order equation with a detection limit of about 10 ng injected (Plattner, 1995). A more recent publication has used ELSD for the more demanding requirements of determining the natural presence of fumonisins in maize and a detection limit of 60 ng per injection was reported (Wang *et al.*, 2008b).

3.5   Electrophoretic separations in mycotoxin analysis of food and feed

Chromatographic techniques for the separation of mycotoxins remain the most frequently used owing to the versatility of HPLC in general, including the sensitivity and specificity of the available detectors, whereas other separation techniques such as electrophoresis have mainly languished in the research laboratory. Capillary electrophoresis (CE) achieves chemical separation by utilizing the charge on the molecules in an electrolyte solution contained in a fused silica capillary, rather than on an interaction between the mycotoxin and the surface layer of the separation medium such as GC column coating or HPLC packing. As such it affords a separation functionally different from that of chromatography with the possibility of solving some separation problems, although it may well generate new ones. A further advantage lies in the almost exclusive use of predominantly aqueous solutions and small amounts of reagent. However, owing to the very small injection volumes, highly sensitive detectors are required, especially if analyses are performed at the low levels frequently found in contaminated foods. This latter problem has been addressed by the application of laser-induced fluorescence (LIF) of the mycotoxin or of appropriately derivatized analytes (Maragos, 1998). Alternatively, an amperometric carbon paste electrode has been used as detector for CE of macrocyclic lactone mycotoxins (ZON, alpha-zearalenol and beta-zearalenol) (Arribas *et al.*, 2009).

Neutral species in aqueous buffers can be separated by the addition of a surfactant such as sodium dodecylsulphate at a sufficiently high concentration so as to form micelles, a technique called micellar electrokinetic capillary chromatography (MECC). Neutral analytes partition with these micelles in a chromatographic fashion and are separated based on their retention factors similar to reversed-phase chromatography. On the other hand, charged species such as anionic cyclodextrins can also be used to separate neutral molecules (Bohs *et al.*, 1995; Maragos *et al.*, 2008). A recent description of the application of cyclodextrins to enhance the response from a pyrene derivative of T-2 toxin, which forms an inclusion complex in the cyclodextrin in which the T-2 derivative exists as an excimer with enhanced fluorescence properties, showed the potential of this system (Maragos *et al.*, 2008).

MECC with a modified commercial LIF detector has been used to analyse aflatoxin B_1 in maize samples (Maragos and Greer, 1997). The detection limit of the method depended on the type of clean up used, with immunoaffinity column purification achieving a limit of 1 μg kg^{-1}, whereas the old CB ('contaminants bureau') method using halogenated solvents and silica column improved this to 0.5 μg kg^{-1}. Fumonisins, which lack a useful chromophore or fluorophore, have been analysed in maize by CE following derivatization with fluorescein isothiocyanate (FITC), which was chosen as a derivatization reagent because the resulting derivatives are compatible with commercial argon ion lasers. An immunoaffinity column clean up allowed the determination of FB_1 with a detection limit of 50 μg kg^{-1} and a migration time of 25 min over a 50 cm length of capillary (Maragos *et al.*, 1996).

Holcomb and Thompson (1995) used reversed-phase (C18) SPE for clean up of a maize extract prior to derivatization of the eluate with FMOC for the determination of FB_1 by CE with fluorescence detection at excitation wavelength 210 nm and emission wavelength 305 nm. The detection limit was approximately 500 µg kg^{-1}. OTA has been determined by CE with LIF using excitation at 325 nm provided by a helium–cadmium ion laser. Quantification was achieved in roasted coffee, maize and sorghum after a two-stage clean up using a silica column and an immunoaffinity column (Corneli and Maragos, 1998). Detection limits in all the matrices were 0.2 µg kg^{-1}. Liquid–liquid extraction has been used to purify small volumes of human blood serum for OTA analysis by CE-LIF with a helium–cadmium laser (Koller *et al.*, 2006). More recently, CE has been coupled to electrospray ionization MS to achieve sensitive and specific detection of OTA (Hong and Chen, 2007). An electrophoresis method related to MECC is microemulsion electrokinetic chromatography (MEECK), in which a micro-emulsion in the electrophoresis capillary is used to provide selectivity. MEECK, using an electrolyte composed of borate buffer, sodium dodecylsulphate, butanol and n-heptane, has been used with diode array detection to analyse PAT in apple juice samples with a limit of detection of 3.2 µg l^{-1} (Murillo-Arbizu *et al.*, 2008).

3.6   Future trends

HPLC has become a mature technique with analytical applications in a wide range of sciences. As such, it is the workhorse of many analytical laboratories for a wide spectrum of research and routine analytical work. Given this array of potential applications, the capital outlay for instrumentation is readily recouped, as opposed to instrumentation that may have a more limited application portfolio. For this reason, electrophoretic techniques for mycotoxins are unlikely to challenge HPLC as the method of general choice. HPLC itself continues to evolve as a technique, although changes are more incremental than revolutionary. The shift to smaller particle size column packings, as in UPLC, and the development of more sensitive detectors are examples of this evolution, which will no doubt continue.

The major changes that are occurring in mycotoxin analysis involve the development of immunological methods for rapid analysis and for multi-toxin analytical arrays and the application of LC–MS/MS as a true multi-toxin analytical method with the capability of simultaneous confirmation. Advances in both these areas are described elsewhere in this volume. Nevertheless, the cost of sophisticated instrumentation may limit the spread of these methods. The current trends in mycotoxin analysis are indicated by a recent report on TRIC determination, which found that although GC remains a method of widespread routine use, current advances in TRIC analysis are being reported for LC–MS/MS techniques (Shephard, 2008a).

3.7   Sources of further information and advice

General information on mycotoxins can be obtained from numerous sources such as the Council for Agricultural Science and Technology Task Force Report number 139 (CAST, 2003) or from the European Mycotoxins Awareness Network website (www.mycotoxins.com).

Further information on mycotoxin analytical methods can be obtained from the review by Shephard (2008b), the fact sheets on the European Mycotoxins Awareness Network website or *AOAC International Official Methods of Analysis* (www.aoac.org). Annual reviews of analytical advances can be found in the series of AOAC International mycotoxin General Referee reports, obtainable in the *Journal of AOAC International* or on the AOAC International website. The most recent advances are reviewed in the *World Mycotoxin Journal* (Shephard *et al.*, 2009).

3.8   References

Akiyama H, Uraroongroj M, Miyahara M, Goda Y and Toyoda M (1998), 'Quantitation of fumonisins in corn by HPLC with *o*-phthalaldehyde postcolumn derivatization and their identification by LC/MS'. *Mycopathologia*, **140**, 157–61.

Arranz I, Sizoo E, Van Egmond H, Kroeger K, Legarda T M, Burdaspal P, Reif K and Stroka J (2006), 'Determination of aflatoxin B_1 in medicinal herbs: Interlaboratory study'. *Journal of AOAC International*, **89**, 595–605.

Arribas A S, Bermejo E, Zapardiel A, Tellez H, Rodriguez-Flores J, Zougagh M, Rios A and Chicharro M (2009), 'Screening and confirmatory methods for the analysis of macrocyclic lactone mycotoxins by CE with amperometric detection'. *Electrophoresis*, **30**, 499–506.

Bennett G A and Richard J L (1995), 'Liquid chromatographic method for analysis of the naphthalene dicarboxaldehyde derivative of fumonisins'. *Journal of AOAC International*, **77**, 501–6.

Bohs B, Seidel V and Lindner W (1995), 'Analysis of selected mycotoxins by capillary electrophoresis'. *Chromatographia*, **41**, 631–7.

Brera C, Debegnach F, Minardi V, Pannunzi E, De Santis B and Miraglia M (2007), 'Immunoaffinity column cleanup with liquid chromatography for determination of aflatoxin B_1 in corn samples: Interlaboratory study'. *Journal of AOAC International*, **90**, 765–72.

CAST (2003), *Mycotoxins: Risks in plant, animal, and human systems*, Council for Agricultural Science and Technology, Ames, Iowa.

Cawood, M E, Gelderblom W C A, Vleggaar R, Behrend Y, Thiel P G and Marasas W F O (1991), 'Isolation of fumonisin mycotoxins: A quantitative approach'. *Journal of Agricultural and Food Chemistry*, **39**, 1958–62.

Corneli S and Maragos C M (1998), 'Capillary electrophoresis with laser-induced fluorescence: Method for the mycotoxin ochratoxin A'. *Journal of Agricultural and Food Chemistry*, **46**, 3162–5.

Cunha S C, Faria M A, and Fernandes J O (2009), 'Determination of patulin in apple and quince products by GC–MS using 13C5-7 patulin as internal standard'. *Food Chemistry*, **115**, 352–9.

Dawlatana M, Coker R D, Nagler M and Blunden G (1995), 'A normal phase HPTLC method for the quantitative determination of fumonisin B_1 in rice'. *Chromatographia*, **41**, 187–90.

Dilkin P, Mallman C A, De Almeida C A A and Correa B (2001), 'Robotic automated cleanup for detection of fumonisins B_1 and B_2 in corn and corn-based feed by high-performance liquid chromatography'. *Journal of Chromatography A*, **925**, 151–7.

Duncan K, Kruger S, Zabe N, Kohn B and Prioli R (1998), 'Improved fluorometric and chromatographic methods for the quantification of fumonisins B_1, B_2 and B_3'. *Journal of Chromatography A*, **815**, 41–7.

Eke Z and Torkos K (2004), '*N,N*-dimethyl-trimethylsilyl-carbamate as a derivatizing agent in gas chromatography of trichothecene mycotoxins'. *Microchemical Journal*, **77**, 43–6.

Eke Z, Kende A and Torkos K (2004), 'Simultaneous detection of A and B trichothecenes by gas chromatography with flame ionization or mass selective detection'. *Microchemical Journal*, **78**, 211–6.

Frenich A G, Vidal J L M, Romero-Gonzalez R and Aguilera-Luiz M d M (2009), 'Simple and high-throughput method for the multimycotoxin analysis in cereals and related foods by ultra-high performance liquid chromatography/tandem mass spectrometry', *Food Chemistry*, **117**, 705–712.

Galaverna G, Dall'Asta C, Corradini R, Dossena A and Marchelli R (2008), 'Cyclodextrins as selectors for mycotoxin recognition'. *World Mycotoxin Journal*, **1**, 397–406.

Holcomb M and Thompson H C Jr (1995), 'Analysis of fumonisin B_1 in corn by capillary electrophoresis with fluorescence detection of the FMOC derivative'. *Journal of Microcolumn Separations*, **7**, 451–4.

Holcomb M, Thompson H C Jr and Hankins L J (1993), 'Analysis of fumonisin B_1 in rodent feed by gradient elution HPLC using precolumn derivatization with FMOC and fluorescence detection'. *Journal of Agricultural and Food Chemistry*, **41**, 764–7.

Hong C-Y and Chen Y-C (2007), 'Selective enrichment of ochratoxin A using serum albumin bound magnetic beads as the concentrating probes for capillary electrophoresis/electrospray ionization-mass spectrometric analysis'. *Journal of Chromatography A*, **1159**, 250–5.

Jelen H H and Wasowicz E (2008), 'Determination of trichothecenes in wheat grain without sample cleanup using comprehensive two-dimensional gas chromatography–time-of-flight mass spectrometry'. *Journal of Chromatography A*, **1215**, 203–7.

Jestoi M, Ritieni A and Rizzo A (2004), 'Analysis of the *Fusarium* mycotoxins fusaproliferin and trichothecenes in grains using gas chromatography–mass spectrometry'. *Journal of Agricultural and Food Chemistry*, **52**, 1464–1469.

Katay G, Szecsi A and Tyihak E (2001), 'Separation of fumonisins by OPLC'. *Journal of Planar Chromatography – Modern TLC*, **14**, 53–6.

Koch P (2004), 'State of the art of trichothecenes analysis'. *Toxicology Letters*, **153**, 109–12.

Kok W Th (1994), 'Derivatization reactions for the determination of aflatoxins by liquid chromatography with fluorescence detection'. *Journal of Chromatography B*, **659**, 127–37.

Koller G, Wichmann G, Rolle-Kampczyk U, Popp P and Herbarth O (2006), 'Comparison of ELISA and capillary electrophoresis with laser-induced fluorescence detection in the analysis of ochratoxin A in low volumes of human blood serum'. *Journal of Chromatography B*, **840**, 94–8.

Lin L M, Zhang J, Sui K and Sung W B (1993), 'Simultaneous thin layer chromatographic determination of zearalenone and patulin in maize'. *Journal of Planar Chromatography*, **6**, 274–7.

Lippolis V, Pascale M, Maragos C M and Visconti A (2008), 'Improvement of detection sensitivity of T-2 and HT-2 toxins using different fluorescent labeling reagents by high-performance liquid chromatography'. *Talanta*, **74**, 1476–83.

Llovera M, Viladrich R, Torres M and Canela R (1999), 'Analysis of underivatized patulin by a GC–MS technique'. *Journal of Food Protection*, **62**, 202–5.

MacDonald S, Long M, Gilbert J and Felgueiras I (2000), 'Liquid chromatographic method for the determination of patulin in clear and cloudy apple juices and apple puree: Collaborative study'. *Journal of AOAC International*, **83**, 1387–94.

Maragos C M (1998), 'Analysis of mycotoxins with capillary electrophoresis'. *Seminars in Food Analysis*, **3**, 353–73.

Maragos C M and Greer J I (1997), 'Analysis of aflatoxin B_1 in corn using capillary electrophoresis with laser-induced fluorescence detection'. *Journal of Agricultural and Food Chemistry*, **45**, 4337–41.

Maragos C M, Bennett G A and Richard J L (1996), 'Analysis of fumonisin B$_1$ in corn by capillary electrophoresis, in *Fumonisins in Food*, Jackson L, De Vries J W and Bullerman L B (eds), Plenum Press, New York, 105–12.

Maragos C M, Appell M, Lippolis V, Visconti A, Catucci L and Pascale M (2008), 'Use of cyclodextrins as modifiers of fluorescence in the detection of mycotoxins'. *Food Additives and Contaminants*, B, 164–71.

Marks H S (2007), 'Rapid gas chromatography/mass spectrometry determination and confirmation of patulin in apple juice'. *Journal of AOAC International*, **90**, 879–83.

Mateo J J, Llorens A, Mateo R and Jimenez M (2001), 'Critical study of and improvements in chromatographic methods for the analysis of type B trichothecenes'. *Journal of Chromatography A*, **918**, 99–112.

Melchert H-U and Pabel E (2004), 'Reliable identification and quantification of trichothecenes and other mycotoxins by electron impact and chemical ionization-gas chromatography-mass spectrometry, using an ion-trap system in the multiple mass spectrometry mode. Candidate reference method for complex matrices'. *Journal of Chromatography A*, **1056**, 195–9.

Moricz A M, Fater Z, Otta K H, Tyihak E and Mincsovics E (2007), 'Overpressured layer chromatographic determination of aflatoxin B$_1$, B$_2$, G$_1$ and G$_2$ in red paprika'. *Microchemical Journal*, **85**, 140–4.

Mortensen G K, Strobel B W and Hansen H C B (2003), 'Determination of zearalenone and ochratoxin A in soil'. *Analytical and Bioanalytical Chemistry*, **376**, 98–101.

Munimbazi C and Bullerman L B (2001), 'Chromatographic method for the determination of the mycotoxin moniliformin in corn', in *Mycotoxin Protocols, Methods in Molecular Biology 157*, Trucksess M W and Pohland A E (eds), Humana Press, Totowa, New Jersey, 131–45.

Murillo-Arbizu M, Gonzalez-Penas E, Hansen S H, Amezqueta S and Ostergaard J (2008), 'Development and validation of a microemulsion electrokinetic chromatography method for patulin quantification in commercial apple juice'. *Food and Chemical Toxicology*, **46**, 2251–7.

O'Riordan M J and Wilkinson M G (2009), 'Comparison of analytical methods for aflatoxin determination in commercial chilli spice preparations and subsequent development of an improved method'. *Food Control*, **20**, 700–5.

Otta K H, Papp E and Bagocsi B (2000), 'Determination of aflatoxins in food by overpressured-layer chromatography'. *Journal of Chromatography A*, **882**, 11–6.

Pagliuca G, Zironi E, Ceccolini A, Matera R, Serrazanetti G P and Piva A (2005), 'Simple method for the simultaneous isolation and determination of fumonisin B$_1$ and its metabolite aminopentol-1 in swine liver by liquid chromatography-fluorescence detection'. *Journal of Chromatography B*, **819**, 97–103.

Pittet A, Parisod M and Schellenberg (1992), 'Occurrence of fumonisins B$_1$ and B$_2$ in corn-based products from the Swiss market'. *Journal of Agricultural and Food Chemistry*, **40**, 1352–1354.

Plattner R D (1995), 'Detection of fumonisins produced in *Fusarium moniliforme* cultures by HPLC with electrospray MS and evaporative light scattering detectors'. *Natural Toxins*, **3**, 294–8.

Preis R A and Vargas E A (2000), 'A method for determining fumonisin B$_1$ in corn using immunoaffinity column clean-up and thin layer chromatography/densitometry'. *Food Additives and Contaminants*, **17**, 463–8.

Ren Y, Zhang Y, Shao S, Cai Z, Feng L, Pan H and Wang Z (2007), 'Simultaneous determination of multi-component mycotoxin contaminants in foods and feeds by ultra-performance liquid chromatography tandem mass spectrometry'. *Journal of Chromatography A*, **1143**, 48–64.

Roach J A G, White K D, Trucksess M W and Thomas F S (2000), 'Capillary gas chromatography/mass spectrometry with chemical ionization and negative ion detection for confirmation of identity of patulin in apple juice'. *Journal of AOAC International*, **83**, 104–12.

Ross P F, Rice L G, Plattner R D, Osweiler G D, Wilson T M, Owens D B, Nelson H A and Richard J L (1991), 'Concentrations of fumonisin B$_1$ in feeds associated with animal health problems'. *Mycopathologia*, **114**, 129–35.

Schothorst R C and Jekel A A (2001), 'Determination of trichothecenes in wheat by capillary gas chromatography with flame ionisation detection'. *Food Chemistry*, **73**, 111–7.

Scott P M and Lawrence G A (1992), 'Liquid chromatographic determination of fumonisins with 4-fluoro7-nitrobenzofurazan'. *Journal of AOAC International*, **75**, 829–34.

Senyuva H Z and Gilbert J (2005), 'Immunoaffinity column cleanup with liquid chromatography using post-column bromination for determination of aflatoxins in hazelnut paste: Interlaboratory study'. *Journal of AOAC International*, **88**, 526–35.

Shephard G S (1998), 'Chromatographic determination of the fumonisin mycotoxins'. *Journal of Chromatography A*, **815**, 31–9.

Shephard G S (2008a), 'Committee on natural toxins and food allergens: General referee report'. *Journal of AOAC International*, **91**, 1B–16B.

Shephard G S (2008b), 'Determination of mycotoxins in human foods'. *Chemical Society Reviews*, **37**, 2468–77.

Shepherd M J and Gilbert J (1984), 'An investigation of HPLC post-column iodination conditions for the enhancement of aflatoxin B$_1$ fluorescence'. *Food Additives and Contaminants*, **1**, 325–35.

Shephard G S and Sewram V (2004), 'Determination of the mycotoxin fumonisin B$_1$ in maize by reversed-phase thin layer chromatography'. *Food Additives and Contaminants*, **21**, 498–505.

Shephard G S, Thiel P G, Sydenham E W, Vleggaar R and Marasas W F O (1991), 'Reversed-phase high-performance liquid chromatography of tenuazonic acid and related tetramic acids'. *Journal of Chromatography Biomedical Applications*, **566**, 195–205.

Shephard G S, Thiel P G, Sydenham E W, Vleggaar R and Alberts J F (1994), 'Determination of the mycotoxin fumonisin B$_1$ and identification of its partially hydrolysed metabolites in the faeces of non-human primates'. *Food and Chemical Toxicology*, **32**, 23–9.

Shephard G S, Thiel P G, Stockenström S and Sydenham E W (1996), 'Worldwide survey of fumonisin contamination of corn and corn-based products'. *Journal of AOAC International*, **79**, 671–87.

Shephard G S, Berthiller F, Dorner J, Krska R, Lombaert G A, Malone B, Maragos C, Sabino M, Solfrizzo M, Trucksess M W, Van Egmond H P and Whitaker T B (2009), 'Developments in mycotoxin analysis: an update for, 2007–2008'. *World Mycotoxin Journal*, **2**, 3–21.

Sizoo E A and Van Egmond H P (2005), 'Analysis of duplicate, 24-hour diet samples for aflatoxin B$_1$, aflatoxin M$_1$ and ochratoxin A'. *Food Additives and Contaminants*, **22**, 163–72.

Stroka J, Van Otterdijk R and Anklam E (2000a), 'Immunoaffinity column clean-up prior to thin-layer chromatography for the determination of aflatoxins in various matrices'. *Journal of Chromatography A*, **904**, 251–6.

Stroka J, Petz M and Anklam E (2000b), 'Analytical methods for the determination of aflatoxins in various food matrices at concentrations regarding the limits set in European Regulations: Development, characteristics, limits'. *Mycotoxin Research*, **16**, 23–42.

Stroka J, Anklam E, Jörissen U and Gilbert J (2000c), 'Immunoaffinity column cleanup with liquid chromatography using post-column bromination for determination of aflatoxins in peanut butter, pistachio paste, fig paste, and paprika powder: Collaborative study'. *Journal of AOAC International*, **83**, 320–40.

Stroka J, Anklam E, Joerissen U and Gilbert J (2001), 'Determination of aflatoxin B$_1$ in baby food (infant formula) by immunoaffinity column cleanup liquid chromatography with post-column bromination: Collaborative study'. *Journal of AOAC International*, **84**, 1116–23.

Stroka J, Capelletti C, Papadopoulou-Bouraoui A, Pallaroni L and Anklam E (2002), 'Investigation of alternative reagents to, 2-mercaptoethanol for the pre-column derivatization of fumonisins with *o*-phthaldialdehyde, for HPLC analysis'. *Journal of Liquid Chromatography and Related Technologies*, **25**, 1821–33.

Stroka J, Von Holst C, Anklam E and Reutter M (2003), 'Immunoaffinity column cleanup with liquid chromatography using post-column bromination for determination of aflatoxin B_1 in cattle feed: Collaborative study'. *Journal of AOAC International*, **86**, 1179–1186.

Sydenham E W and Shephard G S (1996), 'Chromatographic and allied methods of analysis for selected mycotoxins, in *Progress in Food Contaminant Analysis*, Gilbert J (ed.), Blackie Academic, London, 65–146.

Sydenham E W, Vismer H F, Marasas W F O, Brown N, Schlechter M, Van der Westhuizen L and Rheeder J P (1995), 'Reduction of patulin in apple juice samples – influence of initial processing'. *Food Control*, **6**, 195–200.

Sydenham E W, Shephard G S, Thiel P G, Stockenström S, Snijman P W and Van Schalkwyk D J (1996), 'Liquid chromatographic determination of fumonisins B_1, B_2, and B_3 in corn: AOAC-IUPAC collaborative study'. *Journal of AOAC International*, **79**, 688–96.

Thiel P G, Stockenström S and Gathercole P S (1986), 'Aflatoxin analysis by reverse phase HPLC using post-column derivatization for enhancement of fluorescence'. *Journal of Liquid Chromatography*, **9**, 103–12.

Tosch D, Waltking A E and Schlesier J F (1984), 'Comparison of liquid chromatography and high performance thin layer chromatography for determination of aflatoxin in peanut products'. *Journal of the Association of Official Analytical Chemists*, **67**, 337–9.

Trucksess M W (2000) Natural toxins, in *Official Methods of Analysis of AOAC International*, Horwitz W (ed.), 17th edn, AOAC Int, Gaithersburg.

Trucksess M W, Weaver C M, Oles C J, Fry F S Jr, Noonan G O, Betz J M and Rader J I (2008), 'Determination of aflatoxins B_1, B_2, G_1, and G_2 and ochratoxin A in ginseng and ginger by multitoxin immunoaffinity column cleanup and liquid chromatographic quantitation: Collaborative study'. *Journal of AOAC International*, **91**, 511–23.

Velazquez C, Van Bloemendal C, Sanchis V and Canela R (1995), 'Derivative of fumonisins B_1 and B_2 with 6-aminoquinolyl N-hydroxysuccinimidylcarbamate'. *Journal of Agricultural and Food Chemistry*, **43**, 1535–7.

Ventura M, Guillen D, Anaya I, Broto-Puig F, Lliberia J L, Agut M and Comellas L (2006), 'Ultra-performance liquid chromatography/tandem mass spectrometry for the simultaneous analysis of aflatoxins B_1, G_1, B_2, G_2 and ochratoxin A in beer'. *Rapid Communications in Mass Spectrometry*, **20**, 3199–3204.

Visconti A, Pascale M and Centonze G (2001a), 'Determination of ochratoxin A in wine and beer by immunoaffinity column cleanup and liquid chromatographic analysis with fluorometric detection: Collaborative study'. *Journal of AOAC International*, **84**, 1818–27.

Visconti A, Solfrizzo M, De Girolamo A (2001b), 'Determination of fumonisins B_1 and B_2 in corn and corn flakes by liquid chromatography with immunoaffinity column cleanup: Collaborative study'. *Journal of AOAC International*, **84**, 1828–37.

Waltking A E and Wilson D (2006), 'Liquid chromatographic analysis of aflatoxin using post-column photochemical derivatization: Collaborative study'. *Journal of AOAC International*, **89**, 678–92.

Wang Y, Chai T, Lu G, Quan C, Duan H, Yao M, Zucker B-A and Schlenker G (2008a), 'Simultaneous detection of airborne aflatoxin, ochratoxin and zearalenone in a poultry house by immunoaffinity clean-up and high-performance liquid chromatography'. *Environmental Research*, **107**, 139–44.

Wang J, Zhou Y and Wang Q (2008b), 'Analysis of mycotoxin fumonisins in corn products by high-performance liquid chromatography coupled with evaporative light scattering detection'. *Food Chemistry*, **107**, 970–6.

Welke J E, Hoeltz M, Dottori H A and Noll I B (2009), 'Quantitative analysis of patulin in apple juice by thin-layer chromatography using a charge couple device detector'. *Food Additives and Contaminants*, **26**, 754–8.

Wilkes J G, Sutherland J B, Churchwell M I and Williams A J (1995), 'Determination of fumonisins B_1, B_2, B_3 and B_4 by high-performance liquid chromatography with evaporative light-scattering detection'. *Journal of Chromatography A*, **695**, 319–23.

4

Mass spectrometry in multi-mycotoxin and fungal spore analysis

M. C. Spanjer, Food and Consumer Product Safety Authority, The Netherlands

Abstract: The development of liquid chromatography–mass spectrometry (LC–MS) from a single mycotoxin confirmation technique into a multi-mycotoxin routine analytical method is described. LC–MS/MS not only has the advantage of measuring all regulated compounds in one single run, but also the possibility to discover unknown mycotoxins, or to find mycotoxins in matrices where they were never expected to be present or ever found before. After compilation of 45 methods published in the last eight years, the critical performance criteria in liquid chromatography and mass spectrometry are discussed. Typical problems like matrix effect, choice of internal standard and considerations concerning validation are worked out. Finally some future trends like Orbitrap, MALDI, DESI and the combination of multi-methods for different groups of chemical contaminants and residues are considered.

Key words: internal standard, isotopically labelled standard, LC–MS, matrix effect, multi-mycotoxin, validation.

4.1   Introduction

Application of mass spectrometry to mycotoxin analysis goes back as far as 1971, when Haddon *et al.* (1971) published a paper on aflatoxin detection by thin layer chromatography–mass spectrometry (TLC–MS). At that time it was already recognized that fluorescing artifacts could falsely indicate the presence of aflatoxins. This led to the introduction of a mass spectral method to provide unambiguous identification of aflatoxins, which were isolated from individual TLC spots. The extracts were concentrated on a probe, which was heated after being placed in the mass spectrometer, which is obviously a huge sector type instrument. A detection limit of 20 μg kg^{-1} could be reached. It took 13 years more before Plattner *et al.* (1984) published an application of a triple quadrupole instrument for identification of the aflatoxins B$_1$, B$_2$, G$_1$, G$_2$, M$_1$ and M$_2$ (AFB$_1$, AFB$_2$, AFG$_1$, AFG$_2$, AFM$_1$ and AFM$_2$). After electron impact ionization they recorded the molecular M$^+$ ion and the daughter ions from the protonated MH$^+$ molecules. Samples were still introduced via a (direct insertion) probe. Crude extracts could be applied, since the argon in the collision cell of the instrument produced so many daughter ions that clear spectra were obtained. This was accompanied by observation of a decrease in signal strength when more extract was introduced into the ion source, reported as a matrix effect. Nevertheless, amounts as low as 10 μg kg^{-1} could be quantified.

Obviously this method of sample introduction was not suitable for routine analysis. Only with the introduction of microbore columns for high performance liquid chromatography (HPLC), which decreased the necessary eluent volumes, was it technically possible to improve coupling devices between the liquid chromatograph and the mass spectrometer. Tiebach *et al.* (1985) reported transformation of their gas chromatography–mass spectrometry (GC–MS) method for the confirmation of nivalenol (NIV) and deoxynivalenol (DON) towards online microbore liquid chromatography–mass spectrometry (LC–MS). They applied negative chemical ionization in selected ion monitoring mode with hexachlorobiphenyl as internal standard and determined amounts up to 5–10 μg kg^{-1} in naturally contaminated cereals. Rajakyla *et al.* (1987) determined DON, patulin (PAT), diacetoxyscirpenol (DAS), T-2, HT-2, zearalenone (ZEN) and ochratoxin A (OTA) in cereals by thermospray LC–MS and reported the influence of pH and acetate concentration of the eluent on retention and ion spectra. Their efforts however, cannot be considered as a multi-mycotoxin method, as they handled the selected mycotoxins in two separate groups for which they applied two different clean-up steps. Korfmacher *et al.* (1991) reported application of mass spectrometry to determine fumonisin B$_1$ (FB$_1$). Thus, it became possible to confirm all mycotoxins that are regulated in the EU for food (Commission Regulations 2006/1881/EC and 2007/1126/EC) and feed (Commission Directive 2002/32/EC and Commission Recommendation 2006/576/EC). They also compared thermospray, fast-atom bombardment (FAB) and electrospray ionization (ESI).

The introduction of the latter technique meant that there was a breakthrough in the application of LC–MS, as it made high throughput of samples possible and thus turned LC–MS from a confirmation technique for individually single compounds into an option for routine multi-analyte determination. For mycotoxins this was first recognized by Smedsgaard and Frisvad (1996), who determined mould species for taxonomy and secondary metabolite profiling. In crude fungal extracts of *Penicillium* isolates they identified, with a single quadrupole, 13 mycotoxins, of which ochratoxin A, citrinin (CIT), penicillic acid and roquefortin C (Roq C) are the most important for food and feed control. The introduction of ion trap detector (ITD) technology was another boost for routine applications, as it brought benchtop instruments into the laboratory. Tuomi *et al.* (1998) applied LC-ITD with soft positive ESI to detect polar and macrocyclic trichothecenes (e.g. T-2, HT-2, DAS) from indoor environment samples. However, the accuracy of quantitative analysis was limited by the technical characteristics of the ion trap. They also noted that more accurate measurements would require labelled internal standards of all toxins, which were not commercially available at that time.

Berger *et al.* (1999) also applied LC-ITD, with atmospheric pressure chemical ionization (APCI), to investigate NIV, DON, DAS, T-2, HT-2, neosolaniol (NEO), fusarenone X (F-X), 3- and 15-acetyldesoxynivalenol (3- and 15-AcDON) in wheat samples. Griesshaber (2007) further developed this ion trap method by adding citrinin, OTA and ZEN, whereas the aflatoxins were still missing. Tuomi *et al.* (2000) added CIT and OTA two years after their first published ITD method and, even more importantly, added the aflatoxins just one year later (Tuomi *et al.*, 2001). They also emphasized the possible application for food and feed analysis, without testing it themselves. Other researchers picked this up and 25 multi-mycotoxin methods in these matrices were published in the next six years (Table 4.1). Developments in the application of LC–MS for mycotoxin analysis in food and feed by hyphenated chromatographic/mass spectrometric technology were described by Razzazi-Fazeli (2006), reviewed by Sforza *et al.* (2006) in 2004, and Zöllner and Mayer-Helm (2006). The latter authors compiled an impressive detailed literature review, updated up to July 2006, but mainly focussed on single methods. Song-sermsakul and Razzazi-Fazeli (2008) and Schuhmacher *et al.* (2008) published updates, which covered references up to October 2007. Since then another 20 multi-mycotoxin methods have been published in less then three years.

Multi-mycotoxin methods not only have the advantage of measuring all regulated compounds in one single run. The greatest advantage could even be the possibility of discovering unknown mycotoxins, or to find them in matrices where they were not expected to be present. Examples of the latter option were findings of fumonisin B$_1$ in figs and ochratoxin A in pistachios. Bayman *et al.* (2002) reported the isolation of fungi with the ability to produce ochratoxin A in tree nuts and the British Food Standards Agency reported the actual analysis of it in a mycotoxin survey (Matthews, 2002). The Italian Institute of Sciences of Food Production in Bari was the first to report the presence of toxigenic fungi, able to produce other mycotoxins than aflatoxin, in figs (Logrieco *et al.*, 2003). Two years later they reported the results of chemical analysis of rotten figs, showing fumonisin contamination at a low level in some samples (Moretti *et al.*, 2005). Recently Noonim *et al.* (2009) found fumonisins B$_2$ and B$_4$ in coffee beans, which until then were only known to have been contaminated with ochratoxin A. These had already been reported in rice by Abbas *et al.* (1998) and detected again in Japan (Kushiro *et al.*, 2009). Mizutani *et al.* (2009) determined luteoskyrin, a yellow rice toxin, which had been discovered much earlier (Uraguchi *et al.*, 1961). Using LC–MS multi-mycotoxin analysis it can easily be monitored, to evaluate whether it could be a re-emerging risk. These findings are comparable with those of Uhlig *et al.* (2007) when they discovered ergot alkaloids (ERG) in Norwegian wild grasses, which could be dangerous to animal health. They determined a complete set of relevant ergot alkaloids in a single run. Song *et al.* (2008) found beauvericin (BEA) and enniantins (ENNs) in potato while Monbaliu *et al.* (2009) found BEA and FB in sweet pepper, food matrices that are seldom related to mycotoxins. Sørensen *et al.* (2009) analysed wet apple core rot on *Fusarium avenaceum* metabolites and found, amongst others, moniliformin (MON) and ENNs, but no patulin that is routinely controlled in apple and apple products. Ren *et al.* (2007) also unexpectedly found aflatoxin M_1 and citrinin in peanut butter. The presence of aflatoxin M_1 was especially surprising, as it is often used as an internal standard in analysis of non-dairy products.

4.2   Liquid chromatography–mass spectrometry (LC–MS) methods in multi-mycotoxin analysis

Owing to rapid technical developments in the field of multi-mycotoxin analysis, only the literature of the last ten years is presented. Table 4.1 gives an overview of mycotoxin/matrix combinations, additional information and references. Table 4.2 gives the corresponding analytical details. Methods are collected that analyse at least two mycotoxins for which EU legislation exists. They are compiled in chronological order, i.e. according to date as sent to the journal. Regarding the gap of sometimes more than one year until publication, this represents a more realistic overview of the analytical progress in time, when compared to publication date.

For method 2, the extraction was taken from Pallaroni and Von Holst (2003). In some references different extraction solvents were investigated. The final and optimal one is mentioned in the table. In addition to the compiled methods, a study of Asam and Rychlik (2007), who investigated the accuracy of the extraction by means of isotopically labelled DON, 4-acetyl-nivalenol (= fusarenone X), 3- and 15-acetyl-DON, monoacetoxyscirpenol (MAS), DAS, T-2 and HT-2 and determined acetonitrile (ACN):water = 84:16 v/v to be the optimum at an extraction volume of 10 ml g^{-1} of sample. With a slightly modified QUECHERS (QUick, Easy, CHeap, Effective, Rugged and Safe) method, Sospedra *et al.* (2010) obtained better recoveries for DON, NIV, DAS T-2 and HT-2, when applying a mixture of methanol and acetonitrile (85:15, v/v).

A systematic study of the decrease in retention time and improvement in peak shape and thus baseline separation when going from 5 µm to 1.8 µm particle size, was performed by Senyuva *et al.* (2008b). Method 26 was first extended to 106 mycotoxins (Sulyok *et al.*, 2010), later to 186 fungal metabolites (Vishwanath *et al.*, 2009) and validated in different nuts and dried fruits by Varga *et al.* (2009). Method 42 was successfully validated for eight analytes qualitatively and 19 quantitatively (Rasmussen *et al.*, 2010). Both tables give options for solving analytical problems of interest, for which further details can be found in the original references. For example Monbaliu *et al.* (2010) combined sample preparation of method 26 with their LC–MS procedure for sweet pepper (method 33), to make it suitable for different feed matrices (sow feed, wheat and maize). López Grío *et al.* (2010) minimized method 39 to achieve an analysis time for aflatoxins and ochratoxin A in animal feed of only 3.5 min.

Table 4.1   Mycotoxins, matrices, additional information and references for the collected methods

No	AFB$_1$	OTA	DON	FB$_1$	ZEN	PAT	AFM$_1$	T2/HT2	ERG	Other mycotoxins	Matrix	Additional information	Reference
1			x		x						Wheat, maize	Full scan m/z 150–350	Pallaroni et al., 2002
2			x		x					15-AcDON	Fungi & maize	Recovery only reported for blood serum of pig	Bily et al., 2004
3	x	x	x	x	x	x	x	x	x	474 included in database	Fungal culture	Full scan m/z 100–900; relative sensitivity MS/UV list	Nielsen and Smedsgaard, 2003
4	x	x	x		x					NIV, α- and β-ZOL	Liver, kidney	Animal offal (pork)	Driffield et al., 2003
5			x	x	x						Maize, wheat	CID with He; pooled extracts of FB and DON/ZEN	Royer et al., 2004
6	x	x	x		x		x	x		NIV, AcDON, F-X, DAS, ZAN	Food and feed	Flour, bread, pastry, cereal, pasta, corn, soy, molasses	Biselli et al., 2004
7			x		x			x		NIV, F-X, AcDON, DAS	Maize	IS (ZAN) compensated well for ME	Berthiller et al., 2005
8			x		x					NIV, F-X, AcDON, ZAN	Maize	Sample is extracted twice	Cavaliere et al., 2005a
9	x	x					x			MPA, Penicillic acid, Roq C	Cheese	Extract re-dissolved in MeOH	Kokkonen et al., 2005
10		x	x	x	x		x	x		DAS, MAS, AcDON, DOM	Bovine milk	Extract re-dissolved in 20% MeOH	Sørensen and Elbæk, 2005
11			x	x	x			x		NIV, F-X, AcDON, DAS, MAS	Corn meal	3- and 15-AcDON not separated	Cavaliere et al., 2005b
12			x		x			x		13 in total, e.g. AcDON, DAS	Food	Corn (flour), wheat flour and bran, durum, oats, bread	Klötzel et al., 2006
13	x		x		x			x		NIV, F-X, AcDON, DAS, AFG$_{1,\,2}$	Corn, wheat	Also corn flakes and biscuits	Tanaka et al., 2006

No.									Toxins	Matrix	Comments	Reference
14	x	x	x		x				x	16 in total, e.g. NIV, DAS, STE CIT and gliotoxin	Fungal culture	ME eliminated by IS (DOM) and extraction	Delmulle *et al.*, 2006
15	x	x	x	x	x	x					Corn silage	SIM mode, 9 months silage survey	Garon *et al.*, 2006
16	x	x	x	x	x	x			x	39 in total	Wheat, maize	ME in maize, not in wheat, PAT finally excluded	Sulyok *et al.*, 2006
17	x	x									Beer	Quantification by standard addition method	Ventura *et al.*, 2006
18	x	x	x						x	17 in total	Peanut, corn	Peanut butter as food and corn as feed matrix	Ren *et al.*, 2007
19			x	x	x				x	NIV, F-X, AcDON, DAS, MAS	Corn meal	Improvement of method 11 with type A & B runs	Cavaliere *et al.*, 2007
20	x	x	x	x	x					CIT and gliotoxin	Corn silage	11 months silage, better FB extraction as in no. 15	Richard *et al.*, 2007
21		x	x		x				x	8 in total, e.g. CIT, NIV	Wheat, maize	Full scan *m/z* 235–800	Griesshaber, 2007
22	x	x	x	x	x				x	33 in total	Food and feed	FB recovery after extra dilution with water	Spanjer *et al.*, 2008
23	x	x	x	x	x				x	38 in total	Maize, rice	Spelt and barley were also analysed	Sulyok *et al.*, 2007a
24	x	x	x	x	x				x	11 in total	Maize	ME only for aflatoxins and slightly for OTA	Lattanzio *et al.*, 2007
25			x		x				x	NIV, F-X, AcDON, DAS	Wheat, oat	Hardly ME in durum wheat	Suman and Catellani, 2008
26	x	x	x	x	x	x	x		x	87 in total	Bread, fruit	Lower recoveries for ERG due to epimerization	Sulyok *et al.*, 2007b
27	x	x	x	x	x				x	20 in total	Silage	Corn, wheat and grass silage	Driehuis *et al.*, 2008

Table 4.1   *continued*

No	AFB_1	OTA	DON	FB_1	ZEN	PAT	AFM_1	T2/HT2	ERG	Other mycotoxins	Matrix	Additional information	Reference
28	x	x		x	x	x		x		FB_4	Dried figs	IS before extraction, AF and OTA via library search	Senyuva and Gilbert, 2008
29	x	x	x	x	x		x	x		20 in total, e.g. CIT, STE	Fungal culture	Compound identification by database search	Senyuva *et al.*, 2008a
30	x	x	x	x	x			x		NIV	Spices	Also wheat, maize, cookies, feed, peanut butter	Boonzaaijer *et al.*, 2008
31			x		x			x		NIV, F-X, 3-AcDON, NEO, ZAN	Wheat, maize	Ammonium and acetate adducts, except for ZEN/ZAN	Santini *et al.*, 2009
32	x	x	x	x	x			x	x	31 in total, e.g. ENNs, MPA	Grains	Wheat, barley, oats	Kokkonen and Jestoi, 2009a,b
33	x	x	x	x	x			x		24 in total, e.g. NIV, DAS	Sweet pepper	Extract was split, cleaned and recombined before inj.	Monbaliu *et al.*, 2009
34	x	x	x	x	x			x	x	23 in total, e.g. AcDON, DAS	1 feed, 4 food	Feed, maize flour, minced meat, milk, egg, honey	Mol *et al.*, 2008
35	x	x		x	x			x		Alternariol and gliotoxin	Oilseed cakes	Survey during 5 months farm storage	Lanier *et al.*, 2009
36	x	x	x	x	x	x		x	x	32 in total, e.g. CIT, DAS, AOH	Wheat, maize	Triple quadrupole (tQ). LOQ = 2640 µg kg⁻¹ for PAT	Herebian *et al.*, 2009

										Mycotoxins	Matrix	Remarks	Reference
37	x	x	x	x	x	x		x	x	32 in total, e.g. CIT, DAS, AOH	Wheat, maize	Orbitrap full scan m/z 90–900. LOQ PAT = 6600 µg kg^{-1}	Herebian et al., 2009
38	x	x	x	x	x			x		23 in total, e.g. AAL, BEA	Food supplem.	Standard addition was applied for quantification	Di Mavungu et al., 2009
39	x	x	x	x	x		x	x		e.g. 12 in total, AFB$_2$, AFG$_{1,2}$	Food, 4 types	Maize, walnut, biscuit, breakfast cereals	Garrido Frenich et al., 2009
40	x	x	x	x	x			x		e.g. 11 in total, AFB$_2$, AFG$_{1,2}$	Food, 3 types	Maize, dry pasta, babyfood	Beltran et al., 2009
41	x	x	x	x	x		x	x		12 in total, AFB$_2$, AFG$_{1,2}$, FB$_2$	Beer	Method 39 optimized for beer, including clean up step	Romero-Gonzales et al., 2009
42		x	x	x	x	x		x		32 in total, e.g. CIT, gliotoxin	Maize silage	Metabolites from fungi in field and silage storage	Rasmussen et al., 2009
43			x	x	x			x		NIV, F-X, 3-AcDON, FB$_2$, FB$_3$	Wheat, maize	Also in barley. TOF at 11,500–12,500 FWHM	Zachariasova et al., 2010a
44			x	x	x			x		NIV, F-X, 3-AcDON, FB$_2$, FB$_3$	Wheat, maize	Also in barley. TOF at 11,500–12,500 FWHM	Zachariasova et al., 2010a
45			x	x	x			x		NIV, F-X, 3-AcDON, FB$_2$, FB$_3$	Wheat, maize	Also in barley. Orbitrap at 100,000 FWHM (m/z 200)	Zachariasova et al., 2010a

Note: AOH = alternariol, CID = collision induced dissociation, MPA = mycophenolic acid, SIM = selected ion monitoring and STE = sterigmatocystin. Other abbreviations are mentioned in the text.

Table 4.2   Analytical details of extraction, clean up, chromatographic system and mass spectrometric settings

No.	Matrix	Extraction (modifier%)	Clean up	Eluent	Column	μm	Ion mode	MS	Internal standard (IS)	ME	LOQ	Rec.%	Rt
1	Wheat, maize	MeOH-ACN	SPE/filter	MeOH	RP18	3.5	APCI −	ion trap	ZAN	n.r.	12–15	107–118	8
2	Fungi and maize	84% ACN, redis. MeOH	Charcoal	ACN	RP18	5	APCI +	ion trap		n.r.	24–36	72	18
3	Fungal culture	EtOAc, isopropanol	PTFE filter	ACN, TFA in water	RP18	3	ESI +	TOF	tuning with reserpine	n.r.	n.r.	n.r.	50
4	Pork liver, kidney	MeOH-ACN	SPE/filter	MeOH, acetate	RP18	4	+/− switch	tQ		MMC	0.2–2.7	n.r.	14
5	Maize, wheat	Accelerated solvent extr.	SAX/ Mycosep®	ACN, acetate	RP18	5	APCI +/−	ion trap	d_6-FB$_1$, VER, ZOL	MMC	10–50	40–90	9
6	Food and feed	85% ACN	Mycosep®	MeOH	RP18	3	+/− switch	tQ	Verrucarol (VER)	MMC	1–10	53–94	15
7	Maize	84% ACN	Mycosep®	MeOH, acetate	RP18	3	APCI +/−	tQ	ZAN	rep.	1–12	82–98	7
8	Maize	75% ACN	Carbograph®	ACN/MeOH	RP18	5	ESI −	tQ	nafcillin/α-estradiol	MMC	6–36	79–106	25
9	Cheese	ACN-hexane	Re-dissolved	ACN, acetate	RP18	3.5	ESI +	tQ		MMC	0.6–5	96–143	33
10	Bovine milk	ACN-hexane	SPE	MeOH, acetate	RP18	5	+/− runs	tQ		MMC	0.2–10	76–108	16
11	Corn meal	75% ACN	Carbograph®	MeOH, formate	RP18	5	+/− switch	tQ	ZAN	MMC	6–120	81–104	26
12	Food	80% ACN	SPE	ACN, acetate	RP18	4	+/- switch	tQ	DOM (trichothecenes) and ZAN	MMC	1–16	65–116	30
13	Corn, wheat	85% ACN	Mycosep®	MeOH, acetate	RP18	3.5	APCI +	TOF		MMC	0.3–18	71–133	40
14	Fungal culture	EtOAc, CH$_2$Cl$_2$	Re-dissolved	MeOH, acetate	RP18	3.5	ESI +	tQ	DOM	r.a.	1–50	75–97	20
15	Corn silage	80% MeOH	SPE	ACN, acetate	RP18	5	+/− switch	ion trap		MMC	5–20	65–80	22
16	Wheat, maize	79% ACN, acetic acid	1 to 9 dilution	MeOH, acetate	RP18	5	+/− runs	Q-Trap		MMC	0.1–660	41–113	42
17	Beer	Re-dissolution in eluent	SPE	MeOH, formic acid	UPLC	1.7	+/− switch	tQ		rep.	0.2	70, 85	3.2
18	Peanut, corn	84% ACN	SPE and IA	MeOH, acetate	UPLC	1.7	+/− runs	tQ	ZAN and standard addition	rep.	0.01–70	71–119	21
19	Corn meal	75% ACN	Carbograph®	MeOH, formate	RP18	5	+/− runs	tQ	VER/ZAN/Diclofenac	MMC	1–19	80–111	52

No.	Matrix	Extraction	Clean-up	Mobile phase	Column		Ionisation	Analyser	Notes	Quant.	Range	Recovery	RSD
20	Corn silage	80% MeOH	SPE	ACN, acetate	RP18	5	+/– switch	ion trap		MMC	5–20	75–80	22
21	Wheat, maize	84% ACN, re-dis. MeOH	SPE	MeOH, formate	RP18	3	APCI +	ion trap	Hydrocortisone/ VER/OTB	rep.	25–125	46–106	10
22	Food and feed	80% ACN	Filter	ACN, formic acid	RP18	5	ESI +	tQ		MMC	1–200	81–126	30
23	Maize, rice	79% ACN, acetic acid	1 to 2 dilution	MeOH, acetate	RP18	5	+/– runs	Q-Trap		MMC	0.1–660	0.3–531	42
24	Maize	MeOH, phosphate buffer	PBS/multi IA	MeOH, acetate	RP18	5	+/– switch	Q-Trap		MMC	1–12	79–104	50
25	Wheat, oat	84% ACN	Mycosep®	MeOH, acetate	RP18	3	+/– switch	ion trap	^{13}C15 DON and ZAN	MMC	6–30	80–115	25
26	Bread, fruit	79% ACN, acetic acid	1 to 9 dilution	MeOH, acetate	RP18	5	+/– runs	Q-Trap		MMC	0.1–675	70–110	33
27	Silage	80% ACN	Freeze dry	ACN, formic acid	RP18	5	ESI +	tQ		MMC	8–250	73–111	22
28	Dried figs	60% ACN, acetic acid	Filter	ACN, acetate	RP18	1.8	ESI +	TOF	Benzophenone in extract	MMC	0.3–16	83–92	22
29	Fungal culture	ACN/EtOAc	Re-dis. MeOH	ACN, acetate	RP18	1.8	ESI +	TOF	Benzophenone in culture	MMC	n.a.	85–105	17
30	Spices	85% ACN	1 to 4 dilution	MeOH, acetate	RP18	5	ESI +	tQ		MMC	0.5–100	72–129	15
31	Wheat, maize	84% ACN	Filter/re-dis.	MeOH, acetate	RP18	3	APCI +/–	tQ		n.r.	0.6–10	71–114	14
32	Grains	Accelerated solvent extr.	Filter	ACN, formic acid	RP18	5	+/– runs	tQ	ACN, acetate in the second run	MMC	1–1250	51–122	27
33	Sweet pepper	EtOAc, formic acid	NH$_2$/C18/ SAX	MeOH, acetate	UPLC	5	ESI +	tQ	ZAN	MMC	1–127	88–110	14
34	1 feed, 4 food	75% ACN, formic acid	Centrifuged	MeOH, formate	UPLC	1.7	+/– runs	tQ		MMC	30–150	63–123	7
35	Oilseed cakes	80% MeOH, acetic acid	SPE	ACN, acetate	RP18	5	+/– switch	ion trap		n.r.	10–100	65–92	22
36	Wheat, maize	79% ACN, acetic acid	1 to 5 dilution	MeOH, acetate	RP18	3	+/– runs	tQ	DON-d_1, 3-Ac DON-d_3, ZON-d_2	MMC	1.6–115	73–152	46
37	Wheat, maize	79% ACN, acetic acid	1 to 5 dilution	ACN, formic acid	RP18	3	ESI +	Orbitrap	d-herbicides and pharmaceutics	MMC	66–6600	73–152	28
38	Food supplement	EtOAc, formic acid	Hexane/SPE	MeOH, acetate	UPLC	5	ESI +	tQ	ZAN	MMC	1–100	85–109	14
39	Food, 4 types	80% ACN	Centr. & filter	MeOH, formate	UPLC	1.7	ESI +	tQ		MMC	0.03–6.30	70–108	8.5

Table 4.2 *continued*

No.	Matrix	Extraction (modifier%)	Clean up	Eluent	Column	μm	Ion mode	MS	Internal standard (IS)	ME	LOQ	Rec.%	Rt
40	Food, 3 types	80% ACN, formic acid	1 to 2 dilution	MeOH, acetate	UPLC	1.7	+/– switch	tQ		MMC	0.5–20	70–110	4
41	Beer	MeOH/ACN 60:40 v/v	SPE	MeOH, formate	UPLC	1.7	ESI +	tQ		MMC	0.07–0.45	70–106	5
42	Maize silage	Modified quechers	Sulphate/ filtrate	ACN, formate	RP18	3	+/– runs	tQ		MMC	3–2440	37–201	22
43	Wheat, maize	Modified quechers	Filter	MeOH, formate	UPLC	1.8	+/– switch	TOF	^{13}C labelled isomers	MMC	10–50	96–105	10
44	Wheat, maize	Crude extract + standards	Filter	MeOH, formate	UPLC	1.8	+/– switch	TOF	^{13}C labelled isomers	MMC	165–2000	n.r.	10
45	Wheat, maize	Crude sample extract	Filter	MeOH, formate	UPLC	1.8	+/– switch	Orbitrap	^{13}C labelled isomers	MMC	20–50	n.r.	10

Note: ME = matrix effect; n.r. = not reported; MMC = matrix matched calibration, i.c. blank matrix is spiked to construct calibration curves in the matrix; r.a. = reported to be absent; rep. = reported to be present, but not handled; LOQ = limit of quantification, expressed in µg kg^{-1}; Rec. = recovery in %; n.a. = not applicable, as this is a qualitative method for profiling fungal metabolites; Rt = retention time of the chromatogram in minutes. Time to flush the system before a new run is not included. Internal standards tabulated at numbers 36 and 37 were applied in both methods. Other abbreviations are mentioned in the text.

Multi-mycotoxin methods all suffer from the difficulties that are connected with this type of analysis: a compromise between the conflicting different chemical properties of the compounds of interest. For the liquid chromatographic part a solvent that is optimal for acceptable retention and baseline resolution often does not provide the best analyte ionization. Owing to the different chemical properties of the compounds of interest, going from very acidic ones like moniliformin via neutral aflatoxins to basic ergot alkaloids, chromatographic retention needs a gradient that starts with almost pure water, which is the worst possible effluent for the ionization process in the ion source of the mass spectrometer. The best sensitivity in ESI mode is mostly achieved when the analyte is already ionized in the liquid phase, but the best chromatographic performance in reversed-phase LC, with good retention factors and resolution, is achieved by adjusting the pH so that the acidic or basic analytes are non-ionized in the mobile phase. These items will be discussed in more detail.

4.3   Liquid chromatographic aspects of multi-mycotoxin methods

It can be concluded from Table 4.2 that all methods are carried out by reversed-phase liquid chromatography with non-polar C18-bonded silica stationary phases. To separate the different types of mycotoxins in this mode, a gradient is needed which starts with a high water percentage to assure retention for the most polar compound in the mixture, which can be the enniantins, but mostly will be moniliformin. This mycotoxin is not only rather acidic, but has also a low molecular mass, which results in poor retention. Non-polar compounds are retained more strongly and therefore these compounds are eluted with an increasing amount of organic solvent. Better evaporation of organic modifier in the ion source results in higher sensitivity for the less polar compounds. All methods apply methanol or acetonitrile as organic modifier. A clear preference for one of these options cannot be given. For trichothecenes it was reported that methanol enhanced MS sensitivity, especially for type B trichothecenes, when compared to acetonitrile (Berthiller *et al.*, 2005; Biselli *et al.*, 2004), whereas acetonitrile is considered to be more selective, as it enabled separation of 3- and 15-AcDON isomers (Gentili *et al.*, 2007). Apparently methanol is mostly preferred, as it is referred to in 28 out of the 45 collected methods.

Buffers and salts are added to improve resolution and reproducibility. Buffers adjust the pH of the eluent in relation to pKa values of the compounds of interest. The retention of the acidic ochratoxins especially is highly pH dependent, similar to MPA, as it is insoluble in water below its pKa but very soluble at pH above its pKa (Sørensen *et al.*, 2008). Formic and acetic acid are selected, since non-volatile buffers such as phosphate cause increased background and signal suppression, as well as rapid contamination of the ion source, resulting in decreased sensitivity and stability. Therefore acetic acid, formic acid, ammonium acetate and ammonium formate are applied, with a preference for acetate buffering, which is also used in

25 of the 45 tabulated cases. Ammonium formate has a stronger signal suppression effect on NIV, DON, F-X and 3-AcDON at a concentration of 10 mM in the negative ionization mode than formic or acetic acid on ESI response (Lagana *et al.*, 2003). Rundberget and Wilkins (2002) reported similar results for *Penicillium* mycotoxins. Buffer concentrations generally should not be higher than 50 mM, as formate or acetate ions in negative ion mode can form $[M+HCOO]^-$ and $[M+CH_3COO]^-$ adducts.

Dall'Asta *et al.* (2004) increased the sensitivity of type A and B trichothecenes by adding a small concentration (0.1 mM) of sodium chloride to the eluent. In the positive ion mode this creates charged sodium adducts $[M+Na]^+$ and in the negative ion mode $[M+Cl]^-$ adducts. Higher concentrations can lead to background interference, rapid contamination of the ion source and blockages in the liquid chromatographic part, especially in UPLC systems, which contain small silica particles. For the same compounds Santini *et al.* (2009) added ammonium to the mobile phase to promote the formation of ammonium adducts $[M+NH_4]^+$ in the positive mode and acetate adducts $[M+CH_3COO]^-$ in the negative mode. ZEN and zearalanone (ZAN) that were also present in their mixture did not show adduct formation.

Gentili *et al.* (2007) added 0.07 mM ammonium acetate to type A, B and D-trichothecenes to achieve ammonium adducts, to satisfy the identification criteria as laid down in EU (Commission Decision 2002/657/EC) for performing confirmatory analysis. Monitoring all these analytes as ammonium adducts with two selected reaction monitoring (SRM) transitions enabled them to use four identification points rather than the required minimum of three. They determined the optimum concentration of ammonium acetate to be between 0.05 and 0.07 mM by plotting the concentration range 0.01 to 0.5 mM against the *S/N* ratios of three trichothecenes from the three different types.

Nowadays performance criteria require molecular ions only, so the formation of adducts is regularly avoided as much as possible. In this respect it has to be considered that sodium can always be present in the mobile phase at this level owing to impurities derived from sample vials or solvents, or it can originate from the sample. Formation of sodium adducts is decreased by formic acid, which can therefore be a reason to prefer formate buffer above another type. In this respect suppression of sodium adducts by adding ammonium ions to the mobile phase was also reported to lead to better MS sensitivity (Cavaliere *et al.*, 2006).

Apfelthaler *et al.* (2008) were the only ones to study systematically the relationship between nature and the amount of organic modifier, pH and ionic strength of buffer additives in the eluent. They listed pKa values and compiled retention patterns of 79 fungal metabolites on three different types of RP/WAX (mixture of reversed-phase and weak anion exchange material) stationary phases, as well as a conventional C18 and an amino-type NH_2 phase in the supporting information of their study. Many basic compounds could not be identified on the amino bonded phase owing to defective peak shape and ochratoxins and fumonisins that were not eluted at all. The latter two groups of compounds and cyclopiazonic acid (CPA) could also not be eluted from the WAX stationary phases. On the C18 material all mycotoxins could be eluted within 12 min. Since they did not apply their method to samples, these data are not presented in Table 4.2. Their retention pattern plots illustrate clearly the contradictory demands of the necessary properties of the eluent for multi-mycotoxin analysis: where basic compounds are retarded at pH 7–8.5 with an organic modifier concentration of 40%, the acidic MON is almost not retarded, while the more neutral aflatoxins have just enough retention to be separated. Regarding the buffer concentration they showed an increase in relative retention by a factor of 2–4 when going from 10 to 100 mM acetate.

Ji *et al.* (2009) worked out a different approach. They used a quantitative structure-retention relationship (QSRR) model to describe the retention times of mycotoxins and fungal metabolites. Based on the dataset of Sulyok *et al.* (2007a) for the quantification of 87 analytes, they calculated the number and magnitude of the factors (descriptors) that were needed to reproduce the dataset. These model descriptors account for the structural properties of the mycotoxins and thus for the retention behaviour. They concluded these to be hydrogen-bonding interactions, steric interaction and hydro-dynamic friction between the compounds and the stationary and mobile phases. Once the descriptors have been calculated, these parameters can be used to estimate the retention times of unknown compounds or give additional information about possible chemical structures that may be present in a chromatographical peak. For the selected dataset they determined five descriptors to be necessary to reach a multi-parameter regression model with a correlation coefficient (R^2) between the predicted and observed values of 0.8932.

The investigated items show that on one hand chromatographic retention, selectivity and baseline separation can be increased, which may help to minimize matrix effects, when eluting compounds at a relatively high percentage of water in the eluent. However, on the other hand, eluting analytes at higher fractions of organic modifier, at a pH around neutral and using fully volatile buffer additives, will increase the ionization yield in the ion source of the mass spectrometer and thus the sensitivity.

4.4   Mass spectrometric aspects of multi-mycotoxin methods

Four different types of mass spectrometric detectors are used in the methods of Table 4.2, of which the preferred applications and (dis)advantages are summarized in Table 4.3.

Ion traps are mostly used for confirmation purposes in single methods, even when they can detect known and unknown, that is unexpected, metabolites in a sample in full scan mode. Triple quadrupoles can detect selected ions rather sensitively, but will give no more information than the compounds targeted for by the instrument settings. Q-traps combine both advantages. They act as a triple quadrupole when the trap is put in SRM mode, but as ITD when set at full scan. In the latter case it can produce complete product ion mass spectra. In the former setting it delivers accurate quantification and confirmation via pre-selected product

Table 4.3   Overview of different MS detector types with their characteristics

Type	Mode	Application and advantage	Performance
Ion trap	Full scan	Additional structural information	Poor sensitivity
Triple quadrupole	SRM	Quantification and confirmation	Only selected ions
Q-trap	SRM	Quantification and confirmation	Triple quad option
TOF and Orbitrap	Full scan	Accurate mass, fast scan	Lower sensitivity

ions. Likewise in both cases the respective disadvantages are met. Modern time-of-flight (TOF) technology can perform fast scans and still measure accurate masses. TOF, Q-trap in ion trap mode and orbitrap (Makarov *et al.*, 2006) can collect and store full scan spectra, giving the opportunity to reprocess stored data long after finishing chemical analytical procedures in the laboratory. Retrospective data analysis gives the option to identify unknown signals or to search for newly emerging compounds, via a computer in any location.

Whatever the chosen technique, two problems have to be solved at the end of the LC column in order to introduce charged ions into the MS. The eluent has to be removed and the compounds of interest have to be transferred into the gas phase and ionized. Enke (1997) described these processes for ESI. The outlet of the LC column is connected to one of the terminals of a high-voltage direct current power supply. The other terminal is connected to a metal plate, which is located a few millimetres away from the capillary tip. The intense electric field at the capillary tip draws charged ions in the exposed solution to the liquid surface. The interior of the droplet is neutral, containing solvent, other molecules, and salts. The charged liquid surface takes the shape of a cone, the so-called Taylor cone, from which a thin filament of solution extends until it breaks up into droplets. These charged droplets move towards the metal plate, because of its opposite charge. During this movement much of the solvent evaporates from the droplets and solvent-free ions are formed. The metal plate is part of the assembly of the ion source of the MS that separates the atmospheric pressure electrospray area from the vacuum of the MS. A little hole in the metal plate allows a small portion of the vapour-phase ions to pass into the mass analyser. LC flow rate and nebulizing temperature are parameters that influence the vapourization process and thus sensitivity. The connection of the power supply poles to the capillary tip of the LC column and the metal plate determines the ionization mode.

4.4.1   Ionization interfaces

The operation mode for the ionization process is an important factor, since sensitivity between positive or negative electrospray can differ, depending on the mycotoxin. In general, all aflatoxins exhibit good ESI ionization efficiency in the positive ion mode with abundant protonated molecules. This was recently confirmed again by Nonaka *et al.* (2009) who evaluated APCI and ESI for the determination of all aflatoxins in both positive and negative ion modes. They concluded that the ESI positive ion mode was the most effective for ionization of aflatoxins. Tanaka *et al.* (2006) found that APCI in the positive mode for their LC-TOF system provided optimum response for the *Fusarium* toxins and ESI in the positive mode showed slightly better signal intensities for the aflatoxins. As their number of analysed trichothecenes was twice as big as the four aflatoxins, they decided to employ APCI in the positive mode to analyse their selected mycotoxins. From Table 4.2 it can be seen that APCI, if applied, was mostly combined with ion trap detection. Cavaliere *et al.* (2006) compared atmospheric pressure photo ionization (APPI) with ESI sources in a confirmatory method to determine aflatoxin M_1 in cow milk and did not consider it to be advantageous. LOQ was lower for APPI, but for ESI it was still four times lower than the EU regulated maximum levels, so ESI was preferred as being more widespread. Capriotti *et al.* (2010) confirmed this in a multi-mycotoxin application, including AFs, OTA, DON, ZEN, T-2 and HT-2, unfortunately revealing the deficiency that the APPI source did not give a response suitable for FB_1 and FB_2 determination at regulated levels.

For OTA, there are different reports of the best ionization mode. Ventura *et al.* (2006) determined AF and OTA in beer by UPLC by switching from positive mode for AF to negative mode for OTA, despite reports of Degelmann *et al.* (1999) who determined OTA in beer with a triple quadrupole in the positive ESI mode up to a level of 0.05 µg l^{-1}. Driffield *et al.* (2003) also applied the positive mode for OTA, but Sørensen and Elbæk (2005) selected the negative mode. Recently Noonim *et al.* (2008) and Noba *et al.* (2009) performed ESI$^+$ to determine OTA in coffee beans. Nielsen *et al.* (2009a) also chose the positive mode in their review of 145 secondary metabolites and mycotoxins from the *Aspergillus niger* group.

For DON, literature data also switch between the positive and negative mode. Lagana *et al.* (2003) reported data in maize for a water/acetonitrile/formic acid eluent system where the positive ion mode was preferred for type A (e.g. T-2 and HT-2) and the negative mode for type B trichothecenes (e.g. NIV and DON), whereas Bily *et al.* (2004) preferred the positive mode for DON. Pallaroni *et al.* (2002) analysed both DON and ZEN in positive mode, but for ZEN in beer Zöllner *et al.* (2000) experienced more sensitivity in the negative mode. They attributed this to exclusive formation of deprotonated molecules. On the other hand this has the disadvantage of having a rather non-specific fragmentation pattern as loss of carbon dioxide is commonly observed with a lot of carboxyl acids, which is unfavourable for identification. Romero-González *et al.* (2009) reached a LOQ of 0.3 µg l^{-1} for ZEN in the positive mode, when they optimized method 39 to analyse beer. For fumonisins, both modes show almost similar sensitivity, with a slight preference for the positive mode (Lukacs *et al.*, 1996, p. 126). Mohamed *et al.* (2006) concluded that the positive ionization mode also revealed a higher total ion current for all ergot alkaloids, with sensitivity increased up to 50-fold for ergotamine compared to the negative mode. For enniatins, Uhlig and Ivanova (2004) reported the negative ion mode to be 100 times less sensitive than the positive mode, since these compounds easily produce protonated molecular ions.

To make the right choice regarding the preferred ionization mode it is necessary to keep in mind what regulations have to be maintained, as limits might be higher than analytical method capabilities, so it will not always be necessary to go for the technical limits of the equipment. Delmulle *et al.* (2006) ran 16 mycotoxins in both modes on a triple quadrupole mass spectrometer and found out that not all compounds were detectable in the negative mode. In the positive mode however, they achieved LOQ values of 1, 2.3 and 50 pg μl^{-1} for aflatoxin B$_1$, ochratoxin A and DON respectively. Garrido Frenich *et al.* (2009) also evaluated ESI positive and negative modes, observing that whereas all mycotoxins were detectable in the positive mode, not all of them were efficiently ionized in the negative mode. Therefore they also selected the positive mode with which they achieved LOQs far below the EU regulatory limits. Other authors dealt with this by running a sample extract twice, both in positive and negative mode. Because another buffer sometimes is preferred in an opposite mode, this can result in two different optimized runs, which doubles analysis time and gives rise to possible errors, that is combining the wrong runs for one sample. These disadvantages can be avoided by switching between negative and positive ionization mode within a single LC–MS run, as indicated by '+/– switch' in Table 4.2.

4.4.2   Patulin: a special case

Owing to its small molecular size and high polarity, patulin is an exceptional mycotoxin regarding its analysis, generally resulting in a single method. Moniliformin and nivalenol have similar chemical properties, but lack formal regulation, so in most cases qualitative analysis suffices, which can be provided for by including these compounds in a multi-method. Sewram *et al.* (2000) investigated the mass spectrometric capabilities of patulin and found out that positive and negative ion modes were possible, but the negative one was favourable. Garon *et al.* (2006) tried to incorporate it in their method for silage, but failed to extract it under the same conditions as the other mycotoxins. Sulyok *et al.* (2006) ran it in their multi-mycotoxin method, achieving a LOQ as high as 100 μg kg^{-1}, which is 10 times above the EU limit for baby food. Similar problems were experienced by Senyuva and Gilbert (2008), who operated their TOF instrument in positive electrospray mode, except for patulin. In case the experimental data from their TOF run indicated the possible presence of patulin, they confirmed these findings by re-extracting the samples and subjected them to a single LC–MS analysis with a triple quadrupole in the negative mode. Recently Kataoka *et al.* (2009) confirmed these findings, as they evaluated APCI and ESI for the determination of patulin in both positive and negative ion modes and concluded that the negative ESI mode was the most effective. For these reasons it would be more logical to combine patulin and *Alternaria* toxins analysis, as for these mycotoxins higher sensitivity was also observed in the negative mode (Lau *et al.*, 2003) and both compounds can occur in fruit (juices). Another option could be to combine analysis in apple products with pesticides, as published by Christensen *et al.* (2009), even if their LOQ for patulin of 40 μg l^{-1} is high when compared to EU limits of 10–50 μg kg^{-1}.

4.4.3   Optimizing instrument settings

After selecting an appropriate mass analyser, all parameters (cone and capillary voltage; desolvation, nebulizer and cone gas flow; desolvation temperature, mass resolution and collision energies) have to be set for the mycotoxins of interest. This is achieved by continuous infusion of a standard solution of each mycotoxin into the MS/MS. Plotting the settings against the signal areas will result in graphs from which an optimum can be determined. An example of this preliminary work was first presented at the 16th Montreux Symposium on Liquid Chromatography–Mass Spectrometry in 1999 and later published by Razzazi-Fazeli *et al.* (2002). When triple quadrupole is applied in the SRM mode, monitoring ions are selected after analysing the ESI mass spectra obtained by direct liquid injection of standards. When considering the number of mycotoxins of interest to be analysed simultaneously by one method, it will also be necessary to optimize dwell times to find the best detection parameters, as a sufficient number of data points per peak has to be acquired in order to have enough sensitivity and to allow reproducible integration for correct quantitative results.

4.5   LC–MS aspects of multi-mycotoxin analysis

4.5.1   Matrix effect

The ultimate goal of a multi-mycotoxin method is, like for any other comparable analytical problem, to measure every possible compound in any type of food or feed in a single run. This implies that any type of clean up should be avoided, as using clean up indirectly implies a restriction of the types of mycotoxins that can be analysed, since the presence of compounds in the finally extracted aliquot to be injected into a chromatographic system will be modified by a clean up step. For example, application of Mycosep® columns mostly led to methods which were only suitable for handling *Fusarium* mycotoxins like trichothecenes, fumonisins and zearalenone derivatives (Tuomi *et al.*, 1998; Razzazi-Fazeli *et al.*, 1999, 2002; Biselli *et al.*, 2004; Klötzel *et al.*, 2005; Berthiller *et al.*, 2005; Cavaliere *et al.*, 2005a,b; Hartmann *et al.*, 2007; Lattanzio *et al.*, 2008).

Recently preliminary results were presented for 22 different mixtures of Mycosep® column packing materials giving satisfactory recoveries over the full range of mycotoxins of interest (Mitterer *et al.*, 2009b). Injecting a crude extract unavoidably leads to the introduction of matrix components into the LC–MS system and thus into the ion source. As the instrument settings are not focused on them, these compounds will not be measured. Nevertheless they will influence the measurements, as they disturb the ionization processes of the target compounds in the source of the mass spectrometer, a phenomenon well known as the matrix effect. This can result in ion suppression or signal decrease, or enhancement, which results in a higher signal. The latter version is far less common, so generally an undesired decrease in sensitivity has to be expected, resulting in lower limits of quantification.

For aflatoxins Plattner *et al.* (1984) already reported a highly (up to 75%)

variable loss in response for the 312/297 daughter signal on injection of a 25 mg aliquot of contaminated (8 ppb) corn, when compared to a standard solution. Consistent recoveries and minimized matrix effects were only obtained below injection of 1 mg of crude corn extract. This phenomenon was more systematically investigated by Buhrman *et al.* (1996) in a pharmacological study of SR 27417 (2-[*N*-(2-dimethylamino ethyl)-N-(3-pyridinyl methyl) [4- (2, 4, 6- tri-isopropyl-phenyl) thiazol-2-yl] amine]) in human plasma. They ran spiked plasma samples in full scan mode from *m/z* 200–700, compared the different spectra and determined a quantitative correlation with the ion suppression, which was attributed to the amount of matrix.

The term matrix effect was introduced and defined as: matrix effect = (peak area of spiked sample extract/peak area of standard) × 100%. For the extraction procedure and the efficiency of the total procedure, different formulae were derived by Matuszewski *et al.* (2003). Zöllner *et al.* (2000) reported the presence of matrix in beer as a huge peak during the first 5 min of a chromatogram in the SRM mode analysing ZEN, α-zearalenol and β-zearalenol (α-ZOL and β-ZOL), while applying ZAN as an internal standard. The four peaks of interest were retained between 6 and 12 min. Nevertheless, they also reported that calibration curves for β-ZOL, presented as ratio of the internal standard ZAN, were considerably different for individual brands. Plattner and Maragos (2003) also considered the matrix to be present in the void volume of the chromatographic system when they determined DON and NIV in corn and wheat by ESI LC–MS in the SRM mode. When sample extracts were injected, detection of DON (retention time approximately 5 min) was obscured by its co-elution with polar matrix components in the extract, which eluted at the void volume (3 min) and the abundance of the signals was reduced by matrix effects. To avoid this, a gradient was applied to separate DON from these polar matrix components, which also allowed NIV and DON to be clearly separated and measured directly from extracts of wheat, corn, and other small grains without clean up.

These findings did not fit a study of Rundberget and Wilkins (2002), who reported that peak areas of standards were different in the mobile phase or in a food mixture for penitrem A, which under their conditions had a retention time of 15 min, far beyond any void volume. They expressed this phenomenon by presenting calibration curves in the MS/MS mode for standard solutions in a methanol–water mixture versus equivalent levels in a food mixture extract. This procedure is generally referred to as matrix matched calibration. Reductions of signal strength up to 40% were observed. In an attempt to correct for the matrix effect they composed a kind of 'general food' matrix by producing a blank food mixture from bread (30%), cooked rice (20%), cooked pasta (20%), cooked potatoes (15%), vegetables (3%), fruits (3%), cheese (1%), salami (1%), used filter coffee (5%) and minced meat (5%). All components were fresh from a local grocery store. Water (10%) was added, the mixture was homogenized and freeze dried and portions of the blank food mixture were spiked to construct calibration curves in a matrix.

Unfortunately, this universal approach to tackling the matrix effect by introducing a general type of corrective action does not work, as it turned out that every mycotoxin/matrix combination appears to have a different, unpredictable magnitude (Spanjer *et al.*, 2003). They presented calibration curves for aflatoxin B_1, OTA and DON in a peanut and a cornflake sample. These two matrices were selected to investigate the difference in matrix effect for these mycotoxins, despite the fact that it is unusual to expect DON in a peanut sample. In Fig. 4.1 calibration curves are given for aflatoxin B_1, OTA, DON and ZEN in peanut, for HPLC and UPLC systems. The latter data were presented later (Spanjer *et al.*, 2007), after the introduction of UPLC columns and are given here to show the improvement that can be gained by applying UPLC instead of HPLC (refer to Chapter 3 for further details).

The presented graphs clearly indicate that the matrix effect depends on the analyte as well as on the matrix, because, for example DON showed an unpredictable, huge matrix effect in peanut. In investigations of unexpected mycotoxins in food by a multi-mycotoxin method this factor cannot be neglected. The matrix effect of ZEN appeared to be moderate in peanut, but Hartmann *et al.* (2007) experienced 28–58% ion suppression for ZEN, while going from river to waste water samples. Determination of ion suppression for every single separate matrix–mycotoxin combination seems therefore obligatory, even when this would result in an enormous validation task.

It can be questioned whether performing this kind of validation will solve the problem of the nature of the matrix effect completely. Bonfiglio *et al.* (1999) investigated the matrix effect by a different approach. They constructed a mixing tee at the outlet of the chromatographic system and infused a constant flow of a standard solution of the compound of interest into the effluent to the mass spectrometer. After gaining a steady baseline, a blank sample extract (i.e. the matrix) was injected into the system. Any eluted matrix compound that suppresses or increases ionization in the mass spectrometer will cause a drop or rise in this baseline. They tested two compounds and recorded two rather different infusion chromatograms, which led them to conclude that it would not be possible simply to subtract an infusion chromatogram from a sample chromatogram and thus obtain the corrected chromatogram for all analytical compounds. Surprisingly Stahnke *et al.* (2009) simplified this idea after a thorough and extensive study of 129 pesticides in 20 plant matrices. Matrix effects in this variety of matrices appeared to be similar for different analytes. Therefore they suggested making corrections based on the permanent post-column infusion of one single monitor substance. Nevertheless matrix effects up to 25% could still remain. Thus they concluded that their approach is useful, as it would reduce the number of cases in which standard addition would be required to confirm violations of legal limits.

Only Klötzel *et al.* (2005) studied post-column infusion for mycotoxins. They applied this technique to determine the matrix effect for trichothecenes in cereals and presented two infusion chromatograms. The first was obtained after injection of a blank bread extract, cleaned with Mycosep® 227, in a constant flow of NIV with the LC–MS/MS system in the negative ESI mode. A dip in the signal intensity was recorded between 2 and 3 min, whereas the intensity was stable again after

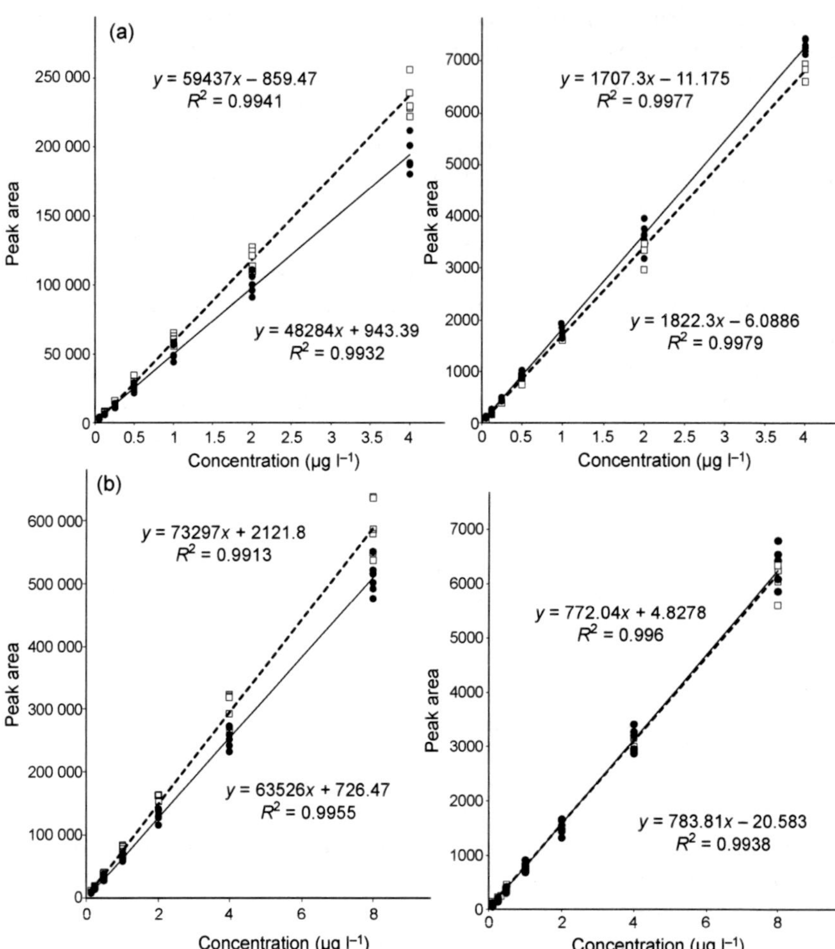

Fig. 4.1   Comparison of calibration curves of standards and spiked peanut extracts with HPLC system (left) and UPLC system (right) for (a) aflatoxin B_1 (0.1–4.0 ng ml⁻¹). HPLC: standards upper line. UPLC: standards lower line; (b) ochratoxin A (0.1–8.0 ng ml⁻¹); (c) DON (0.1–200 ng ml⁻¹); (d) ZEN (1–80 ng ml⁻¹). For (b), (c) and (d) standards are upper lines in both systems. Injections of crude extracts ($n = 7$).

5 min. The retention time in a regular LC–MS/MS run was 4 min, so clearly within that time the matrix influenced the signal. In a second example they infused the system with DAS and recorded in the positive ESI mode three small, but intense peaks between 2.5 to 3 and at 5 min. In addition, a huge dip was recorded between 8 and 15 min. A stable intensity was reached again after 20 min, the retention time of DAS, so its signal was not influenced by bread matrix. Except for nivalenol, no ion suppression or enhancement was observed for the 12 other trichothecenes using this method to determine matrix effects. They therefore concluded that these results demonstrate that the matrix effect could not be eliminated completely and thus minimized it as far as possible by the use of an internal standard (de-epoxy-

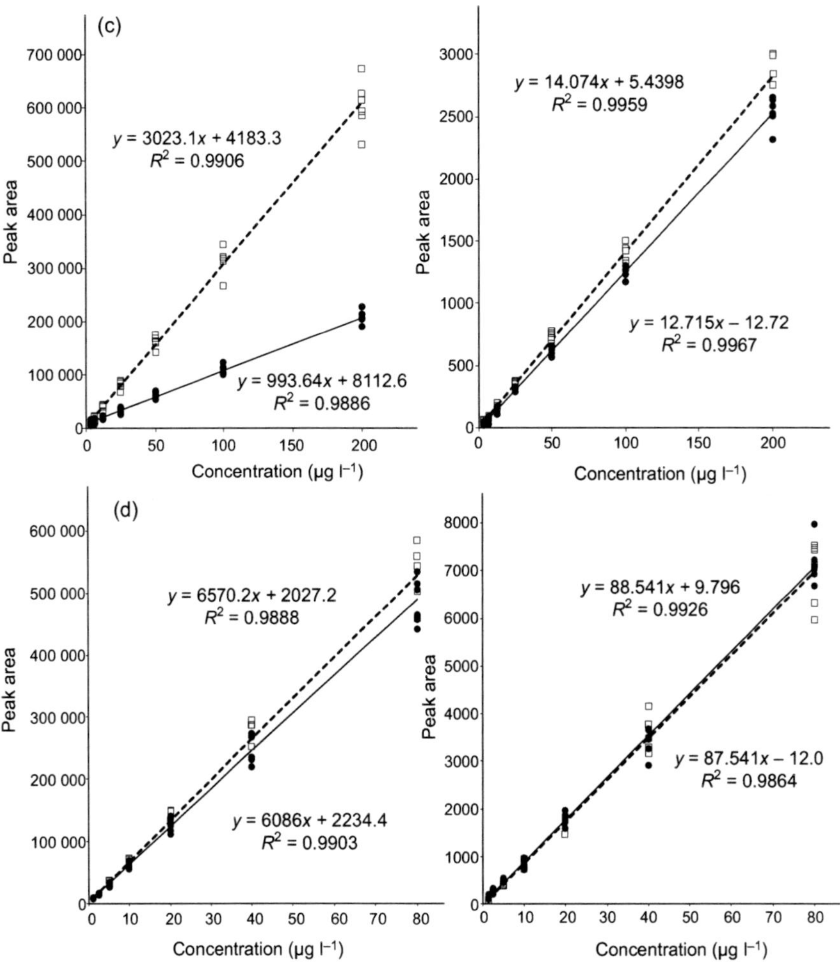

Fig. 4.1 *Continued*

deoxynivalenol). But this gives no evidence for any other mycotoxin, as this experiment clarified that the matrix effect depends on the matrix, the mycotoxin and the ionization mode of the MS. Data from Lattanzio *et al.* (2008) support this view, as they experienced severe matrix effects in different cookies and snacks for NIV and less for DON, T-2 and HT-2, where the opposite was noticed in maize samples.

Another aspect, which complicates the correct treatment of the matrix effect, is the fact that it can differ even in the same type of matrix. Zöllner *et al.* (2000) reported a brand-to-brand difference in beer and Jestoi *et al.* (2004) reported severe effects varying from grain to grain. Matuszewski *et al.* (2003) experienced the same in human plasma samples and concluded that a blank matrix should be a mixture of at least five different plasma lots. Matuszewski (2006) developed this idea further and concluded that comparison of the coefficient of variation (CV) values of MMC curves in five different lots of a biofluid with analogous values obtained by repeated analysis ($n = 5$) in a single lot, may serve as a measure of the relative matrix effect, that is a good *quantitative* indicator of the presence/absence of a relative matrix effect in bioanalytical methods.

Based on calculations of 52 compounds he proposed that precision (CV) values of standard line slopes constructed in five different lots of a biofluid should not exceed 3–4% in order for the method to be considered practically free from the relative matrix effect. To calculate signal suppression or enhancement (SSE) in this way, Sulyok *et al.* (2006) reformulated the equation for the matrix effect into: $SSE = (slope_{spiked\ extract}/slope_{standard}) \times 100\%$. Unexpectedly the matrix effect was not reduced to zero after clean up. Romero-González *et al.* (2009) applied solid-phase extraction to beer samples and found out that a matrix effect remained, certainly for special beers. Lattanzio *et al.* (2007) applied a multi-analyte assay containing antibodies for aflatoxins, OTA, fumonisins, DON, ZEN, T-2 and HT-2, but reported significant ion suppression in maize for aflatoxins and slight effects for OTA. Therefore both authors still used MMC for all mycotoxins of interest. Recently, better results were presented for determination of 11 mycotoxins in corn samples with an improved version (Liu and Powers, 2009), but matrix effects could still be observed.

All together this leads to the conclusion that a clean up procedure cannot completely exclude matrix effects from being present during LC–MS analysis, not even one on an immuno assay basis. A thorough validation of any developed method will have to reveal which additional measures, that is internal standards or standard addition, have to be taken.

4.5.2   Internal standard

The ideal internal standard behaves exactly the same as the compound of interest, during extraction from the sample, chromatographic retention and ionization in the mass spectrometric source, but can still be distinguished from the target compound of analytical interest. This is an almost impossible combination of conflicting demands. The fact that mycotoxins can be found in almost every food or feed sample is another complication in finding an appropriate internal standard. Therefore, Tuomi *et al.* (2000, 2001) applied reserpine, Cavaliere *et al.* (2005a) nafcillin and α-estradiol; Senyuva *et al.* (2008a) benzophenone and Herebian *et al.* (2009) d_6-diuron, d_3-phenanzon, d_3-mecoprop and dihydrocarbamazepine as internal standards (IS). These compounds differ completely from mycotoxins and are intended to avoid intervention of naturally present compounds in a sample. However, the different behaviour of these compounds during extraction, chromatography and ionization is a disadvantage. Therefore Biselli *et al.* (2004) and Royer *et al.* (2004) applied verrucarol for trichothecenes and the latter α-ZOL to determine ZEN. ZAN was used for this purpose by Pallaroni *et al.* (2002), Berthiller *et al.* (2005), Cavaliere *et al.* (2005b), Ren *et al.* (2007), Suman and Catellani (2008), Monbaliu *et al.* (2009) and Di Mavungu *et al.* (2009). Cavaliere *et al.* (2007) applied verrucarol, ZAN and diclofenac.

Griesshaber (2007) applied verrucarol, hydrocortisone and ochratoxin B (OTB).

The chemical structures of these compounds are all more or less related to mycotoxins, which will reveal similar chemical behaviour during extraction and analysis, but has the disadvantage that their complete absence in a naturally contaminated sample cannot be fully guaranteed. This is a drawback to the expected precision of the analysis, even when the methods are focused on *Fusarium* toxins only. It is known that verrucarol is likely to be present in samples of mouldy interior finishes from buildings with moisture problems (Tuomi *et al.*, 1998). Its presence in food was questioned by Cavaliere *et al.* (2005a and b), who for this reason chose 17 α-estradiol and penicillin, which nevertheless can also be present in nature.

Klötzel *et al.* (2005) observed an interfering effect of fusarenon X when using verrucarol, whereas the sensitivity was three times lower than for DOM. This de-epoxy-DON is basically found *in vivo* in animals, but it can also be present in milk as has been shown by Sørensen and Elbæk (2005). A drawback of the application of OTB is the absence of a chlorine atom in its structure, which causes a different retention time, so it will not elute and thus not evaporate in a similar way to OTA in the ion source (Noba *et al.*, 2009). Earlier Bacaloni *et al.* (2005) reported different SPE conditions for OTB, when compared with OTA. Regarding aflatoxins it is sometimes suggested that aflatoxin M_1 be applied as internal standard when B_1, B_2, G_1 and G_2 have to be analysed in nuts (Nonaka *et al.*, 2009), but even this option can be questioned since Ren *et al.* (2007) detected aflatoxin M_1 in peanut butter.

4.5.3   Isotope labelled internal standard

The only way to guarantee application of compounds that show the best compara-ble chemical behaviour during extraction or clean up, chromatography and ionization, and that cannot be found in any food or feed sample caused by natural contamination, is to use isotope labelled isomers. And when analysis has to be scientifically perfect, it implies addition of a labelled compound for every single mycotoxin of interest in a LC–MS method, as already noticed by Tuomi *et al.* (1998). At that time not all these compounds were commercially available, but that deficiency has been completely remedied now.

Cervino *et al.* (2008) synthesized aflatoxins d_2-AFB$_2$ and d_3-AFG$_2$ with which they achieved LOQs of 1 µg kg^{-1} for aflatoxins B_1, G_1 and G_2 and 0.3 µg kg^{-1} for aflatoxin B_2 in almonds and wheat flour. Lindenmeier *et al.* (2004) synthesized $^2H_5$ OTA and reached an LOQ of 1.4 µg kg^{-1} in raisins after immuno assay clean up. Leong (2005) synthesized $^{13}C_{11}$ and $^{13}C_{20}$ OTA to determine OTA in wine and grapes. Bretz *et al.* (2005) synthesized 3-d_3-AcDON, in a later study 15-d_1-DON and developed a method for the analysis of DON and 3-AcDON using HPLC–MS/MS with these isotopes as internal standards (Bretz *et al.*, 2006).

Häubl *et al.* (2006a,b) applied commercially available $^{13}C_{15}$ DON (Biopure Referenzsubstanzen GmbH, Tulln, Austria) and analysed maize and wheat ex-tracts without any clean up, decreasing their CV values from 115% without an IS to 2.5% with a labelled IS. In another publication they improved the slopes of calibration curves for DON in maize from 0.76 without an IS into 1.01 with an IS.

Suman and Catellani (2008) applied $^{13}C_{15}$ DON to analyse trichothecenes in durum wheat, oat flour and flakes. Deuterated fumonisin B_1 [$^2H_6$ FB_1] was synthesized by Lukacs *et al.* (1996) and applied to fumonisin analysis in corn grit, meal, flakes, popcorn and baby food until a level of 1 µg kg^{-1}. Hartl *et al.* (1999) applied the same isotope in a rapid method without any clean up and included hydrolyzed fumonisin in corn products in the same analytical run (Hartl and Humpf, 1999).

Royer *et al.* (2004) tested accelerated solvent extraction for corn samples with this internal standard. Hexadeuterated ZEN [$^2H_6$ ZEN] was prepared by Dall'Asta *et al.* (2005) and successfully applied to the analysis of corn flour. Hartmann *et al.* (2007) obtained the commercial version (Sigma, Buchs, Switzerland) to analyse river and drainage water to a level as low as 5 ng l^{-1} and wastewater effluent at 25 ng l^{-1}. This study was repeated with plant material, soil, manure and sewage sludge (Hartmann *et al.*, 2008).

Cramer *et al.* (2007) produced 3, 5-d_2-zearalenone and analysed corn flakes and tortilla chips. Rychlik and Schieberle (1998, 1999) synthesized $^{13}C_2$ patulin to quantify this mycotoxin in moulded wheat bread, apple, other fruit juices and pulp up to a level of 0.1 µg kg^{-1}. Rundberget and Wilkins (2002) and Razzazi-Fazeli *et al.* (2002) applied commercially obtained d_3-T-2 to determine several *Fusarium* toxins. Asam and Rychlik (2006) synthesized $^{13}C_4$ T-2, $^{13}C_2$ HT-2, $^{13}C_4$ DAS and $^{13}C_2$ MAS and applied these as IS after clean up on multi-functional columns. Beyer *et al.* (2009) produced d_3-T-2 and d_3-HT-2 on a large scale, but only utilized it themselves. Häubl *et al.* (2007) characterized $^{13}C_{24}$ T-2 and its use for the quantification of T-2 toxin in cereals. Trebstein *et al.* (2009) demonstrated d_1-DON, $^{13}C_{34}$ FB_1, $^{13}C_{34}$ FB_2, $^{13}C_{24}$ T-2 and $^{13}C_2$ HT-2 to be applicable for analysing naturally contaminated barley, breakfast cereals, oat cookies, wheat flour, corn flour, corn grits, taco chips and corn flakes. Isotopically labelled IS compensated perfectly for matrix effects and proved to give similar results when the standard addition method was applied, whereas external calibration without correction failed completely.

From this overview it is clear that isotopically labelled compounds for all regulated mycotoxins are available now, even with different labels, as deuterium and ^{13}C atoms. The latter are preferred, since deuterium atoms can equilibrate with hydrogen atoms during extraction, clean up and chromatography, which has a negative effect on the precision of the analysis, as it is essential that the label remains present in the ions to be analysed in the MS. As most carbon atoms are part of the molecular frame these are less sensitive to loss of label, especially when loss of COO^- is avoided. Also, going from ^{12}C to ^{13}C atoms does not change molecular mass too much, when compared to deuterated labels where the atomic mass is doubled from hydrogen to deuterium. Where there are six deuterated atoms the chemical behaviour does not remain unchanged, which can cause slightly different behaviour during extraction, an effect that will be more and more apparent at the LOQ. The mass spectrometer has to differentiate between the mycotoxin and the labelled version, since these compounds enter the ion source simultaneously. To ensure complete resolution, a difference of at least three mass units is preferred. Therefore, Noba *et al.* (2009) applied $^{13}C_{20}$ OTA before extraction of coffee beans and obtained a mean recovery rate of 99.5% for OTA, where it was 82.6% without an IS. Kataoka *et al.* (2009) used $^{13}C_3$ patulin and achieved a LOQ as low as 100 pg ml^{-1} in juice samples with overall recoveries of 92.5 ± 4.2% (relative standard deviation, RSD = 4.5%) and 94.4 ± 0.9% (RSD = 0.95%) at concentrations of 1.0 and 10 ng ml^{-1} in SIM mode, respectively. Nowadays full ^{13}C labelled versions are available, e.g. at Sigma-Alldrich (www.sigmaaldrich.com/), Romerlabs (http://www.romerlabs.com/) or Coring (www.coring.de/) for all regulated mycotoxins and some other important ones (Mitterer *et al.*, 2009a).

All together it can be concluded that the best approach to overcome problems related to matrix effects is the use of isotope labelled ISs, especially since they are not present in nature and have identical chemical properties to the mycotoxins of interest. They can be spiked to the sample before extraction to correct both for losses during sample preparation and ionization in the MS source. All necessary labelled standards are commercially available nowadays, all over the world. Sometimes the costs are discussed, but since Rychlik and Asam (2008, p. 619) calculated only 10 Eurocent additional costs per sample, which can indeed be considered to be negligible compared to labour and equipment costs, even when in a multi-mycotoxin method many labels are required. When a fair comparison costs are also included which are related to taking a wrong decision about a lot based on an incorrect analysis, it might even be cheaper.

4.5.4 Method validation in multi-toxin analysis

Commission Regulation 2006/401/EC lays down the methods of sampling and analysis for the official control of the levels of mycotoxins in foodstuffs. Annex II contains criteria for sample preparation and methods of analysis. For these regulated mycotoxins in food, analytical methods are available that have been fully validated by a collaborative trial (CEN, Comité Européen de Normalisation, the European Committee for Standardization, Brussels, Belgium). These criteria and methods are not yet available for LC–MS analysis in food and feed. For food of animal origin, Commission Decision 2002/657/EC has to be applied. To validate the identity of the analyte, precursor ions are fragmented into product ions in the collision cell of the triple quadrupole. Quantification is carried out on the primary transition. Each mycotoxin is confirmed by a second transition and the ratio between primary and secondary product ion. This fulfils current guidelines about unambiguous compound identification, which prescribe a minimum of two precursor ion/product ion pairs for SRM experiments. Deviations of relative ion intensities related to standard solutions, expressed as a percentage of intensity of the most abundant ion, should for the SRM transitions not be greater than 20% (relative ion intensity > 50%), 25% (relative ion intensity: 21–50%), 30% (relative ion intensity: 11–20%) or 50% (relative ion intensity ≤ 10%). The third criterion is the retention time where the relative retention time, that is the retention time of the analyte to that of the internal standard, should correspond to that of the calibration solution at a tolerance of ± 2.5%.

Sørensen and Elbæk (2005) validated their LC–MS method for α-zearalanol

(α-ZAL), which within the EU is a prohibited veterinary agent in food producing animals, according to this document and determined a decision limit of CCα and a detection capability of CCβ. Monbaliu *et al.* (2010) followed the same procedure in their multi-mycotoxin method for the feed matrices in sow feed, wheat and maize. The latter two matrices can easily be linked to food, but for food of non-animal origin no such criteria document yet exists. A good starting point for such a regulation for contaminants in food could be the article of Lehotay *et al.* (2008), who discussed the proper use of terminology, highlighted the identification power of various MS techniques, demonstrated how MS identifications can fail if precautions are not taken and re-asserted the value of basic confirmation practices, qualitative method validation, information checklists, routine quality-control procedures and blind proficiency-test analyses. Elements of this work were taken into the document for method validation of pesticide residue analysis (Pihlström, 2010). This document is applicable to food and feed, whereas the EC 657 decision is limited to food of non-animal origin, as it originates from the veterinary drug domain. Since mycotoxins are contaminants, and so are neither veterinary drugs, nor pesticides, it can be questioned first which approach should be appropriate. Unless this is definitively decided, the laboratory can go for the 'fit-for-purpose' approach.

Nevertheless, whereas not prescribed, isotopically labelled standards are the first option to choose. If not available, MMC or standard addition are the next best practical options. MMC does not correct perfectly for analyte losses during the extraction step, since the spike is added to blank matrix, which favours extraction when compared to extraction of naturally contaminated samples. Standard addition avoids this disadvantage, but is rather time consuming, as every individual sample has to be fortified on multiple levels before correct calibration is obtained, even when it is only done for relevant samples.

4.5.5   Limitations of LC–MS methods

In 27 of the tabulated 45 methods, triple quadrupole instruments are applied. These have the disadvantage that ion settings have to be made before analysis, so what is searched for has to be known beforehand. To investigate unknown or unexpected compounds, it would be better to apply full scan mass spectrometry. For this purpose TOF is the first option. Recording complete spectra has the exclusive advantage that it enables the possibility to look for unforeseen compounds after the analysis, by reprocessing the data again, preferably on a computer with a relevant database. This procedure is well known in the field of pesticides and veterinary drugs, but is still uncommon in the mycotoxin area, mainly owing to the lack of the required database.

Hancock and D'Agostino (2002) searched with LC-TOF for toxins of biological origin in general, but only included aflatoxin B$_1$ and OTA in their attempts to identify biological and chemical weapons. Liao *et al.* (2008) included only T-2 toxin in their database of 19 438 substances, which was compiled of pesticides, pharmaceutics and chemicals. Senyuva *et al.* (2008a) and Senyuva and Gilbert

(2008) set up a database of 465 fungal metabolites. They identified fumonisin B_2, HT-2 toxin, patulin and zearalenone in dried figs with this system. Polizzi *et al.* (2009) elucidated the possible occurrence of another 42 different fungal metabolites by LC-Q-TOF-MS in samples that had first been subject to triple quadrupole analysis with 20 targeted mycotoxins. Another example of this is given by Andersen *et al.* (2004), who applied LC-TOF to analyse the production of metabolites by *Penicillium expansum* and found not only that patulin was formed in apple juice, but also chaetoglobosins and communesins. The latter two groups were also detected in fruit juices and potato pulp, in which no patulin was found at all. A study of Watanabe (2008) confirmed these findings and concluded that products from decayed apples might contain expansolides A and B in addition to patulin and citrinin. The latest development in this field is an update of the 474 mycotoxins and microbial metabolites database for the fungal screening by Nielsen and Smedsgaard (2003) of up to 800 standards (Nielsen *et al.*, 2009b). The enlargement of the scope led to some loss in sensitivity in the required full scan mode, but it fits well for monitoring purposes.

4.6   Future trends in liquid chromatography–mass spectrometry analysis

4.6.1   Future trends in multi-methods

Since the existence of analytical chemistry, the ultimate aim has been to combine analysis of every possible chemical compound in one single method. Lehotay *et al.* (2005a) integrated sample preparation and extraction for different multi-methods for pesticides, known as the QUECHERS method. Figure 4.2 shows the scheme for this procedure as modified by Pizzutti *et al.* (2007), who combined sample preparation for multi-methods with 221 pesticides and 26 mycotoxins (RSD values 5–15%, recoveries 70–120% and LOQ values from 1–15 µg kg^{-1}).

Their test on rice samples revealed the presence of AFB$_1$, ZEN and carbendazim. The calibration curves for AFB$_1$ and ZEN showed an almost negligible matrix effect for these mycotoxin/matrix combinations. This differed from the results of Moldes-Anaya *et al.* (2009) who found considerable matrix effects for CPA in rice (as well as feed, wheat and peanut), once again giving evidence of the fact that matrix effects are unpredictable and should be investigated explicitly for any mycotoxin/matrix combination. De Kok and Pizzutti (2009) showed that this sample preparation procedure is also applicable to grape juice, wine, maize, soya, wheat and cocoa beans, while increasing analysis up to 35 mycotoxins and 233 pesticides. Trebstein *et al.* (2009) tested the QUECHERS protocol specifically for its practical use for naturally contaminated barley and oat cookies. Using their isotopically labelled ISs for correction, they achieved T-2, HT-2 and ZEN results that agreed very well with the SPE type clean up procedure of Klötzel *et al.* (2006). DON results could not be improved by IS correction. As their IS for DON was d_1-DON, this might be caused by the too small molecular weight difference

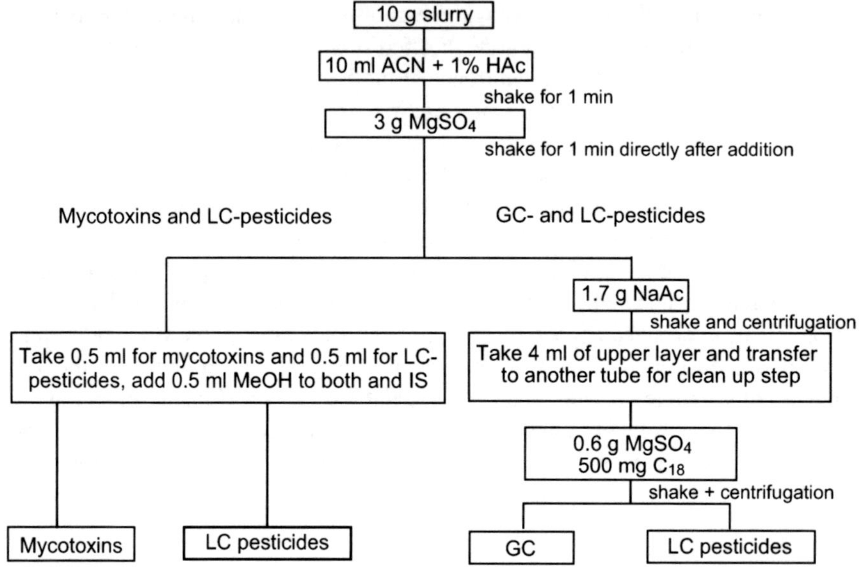

Fig. 4.2   Extraction scheme of combined sample preparation for mycotoxins and pesticides analysis. Slurry is a mixture of matrix with water (Spanjer *et al.*, 2006), so only ACN is added for ACN:water extraction. Acetate buffer and magnesium sulphate (to remove any residual water) are added to improve extraction (Lehotay *et al.*, 2005b).

between both compounds. Mol *et al.* (2008) extended these developments by including veterinary drugs. This revealed promising results for pesticides and veterinary drugs, but unfortunately they reported natural toxins to be more difficult to detect, as demonstrated by the fact that, for example for aflatoxins, their LOQs did not meet the EU regulated limits. Jestoi *et al.* (2009) reported their sample preparation method, developed before for the determination of ionophoric coccidiostats, was applied well to the determination of their mycotoxins of interest – BEA and ENNs – in eggs, which could be a good starting point for further research.

4.6.2   Future trends in mass spectrometers

The introduction of orbitrap mass analysis provides an improvement in collecting high-resolution full spectra. It couples a linear ion trap mass spectrometer, via a trapping quadrupole with a curved axis, to an orbitrap mass analyser wherein ions are trapped at high kinetic energies around an inner electrode. Fourier transformation of the acquired transient radio frequency allows wide mass range detection with high resolving power, mass accuracy and dynamic range. The entire instrument usually operates in LC–MS mode (1 spectrum per second) with nominal mass resolving power of 60 000 FWHM (full width at half-maximum) and automatic gain control to provide high-accuracy mass measurements. Maximum resolving power is 100 000 FWHM. Rapid, automated data-dependent capabilities enable real-time acquisition of up to three high-mass accuracy MS/MS spectra per second. Herebian *et al.* (2009) compared triple quadrupole and orbitrap (at nominal resolving power of 60 000 FWHM) detection of 32 mycotoxins in crude wheat and maize extracts. Tables 4.1 and 4.2 give the results at methods 36 and 37, from which it can be concluded that targeted triple quadrupole analysis is still a more sensitive detection technique. Also remarkable in this study is that both triple quadrupole and orbitrap LOQs are far beyond regulated limits. The missing investigation of the required resolving power was done by Kellmann *et al.* (2009). They spiked 151 pesticides, veterinary drugs, mycotoxins and plant toxins in honey and horse feed. The selected mycotoxins were four aflatoxins, some ergot alkaloids, sterigmatocystin, DAS and T2-toxin. LOQs were 33–825 $\mu g\,kg^{-1}$ in honey, achieved at a resolving power of 25 000 FWHM. The horse feed sample needed a resolving power of 50 000 FWHM (still 50% of the maximum possible value) to achieve comparable LOQs. The study of Zachariasova *et al.* (2010a) showed that QUECHERS/TOF and crude extract/orbitrap methods gave similar results. With the LC-TOF system they were not able to analyse crude extracts, ascribed to the too low resolving power of only 12 000 FWHM. From method 44 in Table 4.2 the difference in LOQ values can be seen when they analysed mixtures of standards dissolved in crude extracts. When comparing their results between corresponding mycotoxins with those of Herebian *et al.* (2009) a clear improvement of about a factor 10 in standards and far more in matrices is noticeable.

All studies reported the presence of matrix effects t, as would be expected, because these were reported earlier by Madalinski *et al.* (2008) who directly introduced yeast cell extracts into a LTQ-orbitrap MS for fast metabolic fingerprinting. When applied to study cadmium toxicity in yeast, direct introduction into this hybrid MS system gave results similar to those of LC/ESI-MS/MS, optimized for a targeted approach. In this case precision was also better for LC/ESI-MS/MS, but LTQ-orbitrap detected most metabolites of interest faster and with the capability to register unexpected metabolites. The presentation of an algorithm for background subtraction of high-resolution LC–MS data by Zhang *et al.* (2009) is also remarkable. In order precisely and thoroughly to remove matrix ions in the analytical data of buspirone and clozapine in human plasma samples, they compiled all kinds of detected ions of metabolites of these compounds throughout the whole chromatogram, obtained by Q-TOF in MS[E] experiments. However, careful study of their supporting information revealed that the structures they proposed for these fragment ions, which were used for background-subtracted CID spectra, were all based on knowledge of the published literature spectra of these metabolites, whereas matrix effects are believed to be caused by anything but the compounds of interest. Developments in the area of post-column infusion look the most promising direction for future research. When full scan equipment reaches the same level of sensitivity as triple quadrupole types, there will be a need for appropriate control of the matrix effect.

4.6.3   Future trends in ionization techniques

MALDI (matrix-assisted laser desorption/ionization) and DESI (desorption electrospray ionization) are soft ionization techniques, usually applied to large molecules such as proteins and peptides to avoid complete fragmentation, which would result in too many charged ions, thus complicating identification. As mycotoxins are small molecules, MALDI cannot be applied directly to a sample, but a matrix solution has to be added. Thereafter it is called matrix assisted, not to be mixed-up with matrix effect. Typical matrix solutions are 3, 5-dimethoxy-4-hydroxycinnamic acid (sinapinic acid, SA), α-cyano-4-hydroxycinnamic acid (CHCA) and 2, 5-dihydroxybenzoic acid (DHB). The sample is spotted onto a metal plate together with a droplet of the matrix solution and the solvent is vaporized, leaving the recrystallized matrix with the analytes on the spot. The laser beam triggers ionization in the spot. The matrix absorbs the laser energy and transfers part of its charge to the analyte molecules to ionize them. Ions observed after this process consist of a neutral molecule [M] and an added or removed ion, which are transferred into a MS, mostly a TOF, owing to its large mass range. DESI lacks the laser, is used under ambient conditions and sometimes can do without any sample preparation or matrix addition. The electrospray is pulsed upon the sample and desorbs particles or molecules, thus producing charged solvent droplets and gas phase solvent ions that collide with analytes on the sample surface. The ions produced are analysed by a MS of choice in both cases.

4.6.4   Identification and differentiation of fungal spores with matrix-assisted laser desorption ionization time-of-flight (MALDI-TOF) mass spectromety

As MALDI-TOF originates from high molecular weight analysis, like peptides and proteins, it did not enter the mycotoxin area by its use to analyse chemical compounds, but by direct analysis of intact fungal spores (Welham *et al.*, 2000). Welham *et al.* studied *Penicillium* spp., *Scytalidium dimidiatum* and *Trichophyton rubrum* and observed profiles of the cellular material over the mass range 2–13 kDa. Li *et al.* (2000) presented MALDI-TOF mass spectra of intact fungal spores of four *Aspergillus* species, *Aspergillus flavus, A. oryzae, A. parasiticus* and *A. sojae*, and demonstrated that aflatoxigenic strains and non-aflatoxigenic strains have different mass peak profiles. Valentine *et al.* (2002) investigated intact spores of *Aspergillus niger, Rhizopus oryzae, Trichoderma reesei* and *Phanerochaete chrysosporium*. Sulc *et al.* (2009) extended this to 24 *Aspergillus* species. The fungal samples were blotted directly from the fungal culture with double-stick tape and ferulic acid or sinapinic acid matrix solution was layered over the dried samples. Fox *et al.* (2004) detected the *Aspergillus fumigatus* mycotoxins gliotoxin and helvolic acid, to confirm that sensitive and specific antisera could be raised against fungal toxins and may have an application in diagnosing fungal infection. Therefore both toxins were conjugated to bovine serum albumin (BSA) and analysed after SDS-PAGE (sodium dodecylsulfate polyacrylamide gel electrophoresis), freeze-drying and deposition in a SA droplet on the plate.

Chen and Chen (2005) characterized the fungal spores of *Penicillium expansum*, *P. chrysogenum*, *P. citrinum*, *P. digitatum*, *P. italicum*, and *P. pinophilum*, which are frequently found in grain and fruit. They directly scratched fungal spores from surfaces of fruit, contaminated by unknown fungi and demonstrated that these can be rapidly identified using MALDI-TOF analysis without any tedious pretreatment. Hettick *et al.* generated highly reproducible mass spectral 'fingerprints' for 12 *Penicillium* species (2008a) and 12 species of fungi of the genus *Aspergillus* and five different strains of *A. flavus* (2008b). Their mass spectra contained abundant peaks in the range of *m/z* 5000–20 000 and allowed unambiguous discrimination between species. In addition a common biomarker in all *Penicillium* mass spectra was observed at *m/z* 13 900.

Discriminant analysis using MALDI-TOF data yielded 100% correct classification for *Penicillium* and *Aspergillus* species and 95–100% accuracy for strains of *A. flavus*. For large datasets with significant differences in MALDI–TOF mass spectra between species, a subset of 'most significant' peaks allows 100% identification accuracy. However, all of the peaks in the dataset must be included to identify accurately highly similar fingerprint mass spectra such as those from different strains of the same species. Kemptner *et al.* (2009) complete this overview of most relevant fungi with the surface characterization and differentiation of *Fusarium* spores. They investigated five different matrices and solvent mixtures. The best peptide/protein profiles, with respect to MS pattern and signal intensities, were obtained with a matrix of ferulic acid (FA) in ACN/0.1% trifluoroacetic acid (7:3, v/v).

4.6.5   MALDI-TOF in direct mycotoxins analysis

MALDI-TOF mass spectrometry has also been proven to be applicable to the analysis of individual mycotoxins. Catharino *et al.* (2005) detected the aflatoxins B_1, B_2, G_1 and G_2 in a single spot, using CHCA as the matrix and addition of NaCl to enhance sensitivity via Na^+ cationization and obtained LOQ values around 50 pg in a 2 µl spot. They also tested spiked peanut samples, but without mentioning the exact levels. Carpentieri *et al.* (2007) reported a simple and fast method for wine fingerprinting based on measurements of alcohols, esters, organic acids, phenolic compounds, nitrogen-containing compounds, flavonoids, anthocyanins, proanthocyanins and two 'others'. This led to effective and characteristic wine 'fingerprints' or 'signatures', not based on only the detection of anthocyanins, but more than 80 molecular species in a single experiment, of which the two 'other' compounds were β-damascenone and OTA. In 18 of 20 wine samples they detected OTA when using DHB as the added matrix, but did not quantify it, as it was not included in the fingerprint profile.

Elosta *et al.* (2007) simultaneously determined DON, NIV, 3- and 15-AcDON in barley and malt extracts. As the usual matrices failed to ionize these trichothecenes properly, they studied saturated DHB, 0.2% hydrazine hydrate, finely powdered synthetic diamond and sodium azide (NaN_3) suspensions in acetonitrile. For DON and NIV the sodium azide suspension was recommended. The highest signal intensity for determining 15-AcDON was obtained with finely powdered synthetic diamond suspension. *Fusarium* infected barley and malt samples, grown organically and treated with fungicides, were analysed. In organically processed malt 507.1 ± 8.6 µg kg^{-1} DON was reported and compared with 779.5 ± 124.1 µg kg^{-1}, as determined by HPLC. In a malt sample infected with *Fusarium*, but field-treated by fungicides, 393.5 ± 8.6 µg kg^{-1} DON was found, compared to 528.0 ± 84.0 µg kg^{-1} by HPLC analysis. Blechová *et al.* (2006) conducted a similar study on DON and NIV, but did not achieve LOQs lower than 600 µg kg^{-1} for DON in barley malt. The first MALDI-TOF application for fumonisin B$_1$ was not a direct analysis of the mycotoxin itself, but the diazotization product of its deamination in contaminated feed, as identified by Lemke *et al.* (2001).

4.6.6 Desorption electrospray ionization (DESI) and direct analysis in real time (DART) in mycotoxins analysis

DESI and DART are techniques that are based on ionization processes at ambient pressure outside the MS, after which the resulting ions are swept into the ion source. DESI utilizes the impact of electrosprayed solvent droplets upon sample surfaces to generate analyte ions. DART ionization is based on bombardment of the sample with electrical discharged metastable and atmospheric gases. DESI was applied by Long *et al.* (2007) who detected trace levels of chemical warfare agents and biological toxins, including aflatoxin B$_1$, fumonisin B$_1$ and ergotamine at the picogram level. Standards of each toxin were prepared in concentrations as low as 100 pg ml^{-1}. Samples were prepared by depositing a 1 µl aliquot of the appropriate standard onto the test surface material and allowing the spot to air dry at room temperature. The spot was analysed by DESI, selecting protonated or sodiated fragment ions. LODs in individual spots were 5.4 pg for AFB$_1$, 30.8 pg for AFG$_2$, 71 pg for FB$_1$ and 6.5 pg for ergotamine. Their study was carried out on textile surfaces, to test its use for protective clothes. This is an interesting point of view for food and feed analysis, as it might lead to non-destructive analysis of mycotoxins. This had already been demonstrated for pesticides by Borsdorf *et al.* (2009), who coupled a pulsed laser desorption device with a differential mobility spectrometer to screen five pesticides on food surfaces, directly on fruit and vegetable skins. They detected pesticides on apples, grapes, tomatoes and pepper in the nanogram range, for example a 0.5 µg carbaryl spike on grape. When mycotoxins could be determined non-destructively on food surfaces it would be possible to analyse kernels in a lot individually and determine heterogeneity of bulk contamination in practice. The results of such investigations might be interesting for the evaluation of existing sampling schemes. Maragos and Busman (2010) set a first step in this direction by depositing 10 µl of 20ng ml^{-1} FB$_1$ on the surface of maize kernels. Upon drying, the kernels were subjected to a DESI-ITD system, which easily detected the fumonisin spike. After the experiment the seeds were germinated, as to proof the ionization process to be non-destructive. As the spike was put on the corn surface, some more investigations will be necessary to be sure whether naturally contaminated samples give comparable results. Zachariasova *et al.*

(2010b) combined DART with an UPLC-Orbitrap MS system to analyse 32 target mycotoxins (i.e. AFs, OTA, DON, ZEN, T2/HT2 and ERG) in beer at levels lower then 10ng ml^{-1}, unfortunately with an exception for OTA which had lowest calibration levels of 30 to 65 µg l^{-1}, depending on the type of beer. When Vaclavik *et al.* (2010) applied this method to wheat and maize it turned out that aflatoxins and T2/HT2 showed poor ionization, whereas OTA and some other mycotoxins such as ergot alkaloids and fumonisins could not be ionized under the employed experimental conditions, leaving 11 target compounds (i.e. DON and ZEN) for the final validation study.

4.7   Conclusions

LC–MS is a powerful technique in the mycotoxin area, both for multi-method and monitoring purposes. Before application, one first has to consider whether it is necessary with respect to the aim of the analysis (e.g. compared to methods in Chapter 3). If so, the right mass spectrometer type has to be chosen, for example ITD to analyse single (or one class of) compounds, triple quadrupole for targeted analysis of known compounds, triple quadrupole with trap to have a scan option or TOF for full scan availability. In this evaluation process the increased costs must be kept in mind, not only of the equipment but also the costs of maintenance and of technicians with the required levels of skills.

4.8   Acknowledgements

Yvonne van der Laan is thanked for setting up the mycotoxin literature database and retrieving many references. Jaap Voogt is thanked for keeping it up to date.

4.9   References

Abbas H K, Cartwright R D, Shier W T, Abouzied M M, Bird C B, Tice L G, Ross P F, Sciumbato G L and Meredith F I (1998), 'Natural occurrence of fumonisins in rice with sheath rot disease', *Plant Dis.*, **82**, 22–5.

Andersen B, Smedsgaard J and Frisvad J C (2004), '*Penicillium expansum*: consistent production of patulin, chaetoglobosins, and other secondary metabolites in culture and their natural occurrence in fruit products', *J. Agric. Food Chem.*, **52**, 2421–8.

Apfelthaler E, Bicker W, Lammerhofer M, Sulyok M, Krska R, Lindner W and Schuhmacher R (2008), 'Retention pattern profiling of fungal metabolites on mixed-mode reversed-phase/weak anion exchange stationary phases in comparison to reversed-phase and weak anion exchange separation materials by liquid chromatography-electrospray ionisation-tandem mass spectrometry', *J. Chromatogr. A*, **1191**, 171–81. Supporting information: 12 pages annex, with chemical structures, log *P* and pK*a* values and classification of solutes as acidic, amphoteric, basic and neutral.

Asam S and Rychlik M (2006), 'Synthesis of four carbon-13-labelled type a trichothecene mycotoxins and their application as internal standards in stable isotope dilution assays', *J. Agric. Food Chem.*, **54**, 6535–46.

Asam S and Rychlik M (2007), 'Studies on accuracy of trichothecene multitoxin analysis using stable isotope dilution assays', *Mycotoxin Res.*, **23**, 191–8.

Bacaloni A, Cavaliere C, Faberi A, Pastorini E, Samperi R and Laganà A (2005), 'Automated on-line solid-phase extraction-liquid chromatography-electrospray tandem mass spectrometry method for the determination of ochratoxin A in wine and beer', *J. Agric. Food Chem.*, **53**, 5518–25.

Bayman P, Baker J L, Doster M A, Michailides T J and Mahoney N E (2002), 'Ochratoxin production by the *Aspergillus ochraceus* group and *Aspergillus alliaceus*', *Appl. Environ. Microbiol.*, **68**, 2326–9.

Beltrán E, Ibáñez M, Vicente Sancho J and Hernández F (2009), 'Determination of mycotoxins in different food commodities by UPLC coupled to triple quadrupole mass spectrometry', *Rapid Commun. Mass Spectrom.*, **23**, 1801–9.

Berger U, Oehme M and Kuhn F (1999), 'Quantitative determination and structure elucidation of type A- and B-trichothecenes by HPLC/ion trap multiple mass spectrometry', *J. Agric. Food Chem.*, **47**, 4240–5.

Berthiller F, Schuhmacher R, Buttinger G and Krska R (2005), 'Rapid simultaneous determination of major type A- and B-trichothecenes as well as zearalenone in maize by high performance liquid chromatography–tandem mass spectrometry', *J. Chromatogr. A*, **1062**, 209–16.

Beyer M, Ferse I and Humpf H-U (2009), 'Large-scale production of selected type A trichothecenes: the use of HT-2 toxin and T-2 triol as precursors for the synthesis of d_3-T-2 and d_3-HT-2 toxin', *Mycotoxin Res.*, **25**, 41–52.

Bily A C, Reid L M, Savard M E, Reddy R, Blackwell B A, Campbell C M, Krantis A, Durst T, Philogène B J R, Arnason J T and Regnault-Roger C (2004), 'Analysis of *Fusarium graminearum* mycotoxins in different biological matrices by LC/MS', *Mycopathologia*, **157**, 117–26.

Biselli S, Hartig L, Wegner H and Hummert C (2004), 'Analysis of *Fusarium* toxins using LC/MS-MS: application to various food and feed matrices', *LC·GC Europe*, **17**, 25–30.

Blechová P, Havlová P, Gajdošová D and Havel J (2006), 'New possibilities of matrix-assisted laser desorption ionization time of flight mass spectrometry to analyze barley malt quality. Highly sensitive detection of mycotoxins', *Environ. Toxicol.*, **21**, 403–8.

Bonfiglio R, King R C, Olah T V and Merkle K (1999), 'The effects of sample preparation methods on the variability of the electrospray ionization response for model drug compounds', *Rapid Commun. Mass Spectrom.*, **13**, 1175–85.

Boonzaaijer G, Osenbruggen W A, Van Kleinnijenhuis A J and Van Dongen W D (2008), 'An exploratory investigation of several mycotoxins and their natural occurrence in flavour ingredients and spices, using a multi-mycotoxin LC–MS/MS method', *World Mycotoxin J.*, **1**, 167–74.

Borsdorf H, Roetering S, Nazarov E G and Weickhardt C (2009), 'Rapid screening of pesticides from fruit surfaces: preliminary examinations using a laser desorption–differential mobility spectrometry coupling', *Int. J. Ion Mobility Spectrom.*, **12**, 15–22.

Bretz M, Beyer M, Cramer B and Humpf H-U (2005), 'Synthesis of stable isotope labelled 3-acetyldeoxynivalenol, 3-d_3-AcDON', *Mol. Nutr. Food Res.*, **49**, 1151–3.

Bretz M, Beyer M, Cramer B and Humpf H-U (2006), 'Stable isotope dilution analysis of the *Fusarium* mycotoxins deoxynivalenol and 3-acetyldeoxynivalenol', *Mol. Nutr. Food Res.*, **50**, 251–60.

Buhrman D L, Price P I and Rudewicz P J (1996), 'Quantitation of SR 27417 in human plasma using electrospray liquid chromatography–tandem mass spectrometry: a study on ion suppression', *J. Am. Soc. Mass Spectrom.*, **7**, 1099–105.

Capriotti A L, Foglia P, Gubbiotti R, Roccia C, Samperi R and Laganà A (2010), 'Development and validation of a liquid chromatography/atmospheric pressure photoionization–tandem mass spectrometric method for the analysis of mycotoxins subjected to commission regulation (EC) No. 1881/2006 in cereals', *J. Chromatogr. A*, **1217**, 6044–51.

Carpentieri A, Marino G and Amoresano A (2007), 'Rapid fingerprinting of red wines by MALDI mass spectrometry', *Anal. Bioanal. Chem.*, **389**, 969–82.

Catharino R R, de Azevedo Marques L, Santos L S, Baptista A S, Gloria E M, Calori-Domingues M A, Facco E M P and Eberlin M N (2005), 'Aflatoxin screening by MALDI-TOF mass spectrometry', *Anal. Chem.*, **77**, 8155–7.

Cavaliere C, D'Ascenzo G, Foglia P, Pastorini E, Samperi R and Laganà A (2005a), 'Determination of type B trichothecenes and macrocyclic lactone mycotoxins in field contaminated maize', *Food Chem.*, **92**, 559–68.

Cavaliere C, Foglia P, Pastorini E, Samperi R and Laganà A (2005b), 'Development of a multiresidue method for analysis of major *Fusarium* mycotoxins in corn meal using liquid chromatography–tandem mass spectrometry', *Rapid Commun. Mass Spectrom.*, **19**, 2085–93.

Cavaliere C, Foglia P, Pastorini E, Samperi R and Laganà A (2006), 'Liquid chromatography/tandem mass spectrometric confirmatory method for determining aflatoxin M_1 in cow milk – Comparison between electrospray and atmospheric pressure photoionization sources', *J. Chromatogr. A*, **1101**, 69–78.

Cavaliere C, Foglia P, Guarino C, Motto M, Nazzari M, Samperi R, Laganà A and Berardo N (2007), 'Mycotoxins produced by *Fusarium* genus in maize: determination by screening and confirmatory methods based on liquid chromatography tandem mass spectrometry', *Food Chem.*, **105**, 700–10.

Cervino C, Asam S, Knopp D, Rychlik M and Niessner R (2008), 'Use of isotope-labelled aflatoxins for LC–MS/MS stable isotope dilution analysis of foods', *J. Agric. Food Chem.*, **56**, 1873–9.

Chen H-Y and Chen Y-C (2005), 'Characterization of intact *Penicillium* spores by matrix-assisted laser desorption/ionization time-of-flight mass spectrometry', *Rapid Commun. Mass Spectrom.*, **19**, 3564–8.

Christensen H B, Poulsen M E, Rasmussen P H and Christen D (2009), 'Development of an LC–MS/MS method for the determination of pesticides and patulin in apples', *Food Addit. Contam.*, **26**, 1013–23.

Commission Decision 2002/657/EC of 14 August 2002 implementing Council Directive 96/23/EC concerning the performance of analytical methods and the interpretation of results. OJ L 221/8–36, 17.08.2002.

Commission Directive 2002/32/EC of 7 May 2002 of the European Parliament and of the Council, on undesirable substances in animal feed. OJ L 140/10-21, 30.05.2002.

Commission Recommendation 2006/576/EC of 17 August 2006 on the presence of deoxynivalenol, zearalenone, ochratoxin A, T-2 and HT-2 and fumonisins in products intended for animal feeding. OJ L 229/7–9, 23.08.2006.

Commission Regulation 2006/401/EC of 23 February 2006 laying down the methods of sampling and analysis for the official control of the levels of mycotoxins in foodstuffs. OJ L 70/12–34, 9.3.2006.

Commission Regulation 2006/1881/EC of 19 December 2006 replacing Regulation (EC) 466/2001 setting maximum levels for certain contaminants in foodstuffs. OJ L 364/5–24, 20.12.2006.

Commission Regulation 2007/1126/EC of 28 September 2007 amending Regulation (EC) 1881/2006 setting maximum levels for certain contaminants in foodstuffs, regarding *Fusarium* toxins in maize and maize products. OJ L 255/14–17, 29.09.2007.

Cramer B, Bretz M and Humpf H-U (2007), 'Stable isotope dilution analysis of the fusarium mycotoxin zearalenone', *J. Agric. Food Chem.*, **55**, 8353–8.

Dall'Asta C, Sforza S, Galaverna G, Dossena A and Marchelli R (2004), 'Simultaneous detection of type A and type B trichothecenes in cereals by liquid chromatography–electrospray ionization mass spectrometry using NaCl as cationization agent', *J. Chromatogr. A*, **1054**, 389–95.

Dall'Asta C, Sforza S, Moseriti A, Galaverna G, Dossena A and Marchelli R (2005), 'An innovative LC/MS approach applied to the determination of zeralenone in maize: Alternate Isotope-coded Derivatization Assay (AIDA)', *Mycotoxin Res.*, **21**, 218–23.

Degelmann P, Becker M, Herderich M and Humpf H U (1999), 'Determination of ochratoxin A in beer by high-performance liquid chromatography', *Chromatographia*, **49**, 543–6.

De Kok A and Pizzutti I (2009), 'Efficiency optimization for pesticides and mycotoxins multi-residue analyses by integrated sample preparation method'. Oral and poster presentations at the *4th International Symposium on Recent Advances in Food Analysis*, 4–6 November, Prague, Czech Republic, book of abstracts, available from: http://rafa2009.eu/pdf/BoA_RAFA_2009_FINAL.pdf, pp. 83, 374, 385 and 387, accessed: 11 March 2010.

Delmulle B, De Saeger S, Adams A, De Kimpe N and Van Peteghem C (2006), 'Development of a liquid chromatography/tandem mass spectrometry method for the simultaneous determination of 16 mycotoxins on cellulose filters and in fungal cultures', *Rapid Commun. Mass Spectrom.*, **20**, 771–6.

Di Mavungu J D, Monbaliu S, Scippo M-L, Maghuin-Rogister G, Schneider Y-J, Larondelle Y, Callebaut A, Robbens J, Van Peteghem C and De Saeger S (2009), 'LC–MS/MS multi-analyte method for mycotoxin determination in food supplements', *Food Addit. Contam. Part A*, **26**, 885–95.

Driehuis F, Spanjer M C, Scholten J M and Te Giffel M C (2008), 'Occurrence of mycotoxins in maize, grass and wheat silage for dairy cattle in the Netherlands', *Food Addit. Contam. Part B*, **1**, 41–50.

Driffield M, Hird S J and Mac Donald S (2003), 'The occurrence of a range of mycotoxins in animal offal food products by HPLC–MS/MS', *Aspects Appl. Biol.*, **68**, 205–10.

Elosta S, Gajdošová D, Hégrová B and Havel J (2007), 'MALDI TOF mass spectrometry of selected mycotoxins in barley', *J. Appl. Biomedicine*, **5**, 39–47.

Enke C G (1997), 'A predictive model for matrix and analyte effects in electrospray ionization of singly-charged ionic analytes', *Anal. Chem.*, **69**, 4885–93.

Fox M, Gray G, Kavanagh K, Lewis C and Doyle S (2004), 'Detection of *Aspergillus fumigatus* mycotoxins: immunogen synthesis and immunoassay development', *J. Microbiol. Methods*, **56**, 221–30.

Garon D, Richard E, Sage L, Bouchart V, Pottier D and Lebailly P (2006), 'Mycoflora and multimycotoxin detection in corn silage: experimental study', *J. Agric. Food Chem.*, **54**, 3479–84.

Garrido Frenich A, Martínez Vidal J L, Romero-González R and Aguilera-Luiz M M (2009), 'Simple and high-throughput method for the multimycotoxin analysis in cereals and related foods by ultra-high performance liquid chromatography/tandem mass spectrometry', *Food Chem.*, **117**, 705–12.

Gentili A, Caretti F, D'Ascenzo G, Rocca L M, Marchese S, Materazzi S and Perret D (2007), 'Simultaneous determination of trichothecenes A, B, and D in maize food products by LC–MS-MS', *Chromatographia*, **66**, 669–76.

Griesshaber D S (2007), *Development of Ion Trap HPLC–MS Detection Methods for the Determination of Prevalent Mycotoxins in Grain and application to Real Samples*, PhD thesis, University of Basel, 90–122.

Haddon W F, Wiley M and Waiss A C (1971), 'Aflatoxin detection by thin-layer chromatography–mass spectrometry', *Anal. Chem.*, **43**, 268–70.

Hancock J R and D'Agostino P A (2002), 'Mass spectrometric identification of toxins of biological origin', *Anal. Chim. Acta*, **457**, 71–82.

Hartl M and Humpf H-U (1999), 'Simultaneous determination of fumonisin B-1 and hydrolyzed fumonisin B-1 in corn products by liquid chromatography/electrospray ionization mass spectrometry', *J. Agric. Food Chem.*, **47**, 5078–83.

Hartl M, Herderich M and Humpf H-U (1999), 'Rapid determination of fumonisin FB_1 in corn products by high-performance liquid chromatography-electrospray mass spectrometry', *Eur. Food Res. Techn.*, **209**, 348–51.

Hartmann N, Erbs M, Wettstein F E, Schwarzenbach R P and Bucheli T D (2007), 'Quantification of estrogenic mycotoxins at the ng/L level in aqueous environmental samples using deuterated internal standards', *J. Chromatogr. A*, **1138**, 132–40.

Hartmann N, Erbs M, Wettstein F E, Hoerger C C, Schwarzenbach R P and Bucheli T D (2008), 'Quantification of zearalenone in various solid agroenvironmental samples using D-6-zearalanone as the internal standard', *J. Agric. Food Chem.*, **56**, 2926–32.

Häubl G, Berthiller F, Krska R and Schuhmacher R (2006a), 'Suitability of a fully C-13 isotope labelled internal standard for the determination of the mycotoxin deoxynivalenol by LC–MS/MS without cleanup', *Anal. Bioanal. Chem.*, **384**, 692–6.

Häubl G, Berthiller F, Rechthaler J, Jaunecker G, Binder E M, Krska R and Schuhmacher R (2006b), 'Characterization and application of isotope-substituted (13C15)-deoxynivalenol (DON) as an internal standard for the determination of DON', *Food Addit. Contam.*, **23**, 1187–93.

Häubl G, Berthiller F, Hametner C, Rechthaler J, Jaunecker G, Freudenschuss M, Krska R and Schuhmacher R (2007), 'Characterization of (13C24) T-2 toxin and its use as an internal standard for the quantification of T-2 toxin in cereals with HPLC–MS/MS', *Anal. Bioanal. Chem.*, **389**, 931–40.

Herebian D, Zühlke S, Lamshöft M and Spiteller M (2009), 'Multi-mycotoxin analysis in complex biological matrices using LC-ESI/MS: experimental study using triple stage quadrupole and LTQ-Orbitrap', *J. Sep. Sci.*, **32**, 939–48.

Hettick J M, Green B J, Buskirk A D, Kashon M L, Slaven J E, Janotka E, Blachere F M, Schmechel D and Beezhold D H (2008a), 'Discrimination of *Penicillium* isolates by matrix-assisted laser desorption/ionization time-of-flight mass spectrometry fingerprinting', *Rapid Commun. Mass Spectrom.*, **22**, 2555–60.

Hettick J M, Green B J, Buskirk A D, Kashon M L, Slaven J E, Janotka E, Blachere F M, Schmechel D and Beezhold D H (2008b), 'Discrimination of *Aspergillus* isolates at the species and strain level by matrix-assisted laser desorption/ionization time-of-flight mass spectrometry fingerprinting', *Anal. Biochem.*, **380**, 276–281.

Jestoi M, Somma M C, Kouva M, Veijalainen P, Rizzo A, Ritieni A and Peltonen K (2004), 'Levels of mycotoxins and sample cytotoxicity of selected organic and conventional grain-based products purchased from Finnish and Italian markets', *Mol. Nutr. Food Res.*, **48**, 299–307.

Jestoi M, Rokka M, Järvenpää E and Peltonen K (2009), 'Determination of *Fusarium* mycotoxins beauvericin and enniatins (A, A$_1$, B, B$_1$) in eggs of laying hens using liquid chromatography–tandem mass spectrometry (LC–MS/MS)', *Food Chem.*, **115**, 1120–7.

Ji C, Li Y, Su L, Zhang X and Chen X (2009), 'Quantitative structure-retention relationships for mycotoxins and fungal metabolites in LC–MS/MS', *J. Sep. Sci.*, **32**, 3967–79.

Kataoka H, Itano M, Ishizaki A and Saito K (2009), 'Determination of patulin in fruit juice and dried fruit samples by in-tube solid-phase microextraction coupled with liquid chromatography–mass spectrometry', *J. Chromatogr. A*, **1216**, 3746–50.

Kellmann M, Muenster H, Zomer P and Mol H (2009), 'Full Scan MS in comprehensive qualitative and quantitative residue analysis in food and feed matrices: How much resolving power is required?', *J Am. Soc. Mass Spectrom.*, **20**, 1464–76.

Kemptner J, Marchetti-Deschmann M, Mach R, Druzhinina I S, Kubicek C P and Allmaier G (2009), 'Evaluation of matrix-assisted laser desorption/ionization (MALDI) preparation techniques for surface characterization of intact *Fusarium* spores by MALDI linear time-of-flight mass spectrometry', *Rapid Commun. Mass Spectrom.*, **23**, 877–84.

Klötzel M, Gutsche B, Lauber U and Humpf H-U (2005), 'Determination of 12 type A and B trichothecenes in cereals by liquid chromatography–electrospray ionization tandem mass spectrometry', *J. Agric. Food Chem.*, **53**, 8904–10.

Klötzel M, Lauber U and Humpf H-U (2006), 'A new solid phase extraction cleanup method for the determination of 12 type A and B trichothecenes in cereals and cereal-based food by LC–MS/MS', *Mol. Nutr. Food Res.*, **50**, 261–9.

Kokkonen M K and Jestoi M N (2009a), 'A multi-compound LC–MS/MS method for the screening of mycotoxins in grains', *Food Anal. Methods*, **2**, 128–40.

Kokkonen M K and Jestoi M N (2009b), Erratum, 'A multi-compound LC–MS/MS method for the screening of mycotoxins in grains', *Food Anal. Methods*, **2**, 239.

Kokkonen M, Jestoi M and Rizzo A (2005), 'Determination of selected mycotoxins in mould cheeses with liquid chromatography coupled to tandem with mass spectrometry', *Food Addit. Contam*, **22**, 449–56.

Korfmacher W A, Chiarelli M P, Lay J O, Bloom J, Holcomb M and McManus K T (1991), 'Characterization of the mycotoxin fumonisin-B$_1$ – comparison of thermospray, fast-atom-bombardment and electrospray mass–spectrometry', *Rapid Commun. Mass Spectrom.*, **5**, 463–8.

Kushiro M, Zheng Y, Nagata R, Nakagawa H and Nagashima H (2009), 'Limited surveillance of fumonisins in brown rice and wheat harvested in Japan', *J. Food Prot.*, **72**, 1327–31.

Laganà A, Curini R, D'Ascenzo G, De Leva I, Faberi A and Pastorini E (2003), 'Liquid chromatography/tandem mass spectrometry for the identification and determination of trichothecenes in maize', *Rapid Commun. Mass Spectrom.*, **17**, 1037–43.

Lanier C, Heutte N, Richard E, Bouchart V, Lebailly P and Garon D (2009), 'Mycoflora and mycotoxin production in oilseed cakes during farm storage', *J. Agric. Food Chem.*, **57**, 1640–5.

Lattanzio V M, Solfrizzo M, Powers S and Visconti A (2007), 'Simultaneous determination of aflatoxins, ochratoxin A and *Fusarium* toxins in maize by liquid chromatography/ tandem mass spectrometry after multitoxin immunoaffinity cleanup', *Rapid. Commun. Mass Spectrom.*, **21**, 3253–61.

Lattanzio V M, Solfrizzo M and Visconti A (2008), 'Determination of trichothecenes in cereals and cereal-based products by liquid chromatography–tandem mass spectrometry', *Food Addit. Contam.*, **25**, 320–30.

Lau B P Y, Scott P M, Lewis D A, Kanhere S R, Cleroux C and Roscoe V A (2003), 'Liquid chromatography–mass spectrometry and liquid chromatography–tandem mass spectrometry of the *Alternaria* mycotoxins alternariol and alternariol monomethyl, ether in fruit juices and beverages', *J Chromatogr. A*, **998**, 119–31.

Lehotay S J, De Kok A, Hiemstra M and Van Bodegraven P (2005a), 'Validation of a fast and easy method for the determination of residues from 229 pesticides in fruits and vegetables using gas and liquid chromatography and mass spectrometric detection', *J. AOAC Int.*, **88**, 595–614.

Lehotay S J, Mastovska K and Lightfield A R (2005b), 'Use of buffering and other means to improve results of problematic pesticides in a fast and easy method for residue analysis of fruits and vegetables', *J. AOAC Int.*, **88**, 615–29.

Lehotay S J, Mastovska K, Amirav A, Fialkov A B, Alon T, Martos P A, Kok A de and Fernández-Alba A R (2008), 'Identification and confirmation of chemical residues in food by chromatography–mass spectrometry and other techniques', *Trends Anal. Chem.*, **27**, 1070–90.

Lemke S L, Ottinger S E, Ake C L, Mayura K and Phillips T D (2001), 'Deamination of fumonisin B(1) and biological assessment of reaction product toxicity', *Chem. Res. Toxicol.*, **14**, 11–5.

Leong S L (2005), *Black* Aspergillus *Species: Implications for Ochratoxin A in Australian Grapes and Wine*, PhD thesis, University of Adelaide, School of Agriculture and Wine, Discipline of Plant and Pest Science, Adelaide, p. 204.

Li T-Y, Liu B-H and Chen Y-C (2000), 'Characterization of *Aspergillus* spores by matrix-assisted laser desorption/ionization time-of-flight mass spectrometry', *Rapid Commun. Mass Spectrom.*, **14**, 2393–400.

Liao W T, Draper W M and Perera S K (2008), 'Identification of unknowns in atmospheric pressure ionization mass spectrometry using a mass to structure Search Engine', *Anal. Chem.*, **80**, 7765–77.

Lindenmeier M, Schieberle P and Rychlik M (2004), 'Quantification of ochratoxin A in foods by a stable isotope dilution assay using high-performance liquid chromatography–tandem mass spectrometry', *J. Chromatogr. A*, **1023**, 57–66.

Liu J and Powers S (2009), 'Simultaneous detection of eleven components of mycotoxin family by high pressure liquid chromatography in corn'. Poster 184 at the *First International Society for Mycotoxicology Conference*, 9–11 September 2009, Tulln, Austria, book of abstracts, available from: http://www.ism2009.at/ISM2009_posters.pdf, p. 209, accessed: 29 May 2010.

Logrieco A, Bottalico A, Mulé G, Moretti A and Perrone G (2003), 'Epidemiology of toxigenic fungi and their associated mycotoxins for some Mediterranean crops', *Eur. J. Plant Path.*, **109**, 645–67.

Long M S, Laughlin B C, Wiseman J M, Pyle T, Boscacci K J, Rothhaar K and Helphingstine K J (2007), *Incubation and Growth of Life Sciences, Medical and Biotechnology Businesses in Proteomics, Genomics, Medicine, and Dentistry,* Advanced Research and Technology Institute Indianapolis. Available from: www.dtic.mil/cgi-bin/GetTRDoc?AD= ADA477669&Location=U2&doc=GetTRDoc, accessed 1 June 2010, 22–7.

López Grío S J, Garrido Frenich A, Martínez Vidal J L and Romero-González R (2010), 'Determination of aflatoxins B_1, B_2, G_1, G_2 and ochratoxin A in animal feed by ultra high-performance liquid chromatography–tandem mass spectrometry', *J. Sep. Sci.*, **33**, 502–8.

Lukacs Z, Schaper S, Herderich M, Schreier P and Humpf H-U (1996), 'Identification and determination of fumonisin FB_1 and FB_2 in corn and corn products by high-performance liquid chromatography–electrospray-ionization tandem mass spectrometry (HPLC-ESI-MS-MS)', *Chromatographia*, **43**, 124–8.

Madalinski G, Godat E, Alves S, Lesage D, Genin E, Levi P, Labarre J, Tabet J-C, Ezan E and Junot C (2008), 'Direct introduction of biological samples into a LTQ-orbitrap hybrid mass spectrometer as a tool for fast metabolome analysis', *Anal. Chem.*, **80**, 3291–303.

Makarov A, Denisov E, Kholomeev A, Balschun W, Lange O, Strupat K and Horning S (2006), 'Performance evaluation of a hybrid linear ion trap/orbitrap mass spectrometer', *Anal. Chem.*, **78**, 2113–20.

Maragos C M and Busman M (2010), 'Rapid and advanced tools for mycotoxin analysis: a review', *Food Addit. Contam. Part A*, **27**, 688–700.

Matthews W (2002), *Survey of nuts, nut products and dried tree fruits for mycotoxins.* Food Survey Information Sheet 21/02. Food Standards Agency, Contaminants Division: London, p. 52, available from: http://www.food.gov.uk/science/surveillance/, accessed: 13 March 2007.

Matuszewski B K (2006), 'Standard line slopes as a measure of a relative matrix effect in quantitative HPLC–MS bioanalysis', *J. Chromatogr. B*, **830**, 293–300.

Matuszewski B K, Constanzer M L and Chavez-Eng C M (2003), 'Strategies for the assessment of matrix effect in quantitative bioanalytical methods based on HPLC–MS/MS', *Anal. Chem.*, **75**, 3019–30.

Mitterer G, Kainz M, Freudenschuss M, Häubl G and Krska R (2009a), 'The use of fully stable isotope labelled mycotoxins as internal standards for mycotoxin analysis with LC–MS/MS'. Poster 193 at the *First International Society for Mycotoxicology Conference*, 9–11 September 2009, Tulln, Austria , book of abstracts, available from: http://www.ism2009.at/ISM2009_posters.pdf, p. 218, accessed: 29 May 2010.

Mitterer G, Kainz M and Sulyok M (2009b), 'Development of a MycoSep based cleanup approach for the simultaneous assessment of all relevant mycotoxins by LC–MS/MS'. Poster 194 at the *First International Society for Mycotoxicology Conference*, 9–11 September 2009, Tulln, Austria, book of abstracts, available from: http://www.ism2009.at/ISM2009_posters.pdf, p. 219, accessed: 29 May 2010.

Mizutani K, Kumagai S, Mochizuki N, Kitagawa Y and Sugita-Konishi Y (2009), 'Determination of a yellow rice toxin, luteoskyrin, in rice by using liquid chromatography–tandem mass spectrometry with electrospray ionization', *J. Food Prot.*, **72**, 1321–6.

Mohamed R, Gremaud E, Richoz-Payot J, Tabet J C and Guy P A (2006), 'Quantitative determination of five ergot alkaloids in rye flour by liquid chromatography–electrospray ionisation tandem mass spectrometry', *J. Chromatogr. A*, **1114**, 62–72.

Mol H G J, Plaza-Bolanos P, Zomer P, De Rijk T C, Stolker A A M and Mulder P P J (2008), 'Toward a generic extraction method for simultaneous determination of pesticides, mycotoxins, plant toxins, and veterinary drugs in feed and food matrixes', *Anal. Chem.*, **24**, 9450–9 and supporting information.

Moldes-Anaya A S, Asp T N, Eriksen G S, Skaar I and Rundberget T (2009), 'Determination of cyclopiazonic acid in food and feeds by liquid chromatography–tandem mass spectrometry', *J. Chromatogr. A*, **1216**, 3812–8.

Monbaliu S, Van Poucke C, Van Peteghem C, Van Poucke K, Heungens K and De Saeger S (2009), 'Development of a multi-mycotoxin liquid chromatography/tandem mass spectrometry method for sweet pepper analysis', *Rapid Commun. Mass Spectrom.*, **23**, 3–11.

Monbaliu S, Van Poucke C, Detavernier C, Dumoulin F, Van de Velde M, Schoeters E, Van Dyck S, Averkieva O, Van Peteghem C and De Saeger S (2010), 'Occurrence of mycotoxins in feed as analyzed by a multi-mycotoxin LC–MS/MS method', *J. Agric. Food Chem.*, **58**, 66–71.

Moretti A, Ritieni A and Logrieco A (2005), 'Toxin profile and morphological identification of *Fusarium* species isolated from fig in Apulia'. Poster presentation at the *Third International Symposium on Fig*, Vilamoura, Portugal, May 16–20.

Nielsen K F and Smedsgaard J (2003), 'Fungal metabolite screening: database of 474 mycotoxins and fungal metabolites for dereplication by standardised liquid chromatography–UV-mass spectrometry methodology', *J. Chromatogr. A*, **1002**, 111–36.

Nielsen K F, Mogensen J F, Johansen M, Larsen T O and Frisvad J C (2009a), 'Review of secondary metabolites and mycotoxins from the *Aspergillus niger* group', *Anal. Bioanal. Chem.*, **395**, 1225–42.

Nielsen K F, Larsen T O and Frisvad J C (2009b), 'LC-HRMS dereplication of mycotoxins and microbial metabolites based on 800 reference standards. Interpretation of ESI spectra and chromatographic retention'. Oral presentation at the *First International Society for Mycotoxicology Conference*, 9–11 September 2009, Tulln, Austria.

Noba S, Uyama A and Mochizuki N (2009), 'Determination of ochratoxin A in ready-to-drink coffee by immunoaffinity cleanup and liquid chromatography–tandem mass spectrometry', *J. Agric. Food Chem.*, **57**, 6036–40.

Nonaka Y, Saito K, Hanioka N, Narimatsu S and Kataoka H (2009), 'Determination of aflatoxins in food samples by automated on-line in-tube solid-phase microextraction coupled with liquid chromatography–mass spectrometry', *J. Chromatogr. A*, **1216**, 4416–22.

Noonim P, Mahakarnchanakul W, Nielsen K F, Frisvad J C and Samson R A (2008), 'Isolation, identification and toxigenic potential of ochratoxin A-producing *Aspergillus* species from coffee beans grown in two regions of Thailand', *Int. J. Food Microbiol.*, **128**, 197–202.

Noonim P, Mahakarnchanakul W, Nielsen K F, Frisvad J C and Samson R A (2009), 'Fumonisin B_2 production by *Aspergillus niger* in Thai coffee beans', *Food Addit. Contam. Part A*, **26**, 94–100.

Pallaroni L and Von Holst C (2003), 'Determination of zearalenone from wheat and corn by pressurized liquid extraction and liquid chromatography–electrospray mass spectrometry', *J. Chromatogr. A*, **993**, 39–45.

Pallaroni L, Bjorklund E and Von Holst C (2002), 'Optimization of atmospheric pressure chemical ionization interface parameters for the simultaneous determination of deoxynivalenol and zearalenone using HPLC/MS', *J. Liq. Chrom. Rel. Technol.*, **25**, 913–26.

Pihlström T (2010), *Method Validation and Quality Control Procedures for Pesticide Residues Analysis in Food and Feed*, Document No. SANCO/10684/2009, available from: http://ec.europa.eu/food/plant/protection/resources/qualcontrol_en.pdf , implemented by 01/01/2010, accessed: 13 January 2010.

Pizzutti I R, Adaime M B, Zanella R, de Kroon M, Rensen P, Prestes O D and de Kok A (2007), 'Fully integrated sample preparation method for the GC–MS (MS) and LC–MS/MS multi-residue analysis of 221 pesticides and 26 mycotoxins in rice'. Oral presentation at the *1st Latin American Pesticide Residue Workshop (LAPRW 2007) – Pesticides in Food and Environmental Samples*, 6–11 May 2007, Federal University of Santa Maria, RS, Brazil, book of abstracts, p. 32.

Plattner R D and Maragos C M (2003), 'Determination of deoxynivalenol and nivalenol in corn and wheat by liquid chromatography with electrospray mass spectrometry', *J. AOAC Int.*, **86**, 61–5.

Plattner R D, Bennett G A and Stubblefield R D (1984), 'Identification of aflatoxins by quadrupole mass spectrometry/mass spectrometry', *J. AOAC*, **67**, 734–8.

Polizzi V, Delmulle B, Adams A, Moretti A, Susca A, Picco A M, Rosseel Y, 't Kindt R, Van Bocxlaer J, De Kimpe N, Van Peteghem C and De Saeger S (2009), 'JEM Spotlight: Fungi, mycotoxins and microbial volatile organic compounds in mouldy interiors from water-damaged buildings', *J. Environ. Monit.*, **10**, 1849–58.

Rajakyla E, Laasasenaho K and Sakkers P J D (1987), 'Determination of mycotoxins in grain by high-performance liquid-chromatography and thermospray liquid-chromatography mass-spectrometry', *J. Chromatogr.*, **384**, 391–402.

Rasmussen R R, Storm I M L D, Rasmussen P H and Nielsen K F (2009), 'Determination of mycotoxins in maize silage by LC–MS/MS using a rapid extraction procedure'. Poster 213 at the *First International Society for Mycotoxicology Conference*, 9–11 September 2009, Tulln, Austria, book of abstracts, available from: http://www.ism2009.at/ISM2009_posters.pdf, p. 238, accessed: 29 May 2010.

Rasmussen R R, Storm I M L D, Rasmussen P H, Smedsgaard J and Nielsen K F (2010), 'Multi-mycotoxin analysis of maize silage by LC–MS/MS', *Anal. Bioanal. Chem.*, **397**, 765–76.

Razzazi-Fazeli E (2006), 'Analyses and identification of mycotoxins in food using LC–MS', in *The Encyclopedia of Mass Spectrometry Vol. 8: Hyphenated Methods*, Niessen W M A (ed.), Elsevier, Amsterdam, 707–21.

Razzazi-Fazeli E, Böhm J and Luf W (1999), 'Determination of nivalenol and deoxynivalenol in wheat using liquid chromatography–mass spectrometry with negative ion atmospheric pressure chemical ionisation', *J. Chromatogr. A*, **854**, 45–55.

Razzazi-Fazeli E, Rabus B, Cecon B and Böhm J (2002), 'Simultaneous quantification of A-trichothecene mycotoxins in grains using liquid chromatography atmospheric pressure chemical ionisation mass spectrometry', *J. Chromatogr. A*, **968**, 129–42.

Ren Y P, Zhang Y, Shao S L, Cai Z X, Feng L, Pan H F and Wang Z G (2007), 'Simultaneous determination of multi-component mycotoxin contaminants in foods and feeds by ultra-performance liquid chromatography tandem mass spectrometry', *J. Chromatogr. A*, **1143**, 48–64.

Richard E, Heutte N, Sage L, Pottier D, Bouchart V, Lebailly P and Garon D (2007), 'Toxigenic fungi and mycotoxins in mature corn silage', *Food Chem. Toxicol.*, **45**, 2420–5.

Romero-González R, Martínez Vidal J L, Aguilera-Luiz M M and Garrido Frenich A (2009), 'Application of conventional solid-phase extraction for multimycotoxin analysis in beers by ultrahigh-performance liquid chromatography–tandem mass spectrometry', *J. Agric. Food Chem.*, **57**, 9385–92.

Royer D, Humpf H-U and Guy P A (2004), 'Quantitative analysis of *Fusarium* mycotoxins in maize using accelerated solvent extraction before liquid chromatography atmospheric pressure chemical ionization tandem mass spectrometry', *Food Addit. Contam.*, 21, 678–92.

Rundberget T and Wilkins A L (2002), 'Determination of *Penicillium* mycotoxins in foods and feeds using liquid chromatography–mass spectrometry', *J. Chromatogr. A*, **964**, 189–97.

Rychlik M and Asam S (2008), Review, 'Stable isotope dilution assays in mycotoxin analysis', *Anal. Bioanal. Chem.*, **390**, 617–28.

Rychlik M and Schieberle P (1998), 'Synthesis of C-13-labelled patulin [4-hydroxy-4*H*-furo[3,2-c]pyran-2(6H)-one] to be used as internal standard in a stable isotope dilution assay', *J. Agric. Food Chem.*, **46**, 5163–9.

Rychlik M and Schieberle P (1999), 'Quantification of the mycotoxin patulin by a stable isotope dilution assay', *J. Agric. Food Chem.*, **47**, 3749–55.

Santini A, Ferracane R, Somma M C, Aragón A and Ritieni A (2009), 'Multitoxin extraction and detection of trichothecenes in cereals: an improved LC–MS/MS approach', *J. Agric. Food Chem.*, **89**, 1145–53.

Schuhmacher R, Sulyok M and Krska R (2008), 'Recent developments in the application of liquid chromatography–tandem mass spectrometry for the determination of organic residues and contaminants', *Anal. Bioanal. Chem.*, **390**, 253–6.

Senyuva H Z and Gilbert J (2008), 'Identification of fumonisin B-2, HT-2 toxin, patulin, and zearalenone in dried figs by liquid chromatography-time-of-flight mass spectrometry and liquid chromatography–mass spectrometry', *J. Food Prot.*, **71**, 1500–4.

Senyuva H Z, Gilbert J and Öztürkoglu S (2008a), 'Rapid analysis of fungal cultures and dried figs for secondary metabolites by LC/TOF-MS', *Anal. Chim. Acta*, **617**, 97–106.

Senyuva H Z, Gilbert J, Özcan S and Gürel N (2008b), 'Rapid LC and LC/MS for routine analysis of mycotoxins in foods', *World Mycotoxin J.*, **1**, 229–35.

Sewram V, Nair J J, Nieuwoudt T W, Leggott N L and Shephard G S (2000), 'Determination of patulin in apple juice by high-performance liquid chromatography–atmospheric pressure chemical ionization mass spectrometry', *J. Chromatogr. A*, **897**, 365–74.

Sforza S, Dall'Asta C and Marchelli R (2006), 'Recent advances in mycotoxin determination in food and feed by hyphenated chromatographic techniques/mass spectrometry', *Mass Spectrom. Rev.*, **25**, 54–76.

Smedsgaard J and Frisvad J C (1996),'Using direct electrospray mass spectrometry in taxonomy and secondary metabolite profiling of crude fungal extracts', *J. Microbiol. Methods*, **25**, 5–17.

Song H-H, Lee H-S, Jeong J-H, Park H-S and Lee C (2008), 'Diversity in beauvericin and enniatins H, I, and MK1688 by *Fusarium oxysporum* isolated from potato', *Int. J. Food Microbiol.*, **122**, 296–301.

Songsermsakul P and Razzazi-Fazeli E (2008), 'A review of recent trends in applications of liquid chromatography–mass spectrometry for determination of mycotoxins', *J. Liq. Chrom. Rel. Technol.*, **31**, 1641–86.

Sørensen L K and Elbæk T H (2005), 'Determination of mycotoxins in bovine milk by liquid chromatography tandem mass spectrometry', *J. Chromatogr. B*, **820**, 183–96.

Sørensen L M, Nielsen K F, Jacobsen T, Koch A G, Nielsen P V and Frisvad J C (2008), 'Determination of mycophenolic acid in meat products using mixed mode reversed phase-anion exchange clean-up and liquid chromatography-high-resolution mass spectrometry', *J. Chromatogr. A*, **1205**, 103–8.

Sørensen J L, Phipps R K, Nielsen K F, Schroers H J, Frank J and Thrane U (2009), 'Analysis of *Fusarium avenaceum* metabolites produced during wet apple core rot', *J. Agric. Food Chem.*, **57**, 1632–9.

Sospedra I, Blesa J, Soriano J M and Mañes J (2010), 'Use of the modified quick easy cheap effective rugged and safe sample preparation approach for the simultaneous analysis of type A- and B-trichothecenes in wheat flour', *J. Chromatogr. A*, **1217**, 1437–40.

Spanjer M C, Rensen P M and Scholten J M (2003), 'Single run LC–MS/MS analysis of mycotoxins subject to actual and upcoming EU legislation in one sample extract'. Poster presentation at the *Second World Mycotoxin Forum*, 17–18 February 2003, Book of Abstracts, p. 95, Noordwijk aan Zee, The Netherlands.

Spanjer M C, Scholten J M, Kastrup S, Jörissen U, Schatzki T F and Toyofuku N (2006), 'Sample comminution for mycotoxin analysis: dry milling or slurry mixing?', *Food Addit. Contam.*, **23**, 73–83.

Spanjer M C, Rensen P M, Scholten J M and De Kok A (2007), 'Multi-mycotoxin analysis by UPLC–MS/MS'. Oral presentation at the *XIIth IUPAC International Symposium on Mycotoxins and Phycotoxins*, Istanbul, Turkey, 21–25 May 2007.

Spanjer M C, Rensen P M and Scholten J M (2008), 'LC–MS/MS multi-method for mycotoxins after single extraction, with validation data for peanut, pistachio, wheat, maize, cornflakes, raisins and figs', *Food Addit. Contam.*, **25**, 472–489.

Stahnke H, Reemtsma T and Alder L (2009), 'Compensation of matrix effects by postcolumn infusion of a monitor substance in multiresidue analysis with LC–MS/MS', *Anal. Chem.*, **81**, 2185–92 and supporting information.

Sulc M, Peslova K, Zabka M, Hajduch M and Havlicek V (2009), 'Biomarkers of *Aspergillus* spores: Strain typing and protein identification', *Int. J. Mass Spectrom. Microbiol.*, **280**, 162–8.

Sulyok M, Berthiller F, Krska R and Schuhmacher R (2006), 'Development and validation of a liquid chromatography/tandem mass spectrometric method for the determination of 39 mycotoxins in wheat and maize', *Rapid Commun. Mass Spectrom.*, **20**, 2649–59.

Sulyok M, Krska R and Schuhmacher R (2007a), 'A liquid chromatography/tandem mass spectrometric multi-mycotoxin method for the quantification of 87 analytes and its application to semi-quantitative screening of moldy food samples', *Anal. Bioanal. Chem.*, **389**, 1505–23.

Sulyok M, Krska R and Schuhmacher R (2007b), 'Application of a liquid chromatography–tandem mass spectrometric method to multi-mycotoxin determination in raw cereals and evaluation of matrix effects', *Food Addit. Contam.*, **24**, 1184–95.

Sulyok M, Krska R and Schuhmacher R (2010), 'Application of an LC–MS/MS based multimycotoxin method for the semi-quantitative determination of mycotoxins occurring in different types of food infected by molds', *Food Chem.*, **119**, 408–16.

Suman M and Catellani D (2008), 'Ion trap LC–MS/MS as a valid multi-method to determine trichothecenes and zearalenone in the food industry', *World Mycotoxin J.*, **1**, 255–62.

Tanaka H, Takino M, Sugita-Konishi Y and Tanaka T (2006), 'Development of a liquid chromatography/time-of-flight mass spectrometric method for the simultaneous determination of trichothecenes, zearalenone and aflatoxins in foodstuffs', *Rapid Commun. Mass Spectrom.*, **20**, 1422–8.

Tiebach R, Blaas W, Kellert M, Steinmeyer S and Weber R (1985), 'Confirmation of nivalenol and deoxynivalenol by online liquid-chromatography mass-spectrometry and gas-chromatography mass-spectrometry – comparison of methods', *J. Chromatogr.*, **318**, 103–11.

Trebstein A, Lauber U and Humpf H-U (2009), 'Analysis of *Fusarium* toxins via HPLC–MS/MS multimethods: matrix effects and strategies for compensation', *Mycotoxin Res.*, **25**, 201–13.

Tuomi T, Saarinen L and Reijula K (1998), 'Detection of polar and macrocyclic trichothecene mycotoxins from indoor environments', *The Analyst*, **123**, 1835–41.

Tuomi T, Reijula K, Johnsson T, Hemminki K, Hintikka E-L, Lindroos O, Kalso S, Koukila-Kähkölä P, Mussalo-Rauhamaa H and Haahtela T (2000), 'Mycotoxins in crude building materials from water-damaged buildings', *Appl. Environ. Microbiol.*, **66**, 1899–904.

Tuomi T, Johnsson T, Hintikka E-L and Reijula K (2001), 'Detection of aflatoxins (G(1-2), B(1-2)), sterigmatocystin, citrinine and ochratoxin A in samples contaminated by microbes', *The Analyst*, **126**, 1545–50.

Uhlig S and Ivanova L (2004), 'Determination of beauvericin and four other enniatins in grain by liquid chromatography–mass spectrometry', *J. Chromatogr. A*, **1050**, 173–8.

Uhlig S, Vikøren T, Ivanova L and Handeland K (2007), 'Ergot alkaloids in Norwegian wild grasses: a mass spectrometric approach', *Rapid Commun. Mass Spectrom.*, **21**, 1651–60.

Uraguchi K, Tatsuno T, Sakai F, Tsukioka M, Sakai Y, Yonemitsu O, Ito H, Miyake M, Enomoto M, Shikata T and Ishiko T (1961), 'Isolation of two toxic agents, luteoskyrin and chlorine-containing peptide from the metabolites of *Penicillium islandicum* spp., with some properties thereof', *Jpn. J. Exp. Med.*, **31**, 19–41.

Vaclavik L, Zachariasova M, Hrbek V and Hajslova J (2010), 'Analysis of multiple mycotoxins in cereals under ambient conditions using direct analysis in real time (DART) ionization coupled to high resolution mass spectrometry', *Talanta*, **82**, 1950–7.

Valentine N B, Wahl J H, Kingsley M T and Wahl K L (2002), 'Direct surface analysis of fungal species by matrix-assisted laser desorption/ionization mass spectrometry', *Rapid Commun. Mass Spectrom.*, **16**, 1352–7.

Varga E, Sulyok M, Schuhmacher R and Krska R (2009), 'Validation and application of an LC–MS/MS based method for multi-mycotoxin analysis in different nuts and dried fruits'. Poster 236 at the *First International Society for Mycotoxicology Conference*, 9–11 September 2009, Tulln, Austria, Book of abstracts, available from: http://www.ism2009.at/ISM2009_posters.pdf, p. 261, accessed: 29 May 2010.

Ventura M, Guillén D, Anaya I, Broto-Puig F, Lliberia J L, Agut M and Comellas L (2006), 'Ultra-performance liquid chromatography/tandem mass spectrometry for the simultaneous analysis of aflatoxins when B_1, B_2, G_1 and G_2 and ochratoxin A in beer', *Rapid Commun. Mass Spectrom.*, **20**, 3199–204.

Vishwanath V, Sulyok M, Labuda R, Bicker W and Krska R (2009), 'Simultaneous determination of 186 fungal and bacterial metabolites in indoor matrices by liquid chromatography/tandem mass spectrometry', *Anal. Bioanal. Chem.*, **395**, 1355–72.

Watanabe M (2008), 'Production of mycotoxins by *Penicillium expansum* inoculated into apples', *J. Food Prot.*, **71**, 1714–9.

Welham K J, Domin M A, Johnson K, Jones L and Ashton D S (2000), 'Characterization of fungal spores by laser desorption/ionization time-of-flight mass spectrometry', *Rapid Commun. Mass Spectrom.*, **14**, 307–10.

Zachariasova M, Lacina O, Malachova A, Kostelanska M, Poustka J, Godula M and Hajslova J (2010a), 'Novel approaches in analysis of Fusarium mycotoxins in cereals employing ultra performance liquid chromatography coupled with high resolution mass spectrometry', *Anal. Chim. Acta*, **662**, 51–61.

Zachariasova M, Cajka T, Godula M, Malachova A, Veprikova Z and Hajslova J (2010b), 'Analysis of multiple mycotoxins in beer employing (ultra)-high-resolution mass spectrometry', *Rapid Commun. Mass Spectrom.*, **24**, 3357–67.

Zhang H, Grubb M, Wu W, Josephs J and Humphreys W G (2009), 'Algorithm for thorough background subtraction of High-Resolution LC/MS data: Application to obtain clean product ion spectra from nonselective collision-induced dissociation experiments', *Anal. Chem.*, **81**, 2695–700.

Zöllner P, Berner D, Jodlbauer J and Lindner W (2000), 'Determination of zearalenone and its metabolites alpha- and beta-zearalenol in beer samples by high-performance liquid chromatography–tandem mass spectrometry', *J. Chromatogr. B*, **738**, 233–41.

Zöllner P and Mayer-Helm B (2006), 'Trace mycotoxin analysis in complex biological and food matrices by liquid chromatography–atmospheric pressure ionisation mass spectrometry', *J. Chromatogr. A*, **1136**, 123–69.

5

Immunochemical methods for rapid mycotoxin detection in food and feed

I. Y. Goryacheva, Saratov State University, Russia and S. De Saeger, Ghent University, Belgium

Abstract: This chapter discusses immunochemical methods for rapid mycotoxin detection. The first section reviews the production and characterization of specific anti-mycotoxin antibodies. The second subchapter is devoted to the specificity of immunochemical methods. Further sections then discuss principles and applications of microtiter plate enzyme-linked immunosorbent assays and non-instrumental rapid tests. A substantial amount of information has been collected regarding immunochemical tests for detection of one or two mycotoxins, while rapid immunochemical tests for multiple mycotoxins are still scarce.

Key words: antibody, ELISA, immunoassay, mycotoxin, rapid test, visual detection.

5.1 Introduction

Mycotoxins can contaminate commodities in the different stages of processing, in the field, at harvest, during storage, transportation and treatment. For this reason it is important to analyse an appropriate number of samples at critical points in the food chain. Screening methods should be fast, simple, low-cost and suitable for high-throughput screening.

Different kinds of rapid methods, based on thin layer chromatography (TLC) or fluorescence detection were developed, but the absolute leaders in the group of analytical methods for screening (rapid detection) of mycotoxins are immunochemical methods with high sensitivity and selectivity caused by specific antibodies. Application of immunoassays for mycotoxin screening is now a common analytical practice. The most widespread method is the enzyme-linked immunosorbent assay (ELISA). For on-site screening, different kinds of membrane-based methods were developed. The main trends for research in this field are sensitivity improvement, matrix effect reduction, simplification, shorter time of analysis and multi-analyte applications.

The advantages of immunochemical methods are as follows:

1. simplicity and speed of operation
2. applicability to routine analyses under field conditions
3. possibility of automation
4. high reliability of determinations
5. simple and fast sample preparation.

Moreover, these methods require simple instruments (the majority of immunochemical methods are based on photometric, fluorimetric, luminescence or electrochemical detection) or do not require instrumentation (rapid tests with visual detection for qualitative/semiquantitative evaluation). Common disadvantages of methods based on the antigen–antibody interaction are:

1. possible non-specific binding of components
2. non-reproducibility of the sorption processes
3. possible cross-reaction with related compounds.

In some cases, especially for groups of related mycotoxins, such as aflatoxins, fumonisins and trichothecenes, an overestimation of levels is reported in comparison with chromatographic methods. Among the drawbacks are time and costs of the antibody production and the possible interference of matrix components.

Immunomethods for rapid detection of mycotoxins are summarized in recent reviews of immunochemical methods (Maragos, 2004; Zheng *et al.*, 2006; Goryacheva *et al.*, 2007a; Krska and Molinelli, 2009; Goryacheva *et al.*, 2009b) and analytical methods in general (Gilbert, 1999; Krska *et al.*, 2005; Krska *et al.*, 2008; Shephard, 2008; Turner *et al.*, 2009). In addition, reviews have been published devoted to the determination of single mycotoxins or groups of related toxins, such as ochratoxin A (OTA) (Visconti and De Girolamo, 2005), trichothecenes (Krska *et al.*, 2001; Koch, 2004; Schneider *et al.*, 2004), aflatoxins (Maragos 2002) and citrinin (Xu *et al.*, 2006).

5.2   Antibody production and characterization

Antibodies are the main driving force of all immunochemical methods, including ELISA. The quality of the antibodies defines the specificity and sensitivity of analysis. Antibodies are proteins from the class of immunoglobulin (with molecular weight over 150 000 Da) with specific binding sites for the antigen. Antibodies are produced as a result of a defence reaction to an immunogen by the immune system of vertebrates. As antigens with molecular weights below 1000 Da (e.g. mycotoxins) are not immunogenic, the suitable immunogen can be synthesized by coupling the analyte of interest or analogue to a carrier high-molecular weight molecule (protein). Usually, bovine serum albumin (BSA), ovalbumin (OVA), or keyhole limpet hemocyanine (KLH) are used as carrier proteins. For antibody production, laboratory animals (rabbits, mice, sheep, goats, horses, chickens, etc) are immunized with the mycotoxin–protein conjugate (immunogen). Recognition of the antibody is based on the spatial complement of specific chemical groups (the epitopes) on the antigen, not on the whole antigen. The affinity of antibody is strongly dependent on the immunogen structure and especially on the hapten derivative design. To develop antibodies with high affinity and specificity, the structure of the mycotoxin should be affected as little as possible by the conjugation procedure. Furthermore, the antibody is likely to recognize the toxin preferentially at the moiety opposite to where it has been attached to the immunogenic protein (Cervino *et al.*, 2007).

Polyclonal antibodies are separated from the blood of animals. Chicken immunization allows eggs to be used as a source of antibody. This method is less traumatic for laboratory animals. Polyclonal antibodies have heterogeneous physicochemical properties, namely, specificity and affinity (Frank, 2002). They are easy to produce, but contain limited amounts in each batch and show significant batch to batch variations.

This limitation was eliminated by introducing the hybridoma technology for antibody production. Monoclonal antibodies have identical specificity and affinity. Screening and selection of antibody clones, which show affinity for particular epitopes, make it possible to select an antibody with suitable physicochemical properties.

Hybridoma screening is a key step in the successful generation of high-affinity analyte-specific monoclonal antibodies. For this purpose ELISA is usually used (Fig. 5.1). The most common way is indirect competitive ELISA (*ic*ELISA) screening with immobilization of a hapten–protein conjugate and use of an enzyme-labelled secondary antibody as tracer. But, direct competitive ELISA (*dc*ELISA) with antibody immobilization and application of a hapten–enzyme conjugate as tracer could also be used for routine hybridoma screening. *dc*ELISA gives useful screening results for the different supernatant dilutions chosen; however, the sensitivity of the direct ELISA screening is generally lower compared to indirect ELISA (Cervino *et al.*, 2008). A modified two-step screening procedure was proposed by Zhang *et al.* (2009) for production of ultrasensitive generic monoclonal antibodies against major aflatoxins. The first step was indirect ELISA and resulted in positive hybridomas and hapten-specific antibodies. A modified *ic*ELISA was the second step, in which aflatoxins B_1, B_2, G_1 and G_2 were all used as competitors.

Recently Reiger *et al.* (2009) proposed an automated screening method that makes use of antibody microarrays. The hybridoma cell supernatant samples were printed on a glass chip initially coated with capture antibodies. The affinity of monoclonal anti-aflatoxin antibodies was displayed by the specific binding of an aflatoxin B_2–horseradish peroxidase conjugate and was performed in an automated fashion using a chemiluminescence readout system. The quality of the

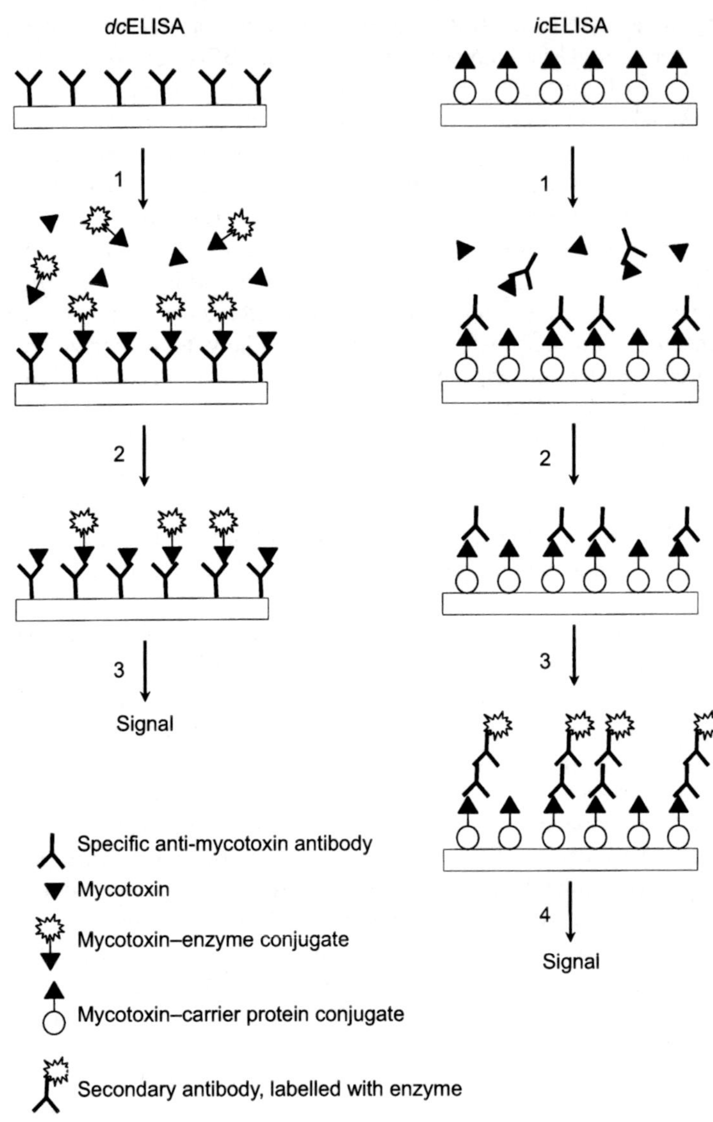

Fig. 5.1 Competitive ELISA principle. Direct format (*dc*ELISA): 1 – application of sample with addition of mycotoxin–enzyme conjugate, 2 – washing step, 3 – substrate application; indirect format (*ic*ELISA): 1 – application of sample with addition of specific antibody, 2 – washing step, 3 – application of secondary antibody, 4 – substrate application.

generated data was comparable to data generated by a previously optimized microplate-based immunoassay method.

For monoclonal antibodies, it may be difficult initially to obtain the correct affinity and specificity, but they are preferred for commercial use as they have uniform affinity and specificity and can be produced repeatedly in sufficient quantities. Both polyclonal and monoclonal antibodies are used for mycotoxin analysis. The applicability of the standard ELISA procedure to individual or groups of related mycotoxins depends on the cross-reactivity of the antibodies.

Modern technologies for antibody production are also applied to mycotoxin detection. The selection of synthetic antibody fragments from large phage libraries has become a common technique for the generation of specific antibodies to antigens. In this way a single variable chain antibody (scFv) against aflatoxin B_1 (AFB_1) (Moghaddam *et al.*, 2001), deoxynivalenol (DON) (Choi *et al.*, 2004; Wang *et al.*, 2007), zearalenone (ZEA) (Wang *et al.*, 2008) and fumonisin B_1 (FB_1) (Lauer *et al.*, 2005) has been produced. A monoclonal anti-anti-idiotype antibody against FB_1 was produced from the hybridoma cell, which was generated by the fusion of myeloma cells with spleen cells isolated from a mouse that had been immunized with the Fab fragments of affinity-purified anti-idiotype antibodies (Yu and Chu, 1999).

5.3   Specificity of immunochemical methods for rapid mycotoxin detection in food and feed

Performance characteristics (first of all specificity) of immunochemical methods are mainly influenced by the antibody quality. As mycotoxins have different chemical structures, non-related compounds usually do not affect each other during the immunoassay. But for related mycotoxins, such as the aflatoxin, fumonisin, or trichothecene family, cross-reactivity could influence results.

The most important question concerning specificity is for related mycotoxins with different toxicity. For example, AFB_1 is the most potent of all aflatoxins and is generally found in the highest concentration in food and feed samples. Regulatory requirements are established for AFB_1, total aflatoxins (sum of AFB_1, aflatoxin B_2 (AFB_2), aflatoxin G_1 (AFG_1) and aflatoxin G_2 (AFG_2)) and also for aflatoxin M_1 (AFM_1) in milk and milk products. According to these requirements ELISAs for total aflatoxins and those specific to AFB_1 and AFM_1 are commercially available. In most cases, ELISAs based on AFB_1-specific antibodies are used for total mycotoxin assay. Most reported anti-aflatoxin antibodies from immunization with AFB_1 conjugates showed some cross-reactivity with AFB_2, AFG_1 and AFG_2. It is interesting that almost all of these antibodies have the lowest cross-reactivity with AFG_2 (Chu *et al.*, 1977; Devi *et al.*, 1999; Li *et al.*, 2009). Cervino *et al.*, 2007 proposed a new approach for immunogen synthesis to obtain antibodies with significantly different cross-reactivities for aflatoxins B and G. Further, anti-AFB_1 antibodies could be used for AFM_1 individual detection in milk and milk products (Goryacheva *et al.*, 2009a) in addition to anti-AFM_1 antibody-based immunoassays. Antibodies were produced and applied for ELISA specific to AFB_1 (Lee and Rachmawati, 2006), AFB_2 (Hastings *et al.*, 1988), AFG_1 (Chu *et al.*, 1985), total aflatoxin (Zhang and Chu, 1989) and some major metabolites such as AFB_{2a} (Gaur *et al.*, 1981), AFQ_1 (Fan *et al.*, 1984) and AFM_1 (Haltzapple *et al.*, 1996).

Similarly, FB_1 specific antibodies were used for total fumonisin assay. Among the fumonisins, FB_1 is the most abundant in foods and is known to be the most potent. Kim *et al.* (1998) reported that the cross-reactivities of the antibody to fumonisin B_2 (FB_2) and B_3 (FB_3) were 70 and 166%, respectively.

Different approaches were applied to modify immunoassay specificity , such as variation of the haptens in the immunogen and enzyme conjugate. To obtain a group-specific assay, different but related haptens (mycotoxins) could be used for immunogen and enzyme conjugate synthesis. For example, anti-fumonisin B_4 (FB_4) polyclonal antibody was used for *dc*ELISA with different enzyme conjugates. In the FB_1-HRP-based *dc*ELISA, the relative cross-reactivities towards FB_1, FB_2, FB_3 and FB_4 were 58.5, 309.5, 58.5 and 100%, respectively. In the FB_3-HRP-based *dc*ELISA, the relative cross-reactivities towards FB_1, FB_2, FB_3 and FB_4 were 74, 280, 70 and 100%, respectively (Christensen *et al.*, 2000). Using an FB_1 monoclonal antibody and FB_1–HRP conjugate resulted in relative cross-reactivities with FB_1, FB_2 and FB_3 of 100, 91.8 and 209, respectively (Barna-Vetro *et al.*, 2000). Modification of various groups in the synthesis of the enzyme hapten conjugate allowed a system to be chosen with lower selectivity for FB_1; this provided an opportunity to determine FB_4 selectively with the use of two antibodies (Christensen *et al.*, 2000).

Antibodies with various specificities to ZEA were obtained by varying the immunogen. This enabled either the simultaneous determination of ZEA and α-zearalenol (Burkin *et al.*, 2002a) or the selective determination of only ZEA (Burkin *et al.*, 2002b).

Quantification of DON by immunoassays is complicated by cross-reaction with 3-acetyl-DON and 15-acetyl-DON (Schuhmacher *et al.*, 1997; Josephs *et al.*, 2001; Cavaliere *et al.*, 2005). To determine DON together with its related derivatives, preliminary acetylation of toxins in the extracts followed by determination of the sum of DON acetylderivatives was suggested (Kononenko *et al.*, 1999a). The specificity of antibodies to nivalenol (NIV), where toxicity can be compared to DON, is also low and this requires an additional derivatization of the toxin (Yoshizawa *et al.*, 2004). Monoclonal antibodies produced with the nivalenol–glycine immunogen made it possible to determine DON and NIV selectively without preliminary modification of the toxins (Maragos *et al.*, 2006).

Ochratoxin A is the most widespread in its group of related ochratoxins. Other related mycotoxins (ochratoxin B (OTB), ochratoxin C, ochratoxin α, OTA methyl ester) are very rare and detected only in low concentrations. Nevertheless, there are data that show that the presence of OTB can influence the results for OTA detection (Visconti and De Girolamo, 2005).

5.4   Microtiter plate enzyme-linked immunosorbent assay for rapid mycotoxin detection in food and feed

5.4.1   Introduction to enzyme-linked immunosorbent assay

The microtiter plate ELISA format occupies a leading position as a rapid test for mycotoxins. ELISA tests are commercially available for the determination of individual mycotoxins (AFB_1, AFM_1, DON, OTA, ZEA, T-2 toxin, citrinin) in some foodstuffs as well as for groups of related mycotoxins (aflatoxins (AFB_1, AFB_2, AFG_1, AFG_2), fumonisins (FB_1, FB_2 and FB_3), trichothecenes); ELISA test kits consist of the required reagents and materials. They are useful tools for screening and quantification and offer benefits with respect to speed and sensitivity.

Examples of ELISA test kit validation for mycotoxin detection were presented by several authors (Sutikno *et al.*, 1996; Bird *et al.*, 2002; Yoshizawa *et al.*, 2004; Zheng *et al.*, 2005a,b; Wang *et al.*, 2007; Rubio *et al.*, 2009; Lee *et al.*, 2005). Validation generally includes evaluation of the analytical range, detection limit, sensitivity, accuracy and precision of the procedure in contaminated samples of different food matrices. Commercial ELISA test kits have been validated by organizations such as the Association of Official Analytical Chemists (AOAC) International and the Grain Inspection, Packers and Stockyards Administration (GIPSA).

ELISA is a heterogeneous technique mainly performed in 96-well polystyrene plates (or 384-well plates). One of the immunoreagents is usually immobilized on the bottom and walls of the wells for separation of bound and free reagents. This heterogeneous principle of ELISA makes it possible to separate effectively the bound and free components of an analytical system after the immunochemical reaction.

Because mycotoxins are monovalent antigens, competitive ELISA techniques are usually used. Possible formats are presented in Fig. 5.1. Direct competitive ELISA (*dc*ELISA) is most commonly used. In this format the analyte competes with an enzyme labelled analyte to bind with a restricted amount of antibodies immobilized on the wells. After incubation unbound compounds are removed by washing and chromogenic substrate is added. The enzymatic activity in each well is inversely proportional to the mycotoxin concentration. In the case of indirect competitive ELISA (*ic*ELISA), the analyte or its analogue, conjugated with a macromolecular carrier (e.g. BSA or OVA) is coated onto the well during incubation. Then the sample extract and specific antibody are added to each well. The analyte present in the sample and the immobilized analyte compete to bind with the antibody in solution. After a washing step, the amount of bound specific antibody is detected by a secondary antibody, labelled with an enzyme. This approach makes it possible to simplify immunoreagents preparation (enzyme-labelled secondary antibodies, e.g. labelled with horseradish peroxidase (HRP) or alkaline phosphatase (AP) are commercially available), but includes an additional assay step. To eliminate this additional step, direct labelling of the specific antibody is also possible.

Application of a non-competitive open sandwich ELISA was described for ZEA detection. The fundamental limitation that the antigen to be measured by sandwich format must be large enough to have at least two epitopes to be captured was circumvented by a new immunoassay approach. This assay exploits reassociation of the generally weak antibody variable region V_H–V_L complex by a bridging antigen. Using immobilized V_L and enzyme-tagged V_H fragments, the ZEA concentration (limit of detection (LOD) ~0.1 ng ml^{-1}) could be determined in

Table 5.1   Main characteristics of developed ELISAs for mycotoxin determination in different food and feed samples

Mycotoxin	Sample	ELISA	Antibody	IC50 (μg l^{-1})	LOD (μg kg^{-1})	Range (μg kg^{-1})	References
Aflatoxins' sum (B$_1$, B$_2$, G$_1$, G$_2$)	Grain and grain products	Direct	n.d.*	n.d.	2.5	4–40	Zheng et al., 2005b
Aflatoxin B$_1$	Standard solutions	Indirect	P** M***	n.d.	0.4 μg l^{-1} 0.1 μg l^{-1}	n.d.	Burkin et al., 2000 Burkin et al., 2000a
	Corn, peanut	n.d.	M	n.d.	4	n.d.	Lipigorngosonyz et al., 2003
	Peanut	Indirect	P	n.d.	0.5	20–650 ng l^{-1}	Asis et al., 2002
	Feed	Direct	P	0.85	0.18	n.d.	Lee and Rachmawati, 2006
	Corn	Direct	n.d.	240	2	2.4–4000	Neagu et al., 2009
Aflatoxin M$_1$	Milk and milk-based confectionery	Indirect	P	0.5	n.d.	0.01–10 μg l^{-1}	Thirumala-Devi et al., 2002
	Milk	Direct	P	n.d.	0.25 μg l^{-1}	n.d.	Kim et al., 2000
Ochratoxin A	Rice, wheat, corn, feed	n.d.	M	n.d.	0.15 μg l^{-1}	0.2–8 μg l^{-1}	Liu et al., 2007
	Barley, oat, wheat, corn, rice, raisins, grape juice, beer	Direct	P	0.07	n.d.	n.d.	Wang X H et al., 2007
	Wheat	Indirect	P	1.018	0.02 μg l^{-1}	0.02–400 μg l^{-1}	Huang et al., 2006
	Agricultural commodities	Direct	P	0.9	n.d.	n.d.	Yu et al., 2005
	Chilli	Direct	P	5	0.1 μg l^{-1}	n.d.	Thirumala-Devi et al., 2000
	Grain	Indirect	P	n.d.	20	0.4–50 μg l^{-1}	Kononenko et al., 2000
	Corn, milo, barley, wheat, soybeans and green coffee	Direct	n.d.	n.d.	2	2–40	Zheng et al., 2005a
Ochratoxin B	Standard solutions	Indirect	M	500 nM	27 nM	n.d.	Heussner et al., 2007
Fumonisin B$_1$	Corn, barley, rice, oat, peanut and sorghum	Direct	P	0.32	0.09 μg l^{-1}	0.14–0.9 μg l^{-1}	Quan et al., 2006
	Wheat	Direct	P	1.1	0.2 μg l^{-1}	n.d.	Wang S H et al., 2007

	Grain	Direct	M	29.1	n.d.	50–2000 µg l^{-1}	Yu and Chu, 1999a
		Indirect		20.1			
	Corn and corn-based products	Indirect	P	n.d.	200	40–1000 µg l^{-1}	Kononenko *et al.*, 1999
T-2 toxin	Grain	Indirect	P	n.d.	30	30–1000	Kononenko *et al.*, 1999b
	Wheat	Sandwich	n.d.	8.2	~1 µg l^{-1}	3–100 µg l^{-1}	Suzuki *et al.*, 2007
	Barley	Indirect	M	n.d.	0.2	25–1000	Tanaka *et al.*, 1995
Zearelenone	Feed	Indirect	P	n.d.	1 µg l^{-1}	1–100 µg l^{-1}	Burkin *et al.*, 2000d
	Standard solution	Indirect	P yolk	n.d.	n.d.	10–200 µg l^{-1}	Pichler *et al.*, 1998
Deoxynivalenol	Standard solutions	Direct	M	15.8	n.d.	n.d.	Maragos *et al.*, 2006
		Indirect		16.5			
	Wheat	Direct	M	18	n.d.	50–5000	Maragos and McCormick, 2000
	Wheat	Direct	scFv ****	8.2	n.d.	n.d.	Wang S H *et al.*, 2007
Nivalenol	Standard solutions	Direct	M	27.5	n.d.	n.d.	Maragos *et al.*, 2006
		Indirect		33.4			
Roridin	Cereal straw	Indirect	P	n.d.	n.d.	0.2–40 µg l^{-1}	Burkin *et al.*, 2000b
Citrinin	Cereals, soybean, sunflower products, feed	Indirect	P	n.d.	n.d.	20–500	Kononenko and Burkin, 2007

*n.d. = not described; **P – polyclonal antibody; ***M – monoclonal antibody; ****scFv – single-chain antibody fragment.

a shorter time period than by using a conventional sandwich assay, owing to the omission of an incubation/washing cycle (Suzuki *et al.*, 2007).

For widespread mycotoxins, such as OTA and aflatoxins, many different ELISAs have been developed, while for less studied mycotoxins only a limited number exist. Table 5.1 summarizes the main characteristics of the developed ELISAs for mycotoxin detection.

The most commonly used enzyme label for ELISA is HRP which is inexpensive, stable and well-studied. Different substrates (e.g. 3,3',5,5'-tetramethylbenzidine (TMB), *o*-phenylendiamine, 2,2'-azinodi(3-ethylbenzothiazoline-6-sulphonate)) can be used with HRP and are combined with spectrophotometric, luminescent, chemiluminescent and electrochemical detection, resulting in sensitive assays. AP is rarely used, nonetheless its conjugates are very stable and their colorimetric and fluorimetric substrates provide sensitive assays (Negu *et al.*, 2009). Other enzyme labels, for example, penicillinase were also used (Paknejad *et al.*, 2008).

Mycotoxin immunoassays were not only applied to food and feed samples, but also to human tissue and body fluids from patients exposed to toxin-producing moulds in their environment. Human urine and methanol extracted tissues and sputum were examined for trichothecenes using competitive ELISA techniques. Trichothecene levels varied in urine, sputum and tissue biopsies (lung, liver, brain) from undetectable ($< 0.2\ \mu g\ kg^{-1}$) to levels up to $18\ \mu g\ kg^{-1}$. Negative control patients had no detectable mycotoxins in their tissues or fluids (Hooper *et al.*, 2009). An ELISA was also developed for satratoxin G and other macrocyclic trichothecenes associated with indoor air (Chung *et al.*, 2003). It is also important to mention that ELISA methods were developed for some mycotoxin adducts AFB_1–*N*7-guanine (Nayak *et al.*, 2001), AFB_1-DNA (Vidyasagar *et al.*, 1997), AFB_1–albumin (Scholl *et al.*, 2006) and sterigmatocystin–DNA (Olson and Chu, 1993).

5.4.2   Extraction and sample pretreatment

In contrast to chromatographic methods, ELISA is usually combined with simple and fast extraction, preconcentration and clean-up procedures, or often without preconcentration and clean up. Traditionally, extraction with aqueous methanol or acetonitrile followed by dilution of the extract with buffers provides sufficient elimination of matrix interferences. The amount of organic solvent directly influences the assay sensitivity and should therefore be optimized in the sample extracts. Usually ≤10% of organic solvent is used. Nevertheless, Lee and Rachmawati (2006) reported a rapid ELISA for AFB_1 screening tolerant to methanol (up to 60%). Non-organic extraction solutions were also used, for example for AfM_1 extraction from milk products (Anfossi *et al.*, 2008) and FB_1 extraction from corn (Yu and Chu, 1999). In the case where matrices are highly coloured or show a strong influence on the assay results, immunoaffinity columns (IAC) were used for clean up and analyte preconcentration resulting in increased sensitivity. For example, for AFB_1 detection in ground red pepper, a LOD of $0.025\ \mu g\ kg^{-1}$ was reached when applying IAC (Ardic *et al.*, 2008).

5.4.3   Assay time and detection systems

Standard ELISA procedures include several incubation steps, some being overnight incubations. This allows equilibrium to be reached between immunoreagents, resulting in good sensitivity and reproducibility of the assays. From this point of view it is difficult to attribute traditional microtiter plate ELISA as a rapid method. In some cases it sacrifices excellent sensitivity to get more rapid tests. Rapid ELISA systems were developed in which the first incubation step (immobilization of mycotoxin–protein conjugate, in the case of *ic*ELISA, or specific antibody, in the case of *dc*ELISA) is still performed overnight, while the following assay steps are reduced in time. A rapid microwell assay could be completed within 20 min with a LOD of 0.5 μg FB_1 per l (Wang *et al.*, 2006). Changing the incubation time influenced the assay characteristics. For example, reducing the time of two incubation steps from $60 + 30$ min to $5 + 10$ min resulted in a larger detection range (from 7.1–55.5 μg aflatoxin per kg sample up to 4.2–99.9 μg kg^{-1}) and a slight decrease insensitivity (Lee *et al.*, 2004). Sample extracts could be prepared within 5 min when using an in-house direct ELISA, while in case of a 96-well ELISA plate, 44 diluted sample extracts could be screened for AFB_1 in 30 min (Lipigorngosonyz *et al.*, 2003). Rapid commercial ELISA kits that can be performed within 15 min are also currently available.

The commonly used detection system in ELISA is based on the use of colorimetric readers. In the presence of an HRP label, TMB-based substrate turns blue. To stop colour development, a sulphuric acid solution is used and the blue colour turns into yellow which is measured at $\lambda = 450$ nm. In the case of on-site use of ELISA test kits, visual colour evaluation could be performed when comparing with standard solutions (Lipigorngosonyz *et al.*, 2003).

Replacement of colorimetric detection by chemiluminescent detection makes it possible to increase the sensitivity of mycotoxin detection, for example up to 10 times for FB_1 (Quan *et al.*, 2006). As the chemiluminescent signal could be immediately read out after addition of substrate, the assay time is less than with colorimetric ELISA. Applying an Eu^{3+} label in an indirect competitive format made it possible to detect fluorescence in the time-resolved mode. This approach results in higher sensitivity of OTA detection and a larger detection range (Huang *et al.*, 2006).

5.4.4   Comparison with other methods

Routinely, ELISA is used as a cost effective and rapid alternative for high performance liquid chromatography (HPLC) where large amounts of samples are screened to a single mycotoxin or group of related mycotoxins (group-specific ELISA). For this reason comparison studies of ELISA (both home-made and test kits) and HPLC results were performed (Kulisek and Hazebroek, 2000; Sydenham *et al.*, 1996; Zachariasova *et al.*, 2008). Some authors indicated a good correlation between both methods, but underestimation (Matrella *et al.*, 2006) or more often overestimation of the mycotoxin content determined by ELISA in comparison to HPLC was also observed. The low regression for HPLC–ELISA methods may be related, for example, to the absence of a clean-up step in the ELISA method (Nilfer and Boyacolu, 2002).

Overestimation is more significant in the presence of structurally related compounds in the case of for example AFB_1, DON or FB_1. The discrepancy may be due to chromatographic methods detecting each mycotoxin separately, while ELISA measures total cross-reacting mycotoxins. For example, with ELISA the fumonisin contents were 15–380% higher than with the HPLC method (Meister, 2001). Ono *et al.* (2000) reported a decreased correlation between ELISA and HPLC for samples with FB_1 levels above 10 µg g^{-1} in comparison with a contamination range of 0.08–10 µg g^{-1}. The relationship between sample handling and DON overestimation was demonstrated; a higher ELISA response was measured in an aqueous extract compared to one prepared by acetonitrile–water (Zachariasova *et al.*, 2008). This study also revealed that in cereal-based matrice, DON-3-glucoside was strongly cross-reacting in all examined commercial ELISA kits showing that masked mycotoxins also could contribute to the overestimation (see Chapter 15). Overestimation of ELISA results was shown not only for instrumental chromatographic methods, but also for TLC (Shelby *et al.*, 1994).

5.5   Non-instrumental rapid tests for mycotoxin detection in food and feed

5.5.1   Introduction to non-instrumental rapid tests

Non-instrumental test methods for the determination of mycotoxins are suitable for performing on-site analysis in the course of manufacture and storage of agricultural products. The possibility of performing screening under out-of-laboratory conditions even in the absence of a power supply will allow contaminated batches of agricultural products to be recognized rapidly and their combination with larger volumes of raw materials to be prevented during their subsequent processing, transportation and storage. The results obtained with the use of non-instrumental test methods are evaluated on a visual basis. In this context, the subjective perception of the person who interprets particular results is one of the main sources of error. Various labels are used to visualize the experimental results: enzymes, colloidal particles (as a rule, colloidal gold particles), fluorescent labels and liposomes containing solubilized dyes (Posthuma-Trumpie *et al.*, 2009). As in instrumental methods, competitive enzyme immunoassay is the main format.

Mostly, non-instrumental tests qualitatively estimate concentrations (yes/no), which characterize the presence (or absence) of the target analyte in a concentration higher than the regulated control level. As a rule, in the development of non-instrumental tests for the determination of mycotoxins, attempts are made to reach a detection limit that corresponds to the legislatively defined maximum permissible concentration of a given mycotoxin in the analysed product. The detection limit is established based on a well-pronounced and considerable decrease in the colour intensity of a test zone or (more often) based on its full colour suppression. A small number of tests imply a semiquantitative assessment based on a comparison between colour intensities. To simplify the interpretation of results, the majority of commercially produced tests include special control zones to confirm the working state of tests and/or to compare the colours of the control and test zones.

Generally, test methods for rapid screening imply a simple procedure for extraction with a mixture of water (a buffer solution) and an organic solvent (acetonitrile or methanol) and, in some cases, filtration followed by dilution with a buffer solution in order to decrease the fraction of the organic solvent.

5.5.2   Immunochromatographic membrane-based tests

Principles, applications and challenges in mycotoxin analysis
Traditional immunochromatographic (IC) strip tests (or strip tests or lateral flow tests or dipstick tests) are a unique and convenient one-step method, which does not require instrumentation and reagents. The analytical procedure consists of the application of sample on the sample pad by dipping a test strip into the test sample to the marked level (or by dropping sample onto the sample pad) with visual evaluation of the results after a specified time (several minutes). The principles of current formats, applications, limitations and perspectives for quantitative monitoring were summarized in recent reviews by Posthuma-Trumpie *et al.* (2009) and Krska and Molinelli (2009). A test strip mainly consists of a porous nitrocellulose membrane, sample pad, conjugate pad and absorbent pad (Fig. 5.2). The test strip can be either a freestanding strip or enclosed within a plastic housing. After dipping the strip into the test solution, the solution components migrate along the membrane to the absorbent pad immobilized at the top part of the membrane. The absorbent pad at the end of the strip allows absorption of excess liquid, ensuring no backflow on to the membrane. Thus, the test solution serves as a mobile phase. Immunoreagents, which were preliminarily coated onto the bottom part of the membrane, migrate together with the solution components. Specific antibodies labelled with colloidal gold are most frequently used for the determination of mycotoxins. Colloidal gold particles about 15–40 nm in diameter form the red zones.

The test line is an analyte conjugated to a protein (for example, BSA). To simplify the interpretation of the results, a control line (anti-species-specific antibodies) is applied to the membrane above the test line. In the presence of the substance to be determined, its molecules are bound to labelled specific antibodies. The resulting immunocomplex passes the test line and binds to the secondary antibodies to colour the control line. If the substance to be determined is absent, the labelled specific antibodies form an immunocomplex with the analyte–protein conjugate at the test line causing the appearance of a colour, whose intensity is inversely proportional to the concentration of the analyte. Unbound antibodies cause the colorization of the control line. The appearance of a colour at the control line is necessary and it suggests the adequacy of the test.

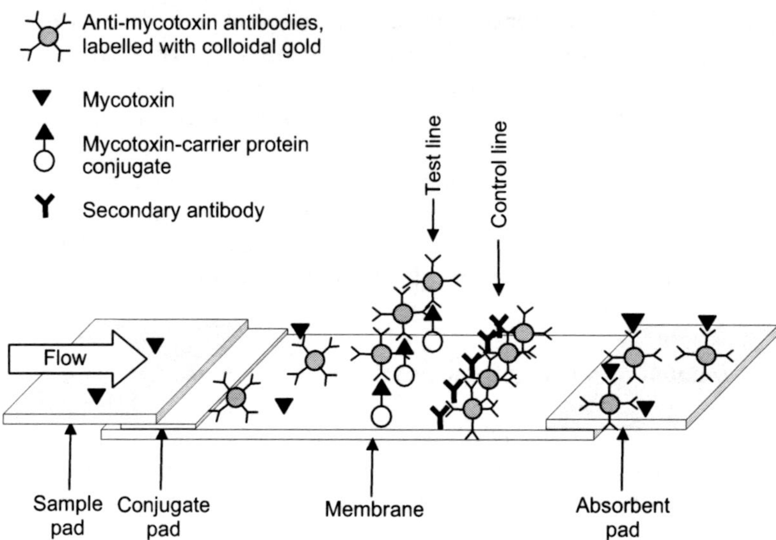

Fig. 5.2   Principle of competitive assay in colloidal-gold based immunochromatographic test strip format.

So, these tests are a unique, dry-chemistry format, which have been in use for decades for rapid diagnostic assays in the clinical and medical sector. The pregnancy test strip is in widespread use. In the last decade, rapid immunoassay-based tests have also increasingly been used in the food and feed sector, where applications range from the screening for drug residues, antibiotics, mycotoxins, foodborne pathogens to allergens and, recently, genetically modified organisms (Krska and Molinelli, 2009).

IC tests for the determination of aflatoxins, fumonisins, DON, OTA and ZEA have been developed and commercialized. Some manufacturers also propose special portable devices (readers) for processing analytical results (Salter *et al.*, 2006). The use of these devices makes it possible to remove the subjective perception of colour by human beings and to improve the sensitivity of the determination. For example, for AFB_1 detection with visual observation, the limit was found to be around 2.5 µg l^{-1}; for quantitative analysis (a photometric strip reader) the LOD was found to be around 0.05–0.1 µg l^{-1} (Sun *et al.*, 2006).

One of the main challenges lies in the reduction of the matrix effect. Therefore, multiple parameters should be optimized including:

1. type and pore size of analytical membrane
2. type and concentration of blocking agent for blocking remaining membrane binding sites after immobilization of immunoreagents
3. type of buffer, pH range and ionic strength;
4. use of surfactants and modifiers for pre- or post-treatment of test strip materials (Krska and Molinelli, 2009).

Current publications are oriented to expand the range of mycotoxins to be determined, to use tests for the analysis of new matrices and to increase assay sensitivity (Table 5.2).

Assay sensitivity and specificity
Various approaches to the determination of the LOD by these test methods have been reported. Zhang *et al.* (2006) considered a concentration of 1 µg l⁻¹, which caused an insignificant but perceptible weakening of the colour of the test line, like the LOD of FB$_1$. For the determination of AFB$_1$ in feeds for pigs, the LOD obtained under optimum conditions (which corresponded to complete colour suppression) was 5 µg kg⁻¹. However, the presence of 3 µg kg⁻¹ of AFB$_1$ had already caused a considerable decrease in the colour intensity of the test line (Delmulle *et al.*, 2005). Assay sensitivity is also often expressed as a cut-off limit (Kolosova *et al.*, 2007b, 2008; Shim *et al.*, 2009a). Because guidelines and recommendations for mycotoxins in food and feed can vary in different countries, IC tests with different indicator ranges were developed. For example, Kolosova *et al.* (2008) designed tests for DON determination with two different ranges, 250–500 µg kg⁻¹ and 1000–2000 µg kg⁻¹.

IC tests are usually combined with a simple extraction with water or buffer (20–40%)–organic solvent (80–60%). As a rule, the dilution of this extract is used to decrease the matrix effect and to reduce the organic solvent concentration. It was found that the matrix effects of barley, peanuts, corn, rice and sorghum in the determination of FB$_1$ could be removed by a 15-fold dilution of an aqueous methanol extract (Zhang *et al.*, 2006). For AFB$_1$ extraction with methanol–water (60:40, v/v) a two-fold and four-fold dilution with PBS for grain and feed extracts was enough (Shim *et al.*, 2007 both factors), while for Sun *et al.*, (2006; methanol influence) a 6-fold dilution of a 60% methanol extract was necessary.

A comparison between the sensitivities of a colloidal gold-based strip test with visual evaluation and ELISA with an HRP label demonstrated that the sensitivity of the strip test (LOD of 1.0 µg l⁻¹)) was only slightly lower than that of ELISA (0.2–0.5 µg l⁻¹) (Wang *et al.*, 2006). Shim *et al.* (2009b) reported a seven times decrease in sensitivity compared to ELISA. At the same time, Cho *et al.* (2005) found that the sensitivity of an IC test was lower than that of ELISA by three orders of magnitude. As prevention of false negative results is the main goal of rapid test optimization, Shim *et al.* (2009b) compared the outcomes of IC tests and *dc*-ELISA with HPLC results for ZEA detection in naturally contaminated corn samples. They reported 8.5% false negative results for strips and concluded that for representative results, strips should be tested more than twice. ELISA showed an underestimation of the ZEA concentration for the same samples. For related compounds, IC tests are also influenced by the antibody's cross-reaction. For example, an IC test for ZEA was also sensitive to α-zearalenol and β-zearalenol (Shim *et al.*, 2009b). Similarly, a strip for AFB$_1$ detection showed cross-reaction to aflatoxins B$_2$, G$_1$, and G$_2$ (Shim *et al.*, 2007).

Alternative labels
In addition to colloidal gold particles, other colloidal labels were used. In the

Table 5.2   Immunochromatographic membrane-based tests for mycotoxin detection

Mycotoxin	Sample matrix	LOD (μg kg^{-1})	Label	Reference
		Individual mycotoxin detection		
Aflatoxin B$_1$	Pig feed	5 3*	Colloidal gold	Delmulle *et al.*, 2005
	Standard solution	2.5 μg l^{-1}	Colloidal gold	Sun *et al.*, 2006
	Standard solution grain feed	0.5 ng l^{-1} 10 20	Colloidal gold	Shim *et al.*, 2007
	Standard solution	18 ng	Sulphorhodamine	Ho and Wauchope, 2002
Ochratoxin A	Standard solution	500 μg l^{-1}	Colloidal gold	Cho *et al.*, 2005
	Standard solution	1* μg l^{-1}	Colloidal gold	Wang *et al.*, 2007
Zearalenone	Standard solution corn	2.5 μg l^{-1} 30	Colloidal gold	Shim *et al.*, 2009
Deoxynivalenol	Wheat Pig feed	400 1500	Colloidal gold	Kolosova *et al.*, 2008
15-Acetyl-deoxynivalenol	Wheat	5 μg l^{-1}	HRP	Usleber *et al.*, 1993
T-2 toxin	Wheat oat	120 100	Colloidal gold	Molinelli *et al.*, 2008
	Standard solution wheat	3 μg l^{-1} 12	HRP	De Saeger and Van Petegham, 1996
Fumonisin B$_1$	Standard solution	1* mg l^{-1}	Colloidal gold	Wang *et al.*, 2006
	Standard solution corn-based food	7.5–10 μg l^{-1} 40–60	HRP	Schneider *et al.*, 1995
Sporidesmin A	Standard solution	4 μg l^{-1} 25 μg l^{-1}	Colloidal gold Colloidal carbon	Collin *et al.*, 1998

Simultaneous detection of several mycotoxins

Zearalenone	Wheat	100	Colloidal gold	Kolosova *et al.*, 2007b
Deoxynivalenol		1500		
Ochratoxin A	Corn	5	Colloidal gold	Shim *et al.*, 2009
Zearalenone		10		
Aflatoxin B_1	Grain	30	HRP	Schneider *et al.*, 1995b
T-2 toxin		100		
3-Acetyl-deoxynivalenol		600		
Roridin A		500		
Zearalenone		60		
Fumonisin B_1	Standard solution	500 µg l^{-1}	HRP	Abouzied and Pestka, 1994
Aflatoxin B_1		0.5 µg l^{-1}		
Zearalenone		3 µg l^{-1}		

*LOD determined as clearly distinguishable colour reduction in comparison with negative control.

determination of sporidesmin as an example, it was found that the LOD with the use of colloidal carbon as a label was higher than that with the use of colloidal gold (Collin *et al.*, 1998). Attempts to use polystyrene microspheres as labels in an IC test were unsuccessful because of the non-specific binding of microspheres to proteins in the test and control strips (Collin *et al.*, 1998). With the use of a sulphorhodamine dye encapsulated in liposomes as a label, a two-step procedure for the determination of AFB_1 was implemented: initially, a test strip was placed in the solution containing the analyte; then, it was transferred to a solution containing AFB_1 labelled with liposomes. The necessity consecutive contact of the test strip was related to the instability of liposomes in the presence of organic solvents, which are commonly used for the extraction of mycotoxins. Measuring the red colour with a densitometer, the LOD of AFB_1 in standard solution was 18 ng (Ho and Wauchope, 2002). Magnetic nanogold microspheres with nano-Fe_2O_3 particles as core and gold nanoparticles as shell were applied as a label for lateral flow immunoassay for AFB_2 detection in nuts (Tang *et al.*, 2009).

Although the use of enzymes implies additional operations (washing and substrate addition), the higher sensitivity of enzyme labels makes it possible to increase (by a factor of 10 or higher) the sensitivity of a strip test considerably (Zhang *et al.*, 2006). This type of test usually contains only the test zone specific antibodies and implies the dipping of the test zone into a sample solution, which also contains the conjugate of the analyte with an enzyme label, commonly with HRP. The second stage of analysis is the dipping of the test strip into a chromogenic substrate. To check the result, a second test strip is commonly used; this strip is dipped into a solution containing no analyte (negative control test). This test format was developed for the determination of 15-acDON with a LOD of 5 µg l^{-1} (a considerable decrease in the colour intensity of the test zone compared with the negative control test). Complete colour suppression occurred at a 15-acDON concentration of 20–25 µg l^{-1}. The use of this test for the determination of 15-acDON in wheat samples allowed the matrix effect of the sample and the effect of methanol to be decreased, compared with ELISA (Usleber *et al.*, 1993). An analogous strip test was proposed for the determination of T-2 in grain. A pronounced difference between the results of a negative control test and the analysis of a grain extract was measured at a concentration of 12 µg kg^{-1} (De Saeger and Van Peteghem, 1996).

Multi-analyte applications
Recently, the results of the development of IC strip tests for the simultaneous determination of two mycotoxins have been published. The membrane consisted of two test zones (with mycotoxin1–protein and mycotoxin2–protein) and one control zone with secondary antibodies. Kolosova *et al.* (2007b) developed a test for DON and ZEA in wheat grains with two test zones (DON–BSA and ZEA–BSA conjugates), while Shim *et al.* (2009a) published the simultaneous detection of OTA and ZEA with OTA–BSA and ZEA–OVA test zones.

Strip tests with enzyme labels were also developed for the simultaneous determination of mycotoxins. In particular, the detection limits of AFB_1, T-2,

3-acDON, roridin A and ZEA in artificially contaminated grain samples were 30, 100, 600, 500 and 60 µg kg^{-1}, respectively (Schneider *et al.*, 1995b). Abouzied and Pestka (1994) developed a test for the simultaneous determination of FB$_1$, AFB$_1$, and ZEA with detection limits of 500, 0.5 and 3 µg l^{-1} in standard solutions, respectively. Pestka (1991) proposed a test method for the determination of structurally related mycotoxins (aflatoxins B$_1$, B$_2$, G$_1$, and G$_2$, as well as ZEA and α-zearalenol); this method combined TLC separation and transfer of the separated mycotoxin zones to a membrane with grafted specific antibodies followed by immunochemical determination.

5.5.3   Immunofiltration membrane-based tests

Principles, applications and challenges in mycotoxin analysis
Immunofiltration (IF) test methods (flow-through or immunofiltration assay or enzyme-linked immunofiltration assay) are based on the use of polymer membranes with bound specific antibodies. The membrane is placed onto an absorbent layer, which absorbs the liquid that passes through the membrane. When the test solution passes through the membrane, the mycotoxin present in this solution forms an immunocomplex with antibodies applied to the test zone of the membrane. Then, a solution of the analyte–enzyme conjugate is added; the final step is the introduction of a chromogenic substrate, which causes colour development in the presence of enzyme. Between each step a washing buffer should be applied. Thus, this format almost completely reproduces the *dc*ELISA sequence of steps with the exception of the consecutive order of sample and mycotoxin–enzyme application.

An IF method based on the use of an HRP label was applied in the determination of OTA in wheat (De Saeger and Van Peteghem, 1999) and T-2 in wheat, maize, oats, and rye (De Saeger *et al.*, 2002). An important step in the development of this method, which allowed the use of an additional negative control test to be excluded, was the implementation of a control zone of antibodies specific to HRP on the membrane (Fig. 5.3). The dilution of the antibodies specific to the enzyme was chosen so that the colour intensity of the control zone was nearly the same as the colour intensity of the test zone in the case of a negative result. This approach was applied to the determination of T-2 in wheat, maize, oat and rye (Sibanda *et al.*, 2000), OTA in roasted (Sibanda *et al.*, 2002) and green coffee beans (Sibanda *et al.*, 2001), and fumonisins in maize (Paepens *et al.*, 2004). In these cases, the LOD was determined based on complete colour suppression. The detection limit estimated from a considerable colour reduction of the test zone was used to determine FB$_1$ in food products (Schneider *et al.*, 1995) and sporidesmin A in standard solutions (Collin *et al.*, 1998).

For some matrices IF tests need additional clean-up steps, such as an immunoaffinity clean-up for AfM$_1$ detection in milk (Sibanda *et al.*, 1999). Indeed, the high concentration of solid substances and the low control level (0.05 µg l^{-1} in accordance with EU regulations), made the analysis of milk without sample

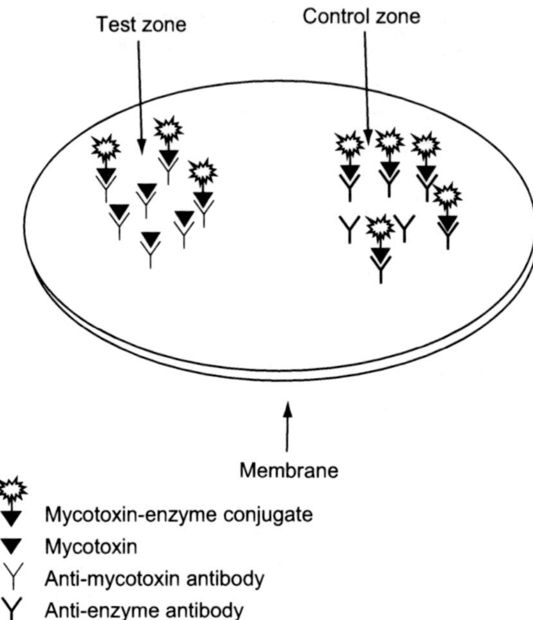

Fig. 5.3   Competitive assay in an immunofiltration test format.

preparation impossible. For OTA detection in roasted coffee Sibanda *et al.* (2002) used solid-phase clean-up columns. To decrease the matrix effect and to increase the sensitivity (because of a lower extract dilution) in the determination of AFB_1 in peanut, wheat, maize, soybean and red pepper, the membrane was washed with a solution containing trifluoroacetic or propionic acid or sodium hydrocarbonate after sample addition (Pal *et al.*, 2005).

To decrease the costs and to shorten the analysis time, different approaches were proposed. The simultaneous determination of AFB_1 in several samples was performed by including four individual antibody zones on a membrane (Pal and Dhar, 2004). A similar test with 36 zones was used to determine T-2 (Pal *et al.*, 2004). Table 5.3 summarizes data on IF tests for mycotoxin detection.

Assay sensitivity
Limits of detection of various tests were compared. With the determination of FB_1 as an example, it was demonstrated that LODs were similar when using a strip test with an enzyme label and an IF test in buffer solutions (Schneider *et al.*, 1995a). However, the tests were approximately 50 times less sensitive than the corresponding ELISA. Sporidesmin A detection in buffer solutions resulted in a lower LOD with ELISA (0.2 µg l^{-1}) than with IF tests (1 µg l^{-1}), IC strip tests (4 µg l^{-1}), and dipstick (1 µg l^{-1}). This lower sensitivity with visual detection was due to the use of a high concentration of antibodies to reach a visually detectable colour in the control sample. The importance of the washing step is reaching a high sensitivity was also demonstrated. The absence of washing before the use of chromogenic

Table 5.3 Immunofiltration membrane tests for mycotoxin detection

Mycotoxin	Sample matrix	LOD ($\mu g\,kg^{-1}$)*	Reference
	Individual mycotoxin detection		
Aflatoxin B_1	Groundnut, corn, wheat, cheese, chilli	$0.01\,\mu g\,l^{-1}$	Pal and Dhar., 2004
Aflatoxin M_1	Milk	0.5	Salter and Douglas, 2006
		0.05	Sibanda et al., 1999
Ochratoxin A	Wheat	4	De Saeger and Van Peteghem, 1999
	Roasted coffee	4	Sibanda et al., 2002
	Green coffee	8	Sibanda et al., 2001
T-2 toxin	Wheat, rye, maize, barley	50	De Saeger et al., 2002 Sibanda et al., 2000
Fumonisin B_1	Corn-based food	40–60	Schneider et al., 1995
Fumonisins	Maize	1000	Paepens et al., 2004
Sporidesmin A	Standard solutions	$1\,\mu g\,l^{-1}$	Collin et al., 1998
	Simultaneous detection of several mycotoxins		
Aflatoxin B_1	Chilli	2	Saha et al., 2007
Ochratoxin A		10	
Aflatoxin B_1	Wheat	10	Schneider et al., 2004
Ochratoxin A		50	
Deoxinivalenol		3500	
T-2 toxin		100	
Diacetoxyscirpenol		5	
Roridin A		250	
Fumonisin B_1		50	

*Label is horseradish peroxidase

substrate resulted in an increase in the detection limit from 1 to 5 $\mu g\,l^{-1}$ (Collin *et al.*, 1998). In addition, for simultaneous determination of several mycotoxins, sensitivity decreased more than when determining individual analytes because of the necessity of reaching comparable colour spot intensities to simplify the visual interpretation of the results. A test design was developed in which the adsorbents were changed during the analysis, hence circumventing the traditional limitations of sample and washing buffer volumes. Although the analytical procedure became more complicated, the LOD was lower. The sensitivity was additionally increased (to 0.25 pg or 0.01 $\mu g\,l^{-1}$ AFB$_1$ in groundnut, corn, wheat, cheese and chilli) by the use of biotinylated tyramine and an avidin–HRP conjugate (Pal and Dhar, 2004).

Different reagents, including polymers and surfactants, were tested for their possible effects on signal generation when an analyte–HRP conjugate was used. Polymers such as poly(vinyl alcohol) (PVA) and poly(ethylene glycol) (PEG) 6000 exerted favourable effects on signal amplification, whereas surfactants negatively affected the assay performance. The highest signal amplification was achieved with the use of 0.5% (w/w) PVA in TMB Colorburst solution (Kolosova *et al.*, 2007a). These authors also compared two enzymes as labels for OTA detection, HRP and AP, both in combination with different substrate systems. Calibrations using OTA–HRP and OTA–AP conjugates were similar, but lower concentrations of immunoreagents were needed for the assay with OTA–AP (Kolosova *et al.*, 2007a). The use of colloidal gold as a label makes it possible to exclude the use of a chromogenic substrate (Wang *et al.*, 2006); however, unlike IC strip tests, this label is not widely used in IF tests because of its low sensitivity.

Multi-analyte applications
An IF prototype with eight wells was developed for the simultaneous determination of the following seven mycotoxins in a sample: AFB_1, FB_1, T-2, roridin A, DON, diacetoxyscirpenol and OTA. Membranes with immobilized specific antibodies were fixed under each well and an additional membrane was used as a negative control test. An absorbent layer was placed under the membranes. The analytical procedure was identical to that for the determination of one analyte, except that a mixture of conjugates was introduced into the wells (Schneider *et al.*, 2004). Spotting two zones of specific antibodies onto a single membrane made it possible to develop a method for the simultaneous determination of AFB_1 and OTA in red pepper samples (Saha *et al.*, 2007).

5.5.4   Gel-based tests
Membrane-based IF tests have advantages such as simplicity and speed (the analysis time is 5–35 min). However, they do not always provide adequate sensitivity; moreover, the matrix effect of the sample often interferes with the results. The replacement of a carrier membrane with a specially prepared gel based on Sepharose (agarose beads) with bound specific antibodies makes it possible to increase the sensitivity. The analysis is performed in a column for solid-phase extraction and it includes the stages of consecutively passing the sample, the antigen conjugated with HRP, and a chromogenic substrate through the bulk of the gel with application of washing buffers between each step. So the procedure of the assay is similar to that for the IF membrane-based technique. As in the case of membrane-based immunochemical methods, the assay sensitivity depends on the concentrations of applied antibodies and analyte–enzyme conjugate. The incorporation of antibodies into the bulk of the gel makes it possible to concentrate mycotoxins, which is similar to the action of immunoaffinity columns; thus, the gel with immobilized antibodies simultaneously serves for separation, preconcentration and determination.

The use of columns also made it possible to include an additional clean-up layer in the assay column, or to use an additional separate column with a clean-up layer, which is required for intensely coloured matrices. Using OTA and aflatoxins determination as an example, it was demonstrated that silica gel with aminopropyl groups provided an optimum sorbent for the clean-up layer. This sorbent minimized colour and matrix effects of the sample and exhibited the lowest sorption capacity for mycotoxins (Lobeau *et al.*, 2005). The clean-up layer could be placed either above or under the detection layer (Fig. 5.4A and B). For analysis of food matrices without an intense colour, such as milk and beer, the clean-up layer could

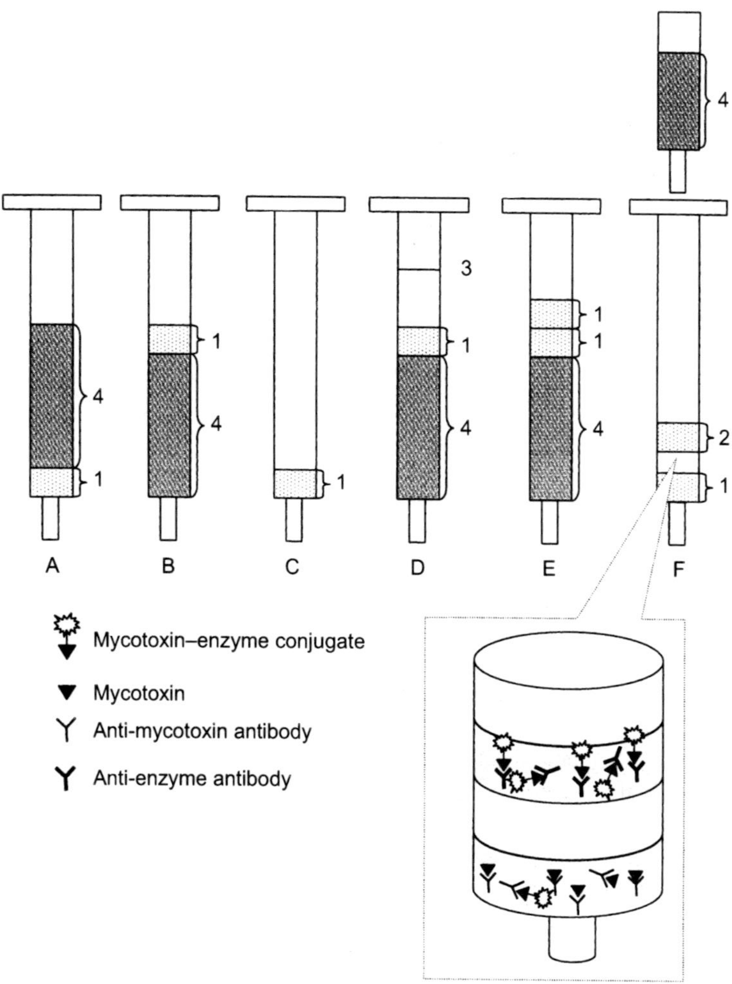

Fig. 5.4   Columns for gel-based immunoassay. 1, Test layer; 2, control layer; 3, conjugate frit; 4, clean-up layer.

be omitted (Fig. 5.4C). To simplify the analytical procedure, a conjugate on an additional frit was introduced into the column (Fig. 5.4D). The simultaneous determination of two mycotoxins was performed upon the introduction of two detection immunolayers with immobilized antibodies specific to the corresponding mycotoxins in the column (Fig. 5.4E). This approach was implemented for the simultaneous determination of OTA and AFB_1 in spices (Goryacheva *et al.*, 2007b). As in the case of membrane tests, the time of gel colour development and the colour intensity were chosen to be similar for both analytes in order to simplify the interpretation of the results. HRP was used as a label for all developed gel-based tests.

This column design had no limitations for volumes of liquid sample or extract,

Table 5.4   Gel-based tests for mycotoxin detection

Mycotoxin	Sample matrix	V (ml)	Cut-off level ($\mu g\ kg^{-1}$ or $\mu g\ l^{-1}$)*	Reference
Ochratoxin A	Beer	6	0.2	Goryacheva *et al.*, 2008a
	Wine	1	2	Rusanova *et al.*, 2009
	Spices	0.5	10	Goryacheva *et al.*, 2006; Goryacheva *et al.*, 2007b
	Grains	2	2	**
	Roasted coffee	1.5	6	Lobeau *et al.*, 2005
	Cocoa	1.5	2	Lobeau *et al.*, 2007
Aflatoxin B_1	Spices	0.5	5	Goryacheva *et al*, 2007
Aflatoxin M_1	Milk	10	0.04	Goryacheva *et al.*, 2009a
T-2 toxin	Grains and feed	2	100	Basova *et al.*, 2010
Zearalenone	Grains and feed	2	100	Burmistrova *et al.*, 2009; Basova *et al.*, 2010

*Label is horseradish peroxidase; **non-published results of our group.

reagent and washing buffers. It allowed use of high sample volumes to improve assay sensitivity. In particular, to obtain a cut-off level of 0.2 $\mu g\ l^{-1}$ for OTA in beer, a beer sample volume of 12 ml was used (Goryacheva *et al.*, 2008a), whereas 10 ml of centrifuged skimmed milk was used to reach a cut-off level of 0.04 $\mu g\ l^{-1}$ in the determination of AfM$_1$ in milk and milk products (Goryacheva *et al.*, 2009a). Table 5.4 summarizes data on developed gel-based tests for the determination of mycotoxins.

To simplify the interpretation of the results and to show the functionality of the tests, a control layer with an antibody specific to the enzyme was introduced, analogous to the membrane-based tests (Fig 5.4F). Mycotoxin–enzyme conjugate, applied to the column, first made contact with the detection layer and was bound to the anti-mycotoxin specific antibody; then it made contact with the control layer and was bound to the HRP-specific antibody. So, the successive distribution of the components between different layers was realized, similar to immunochromatographic techniques. This approach was applied to OTA detection in beer (Goryacheva *et al.*, 2008a) and simultaneous detection of ZEA and T-2 in feed samples (Basova *et al.*, 2010).

5.6   Conclusions and future trends

Generally, immunoassay characteristics (sensitivity, selectivity, throughput, matrix influence) can be improved by the development of high-quality antibodies, new formats and labels.

To date, enzyme labels in the presence of appropriate chromogenic substrates continue to be the most sensitive labels for tests with visual or colourimetric detection. But there is a growing interest in the development of a new generation of particle-based labels, like luminescent labels, fluorescent dyes and metal chelates. A novel class of europium chelate-loaded silica nanoparticles was applied as a label for lateral flow immunoassay by Xia *et al.* (2009). These nanoparticles provide a large fluorescence emission owing to multilayer chemical loading of lanthanide chelates onto the porous silica nanoparticles. Another direction involves increasing the brightness of non-fluorescent nanoparticles. Khlebtsov and Khlebtsov (2008) showed by theoretical estimation that the dot extinction of functionalized gold nanoshells consisting of a silica core and a thin gold layer can be 1000 times higher than that for the same number of traditionally used gold nanoparticles with a diameter of 15–40 nm.

Cyclodextrins (mainly β-cyclodextrin), which have a cavity (pore) that may accommodate small molecules as 'guests', forming inclusion complexes, are of interest (Maragos *et al.*, 2008; Galaverna *et al.*, 2008). This interaction results in a significant enhancement of the fluorescence intensity of aflatoxins (Hashemi *et al.*, 2008; Goryacheva *et al.*, 2008b) and ZEA (Maragos and Appell 2007).

Molecular imprinted polymers (MIPs) as alternative biomimetic receptors are also currently being researched. MIPs were not only used for solid-phase extraction of mycotoxins (De Smet *et al.*, 2009), but also in detection systems, for example in a fluorescence displacement assay for ZEA (Navarro-Villoslada *et al.*, 2007). Further, aptamers are single-stranded oligonucleotides selected *in vitro* to bind to molecular targets. Cruz-Aguado and Penner (2008) describe the identification of an aptamer that binds with high affinity and specificity to OTA and its application to an aptamer affinity column.

Future developments in mycotoxin screening methods could possibly be found in the clinical assays (point-of-care testing) of today. These include new complex microfluidic, lab-on-chip systems and new challenging micro- and nanoparticle labels like magnetic, latex, metal and semiconductor particles with unique optical, electronic, and structural properties.

5.7   References

Abouzied M and Pestka J J (1994) 'Simultaneous screening of fumonisin B_1, aflatoxin B_1 and zearalenone by line immunoblot a computer-assisted multianalyte assay system' *J AOAC Int*, **77**, 495–500.

Anfossi L, Calderara M, Baggiani C, Giovannoli C, Arletti E and Giraudi G (2008) 'Development and application of solvent-free extraction for the detection of aflatoxin M_1 in dairy products by enzyme immunoassay' *J Agric Food Chem*, **56**, 1852–7.

Ardic M, Karakaya Y, Atasever M and Durmaz H (2008) 'Determination of aflatoxin B_1 levels in deep-red ground pepper (isot) using immunoaffinity column combined with ELISA' *Food Chem Toxicol*, **46**, 1596–9.

Asis R, Di Paola R D and Aldao M A J (2002) 'Determination of aflatoxin B_1 in highly contaminated peanut samples using HPLC and ELISA' *Food Agr Immunology*, **14**, 201–8.

Barna-Vetro I, Szabo E, Fazekas B and Solti L (2000) 'Development of a sensitive ELISA for the determination of fumonisin B_1 in cereals' *J Agric Food Chem*, **48**, 2821–5

Basova E Y, Goryacheva I Y, Rusanova T Y, Burmistrova N A, Dietrich R, Maertlbauer E, Detavernier C, Van Peteghem C and De Saeger S (2010) 'An immunochemical test for rapid screening of two Fusarium toxins' *Anal Bioanal Chem*, **397**, 55–62.

Bird C B, Malone B, Rice L G, Ross P F, Eppley R and Abouzied M M (2002) 'Determination of total fumonisins in corn by competitive direct enzyme-linked immunosorbent assay: Collaborative study' *J AOAC Int*, **85**, 404–10.

Burkin M A, Yakovleva I V, Sviridov V V, Burkin A A, Kononenko G P and Soboleva NA (2000a) 'Comparative immunochemical characterization of polyclonal and monoclonal antibodies to aflatoxin B-1' *Appl Biochem Microbiol*, **36**, 496–501.

Burkin A A, Kononenko G P, Soboleva N A and Zotova E V (2000b) 'An enzyme immunoassay system for aflatoxin B-1' *Appl Biochem Microbiol*, **36**, 80–4.

Burkin A A, Kononenko G P, Zoryan V G, Soboleva N A and Zotova E V (2000c) 'Production of antibodies to roridin and their applications' *Appl Biochem Microbiol*, **36**, 368–71.

Burkin A A, Kononenko G P, Soboleva N A and Zotova E V (2000d) 'Preparation of conjugated antigens based on zearalenone carboxymethyloxime and their use in enzyme immunoassay' *Appl Biochem Microbiol*, **36**, 282–8.

Burkin A A, Kononenko G P and Soboleva N A (2002a) 'Group-specific antibodies against zearalenone and its metabolites and synthetic analogs' *Appl Biochem Microbiol*, **38**, 169–176.

Burkin A A, Kononenko G P and Soboleva N A (2002b) 'Production and analytical properties of antibodies with high specificity to zearalenone' *Appl Biochem Microbiol*, **38**, 263–268.

Burmistrova N A, Goryacheva I Yu, Basova E Yu, Franki A S, Elewaut D, Van Beneden K, Deforce D, Van Peteghem C and De Saeger S (2009) 'Application of a new anti-zearalenone monoclonal antibody in different immunoassay formats' *Anal Bioanal Chem*, **395**, 1301–7.

Cavaliere C, D'Ascenzo G, Foglia, P., Pastorini E, Samperi R and Lagana A (2005) 'Determination of type B trichothecenes and macrocyclic lactone mycotoxins in field contaminated maize' *Food Chem*, **92**, 559–68.

Cervino C, Knopp D, Weller M G and Niessner R (2007) 'Novel aflatoxin derivatives and protein conjugates' *Molecules*, **12**, 641–653.

Cervino C, Weber E, Knopp D and Niessner R (2008) 'Comparison of hybridoma screening methods for the efficient detection of high-affinity hapten-specific monoclonal antibodies' *J Immunol Methods*, **329**, 184–93.

Cho Y J, Lee D H, Kim D O, Min W K, Bong K T, Lee G G and Seo J H (2005) 'Production of a monoclonal antibody against ochratoxin A and its application to immunochromatographic assay' *J Agric Food Chem*, **53**, 8447–51.

Choi G H, Lee D H, Min W K, Cho Y J, Kweon D H, Son D H, Park K and Seo J H (2004) 'Cloning, expression, and characterization of single-chain variable fragment antibody against mycotoxin deoxynivalenol in recombinant *Escherichia coli*' *Protein Expr Purif*, **35**, 84–92.

Christensen H R, Yu F Y and Chu F S (2000) 'Development of a polyclonal antibody-based sensitive enzyme-linked immunosorbent assay for fumonisin B$_4$' *J Agric Food Chem*, **48**, 1977–84.

Chu F S, Hsia M T S and Sun P S (1977) 'Preparation and Characterization of Aflatoxin B$_1$-1-(Ocarboxymethyl)-oxime' *J AOAC Int*, **60**, 791–4.

Chu F S, Steinert B W and Gaur P K J (1985) 'Production and characterization of antibody against aflatoxin G$_1$' *Food Safety*, **7**, 161–70.

Chung Y J, Jarvis B B, Tak H and Pestka J J (2003) 'Immunochemical assay for satratoxin G and other macrocyclic trichothecenes associated with indoor air contamination by *Stachybotrys chartarum*' *Toxicol Mech Meth*, **13**, 247–52.

Collin R, Schneider E, Briggs L and Towers N (1998) 'Development of immunodiagnostic field tests for the detection of mycotoxin, sporidesmin A' *Food Agric Immunol*, **10**, 91–104.

Cruz-Aguado J A and Penner G (2008) 'Determination of ochratoxin A with a DNA Aptamer' *J Agric Food Chem*, **56**, 10456–61.

Delmulle B, De Saeger S, Sibanda L, Barna-Vetro I and Van Peteghem C (2005) 'Development of an immunoassay-based lateral flow dipstick for the rapid detection of aflatoxin B_1 in pig feed' *J Agric Food Chem*, **53**, 3364–8.

De Saeger S and Van Peteghem C (1996) 'Dipstick enzyme immunoassay to detect Fusarium T-2 toxin in wheat' *Appl Environ Microbiol*, **62**, 1880–4.

De Saeger S and Van Peteghem C (1999) 'Flow-through membrane-based enzyme immunoassay for rapid detection of ochratoxin A in wheat' *J Food Protect*, **62**, 65–9.

De Saeger S, Sibanda L, Desmet A and Van Peteghem C (2002) 'A collaborative study to validate novel field immunoassay kits for rapid mycotoxin detection' *Int J Food Microbiol*, **75**, 135–42.

De Smet D, Dubruel P, Van Peteghem C, Schacht E and De Saeger S (2009) 'Molecularly imprinted solid-phase extraction of fumonisin B analogues in bell pepper, rice and corn flakes' *Food Addit Contam*, **26**, 874–84.

Devi K T, Mayo M A, Reddy K L N, Delfosse P, Reddy G, Reddy S V and Reddy D V R (1999) 'Production and characterization of monoclonal antibodies for aflatoxin B_1' *Lett Appl Microbiol*, **29**, 284–8.

Fan T S L, Zhang G S and Chu F S (1984) 'Production and characterization of antibody against aflatoxin Q1' *Appl Environ Microbiol*, **47**, 526–32.

Frank S A (2002) 'Specificity and cross-reactivity', in *Immunology and Evolution of Infectious Disease*, Princeton University, 33–56.

Galaverna G, Dall'Asta C, Corradini R, Dossena A and Marchelli R (2008) 'Cyclodextrins as selectors for mycotoxin recognition' *World Mycotoxin Journal*, **1**, 397–406.

Gaur P K, Lau H P, Pestka J J and Chu F S (1981) 'Production and characterization of aflatoxin B_{2a} antiserum' *Appl Environ Microbiol*, **41**, 478–82

Gilbert J (1999) 'Overview of mycotoxin methods, present status and future needs' *Natural Toxins*, **7**, 347–52.

Goryacheva I Yu, De Saeger S, Lobeau M, Eremin S A, Barna-Vetró I and Van Peteghem C (2006) 'Approach for ochratoxin A fast screening in spices using clean-up tandem immunoassay columns with confirmation by high performance liquid chromatography–tandem mass spectrometry (HPLC–MS/MS)' *Anal. Chim. Acta*, **577**, 38–45.

Goryacheva I Yu, De Saeger S, Eremin S A and Van Peteghem C (2007a) 'Immunochemical methods for rapid mycotoxin detection: evolution from single to multiple analyte screening. A review.' Food Addit Contam, 24, 1169–1183.

Goryacheva I Yu, De Saeger S, Lobeau M, Delmulle B, Eremin S A, Barna-Vetró I and Van Peteghem C (2007b) 'Simultaneous non-instrumental detection of aflatoxin B_1 and ochratoxin A using a clean-up tandem immunoassay column' *Anal Chim Acta*, **590**, 118–24.

Goryacheva I Y, Basova E Y, Van Peteghem C, Eremin S A, Pussemier L, Motte J-C and De Saeger S (2008a) 'Novel gel-based rapid test for non-instrumental detection of ochratoxin A in beer' *Anal Bioanal Chem*, **390**, 723–77.

Goryacheva I Y, Rusanova T Y and Pankin K E (2008b) 'Fluorescent properties of aflatoxins in organized media based on surfactants, cyclodextrins, and calixresorcinarenes' *J Anal Chem*, **63**, 751–5.

Goryacheva I Yu, Karagusheva M A, Van Peteghem C, Sibanda L and De Saeger S (2009a) 'Gel-based immunoassay for non-instrumental screening of aflatoxin M_1 in milk' *Food Control*, **20**, 802–6.

Goryacheva I Y, Rusanova T Y, Burmistrova N A and De Saeger S (2009b) 'Immunochemical methods for mycotoxin detection' *J Anal Chem*, **64**, 768–85.

Haltzapple C K, Carlin R J, Rose B G, Kubena L F and Stanker L H (1996) 'Characterization of monoclonal antibodies to aflatoxin M_1 and molecular modeling studies of related aflatoxins' *Mol Immunol*, **33**, 939–46.

Hashemi J, Kram G A and Alizadeh N (2008) 'Enhanced spectrofluorimetric determination of aflatoxin B_1 in wheat by second-order standard addition method' *Talanta*, **75**, 1075–81.

Hastings K L, Tulis J J and Dean J H (1998) 'Production and characterization of a monoclonal antibody to aflatoxin B_2' *J Agric Food Chem*, **36**, 404–8.

Heussner A H, Moeller I, Day B W, Dietrich D R and O'Brien E (2007) 'Production and characterization of monoclonal antibodies against ochratoxin B' *Food Chem Toxicol*, **45**, 827–33.

Ho J A A and Wauchope R D (2002) 'A strip liposome immunoassay for aflatoxin B_1' *Anal Chem*, **74**, 1493–6.

Hooper D G, Bolton V E, Guilford F T and Straus D C (2009) 'Mycotoxin detection in human samples from patients exposed to environmental molds' *Int J Mol Sci*, **10**, 1465–75.

Huang B, Tao W Y, Shi J, Tang L and Jin J (2006) 'Determination of ochratoxin A by polyclonal antibodies based sensitive time-resolved fluoroimmunoassay' *Arch Toxicol*, **80**, 481–5.

Josephs R D, Schuhmacher R and Krska R (2001) 'International interlaboratory study for the determination of the Fusarium mycotoxins zearalenone and deoxynivalenol in agricultural commodities' *Food Addit Contam*, **18**, 417–30.

Khlebtsov B and Khlebtsov N (2008) 'Enhanced solid-phase immunoassay using gold nanoshells: effect of nanoparticle optical properties' *Nanotechnology*, **19**, article number 435703.

Kim E K, Kim Y B, Shon D H, Ryu D and Chung S H (1998) 'Natural occurrence of fumonisin B_1 in Korean rice and its processed foods by enzyme-linked immunosorbent assay' *Food Sci Biotechnol*, **7**, 221–4.

Kim E K, Shon D H, Ryu D, Park J W, Hwang H J and Kim Y B (2000) 'Occurrence of aflatoxin M_1 in Korean dairy products determined by ELISA and HPLC' *Food Addit Contam*, **17**, 59–64.

Kolosova A Yu, De Saeger S, Eremin S A and Van Peteghem C (2007a) 'Investigation of several parameters influencing signal generation in flow-through membrane-based enzyme immunoassay' *Anal Bioanal Chem*, **387**, 1095–104.

Kolosova A Y, De Saeger S, Sibanda L, Verheijen R and Van Peteghem C (2007b) 'Development of a colloidal gold-based lateral-flow immunoassay for the rapid simultaneous detection of zearalenone and deoxynivalenol' *Anal Bioanal Chem*, **389**, 2103–7.

Kolosova A Y, Sibanda L, Dumoulin F, Lewis J, Duveiller E, Van Peteghem C and De Saeger S (2008) 'Lateral-flow colloidal gold-based immunoassay for the rapid detection of deoxynivalenol with two indicator ranges' *Anal Chim Acta*, **616**, 235–44.

Koch P (2004) 'State of the art of trichothecenes analysis' *Toxicol Lett*, **153**, 109–12.

Kononenko G P and Burkin A A (2007) 'Immunoenzyme method for the determination of citrinin' I, 62, 769–774.

Kononenko G P, Burkin A A, Soboleva N A and Zotova EV (1999b) 'Enzyme immunoassay for determination of T-2 toxin in contaminated grain' Appl Biochem Microbiol, 35, 411–416 .

Kononenko G P, Burkin A A, Zotova E V and Soboleva N A (1999a) 'Application of competitive ELISA for determining group B fumonisins' *Appl Biochem Microbiol*, **35**, 183–7.

Kononenko G P, Burkin A A, Zotova E V and Soboleva NA (2000) 'Ochratoxin A: Contamination of grain' *Appl Biochem Microbiol*, **36**, 177–80.

Krska R and Molinelli A (2009) 'Rapid test strips for analysis of mycotoxins in food and feed' *Anal Bioanal Chem*, **393**, 67–71.

Krska R, Welzig E, Berthiller F, Molinelli A and Mizaikoff B (2005) 'Advances in the analysis of mycotoxins and its quality assurance' *Food Addit Contam*, **22**, 345–53.

Krska R, Baumgartner S and Josephs R (2001) 'The state-of-the-art in the analysis of type-A and -B trichothecene mycotoxins in cereals' *Fresenius J Anal Chem*, **371**, 285–99.

Krska R, Schubert-Ullrich P, Molinelli A, Sulyok M, Macdonald S and Crews C (2008) 'Mycotoxin analysis: An update' *Food Addit Contam*, **25**, 152–63.

Kulisek E S and Hazebroek J P (2000) 'Comparison of extraction buffers for the detection of fumonisin B_1 in corn by immunoassay and high-performance liquid chromatography' *J Agric Food Chem*, **48**, 65–9.

Lauer B, Ottleben I, Jacobsen H J and Reinard T (2005) 'Production of a single-chain variable fragment antibody against fumonisin B$_1$' *J Agric Food Chem*, **53**, 899–904.

Lee N A and Rachmawati S (2006) 'A rapid ELISA for screening aflatoxin B$_1$ in animal feed and feed ingredients in Indonesia' *Food Agric Immunol*, **17**, 91–104.

Lee N A, Wang S, Allan R D and Kennedy I R (2004) 'A Rapid aflatoxin B$_1$ ELISA: development and validation with reduced matrix effects for peanuts, corn, pistachio, and soybeans' *J Agric Food Chem*, **52**, 2746–55.

Lee N A, Rachaputi N C, Wright G C, Krosch S, Norman K, Anderson J, Ambarwati S, Retnowati I, Dharmaputra O S and Kennedy I R (2005) 'Validation of analytical parameters of a competitive direct ELISA for aflatoxin B$_1$ in peanuts' *Food Agric Immunol*, **16**, 149–63.

Li P, Zhang Q, Zhang W, Zhang J, Chen X, Jiang J, Xie L and Zhang D (2009) 'Development of a class-specific monoclonal antibody-based ELISA for aflatoxins in peanut' *Food Chem*, **115**, 313–7.

Lipigorngosonyz S, Limtrakulz P, Suttajitz M and Yoshizaway T (2003) 'In-house direct cELISA for determining aflatoxin B$_1$ in Thai corn and peanuts' *Food Addit Contam*, **20**, 838–845.

Liu R R, Yu Z, He Q H and Xu Y (2007) 'An immunoassay for ochratoxin A without the mycotoxin' *Food Control*, **18**, 872–7.

Lobeau M, De Saeger S, Sibanda L, Barna-Vetro I and Van Peteghem C (2005) 'Development of a new clean-up tandem assay column for the detection of ochratoxin A in roasted coffee' *Anal Chim Acta*, **538**, 57–61.

Lobeau M, De Saeger S, Sibanda L, Barna-Vetro I and Van Peteghem C (2007) 'Application and validation of a clean-up tandem assay column for screening ochratoxin A in cocoa powder' *Food Addit Contam*, **24**, 398–405.

Maragos C M (2002) 'Novel assays and sensor platforms for the detection of aflatoxins' *Adv Exper Med Biol*, **504**, 85–93.

Maragos C M (2004) 'Emerging technologies for mycotoxin detection' *J Toxicol Toxin Rev*, **23**, 317–44.

Maragos C M and McCormick S P (2000) 'Monoclonal antibodies for the mycotoxins deoxynivalenol and 3-acetyl-deoxynivalenol' *Food Agric Immunol*, **12**, 181–192.

Maragos C M and Appell M (2007) 'Capillary electrophoresis of the mycotoxin zearalenone using cyclodextrin-enhanced fluorescence' *J Chromatogr A*, **1143**, 252–7.

Maragos C, Busman M and Sugita-Konishi Y (2006) 'Production and characterization of a monoclonal antibody that cross-reacts with the mycotoxins nivalenol and 4-deoxynivalenol' *Food Addit Contam*, **23**, 816–25.

Maragos C M, Appell M, Lippolis V, Visconti A, Catucci L and Pascale M (2008) 'Use of cyclodextrins as modifiers of fluorescence in the detection of mycotoxins' *Food Addit Contam*, **25**, 164–71.

Matrella R, Monaci L, Milillo M A, Palmisano F and Tantillo M G (2006) 'Ochratoxin A determination in paired kidneys and muscle samples from swines slaughtered in southern Italy' *Food Control*, **17**, 114–7.

Meister U (2001) 'Investigations on the change of fumonisin content of maize during hydrothermal treatment of maize. Analysis by means of HPLC methods and ELISA' *Eur Food Res Technol*, **213**, 187–93.

Moghaddam A, Løbersli I, Gebhardt K, Braunagel M and Marvik OJ (2001) 'Selection and characterisation of recombinant single-chain antibodies to the hapten aflatoxin-B$_1$ from naive recombinant antibody libraries' *J Immunol Methods*, **254**, 169–81.

Molinelli A, Grossalber K, Fuehrer M, Baumgartner S, Sulyok M and Krska R (2008) 'Development of qualitative and semiquantitative immunoassay-based rapid strip tests for the detection of T-2 toxin in wheat and oat' *J Agric Food Chem*, **56**, 2589–94.

Navarro-Villoslada F, Urraca J L, Moreno-Bondi M C and Orellana G (2007) 'Zearalenone sensing with molecularly imprinted polymers and tailored fluorescent probes' *Sensor Actuat B-Chem*, **121**, 67–73.

Nayak S, Sashidhar R B and Bhat R V (2001) 'Quantification and validation of enzyme immunoassay for urinary aflatoxin B_1-N7-guanine adduct for biological monitoring of aflatoxins' *Analyst*, **126**, 179–83.

Neagu D, Capodilupo A, Vilkanauskyte A, Micheli L, Palleschi G and Moscone D (2009) 'AFB$_1$–AP conjugate for enzyme immunoassay of aflatoxin B_1 in corn samples' *Anal Lett*, **42**, 1170–86.

Nilfer D and Boyacolu D (2002) 'Comparative study of three different methods for the determination of aflatoxins in tahini' *J Agric Food Chem*, **50**, 3375–9.

Olson J J and Chu F S (1993) 'Urinary-excretion of sterigmatocystin and retention of DNA adducts in liver of rats exposed to the mycotoxin – an immunochemical analysis' *J Agric Food Chem*, **41**, 602–6.

Ono E Y S, Kawamura O, Ono M A, Ueno Y and Hirooka E Y (2000) 'A comparative study of indirect competitive ELISA and HPLC for fumonisin detection in corn of the state of Parana, Brazil' *Food Agric Immunol*, **12**, 5–14.

Paepens C, De Saeger S, Sibanda L, Barna-Vetró I, Leglise I, Van Hove F and Van Peteghem C (2004) 'A flow-through enzyme immunoassay for the screening of fumonisins in maize' *Anal Chim Acta*, **523**, 229–35.

Peknejad M, Rasaee M J, Mohammadnejad J, Pouramir M, Rajabibazl M and Kakhki M (2008) 'Development and characterization of enzyme-linked immunosorbent assay for aflatoxin B-1 measurement in urine sample using penicillinase as label' *J Toxic Sci*, **33**, 565–73.

Pal A and Dhar T K (2004) 'An analytical device for on-site immunoassay. Demonstration of its applicability in semiquantitative detection of aflatoxin B_1 in a batch of samples with ultrahigh sensitivity' *Anal Chem*, **76**, 98–104.

Pal A, Acharya D, Saha D and Dhar T K (2004) 'Development of a membrane-based immunofiltration assay for the detection of T-2 toxin' *Anal Chem*, **76**, 4237–40.

Pal A, Acharya D, Saha D, Roy D and Dhar T K (2005) 'In situ sample cleanup during immunoassay: a simpe method for rapid detection of aflatoxin B_1 in food samples' *J Food Protect*, **68**, 2169–77.

Pestka J J (1991) 'High performance thin layer chromatography ELISAGRAM. Application of a multi-hapten immunoassay to analysis of zearalenone and aflatoxin mycotoxin families' *J Immunol Methods*, **136**, 177–83.

Pichler H, Krska R, Szekacs A and Grasserbauer M (1998) 'An enzyme-immunoassay for the detection of the mycotoxin zearalenone by use of yolk antibodies' *Fresenius J Anal Chem*, **362**, 176–177.

Posthuma-Trumpie G A, Korf J and van Amerongen A (2009) 'Lateral flow (immuno)assay: its strengths, weaknesses, opportunities and threats. A literature survey' *Anal Bioanal Chem*, **393**, 569–82.

Rieger M, Cervino C, Sauceda JC, Niessner R and Knopp D (2009) 'Efficient hybridoma screening technique using capture antibody based microarrays' *Anal Chem*, **81**, 2373–7.

Rubio R, Berruga M I, Roman M and Molina A (2009) 'Evaluation of immunoenzymatic methods for the detection of aflatoxin M_1 in ewe's milk' *Food Control*, **20**, 1049–52.

Rusanova T Y, Beloglazova N V, Goryacheva I Y, Lobeau M, Van Peteghem C and De Saeger S (2009) 'Non-instrumental immunochemical tests for rapid ochratoxin A detection in red wine' *Anal Chim Acta*, **656**, 97–102.

Quan Y, Zhang Y, Wang S, Lee N and Kennedy I R (2006) 'A rapid and sensitive chemiluminescence enzyme-linked immunosorbent assay for the determination of fumonisin B_1 in food samples' *Anal Chim Acta*, **580**, 1–8.

Saha D, Acharya D, Roy D, Shrestha D and Dhar T K (2007), 'Simultaneous enzyme immunoassay for the screening of aflatoxin B_1 and ochratoxin A in chili samples' *Anal Chim Acta*, **584**, 343–9.

Salter R, Douglas D, Tess M, Markovsky B and Saul S (2006) 'Interlaboratory study of the charm ROSA safe level aflatoxin M_1 quantitative lateral flow test for raw bovine milk' *J AOAC Int*, **89**, 1327–34.

Schneider E, Usleber E and Märtlbauer E (1995a) 'Rapid detection of fumonisin B$_1$ in corn-based food by competitive direct enzyme immunoassay/enzyme-linked immunofiltration assay with integrated negative control reaction' *J Agric Food Chem*, **43**, 2548–52.

Schneider E, Usleber E, Märtlbauer E, Deitrich R and Terplan G (1995b) 'Multimycotoxin dipstick enzyme immunoassay applied to wheat' *Food Addit Contam*, **12**, 387–93.

Schneider E, Curtui V, Seidler C, Dietrich R, Usleber E and Märtlbauer E (2004) 'Rapid methods for deoxynivalenol and other trichothecenes' *Toxicol Lett*, **153**, 113–121.

Scholl P F, Turner P C, Sutcliffe A E, Sylla A, Diallo M S, Friesen M D, Groopman J D and Wild C P (2006) 'Quantitative comparison of aflatoxin B$_1$ serum albumin adducts in humans by isotope dilution mass spectrometry and ELISA' *Cancer Epidemiol Biomarkers Prev*, **15**, 823–6.

Schuhmacher R, Krska R, Weingaertner J and Grasserbauer M (1997) 'Interlaboratory comparison study for the determination of the Fusarium mycotoxins deoxynivalenol in wheat and zearalenone in maize using different methods' *Fresenius' J Anal Chem*, **359**, 510–5.

Shelby R A, Rottinghaus G E and Minor H C (1994) 'Comparison of thin-layer chromatography and competitive immunoassay methods for detecting fumonisin on maize" *J Agric Food Chem*, **42**, 2064–7.

Shephard G S (2008) 'Determination of mycotoxins in human foods' *Chem Soc Rev*, **37**, 2468–77.

Shim W B, Yang Z Y, Kim J S, Kim J Y, Kang S J, Woo G J, Chung Y, C, Eremin S A and Chung D H (2007) 'Development of immunochromatography strip-test using nanocolloidal gold-antibody probe for the rapid detection of aflatoxin B$_1$ in grain and feed samples' *J Microbiol Biotechnol*, **17**, 1629–37.

Shim W B, Dzantiev B B, Eremin S A, Chung D H (2009a) 'One-step simultaneous immunochromatographic strip test for multianalysis of ochratoxin A and zearalenone' *J Microbiol Biotechnol*, **19**, 83–92.

Shim W B, Kim K Y and Chung D H (2009b) 'Development and validation of a gold nanoparticle immunochromatographic assay (ICG) for the detection of zearalenone' *J Agric Food Chem*, **57**, 4035–41.

Sibanda L, De Saeger S and Van Peteghem C (1999) 'Development of a portable field immunoassay for the detection of aflatoxin M in milk' *Int J Food Microbiol*, **48**, 203–9.

Sibanda L, De Saeger S, Van Peteghem C, Grabarkiewicz-Szczesna J and Tomczak M (2000) 'Detection of T-2 toxin in different cereals by flow-through enzyme immunoassay with a simultaneous internal reference' *J Agric Food Chem*, **48**, 5864–7.

Sibanda L, De Saeger S, Bauters T G M, Nelis H J and Van Peteghem C (2001) 'Development of a flow-through enzyme immunoassay and application in screening green coffee samples for ochratoxin A with confirmation by high-performance liquid chromatography' *J Food Protect*, **64**, 1597–602.

Sibanda L, De Saeger S, Barna-Vetro I and Van Peteghem C (2002) 'Development of a solid-phase cleanup and portable rapid flow-through enzyme immunoassay for the detection of ochratoxin A in roasted coffee' *J Agric Food Chem*, **50**, 6964–7.

Sun X L, Zhao X L, Tang J, Xiaohong G, Jun Z and Chu F S (2006) 'Development of an immunochromatographic assay for detection of aflatoxin B$_1$ in foods' *Food Control*, **17**, 256–62.

Sutikno, Abouzied M M., Azcona Olivera J I, Hart L P and Pestka J J (1996) 'Detection of fumonisins in *Fusarium* cultures, corn, and corn products by polyclonal antibody-based ELISA: relation to fumonisin B$_1$ detection by liquid chromatography' *J Food Protect*, **59**, 645 –51.

Suzuki T, Munakata Y, Morita K, Shinoda T and Ueda H (2007) 'Sensitive detection of estrogenic mycotoxin zearalenone by open sandwich immunoassay' *Anal Sci*, **23**, 65–70.

Sydenham E W, Shephard G S, Thiel P G, Bird C and Miller B M (1996) 'Determination of fumonisins in corn: evaluation of competitive immunoassay and HPLC techniques' *J Agric Food Chem*, **44**, 159–64.

Tanaka T, Teshima R, Ikebuchi H, Sawada J and Ichinoe M (1995) 'Sensitive enzyme-linked-immunosorbent-assay for the mycotoxin zearalenone in barley and jobs-tears' *J Agric Food Chem*, **43**, 946–50.

Tang D, Sauceda J C, Lin Z, Ott S, Basova E, Goryacheva I, Biselli S, Lin J, Niessner R and Knopp D (2009) 'Magnetic nanogold microspheres-based lateral-flow immunodipstick for rapid detection of aflatoxin B_2 in food' *Biosens Bioelectron*, **25**, 514–8.

Thirumala-Devi K, Mayo M A, Delfosse P, Reddy G, Reddy S V and Reddy D V R (2000) 'Production of polyclonal antibodies against ochratoxin A and its detection in chillies by ELISA' *J Agric Food Chem*, **48**, 5079–82.

Thirumala-Devi K, Mayo M A, Hall A J, Craufurd P Q, Wheeler T R, Waliyar F, Subrahmanyam A and Reddy D V R (2002) 'Development and application of an indirect competitive enzyme-linked immunoassay for aflatoxin M_1 in milk and milk-based confectionery' *J Agric Food Chem*, **50**, 933–7.

Turner N W, Subrahmanyam S and Piletsky S A (2009) 'Analytical methods for determination of mycotoxins: A review' *Anal Chim Acta*, **632**, 168–80.

Usleber E, Schneider E, Maertlbauer E and Terplan G (1993) 'Two formats of enzyme immunoassay for 15-acetyldeoxynivalenol applied to wheat' *J Agric Food Chem*, **41**, 2019–23.

Vidyasagar T, Sujatha N and Sashidhar R B (1997) 'Determination of Aflatoxin B_1–DNA Adduct in Rat Liver by Enzyme Immunoassay' *Analyst*, **122**, 609–13.

Visconti A and De Girolamo A (2005) 'Fitness for purpose – ochratoxin A analytical developments' *Food Addit Contam*, **Supplement 1**, 37–44.

Wang S, Quan Y, Lee N and Kennedy I R (2006) 'Rapid determination of fumonisin B_1 in food samples by enzyme-linked immunosorbent assay and colloidal gold immunoassay' *J Agric Food Chem*, **54**, 2491–5.

Wang S H, Du X Y, Huang Y M, Lin D S, Hart P L and Wang Z H (2007) 'Detection of deoxynivalenol based on a single-chain fragment variable of the antideoxynivalenol antibody' *FEMS Microbiol Lett*, **272**, 214–9.

Wang S H, Du X Y, Lin L, Huang Y M and Wang Z H (2008) 'Zearalenone (ZEN) detection by a single chain fragment variable (scFv) antibody' *World J Microbiol Biotechnol*, **24**, 1681–5.

Wang X H, Liu T, Xu N, Zhang Y and Wang S (2007) 'Enzyme-linked immunosorbent assay and colloidal gold immunoassay for ochratoxin A: investigation of analytical conditions and sample matrix on assay performance' *Anal Bioanal Chem*, **389**, 903–11.

Xia X, Xu Y, Zhao X and Li Q (2009) 'Lateral flow immunoassay using europium chelate–loaded silica nanoparticles as labels' *Clin Chem*, **55**, 179–82.

Xu B J, Jia X Q, Gu L J and Sung C K (2006) 'Review on the qualitative and quantitative analysis of the mycotoxin citrinin' *Food Control*, **17**, 271–85.

Yoshizawa T, Kohno H, Ikeda K, Shinoda T, Yokohama H, Morita K, Kusada O and Kobayashi Y (2004) 'A practical method for measuring deoxynivalenol, nivalenol and T-2+HT-2 toxin in foods by an enzyme-linked immunosorbent assay using monoclonal antibodies' *Biosci Biotechnol Biochem*, **68**, 2076–85.

Yu F Y and Chu F S (1999a) 'Production and characterization of a monoclonal anti-anti-idiotype antibody against fumonisin B_1' *J Agric Food Chem*, **47**, 4815–20.

Yu F Y and Chu F S (1999b) 'Production and characterization of monoclonal antibodies against fumonisin B_1' *Food Agric Immunol*, **11**, 297–306.

Yu F Y, Chi T F, Liu B H and Su C C (2005) 'Development of a sensitive enzyme-linked immunosorbent assay for the determination of ochratoxin A' *J Agric Food Chem*, **53**, 6947–53.

Zachariasova M, Hajslova J, Kostelanska M, Poustka J, Krplova A, Cuhra P and Hochel I (2008) 'Deoxynivalenol and its conjugates in beer: A critical assessment of data obtained by enzyme-linked immunosorbent assay and liquid chromatography coupled to tandem mass spectrometry' *Anal Chim Acta*, **625**, 77–86.

Zhang C, Zhang Y and Wang S (2006), 'Development of multianalyte flow-through and lateral-flow assays using gold particles and horseradish peroxidase as tracers for the rapid determination of carbaryl and endosulfan in agricultural products' *J Agric Food Chem*, **54**, 2502–7.

Zhang D, Li P, Zhang Q, Zhang W, Huang Y, Ding X and Jiang J (2009) 'Production of ultrasensitive generic monoclonal antibodies against major aflatoxins using a modified two-step screening procedure' *Anal Chim Acta*, **636**, 63–9.

Zhang G and Chu F S (1989) 'Production and characterization of antibodies cross-reactive with major aflatoxins' *Experientia*, **45**, 182–4.

Zheng Z M, Hanneken J, Houchins D, King RS, Lee P and Richard JL (2005a) 'Validation of an ELISA test kit for the detection of ochratoxin A in several food commodities by comparison with HPLC' *Mycopathologia*, **159**, 265–72.

Zheng Z, Humphrey C W, King R S and Richard J L (2005b) 'Validation of an ELISA test kit for the detection of total aflatoxins in grain and grain products by comparison with HPLC' *Mycopathologia*, **159**, 255–63.

Zheng M Z, Richard J L and Binder J (2006) 'A review of rapid methods for the analysis of mycotoxins' *Mycopathologia*, **161**, 261–273.

Part II

Quality assurance and official methods for determining mycotoxins in food and feed

6

Official methods and performance criteria for determining mycotoxins in food and feed

H. Z. Senyuva, FoodLife International Ltd, Turkey and J. Gilbert, FoodLife International Ltd, UK

Abstract: The concept of a criteria-based approach for regulatory methods has been adopted in Europe by the European Commission and worldwide by Codex Alimentarius. For mycotoxins in food and feed, EU Directives stipulate that only validated methods can be employed for enforcement purposes and stipulate the minimum performance characteristics such methods must meet. Thus recovery ranges, limits of detection and quantification, repeatability and reproducibility are stipulated for each mycotoxin which is regulated and these minimum performance characteristics are given as a function of the regulatory limit. Notwithstanding the general acceptance of the benefits of adopting a criteria-based approach, some countries outside the EU still have a regulatory framework which requires the publication of 'Official methods' in their own Regulations, and in these countries endorsement of methods by bodies such as AOAC International (Official Methods) or CEN is taken as mandatory. Unfortunately, both AOAC and CEN are very slow to adopt and standardise new methods (frequently taking many years), which can hamper access to new technologies being used for enforcement purposes. In this chapter the current 'state-of-the-art' in terms of validated methods for mycotoxins for food and feed is reviewed and the extent to which these methods meet minimum EU performance criteria or have been adopted elsewhere as official is addressed. Whether a decision based entirely on non-specific detection methodology (e.g. HPLC fluorescence detection) is sufficient for enforcement action without further confirmation is discussed, and thus the place of LC/MS/MS in the regulatory process is reviewed.

Key words: AOAC International, CEN, method performance, method validation, official control of mycotoxins, official methods.

6.1   Introduction

The main driving force behind the development of official methods for mycotoxins in food and feed has been the progressive introduction of regulatory limits particularly by the European Union (EU). Several detailed compilations of world-wide regulations for mycotoxins have been published (FAO, 2004; van Egmond and Jonker, 2004) as well as some explanations of the rationale for these controls (van Egmond *et al.*, 2007). The Codex Alimentarius Commission currently has maximum levels of $10\,\mu g\,kg^{-1}$ for total aflatoxins in 'ready-to-eat' peanuts, almonds, hazelnuts and pistachios, $0.5\,\mu g\,kg^{-1}$ for aflatoxin M_1 in milk, $5\,\mu g\,kg^{-1}$ for ochratoxin A in raw wheat, barley and rye, and $50\,\mu g\,kg^{-1}$ for patulin in apple juice (Codex 1995).

Whilst it is not appropriate in this chapter to discuss the detail of these regulatory limits, it is worthwhile pointing out the main differences in limits globally are between the group of countries (principally USA and MERCUSUR, The Common Market of South America) where the limit for total aflatoxins in foods for human consumption is set at $20\,\mu g\,kg^{-1}$, and the EU and others where the limit for total aflatoxins is set at $4\,\mu g\,kg^{-1}$. The EU and a number of other countries has a combined limit where a limit of $2\,\mu g\,kg^{-1}$ is also applied for aflatoxin B_1, so neither the $4\,\mu g\,kg^{-1}$ total aflatoxin nor $2\,\mu g\,kg^{-1}$ aflatoxin B_1 limit should be exceeded. There are however some moves towards harmonising limits and discussions taking place facilitated by Codex Alimentarius to bring limits in line with one another.

For aflatoxin M_1 in milk there are similar differences, with the USA regulating at $0.5\,\mu g\,kg^{-1}$ whilst the EU has a limit some 10-fold lower ($0.05\,\mu g\,kg^{-1}$). The EU also regulates for ochratoxin A in cereals with limits ranging from 0.5 to $10.0\,\mu g\,kg^{-1}$ depending on whether the cereal in question is intended for infants and young children (lower limit) or whether a processed cereal for general food use. EU limits for ochratoxin A also apply to dried vine fruit, coffee, beer, wine, cocoa, meat products spices and liquorice with limits targeted at each specific product (European Commission, 2006a).

The EU has recently regulated levels of fumonisins in corn and corn-based foods (European Commission, 2006a) basing regulatory limits on the sum of fumonisin B_1 (FB_1) + fumonisin B_2 (FB_2). For unprocessed corn, with the exception of unprocessed corn intended to be processed by wet milling the sum of $FB_1 + FB_2$ should not exceed $4000\,\mu g\,kg^{-1}$. In maize-based foods for direct human consumption, with the exception of breakfast cereals, snacks and baby foods, levels of $FB_1 + FB_2$ should not exceed $1000\,\mu g\,kg^{-1}$. For corn-based breakfast cereals and corn-based snacks, a limit of $800\,\mu g\,kg^{-1}$ for $FB_1 + FB_2$ is applied and for processed corn-based foods and baby foods for infants and young children there is a lower regulatory limit of $200\,\mu g\,kg^{-1}$ (European Commission, 2006a).

In contrast, in the USA there are recommended maximum levels (as opposed to regulatory limits) for fumonisins in human foods and in animal feeds (FDA, 2001) that FDA considers achievable with the use of good agricultural and good manufacturing practices. The recommended FDA limits are expressed as total fumonisins, but in this case total is $FB_1+FB_2+FB_3$ compared to the EU limits, which are based only on the total of FB_1+FB_2. The advisory limits in the USA are also significantly higher than EU limits. For de-germed dry milled corn products the limit is 2000 µg kg^{-1} and for whole or partially de-germed dry milled corn products the limit is 4000 µg kg^{-1}. For dry milled corn bran and cleaned corn intended for mass production the limit is 4000 µg kg^{-1}. For cleaned corn intended for popcorn, the limit is 3000 µg kg^{-1}. (FDA, 2001). The FDA has advisory upper levels for deoxynivalenol in cereals but no statutory maximum limits.

The EU also has limits that apply to patulin in apple juice and apple-based products (ranging from 10–50 µg/kg), deoxynivalenol in cereals (ranging from 200–750 µg kg^{-1} for products intended for direct human consumption) zearalenone (ranging from 20–400 µg kg^{-1}). The intention is also to regulate T-2 toxin and HT-2 toxin with a limit based on the sum total of the two toxins (European Commission, 2006a). There is a considerable complexity in the EU limits, as different upper limits apply to unprocessed commodities. Where it is anticipated there will be a reduction in processing, limits apply to finished ready-to-eat products and significantly lower limits apply to foods intended for infants where additional health protection is provided.

The relationship between the regulatory limits and official methods is that the later must be demonstrated to provide a suitable tool for enforcement of these limits. This means that the official method must have adequate sensitivity so that unequivocal measurement can be made at and below the regulatory limit, the method must have been tested on the various commodities stipulated in the regulations and must have been demonstrated to be sufficiently robust to perform well in the hands of those in different food control laboratories and generate comparable results. Finally as the import and export of foods involves testing across borders, it is also helpful if the official method has international standing, that is it has been endorsed by an international body.

6.2   Official control laboratories for determining mycotoxins in food and feed

The Codex Alimentarius Commission (Codex) was created in 1963 by FAO and WHO to develop food standards, guidelines and related texts such as codes of practice under the Joint FAO/WHO Food Standards Programme. The main purpose of this programme is to protect the health of consumers, ensure fair trade practices in the food trade and promote coordination of all food standards work undertaken by international governmental and non-governmental organizations. Codex has proposed guidelines to provide a framework for the implementation of quality assurance measures to ensure the competence of testing laboratories involved in the import and export control of foods. These guidelines (Codex, 1997) are intended to assist countries in the application of requirements for trade in foodstuffs in order to protect the consumers and to facilitate fair trade. Laboratories involved in the import and export control of foods should comply with four main requirements; first they should meet the general criteria for testing laboratories (ISO/IEC 1993), that is they should be accredited for the analysis of specified mycotoxins. Second they should participate in appropriate proficiency testing schemes for food analysis which conform to the specified requirements (IUPAC/ AOAC, 2006), that is demonstrate their analytical competence, for example through participation in FAPAS®. Third, whenever available, control laboratories should use methods of analysis which have been validated according to the principles laid down by the Codex Alimentarius Commission. Fourth, they should use prescribed internal quality control procedures (IUPAC/AOAC, 1995). The bodies assessing import/export control laboratories should comply with the general criteria for laboratory accreditation, such as those laid down by ISO/IEC (ISO/ IEC, 1993).

These general Codex requirements for import/export testing laboratories including the use of validated methods of analysis are essentially the same as those required for food control laboratories within the EU, and implicit in this requirement is that such validated methods must perform to minimum standards. Elsewhere in the world these principles have not necessarily been applied universally and in a number of countries the approach still remains that only a designated method ('the official method') can be used for the purpose of food control. In these cases such official methods are often enshrined in legislation which makes a rigid system lacking flexibility.

6.3   Establishment of method performance criteria for determining mycotoxins in food and feed

Where an official method is either a single prescribed method for the analysis of a mycotoxin or whether it is one of several methods meeting minimum performance criteria, it nevertheless must be 'fit-for-purpose'. This means that the performance of the method must be such as to enable a clear decision to be taken about whether the mycotoxin being tested is present above or below the regulatory limit.

The method must have sufficient specificity so that there is some certainty that the mycotoxin being determined has been correctly identified and that there are no interfering components which might lead to an overestimate of concentration. Apart from ensuring correct identification, the official method must also have a limit of quantification (LOQ) that is below the regulatory limit by an adequate margin of safety to minimise the uncertainty of measurement near the LOQ. The method must also have an established precision that will need to be taken into account when a decision is taken, about whether a sample contains the mycotoxin at a level above or below the regulatory limit. The method recovery is also important and it is customary to require methods to have demonstrable recoveries within a certain range. In the mycotoxin field by convention, results are corrected for recovery and a batch average recovery figure needs to be established in individual control laboratories. This recovery figure is usually employed for correction purposes.

The five parameters that describe the performance of a method are the limit of detection (LOD), limit of quantification (LOQ), recovery, repeatability (intra-laboratory precision) and reproducibility (inter-laboratory precision). Four of these five parameters can be determined in one laboratory (single-laboratory validation) but the reproducibility can only be established by conducting a full collaborative study with valid (statistically acceptable) results being obtained for a minimum of eight participants.

6.3.1   Single laboratory (in-house or intra-laboratory) validation

A harmonised protocol has been established (Thompson *et al.*, 2002) for single laboratory method validation. This protocol stipulates that duplicate analyses of naturally contaminated samples must be carried out at four different concentration levels to cover the range of interest (normally bracketing the regulatory limit), and that the method recovery must be established by spiking a blank sample of the matrix of interest. The intra-laboratory precision method known as the method repeatability, is directly determined from the four sets of duplicate measurements and is reported as a relative standard deviation (% RSD_R). The LOD can be estimated as being the amount which would give a signal three times the noise level of a blank sample. Although this can be estimated only from analysing a blank sample, it is more meaningful to demonstrate that the theoretical LOD of 3 × signal-to-noise is actually achievable in practice, by carrying out a spiking of the blank sample close to the anticipated LOD and then measuring the peak size. Single laboratory method validation is an essential minimum requisite for any laboratory wishing to gain ISO17025 accreditation and is also a useful preliminary step before undertaking a full inter-laboratory method validation.

6.3.2   Inter-laboratory method validation

An IUPAC/AOAC harmonised protocol has been established (Horwitz, 1995) for inter-laboratory method validation. The parameters determined in a full inter-laboratory study are essentially the same as those described above for a single laboratory validation, except that it is also possible to establish the inter-laboratory method precision known as method reproducibility, which is reported as a relative standard deviation (% RSD_R). The nature of an inter-laboratory study also means that blind duplicate samples can be distributed for analysis and this can include blind 'blank' samples. It is normal to supply additionally a designated blank sample for spike recovery measurements and this can be done preferably using a common spike solution provided to all participants or it may be that participants are permitted to use their own standard solutions for spiking.

An inter-laboratory method validation study requires a study director who takes overall responsibility for design of the study, selection and invitation of participants, drafting of the study protocol and standard operating procedure (SOP), sample preparation including demonstration of sample homogeneity, sample distribution, receipt of all results from participants, statistical analysis and finally drafting a report for publication and/or submission to CEN and/or AOAC International for adoption as an official method. Organisation of an inter-laboratory study is an expensive and time-consuming process which should not be undertaken lightly. It may take as long as 12 months from initiation to completion of a study and a substantial subsequent time for the method to gain official status.

An inter-laboratory method validation study provides an additional dimension in terms of method testing in that it rigorously examines how well the SOP can be followed by participants in different countries using a range of different laboratory equipment and chemicals. Although this is not formally a testing of method robustness, this process inevitably tests how well the method is described and how well the prescribed parameters for example HPLC and LC/MS conditions, can be applied to different instruments.

It is customary to involve around 12 participants in an inter-laboratory validation study based on an assumption that some may not complete the study and some results may be classified as outliers. For satisfactory statistical analysis, a full set of valid results must be obtained from a minimum of eight participants which is normally achievable starting with 12 participants. By convention although the identity of the participants in a study is published, their individual performance in the study remains confidential.

After completion of the study, the study director will carry out the statistical analysis and then, after tabulation of the results, will allow participants the opportunity to check that data have been accurately transcribed. A report will then be written which may be in the form of a peer-reviewed publication, for example submitted to JAOAC International and/or may be submitted to AOAC for consideration for adoption as an Official Method. Alternatively or additionally the completed study may be submitted to CEN (see below) or may be adopted as a national official method.

6.4   Official methods for determining mycotoxins in food and feed

There are a number of international bodies who adopt and endorse methods of analysis for mycotoxins and these are listed in Table 6.1. These organisations are either interested in methods of analysis *per se* (AOAC International, IUPAC, ISO and CEN) or they are international bodies representing sectors of the food chain such as cereals (ICC and AACC), dairy products (IDF), oil-seed products including animal feed (AOCS) and fruit juices (IFU). AOAC International, IUPAC, ISO and CEN do little to initiate collaborative studies and are relatively passive, whereas the food chain sector organisations do initiate collaborative studies themselves where they see a gap in terms of accepted methods. Thus a Joint IDF-IUPAC-IAEA (FAO) interlaboratory study (Grosso *et al.*, 2004) of a method for aflatoxin M_1 in powdered milk using thin layer chromatography (TLC) was conducted, which met IDF needs worldwide, although being TLC-based was unlikely to be of much interest to other method-orientated organisations. Many of

Table 6.1 International bodies adopting, endorsing or standardising methods of analysis for mycotoxins

International body	Total no methods	Mycotoxins	Web reference
Codex Alimentarius Commission (CAC)	Criteria	B_1, M_1, Pat, OTA	http://www.codex alimentarius.net
AOAC International	54	B_1, M_1, Pat, OTA, DON, FB_1, FB_2, FB_3	http://www.aoac.org
European Commission (DG SANCO)	Criteria	B_1, M_1, Pat, OTA, DON, FB_1, FB_2	http://ec.europa.eu/dgs/ health_consumer/
International Standards Organisation (ISO)	11	B_1, M_1, OTA, Pat, ZON	http://www.iso.org
Comité Européen de Normalisation (CEN)	15	B_1, M_1, Pat, OTA, DON, FB_1, FB_2	http://www.cen.eu
International Union of Pure & Applied Chemistry (IUPAC)	5	B_1, Pat, FB_1, FB_2, FB_3	http://www.iupac.org
American Association of Cereal Chemists (AACC)	6	B_1, DON, FB_1, FB_2, FB_3	http://www.aaccnet.org
American Oil Chemists Association (AOCS)	14	B_1, ZON, FB_1, FB_2	http://www.aocs.org
International Dairy Federation (IDF)	3	M_1	http://www.fil-idf.org
International Fruit Juice Union (IFU)	2	OTA, Pat	http://www.ifu-fruit juice.com
International Association for Cereal Chemists (ICC)	1	OTA	http://www.icc.or.at/

B_1 = aflatoxin B_1; M_1 = aflatoxin M_1; Pat = patulin; OTA = ochratoxin A; DON = deoxynivalenol; FB_1 = fumonisin B_1; FB_2 = fumonisin B_2; ZON = zearalenone.

the official methods have been either the result of a joint activity, for example AOAC-IUPAC, AACC-AOAC, or they relate to the same original collaborative study publication.

For mycotoxins in food and feed, EU Directives stipulate that only validated methods can be employed for enforcement purposes and, consistent with the Codex approach, the EU stipulates the minimum performance characteristics such methods must meet (European Commission, 2006b). Thus recovery ranges, limits of detection and quantification, repeatability and reproducibility are stipulated for each mycotoxin which is regulated and these minimum performance characteristics are given as a function of the regulatory limit. Notwithstanding the general acceptance of the benefits of adopting a criteria-based approach, some countries outside the EU still have a regulatory framework which requires the publication of 'official methods' in their own regulations, and in these countries endorsement of official methods by bodies such as AOAC International (Official Methods) or CEN is taken as mandatory. Unfortunately, both AOAC and CEN are very slow to adopt and standardise new methods (frequently taking many years), which can hamper access to new technologies being used for enforcement purposes.

Neither CEN nor the EU in setting method performance criteria for mycotoxins have made any stipulations concerning the criteria for establishing the correct identification. This is in marked contrast, for example, with the area of veterinary drug residues where a points system is employed for identification, points being awarded for different detection systems and a minimum number of points being needed to have rigorously demonstrated correct identification (European Commission, 2002). The reason behind this difference is historical rather than having any rational basis and derives from an assumption that in the mycotoxin field the fluorescence characteristics of some mycotoxins will provide a sufficient degree of specificity in detection. If this property is coupled to the use of highly specific clean up such as employing immunoaffinity columns, there is probably no need for further confirmation as co-extractives are completely removed (Senyuva and Gilbert, 2010). However, this is not the case for some mycotoxins such as deoxynivalenol, T-2 and HT-2 toxins and patulin, and arguably in these cases more rigor should be required in official methods to establish some criteria to ensure correct identification.

6.4.1   Criteria-based approach

The concept of a criteria-based approach for regulatory methods for contaminants and toxins in foods was proposed by the Codex Alimentarius Commission (Codex, 1995). For the analysis of aflatoxins, Codex stipulates that for control limits in the range 1–15 μg kg^{-1} the method must have a recovery of from 70 to 110%, an RSD_R derived from the Horwitz equation (Horwitz and Albert, 2006) and an RSD_r calculated as 0.66 times the precision RSD_R at the concentration of interest. For aflatoxin M_1, ochratoxin A and patulin, which are also designated toxins in Codex standards, recoveries in the concentration range 1–15 μg kg^{-1} must be between 70 to 110% and at >15 μg kg^{-1} must be in the range 80 to110%. The RSD_R in the concentration range 1–120 μg/kg should be derived from the modified Horwitz equation (Thompson, 2000), and above 120 μg kg^{-1} from the Horwitz equation (Horwitz and Albert, 2006). For all concentration ranges the RSD_r should be calculated as 0.66 times the RSD_R (Codex, 1995).

In Table 6.2 the minimum method performance criteria that have been established by both CEN and the European Commission are summarised. The CEN criteria (European Commission for Standardization, CEN, 1999a) were drafted prior to the current EU limits being in place and thus there are some inconsistencies particularly in the concentration ranges not being aligned with legislation. CEN is currently (as of 2009) in the process of producing new criteria guidelines which will presumably overcome these anomalies.

6.4.2   CEN official methods

CEN is the abbreviation used for the Comité Européen de Normalisation and is the European body responsible for standardisation. CEN has a committee structure comprising representatives of each of the 30 national standardisation bodies and

Table 6.2 Minimum performance criteria of methods to be used for food control purposes in the EU

Official body/ref	Mycotoxin	Regulatory limit ($\mu g\ kg^{-1}$)	RSD_r (%)	RSD_R (%)	Recovery (%)
CEN 1999a	Aflatoxin B_1	<1.0	≤40	≤60	50–120
CEN 1999a	Aflatoxin B_1	1–10	≤20	≤30	70–110
EC 2006	Total aflatoxins	<1.0	$0.66 \times RSD_R$	Horwitz*	50–120
EC 2006	Total aflatoxins	1–10	$0.66 \times RSD_R$	Horwitz*	70–110
CEN 1999a	Total aflatoxins	1–10	≤40	≤60	70–110
CEN 1999a	Aflatoxin M_1	0.01–0.05	≤30	≤50	60–120
EC 2006	Aflatoxin M_1	0.01–0.05	$0.66 \times RSD_R$	Horwitz*	60–120
CEN 1999a	Ochratoxin A	<1.0	≤40	≤60	50–120
CEN 1999a	Ochratoxin A	1–10	≤20	≤30	70–110
EC 2006	Ochratoxin A	<1.0	≤40	≤60	50–120
EC 2006	Ochratoxin A	1–10	≤20	≤30	70–110
CEN 1999a	Patulin	20–50	≤20	≤30	70–105
EC 2006	Patulin	<20	≤40	≤60	50–120
EC 2006	Patulin	20–50	≤20	≤30	70–110
CEN 1999a	DON & NIV	>100	≤20	≤40	70–110
EC 2006	DON	>500	≤20	≤40	70–110
CEN 1999a	Zearalenone	>100	≤25	≤40	70–100
EC 2006	Zearalenone	>50	≤25	≤40	70–120
CEN 1999a	T-2 toxin	50–250	≤40	≤60	60–120
EC 2006	T-2 toxin	50–250	≤40	≤60	60–130
CEN 1999a	HT-2 toxin	100–200	≤40	≤60	60–120
EC 2006	HT-2 toxin	100–200	≤40	≤60	60–130
CEN 1999a	FB_1 or FB_2	<500	≤30	≤60	60–120
CEN 1999a	FB_1 or FB_2	500–5000	≤20	≤30	70–110
EC 2006	$FB_1 + FB_2$	<500	≤30	≤60	60–120
EC 2006	$FB_1 + FB_2$	>500	≤20	≤330	70–110

Total aflatoxins = sum of $B_1 + B_2 + G_1 + G_2$; DON = deoxynivalenol; NIV = nivalenol; FB_1 = fumonisin B_1; FB_2 = fumonisin B_2
* Horwitz means that the RSD_R should be derived from the Horwitz equation (Horwitz and Albert, 2006)

also has some 16 affiliate members. Technical Committee 275 (TC275) deals with horizontal methods of analysis for food and has a number of working groups of which WG5 has responsibility for methods of determining biotoxins in food. Biotoxins is the term used in this context to mean both mycotoxins and phycotoxins. The convener for WG5 is the Dutch standardization body Nederlands Normalisatie Instituut (NEN)

At an early stage in its development, TC275 WG5 established a protocol much like that of the European Commission whereby the minimum performance criteria for methods to be considered by CEN for adoption as standards are stipulated (European Committee for Standardization, CEN, 1999a). This criteria document is currently being revised and updated (European Committee for Standardization, CEN, 2009a). Any method for the analysis of mycotoxins in foods which meets these minimum performance criteria can be taken forward by WG5, although it requires a member of the WG to act as a sponsor of the method and needs to be formally adopted by TC275 as part of its CEN mandate. The method sponsor needs to ensure that the method is drafted in CEN format (similar to ISO format) and takes responsibility for 'steering' the method through the committee in terms of any re-drafting that might be required to respond to comments from fellow members of WG5.

In general WG5 does not scrutinise the way in which the collaborative trial data for the proposed standard was obtained, but tends to be more concerned with presentational issues, that is drafting style. Using this approach there are multiple CEN standards (based on different method principles) for the same mycotoxin in the same food matrix. It is also possible to find a method which was validated by an inter-laboratory collaborative study and was subsequently adopted as a CEN standard to be the same method as was also accepted by AOAC as an Official Method. Thus, the same method with identical method performance characteristics may be available as both a CEN standard and as an AOAC First or Final Action Official Method, although the presentational styles may be so different that it is not always apparent that the methods are identical. After acceptance of a method by WG5 it is taken forward to the parent TC275 and is then formally voted upon before being ultimately accepted as a CEN standard. All CEN standards are available in English, French and German which are the three official languages of CEN.

TC275 WG5 has so far adopted some 15 methods of analysis as CEN standards for the determination of mycotoxins covering aflatoxins B_1, B_2, G_1 and G_2 in food (European Committee for Standardization, CEN, 1999b; 2007b; 2008c) and feed (European Committee for Standardization, CEN, 2006c), aflatoxin M_1 (European Committee for Standardization, CEN, 2003a; 2007a), ochratoxin A (European Committee for Standardization, CEN, 1998a; 1998b; 2006a; 2006b; 2008a; 2008b), fumonisins (European Committee for Standardization, CEN, 2001; 2004) and patulin (European Committee for Standardization, CEN, 2003b). Methods for patulin, deoxynivalenol and zearalenone are finalised and awaiting formal approval (European Committee for Standardization, CEN, 2009b; 2009c; 2009d).

To a large extent CEN standards are driven by EU regulations covering mycotoxins in food and feed, although this is not a formal requirement and methods are sometimes adopted ahead of regulations being put in place or might have been driven by a need to support the food and feed industry where standards for levels of mycotoxins are used for trading purposes. For example, with the proliferation of mycotoxin test kits in the market place, CEN has published guidelines which aim to standardise the description of ELISA kits for aflatoxin M_1 analysis (European Committee for Standardization, CEN, 2003a). In fact this is an area where a wider initiative is really needed and consistent standards to be applied by manufacturers and suppliers of antibody-based products to describe the performance of their products would provide a useful basis for enabling comparisons to be made by potential purchasers of these products.

6.4.3   AOAC Official Methods

AOAC International has a long history of production of Official Methods for a range of parameters in food and feed including those of mycotoxins. AOAC International has historically produced an *Official Book of Analytical Methods* (OMA) every 3–5 years which was sold as a hard copy (ring-bound) publication. Around 2003 the electronic format online OMA was introduced as a 'continuous edition', that is new and revised methods are posted as soon as approved and ready. The preferred approach by AOAC International has historically been to be closely involved in method validation, at an early stage in the process, starting by approval of the collaborative study protocol before the study has even been conducted. AOAC International has a technical committee of experts in mycotoxin analysis who are involved in critical assessment of the conduct of the collaborative study (Methods Committee on Natural Toxins and Allergens). Any candidate method for status as an AOAC Official Method must have been validated in strict accordance with the AOAC/IUPAC Harmonised Protocol for Interlaboratory Studies (Horwitz, 1995). This scrutiny of the conduct of the collaborative trial and the resulting data is the main difference between AOAC and CEN. CEN is not prescriptive concerning the conduct of the study and really only focuses on whether the performance characteristics meet their stipulated minimum standards. In contrast AOAC does not have any minimum standards for methods and adopts methods on a case-by-case basis. However, in general AOAC International assesses method performance on the basis of the HORRAT (Horwitz) ratio being taken as a measure of fitness-for-purpose. HORRAT values of below 2 would be seen as a demonstration of adequate method performance (Horwitz and Albert, 2006).

The Methods Committee on Natural Toxins and Allergens (made-up of analytical specialist volunteers) initially considers proposed methods and if deemed acceptable, these are then passed to the Official Methods Board who formally accept methods as AOAC First Action. First Action methods have a provisional status and the philosophy is that after adoption there is a need for a period of time during which the method is been used in practical situations to test its robustness further. If no problems emerge during five years or more of practical application the method then moves from First Action to Final Action status.

In the past there were no limitations as to the number of mycotoxin methods that could be taken forward for adoption as AOAC Official Methods and the only rate-limiting step was the workload of the volunteer Methods Committee on Natural Toxins and Allergens. However, from around 2003/2004 AOAC International moved away from its traditional approach and instead required a payment of around US$30 000 for methods to be submitted to be considered for official adoption. This has had the effect of severely limiting the flow of methods being put forward and where methods are now proposed there has been a strong push from those with commercial interests, for example for adoption of proprietary methods such as the use of mycotoxin test kits.

In Table 6.3 the AOAC Official Methods for mycotoxins are listed and it can be seen that only two mycotoxin methods have been adopted by AOAC International since 2003, which contrasts with CEN which has standardised or had under approval some 13 mycotoxin methods between 2003 and 2009. There is some commonality between AOAC Official Methods and CEN standards with eight of

Table 6.3   AOAC Official Methods for mycotoxins

Mycotoxin	Matrix	Method principle	Date*	AOAC Ref
Aflatoxins	Methanol	Molar absorbances	1970	970.44
Aflatoxins	Mixed solvents	Molar absorbances	1971	971.22
Aflatoxin B_1	Identification	TLC spots extracted	1970	970.47
Aflatoxins	Coconut, copra	Defat, column clean up & TLC	1971	971.24
Aflatoxins	Cocoa beans	Defat, column clean up & TLC	1971	971.23
Aflatoxins	Soybeans	Defat, column clean up & TLC	1972	972.27
Aflatoxins	Corn	Defat, column clean up & TLC	1972	972.26
Aflatoxins	Corn & peanuts	Silica gel col. clean up & TLC	1993	993.17
Aflatoxins	Food and feed	Romer minicolumn	1975	975.36
Aflatoxin B_1	Eggs	Silica gel clean up & TLC	1978	978.15
Aflatoxins	Corn & peanuts	Holaday–Velasco minicolumn	1979	979.18
Aflatoxins	Peanuts & prod	BF method (TLC)	1970	970.45
Aflatoxins	Peanuts	Alternative BF method (TLC)	1998	998.03
Aflatoxins	Cottonseed	Silica gel clean up & TLC/LC	1980	980.20
$B_1, B_2 G_1$	Corn, peanuts etc	ImmunoDot Screen Cup	1990	990.34
$B_1, B_2 G_1$	Corn	Afla-20 Cup (ELISA)	1993	993.16
Aflatoxin B_1	Corn & peanuts	Agri-Screen (ELISA)	1990	990.32
Aflatoxins	Corn, peanut butter	Silica gel clean up & LC	1990	990.33
Aflatoxins	Corn, peanuts	IAC (Aflatest) clean up- LC	1991	991.31
Aflatoxins	Corn & nuts	Mycosep clean up – 2D TLC	1994	994.08
Aflatoxins	Green coffee	Florisil clean up – TLC	1970	970.46
Aflatoxins	Pistachios	TLC	1974	974.16
Aflatoxin B_1	Identification	Derivative formation –TLC	1975	975.37
Aflatoxin B_1	Confirmation	TLC spots – MS (NICI)	1985	985.17
Aflatoxins	Nuts, figs & paprika	IAC clean up HPLC (Br)	1999	999.07[a]
Aflatoxin B_1	Baby food	IAC clean up HPLC (Br)	2000	2000.16[b]
Aflatoxin B_1	Cattle feed	IAC clean up HPLC (Br)	2003	2003.02[c]
Aflatoxins	Corn & peanuts	HPLC – photochemical deriv.	2005	2005.08
Aflatoxins	Ginseng & ginger	Multi-IAC – LC (Br)	2008	2008.02
Aflatoxin M_1	Dairy products	Silica gel clean up & TLC	1974	974.17
aflatoxin M_1	Milk & cheese	Silica gel clean up & TLC	1980	980.21
B_1 & M_1	Liver	Silica gel clean up & TLC	1982	982.24
B_1 & M_1	Liver	Silica gel clean up & 2D TLC	1982	982.25
Aflatoxin M_1	Liver	TLC confirmation	1982	982.26
M_1 & M_2	Liquid milk	C-18 clean up & LC (TFA)	1986	986.16
Aflatoxin M_1	Liquid milk	IAC clean up & LC	2000	2000.08[d]
Ochratoxin A	Barley	Column clean up & TLC	1973	973.37
Ochratoxin A	Green coffee	Column clean up & TLC	1975	975.38

Ochratoxin A	Barley	IAC cleanup & LC (Fl)	2000	2000.03[e]
Ochratoxin A	Roast coffee	SPE+IAC cleanup & LC	2000	2000.09[f]
Ochratoxin A	Corn & barley	Liq/liq extr. + SPE & LC	1991	991.44
Ochratoxin A	Wine & beer	IAC cleanup & LC (Fl)	2001	2001.01
Ochratoxin A	Green coffee	IAC cleanup & LC (Fl)	2004	2004.10
Ochratoxin A	Ginseng & ginger	Multi-IAC cleanup & LC	2008	2008.02
Patulin	Apple juice	Silica gel cleanup & TLC	1974	974.18
Patulin	Apple juice	Liquid extr. & LC (UV)	1995	995.10
Patulin	Apple juice & puree	Liquid extr. & LC (UV)	2000	2000.02[g]
Deoxynivalenol	Wheat	Charcoal/alumina & TLC	1986	986.17
Deoxynivalenol	Wheat	Silica gel cleanup & GC	1986	986.18
Zearalenone	Corn	Column cleanup & TLC	1976	976.22
α-zear & zea	Corn	Liq-liq partition & LC (Fl)	1985	985.18
Zearalenone	Corn, wheat, feed	Agri-screen (ELISA)	1994	994.01
$FB_1+FB_2+FB_3$	Corn	Competitive direct ELISA	2001	2001.06
FB_1+FB_2	Cornflakes	IAC cleanup & LC/OPA	2001	2001.04[h]
$FB_1+FB_2+FB_3$	Corn	SAX cleanup & LC/OPA	1995	995.15

*date method was adopted as First Action.
aflatoxins = B_1 + B_2+G_1+G_2; α-zear = α-zearalenol; zea = zearalenone; FB_1= fumonisin B_1; FB_2 = fumonisin B_2; FB_3 = fumonisin B_3; TLC = thin layer chromatography; LC = high performance liquid chromatography; ELISA = enzyme linked immunosorbent assay; MS = mass spectrometry; NICI = negative ion chemical ionisation; IAC = immunoaffinity column; Br = post-column bromination; TFA = trifluoracetic acid derivatization; SPE = solid phase extraction; Fl = fluorescence detection; SAX = strong anion exchange column; OPA = o-phthaldialdehyde derivatisation.
Shaded rows indicate method corresponds to equivalent CEN standard – European Committee for Standardization [a]1999b & 2007b; [b]2008c; [c]2006c; [d]2007a; [e]2006b; [f]2006b; [g]2003b; [h]2004.

the AOAC First Action methods being identical to CEN standards, that is originating from the same collaborative study although their presentation is very different reflecting the difference between the AOAC and CEN formats.

The on-line OMA contains some 55 First Action methods for mycotoxins of which 47 methods subsequently became Final Action. Two First Action mycotoxin methods were withdrawn as they were based on test kits which were no longer available in the marketplace. The earliest validated methods date back to 1975 using TLC and these methods still remain as Official Methods, although few control laboratories would use such methods for enforcement purposes. AOAC International has two methods which date back to 1970/71 (970.44 and 971.22) which report the molar extinction coefficients of aflatoxin standards in a number of solvents. Although these two methods were originally written in relation to preparing standards for TLC, they are still widely used as the basis for preparing aflatoxin standards and are widely cited both in the literature and in other official methods (e.g. CEN standards).

AOAC Official methods do not have as strong a link to official control as do CEN methods and a number of AOAC methods are intended for screening purposes rather than enforcement. Thus, early methods such as the Romer minicolumn (975.36), and Holaday–Velasco minicolumn from the 1970s through to more recent official methods based on proprietary products such as ImmunoDot screen cup (990.34), Afla-20 Cup (991.16), Agri-Screen (990.32) and a competitive direct ELISA for fumonisins (2001.06) meet industry needs rather than official purposes. There are strong arguments for the CEN position of restricting endorsement of methods to those required for regulatory purposes, as listing test kit methods as 'official' effectively endorses a product. Also in a rapidly changing market the situation can arise where methods need to be withdrawn when the product is no longer available. AOAC International also operates the system of 'Performance Tested Methods[SM]' where the performance claims of a product are verified and endorsed by AOAC which is much more appropriate for such test kits.

6.5   Literature publications of mycotoxin validation studies

A number of inter-laboratory collaborative studies have been coordinated by independent laboratories and these are often published initially in the peer-reviewed literature. From a Codex and EU perspective, provided these methods have been shown to fulfil the performance criteria, they can be selected by any food control or import/export laboratories for use for official purposes. However, these methods in the open literature are not always well publicised and gaining recognition by AOAC International, CEN or another body has attractions in ensuring longevity of the method. In Table 6.4 those methods of analysis for mycotoxins are compiled which have been either single-laboratory or inter-laboratory validated. The principles of these methods are indicated and the method performance characteristics are tabulated. Where these methods have subsequently been adopted as Official methods by AOAC International or adopted by CEN this has been highlighted in Table 6.4.

AOAC International methods do not in general report the method performance characteristics in detail, but the OMA does give a reference citation to the original *JAOAC International* publication. CEN standards tabulate in considerable detail the method performance data in full within the standard and also give some bibliographic information. There is no definitive way to cross-reference one official method to another other than referring back to the cited publication. CEN has also in some instances taken methods from more than one study and these have been combined into a single standard, for example CEN Standard EN 14123 (European Committee for Standardization, CEN, 2007b) combines the results of validation studies from Stroka *et al.* (2003) and Senyuva and Gilbert (2005) into a single standard covering aflatoxins in hazelnuts, peanuts, pistachios, figs and paprika powder. This has been done for administrative reasons to avoid a proliferation of standards for the same mycotoxin each covering a different matrix. Although Stroka *et al.* (2003) and Senyuva and Gilbert (2005) methods both use immunoaffinity column clean up and HPLC with post-column derivatization, there are nevertheless differences in the methods and care needs to be taken when applying the standard depending on the matrix.

6.6   Enforcement of mycotoxin regulations

Official methods serve the purpose of establishing a common measure for trading purposes or being used as the referee method for resolution of disputes, for example in enforcement situations. Thus, some mycotoxin methods originating from bodies such as ICC, AACC, IDF, AOCS or IFU may meet specific needs such as the requirement for screening or on-site testing which is necessary when commodities are bought and sold. The reliability of these screening tests needs to be established particularly in terms of rates of 'false positives' and 'false negatives' but they would never be suitable for enforcement purposes.

In contrast to screening methods for enforcement purposes, unequivocal measurement in terms of identity and quantification is sought at or close to the regulatory limit. In general for a decision to be taken that a sample exceeds a regulatory limit, the measured value must exceed the regulatory limit plus the uncertainty. Thus, for a regulatory limit of 2 µg kg^{-1} for aflatoxin B$_1$, if the official method when validated was shown to have an RSD$_R$ of 25%, then the determined level needs to be greater than 2.5 µg kg^{-1} after correction for recovery to provide certainty that the sample exceeds the limit. In reality the 25% is only the uncertainty which relates to the precision of the analytical method itself and in practice there will also be considerable uncertainty associated with the sampling even when the official sampling and homogenisation protocols are strictly followed. It may thus be necessary to build in a larger margin of safety in decision making to feel confident that a sample which appears to contain mycotoxins above the limit is indeed a sample requiring enforcement action. For regulatory purposes official methods are sometimes also referred to as 'referee methods' and it may be that the official method is not followed on a day-to-day basis and is only strictly applied when samples are detected as being non-compliant and enforcement action is envisaged.

6.7   Confirmation of results

From the early days of AOAC Official Methods going back to the 1970s when TLC was employed which lacks separating power and visual/densitometric measurement lacks specificity, it was recognised that there was a need for some additional confirmation. Thus, AOAC Official Method 975.37 describes derivative formation of aflatoxin B$_1$ on the TLC plate, AOAC Official Method 970.47 describes scraping TLC spots and eluting aflatoxin B$_1$ for confirmation and AOAC Official Method 985.17 describes an additional 2D-TLC step for further clean up of extracts followed by mass spectrometric confirmation (NICI) by putting the scraped spot from the TLC directly into the MS probe. Subsequent developments in both sample clean up such as the widespread use of immunoaffinity columns and improvements in chromatographic separation and detection (use of LC with fluorescence detection) have led to assumptions that no further confirmation is necessary. Whilst generally speaking extracts passed through immunoaffinity columns tend to be very 'clean', there are nevertheless instances with a co-extractive, for example roasted coffee and some spices, where some thought needs to be directed to confirmation of results.

Table 6.4   Method performance characteristics for inter-laboratory validated methods for mycotoxins between 2000 and 2009

Analyte	Matrix	Level $\mu g\ g^{-1}$	RSD_r (%)	RSD_R (%)	Recovery (%)	Analytical technique	Detection	References
Aflatoxin B_1	Peanut, pistachio, fig, paprika	0.9–3.6	3.1–20	9.1–32.2	82–109	HPLC	Fluorescence	(Stroka *et al.*, 2000)
Total aflatoxins		0.8–7.9	4.6–23.3	14.1–34.2	71–92			
Aflatoxin B_1 Total aflatoxins	Hazelnut	1.36–3.82 4.17–12.08	2.2–3.2 2.3–3.4	7.3–7.8 6.1–7.0	85.9–88.8 86.9–89.2	HPLC	Fluorescence	(Senyuva and Gilbert, 2005)
Aflatoxin B_1	Baby food	<0.02, 0.07, 0.09, 0.17	3.5–14	9–23	92–101	HPLC	Fluorescence	(Stroka *et al.*, 2001)
Aflatoxin B_1	Corn	2.20–17.60	11.03–28.71	24.42–36.83	82–84	HPLC	Fluorescence	(Brera *et al.*, 2007)
Aflatoxin B_1	Senna pods Ginger root Devil's claw	0.8–14.9 0.9–2.1 0.8–0.9	5.1–22.0 4.2–10.0 5.5–7.8	18.1–35.2 5.8–30.2 7.6–10.5	78 91–92 83–92	HPLC	Fluorescence	(Arranz *et al.*, 2006)
OTA Total aflatoxin	Ginseng & ginger	0.25–16 0.25–8.0	2.5–10.7 2.6–8.3	5.5–10.7 5.7–28.6	86–113 70–87	HPLC	Fluorescence	Trucksess *et al.*, 2008
Aflatoxin B_1	Cattle feed	0.87–4.19	5.9–8.7	17.5–19.6	74–157	HPLC	Fluorescence	(Stroka *et al.*, 2003)
Aflatoxin M_1	Milk					Thin-layer chromato-graphy	Densitometer	(Grosso *et al.*, 2004)
Aflatoxin M_1	Milk	0.02–0.103	8–18	21–31	74–107	HPLC	Fluorescence	(Dragacci *et al.*, 2001)
OTA	Barley	1.3–4.5	15–33	12–17	93	HPLC	Fluorescence	(Entwisle *et al.*, 2000)
OTA	Green coffee	<0.12–13.5	7.4–20.9	16.3–29.2	92.8	HPLC	Fluorescence	(Vargas *et al.*, 2005)
OTA	Roasted coffee	1.2–5.4	2–22	14–26	65–97	HPLC	Fluorescence	(Entwisle *et al.*, 2001)
OTA	Baby food	0.05–0.22	18–36	29–63	108	HPLC	Fluorescence	(Burdaspal *et al.*, 2001)

OTA	White wine	0.105–1.76	6.6–10.8	13.1–15.9	88.2–105	HPLC	Fluorescence	(Visconti *et al.*, 2001a)
	Red wine	0.19–1.69	6.5–10.9	11.9–13.6	84.3–93.1			
	Beer	0.07–1.4	4.7–16.5	15.2–26.1	87–95			
OTA	Cocoa beans	0.4–12.0	1–4	n/a	89	HPLC	Fluorescence	(Amezqueta *et al.*, 2004)
OTA	Cocoa powder	0.18–0.95	15–31	29–40	80	HPLC	Fluorescence	(Brera *et al.*, 2005)
OTA	Currants	4.5	5.7	28	73.6	HPLC	Fluorescence	(MacDonald *et al.*, 2003)
	Sultanas	11.4	5.6	14	74.5			
	Raisins	7.5	4.9	14	72.2			
	Mixed fruits	1.1	8.6	14	69.2			
	Figs	2.5	8.7	18	72.9			
DON	Cereals	85–1768	3.1–14.1	11.5–26.3	78–87	HPLC	UV	(MacDonald *et al.*, 2005a)
ZON	Baby food	9.1–44	2.8–9.0	8.3–13.3	78–119	HPLC	Fluorescence	(Arranz *et al.*, 2007)
	Animal feed	<20–307	5.7–9.5	15.5–21.4	51–122			
ZON	Baby food	10.9	35.8	38.2	100	HPLC	Fluorescence	(MacDonald *et al.*, 2005b)
	Barley	143.0	6.9	17.9	92			
	Corn	87.2	14.2	20.6	91			
	Polenta	66.5	8.9	16.4	91			
	Wheat	226.6	8.3	17.0	95			
ZON	Dairy feed	134	9.8	16.6	–	HPLC	Fluorescence	(Campbell and Armstrong, 2007)
	Distillers grain	250	5.8	13.4	–			
	Wheat	189	9.7	12.5	107.3			
FB$_1$ & FB$_2$	Corn	200–500	2.8–7.1	n/a	90–101	LC/MS	SIM	(Senyuva *et al.*, 2008)
FB$_1$ & FB$_2$	Corn	500–1950	1.4–11.5	18.8–23.2	–	LC/MS	SIM	(Senyuva *et al.*, 2009)
FB$_1$ & FB$_2$	Corn	650–1410	18.5–26.8	22.1–28.2	72–76	HPLC	Fluorescence	(Visconti *et al.*, 2001b)
	Cornflakes	130–1050	9.2–21.7	26.1–34.8	97–110			

Shaded rows in above table indicate that these methods have been adopted by AOAC International and/or CEN.

Whilst there is a steady increase in the application of LC–MS and LC–MS/MS for multi-mycotoxin analysis (see Section 6.8 below), these tools have mainly been developed in research laboratories and there has been little attention to method validation. One exception has been the recent single laboratory method validation of an LC–MS method for fumonisins B_1 and B_2 in corn (Senyuva *et al.*, 2008), which was subsequently extended to become the first full collaborative study of an LC–MS method for mycotoxins (Senyuva *et al.*, 2010). LC–MS selected ion monitoring was carried out for *m/z* 722, 723 (FB_1) and *m/z* 706, 707 (FB_2) with a dwell time of 218 msec. The identification of FB_1 was based on a retention time of 3.8 ± 0.05 min and an ion ratio of $723/722 = 2.5 +/- 0.1$. The identification of FB_2 was similarly based on a retention time of 8.4 ± 0.05 min and an ion ratio of $706/707 = 2.6 \pm 0.05$.

Royer *et al.* (2004) employed accelerated solvent extraction followed by two solid phase clean up steps using strong ion exchange resin and Mycosep® columns for the simultaneous analysis of 4-deoxynivalenol, zearalenone and FB_1 in corn, using LC–MS/MS in APCI mode and using synthesised d_6-FB_1 as an internal standard. Royer *et al.* (2004) similarly focused on the need for rigorous identification and proposed that three criteria must be met: (1), two transition reactions (*m/z* $722 \rightarrow 546$ and $722 \rightarrow 704$ for FB_1) need to co-elute at the same retention time; (2) that the retention time should be within $\pm 2.5\%$ of the retention time for the standard and (3) the transition reaction ratios must to be within a prescribed coefficient of variation (specified as 15% for FB_1). Both of these examples proposing criteria for identification of mycotoxins originate from individual laboratories but there is really a need for some internationally agreed standards on what constitutes unequivocal identification of mycotoxins.

6.8   Conclusions and future trends

The most obvious future trend in mycotoxin analysis is the increasing use of LC–MS and LC/MS/MS, particularly where the technique is being used to provide a multi-toxin screen. This approach is frequently conducted without sample clean-up making the determination directly on a crude sample extract (Ren *et al.*, 2007; Sulyok *et al.*, 2007; Royer *et al.*, 2004; Cavaliere *et al.*, 2005, 2007). Although there are a number of such published multi-mycotoxin procedures claiming capability for simultaneous determination of as many as 38 mycotoxins, none of the papers have reported the results of single laboratory validation, nor has the robustness of these methods been adequately tested. LC–MS/MS has also been combined with affinity column clean up using multi toxin columns for simultaneous determination of 11 toxins including FB_1 and FB_2 (Lattanzio *et al.*, 2007), but again validation is not reported. Clearly for any of these LC–MS or LC–MS/MS methods, for mycotoxins in food or feed, to be used in an accredited laboratory or for regulatory purposes will require inter-laboratory method validation. More generally, despite the widespread use of LC–MS/MS in food contaminant analysis, few inter-laboratory studies have been conducted, as it is very difficult to standardise a study when participants are using a variety of instruments operating under widely differing conditions.

For the future there is a need for official methods for mycotoxins to address the whole question of identification criteria, which will need an international body to give an authoritative view on whether LC–MS confirmation should be demanded as a minimum for mycotoxins or whether a combination of other tests can be used in HPLC for confirmatory purposes. The existing range of official methods using affinity columns for clean up has dramatically increased the range of mycotoxins and the variety of matrices for which validated methods are available. The most recent of the AOAC Official Methods 2008.02 (Trucksess *et al.*, 2008) involves the use of a multi-affinity column for simultaneous analysis of aflatoxins and ochratoxin A and in the future one might anticipate new validation studies being conducted on a variety of such multi-toxin columns. A 2009 collaborative study of an LC–MS method for fumonisins in corn (Senyuva *et al.*, 2010) provides good evidence that LC–MS can offer significant benefits (in this case avoiding the need for derivatisation) and that, with care, good results can be obtained notwithstanding the difficulties of participants using different instrumentation. This initiative needs to be extended in future work to validate LC/MS/MS multi-toxin methods for adoption as Official Methods. However, this needs some careful thought so that a suitable combination of mycotoxins is chosen which co-occur naturally such as the *Fusarium* toxins, rather than validating an *ad hoc* collection of toxins simply for the purpose of demonstrating a capability for simultaneous monitoring of a large number of fungal metabolites.

6.9   References

Amezqueta, S., Gonzaülez-Penas, E., Murillo, M. and Lopez De Cerain, A. (2004), 'Validation of a high-performance liquid chromatography analytical method for ochratoxin A quantification in cocoa beans', *Food Additives and Contaminants*, **21**, 1096–106.

Arranz, I., Sizoo, E., Van Egmond, H., Kroeger, K., Legarda, T. M., Burdaspal, P., Reif, K. and Stroka, J. (2006), 'Determination of aflatoxin B_1 in medical herbs: Interlaboratory study', *Journal of AOAC International*, **89**, 595–605.

Arranz, I., Mischke, C., Stroka, J., Sizoo, E., Van Egmond, H. and Neugebauer, M. (2007), 'Liquid chromatographic method for the quantification of zearalenone in baby food and animal feed: Interlaboratory study', *Journal of AOAC International*, **90**, 1598–609.

Brera, C., Grossi, S. and Miraglia, M. (2005), 'Interlaboratory study for ochratoxin A determination in cocoa powder samples', *Journal of Liquid Chromatography and Related Technologies*, **28**, 35–61.

Brera, C., Debegnach, F., Minardi, V., Pannunzi, E., De Santis, B. and Miraglia, M. (2007), 'Immunoaffinity column cleanup with liquid chromatography for determination of aflatoxin B_1 in corn samples: Interlaboratory study', *Journal of AOAC International*, **90**, 765–72.

Burdaspal, P., Legarda, T. M. and Gilbert, J. (2001), 'Determination of ochratoxin A in baby food by immunoaffinity column cleanup with liquid chromatography: Interlaboratory study', *Journal of AOAC International*, **84**, 1445–52.

Campbell, H. M. and Armstrong, J. F. (2007), 'Determination of zearalenone in cereal grains, animal feed, and feed ingredients using immunoaffinity column chromatography and liquid chromatography: Interlaboratory study', *Journal of AOAC International*, **90**, 1610–22.

Cavaliere, C., Foglia, F., Pastorini, E., Samperi, R. and Lagana, A. (2005), 'Development of a multresidue method for analysis of major Fusarium mycotoxins in corn meal using liquid chromatography/tandem mas spectrometry', *Rapid Communications in Mass Spectrometry*, **19**, 2085–93.

Cavaliere, C., Foglia, Guarino, C., Motto, M., Nazzari, M., Samperi, R., Lagana, A. and Berardo, N. (2007), 'Mycotoxins produced by Fusarium genus in maize: determination by screening and confirmatory methods based on liquid chromatography tandem mass spectrometry', *Food Chemistry*, **105**, 700–10.

Codex (1995), *Codex General Standard for Contaminants and Toxins in Foods*. CODEX STAN 193–195 Schedule I, 26.

Codex (1997), *Guidelines for the Assessment of the Competence of Testing Laboratories involved in the Import and Export Control of Food*. CAC/GL 27-1997

Dragacci, S., Grosso, F. and Gilbert, J. (2001), 'Immunoaffinity column cleanup with liquid chromatography for determination of aflatoxin M_1 in liquid milk: Collaborative study', *Journal of AOAC International*, **84**, 437–43.

Entwisle, A. C., Williams, A. C., Mann, P. J., Slack, P. T. and Gilbert, J. (2000), 'Liquid chromatographic method with immunoaffinity column cleanup for determination of ochratoxin A in barley: Collaborative study', *Journal of AOAC International*, **83**, 1377–86.

Entwisle, A. C., Williams, A. C., Mann, P. J., Russell, J., Slack, P. T. and Gilbert, J. (2001), 'Combined phenyl silane and immunoaffinity column cleanup with liquid chromatography for determination of ochratoxin A in roasted coffee: Collaborative study', *Journal of AOAC International*, **84**, 444–50.

European Commission (2002), 'Commission Decision 2002/657/EC of 12 August 2002 implementing Council Directive 96/23/EC concerning the performance of analytical methods and the interpretation of results'. *Official Journal of the European Union* L 221/8.

European Commission (2006a), 'Commission Regulation (EC) No 1881/2006 of 19 December 2006 setting out maximum levels for certain contaminants in foodstuff'. *Official Journal of the European Union* L 364/5-23.

European Commission (2006b), 'Commission Regulation (EC) No 401/2006 of 23 February 2006 laying down the methods of sampling and analysis for the official control of the levels of mycotoxins in foodstuffs'. *Official Journal of the European Union*. L 70/12-34.

European Committee for Standardization (CEN) (1998a), *Foodstuffs – Determination of Ochratoxin A in Cereals and Cereal Products. Part 1: – High performance liquid chromatographic method with silica gel clean up* (ISO 15141-1:1998). EN ISO 15141-1:1998 E.

European Committee for Standardization (CEN) (1998b), *Foodstuffs – Determination of Ochratoxin A in Cereals and Cereal Products. Part 2: High performance liquid chromatographic method with bicarbonate clean up*. EN ISO 15141-2:1998 E

European Committee for Standardization (CEN) (1999a), *Food Analysis – Biotoxins – Criteria of Analytical Methods of Mycotoxins*. CR 13505:1999:E.

European Committee for Standardization (CEN) (1999b), *Foodstuffs – Determination of Aflatoxin B_1, and the Sum of Aflatoxin B_1, B_2, G_1 and G_2 in Cereals, Shell-fruits and Derived Products – High performance liquid chromatographic method with post-column derivatization and immunoaffinity column clean-up*. EN 12955:1999:E.

European Committee for Standardization (CEN) (2001), *Foodstuffs – Determination of Fumonisins B_1 and B_2 in Maize – HPLC method with solid phase extraction clean-up* EN 13585: 2001:E.

European Committee for Standardization (CEN) (2003a), *Milk and Milk Products – Guidelines for a Standardized Description of Competitive Enzyme Immunoassays – Determination of aflatoxin M1 content* EN ISO 14675:2003 E

European Committee for Standardization (CEN) (2003b), *Foodstuffs – Determination of Patulin in Clear and Cloudy Apple Juice and Puree – HPLC method with liquid/liquid partition clean-up*. EN 14177:2003

European Committee for Standardization (CEN) (2004), *Foodstuffs – Determination of Fumonisins B₁ and B₂ in Maize based Foods – HPLC method with immunoaffinity column clean-up*. EN 14352: 2004:E

European Committee for Standardization (CEN) (2006a), *Foodstuffs – Determination of Ochratoxin A in Wine and Beer – HPLC method with immunoaffinity column clean-up*. EN 14133:2003/AC:2006:D/E/F.

European Committee for Standardization (CEN) (2006b), *Foodstuffs – Determination of Ochratoxin A in Barley and Roasted Coffee – HPLC method with immunoaffinity column clean-up*. EN 14132:2003/AC:2006:D/E/F.

European Committee for Standardization (CEN) (2006c), *Animal feeding stuffs – Determination of Aflatoxin B₁*. EN ISO 17375:2006: E

European Committee for Standardization (CEN) (2007a), *Milk and milk powder – Determination of Aflatoxin M1 Content – Clean-up by immunoaffinity chromatography and determination by high-performance liquid chromatography*. EN ISO 14501:2003/AC:2007:E

European Committee for Standardization (CEN) (2007b), *Foodstuffs – Determination of Aflatoxin B₁, and the Sum of Aflatoxin B₁, B₂, G₁ and G₂ in Hazelnuts, Peanuts, Pistachios, Figs and Paprika Powder – High performance liquid chromatographic method with post-column derivatization and immunoaffinity column clean-up*. EN 14123:2007:E

European Committee for Standardization (CEN) (2008a), *Foodstuffs – Determination of Ochratoxin A in Currants, Raisins, Sultanas, Mixed Dried Fruit and Dried Figs – HPLC method with immunoaffinity column clean-up and fluorescence detection*. prEN 15829:2008: E

European Committee for Standardization (CEN) (2008b), *Foodstuffs – Determination of Ochratoxin A in Cereal based Foods for Infants and Young Children – HPLC method with immunoaffinity column clean-up and fluorescence detection*. prEN 15835:2008:E

European Committee for Standardization (CEN) (2008c), *Foodstuffs – Determination of Aflatoxin B₁ in Cereal based Foods for Infants and Young Children – HPLC method with immunoaffinity column clean-up*. prEN 15851:2008:E

European Committee for Standardization (CEN) (2009a), *Foodstuffs – Performance Criteria for Methods of Analysis of Mycotoxins*. prCEN/TR 13505. Under development

European Committee for Standardization (CEN) (2009b). *Foodstuffs – Determination of Patulin in Fruit Juice and Fruit based Puree for Young Children – HPLC method with liquid/liquid partition cleanup and solid phase extraction and UV detection*. prEN 15890 Under Approval

European Committee for Standardization (CEN) (2009c). *Foodstuffs – Determination of Deoxynivalenol in Cereals, Cereal Products and Cereal based Foods for infants and Young Children – HPLC method with immunoaffinity column cleanup and UV detection*. prEN 15891 Under approval.

European Committee for Standardization (CEN) (2009d). *Foodstuffs – Determination of Zearalenone in Barley, Maize and Wheat flour, Polenta and Cereal based Foods for Infants and Young children – HPLC method with immunoaffinity column cleanup and fluorescence detection*. prEN 15850 Under approval.

FAO (2004), *Worldwide Regulations for Mycotoxins in Food and Feed in 2003*. FAO Food and Nutrition Paper 81. Food and Agriculture Organization of the United Nations, Rome, Italy.

FDA (2001). *Guidance for Industry – Fumonisin Levels in Human Foods and Animal Feeds*. US Food and Drug Administration. Final Guidance November 9, 2001. http://www.cfsan.fda.gov/~dms/fumongu2.html

Grosso, F., Fremy, J. M., Bevis, S. and Dragacci, S. (2004), 'Joint IDF-IUPAC-IAEA(FAO) interlaboratory validation for determining aflatoxin M₁ in milk by using immunoaffinity clean-up before thin-layer chromatography', *Food Additives and Contaminants*, **21**, 348–57.

Horwitz, W. (1995), 'Protocol for the design, conduct and interpretation of method-performance studies: Revised 1994' (Technical Report) *Pure and Applied Chemistry*, **67**, 331–43.

Horwitz, W. and Albert, R. (2006), 'The Horwitz ratio (HorRat): A useful index of method performance with respect to precision'. *Journal of Association of Official Analytical Chemists, International*, **89**, 1095–109.

ISO/IEC (1993), Guide 58:1993: *Calibration and Testing Laboratory Accreditation Systems – General requirements for operation and recognition*.

IUPAC/AOAC (1995), 'Harmonized guidelines for internal quality control in analytical chemistry laboratories', *Pure and Applied Chemistry*, **67**, 649–66.

IUPAC/AOAC (2006), 'The international harmonized protocol for the proficiency testing of (chemical) analytical laboratories', *Pure and Applied Chemistry*, **78**, 145–96.

Lattanzio, V. M. T., Solfrizzo, M., Powers, S. and Visconti, A. (2007), 'Simultaneous determination of aflatoxins, ochratoxin A and Fusarium toxins in maize by liquid chromatography/tandem mass spectrometry after multitoxin immunoaffinity cleanup', *Rapid Communications in Mass Spectrometry*, **21**, 3253–61.

MacDonald, S. J., Anderson, S., Brereton, P. and Wood, R. (2003), 'Determination of ochratoxin A in currants, raisins, sultanas, mixed dried fruit, and dried figs by immunoaffinity column cleanup with liquid chromatography: Interlaboratory study', *Journal of AOAC International*, **86**, 1164–71.

MacDonald, S. J., Chan, D., Brereton, P., Damant, A. and Wood, R. (2005a), 'Determination of deoxynivalenol in cereals and cereal products by immunoaffinity column cleanup with liquid chromatography: Interlaboratory study', *Journal of AOAC International*, **88**, 1197–204.

MacDonald, S. J., Anderson, S., Brereton, P., Wood, R. and Damant, A. (2005b), 'Determination of zearalenone in barley, maize and wheat flour, polenta, and maize-based baby food by immunoaffinity column cleanup with liquid chromatography: Interlaboratory study', *Journal of AOAC International*, **88**, 1733–40.

Ren, Y., Zhang, Y., Shao, S., Cai, Z., Feng, L., Pan, H. and Wang, Z. (2007), 'Simultaneous determination of multi-component mycotoxin contaminants in foods and feeds by ultra-performance liquid chromatography tandem mass spectrometry' *Journal of Chromatogr. A.*, **1143**, 48–64.

Royer, D., Humpf, H.-U. and Guy, P.A. (2004) 'Quantitative analysis of Fusarium mycotoxins in maize using accelerated solvent extraction before liquid chromatography/atmospheric pressure chemical ionization tandem mass spectrometry', *Food Additives and Contaminants*, **21**, 678–92.

Senyuva, H. Z. and Gilbert, J. (2005), 'Immunoaffinity column cleanup with liquid chromatography using post-column bromination for determination of aflatoxins in hazelnut paste: Interlaboratory study', *Journal of AOAC International*, 88, 526–535.

Senyuva, H, Z. and Gilbert, J. (2010), 'Immunoaffinity column clean-up techniques in food analysis: A review', *Journal of Chromatography B*, **878**, 115–32.

Senyuva, H. Z., Ozcan, S., Cimen, D. and Gilbert, J. (2008), 'Determination of fumonisins B_1 and B_2 in corn by liquid chromatography/mass spectrometry with immunoaffinity column cleanup: Single-laboratory method validation', *Journal of AOAC International*, **91**, 598–606.

Senyuva, H. Z., Gilbert, J. and Stroka, J. (2010), 'Determination of fumonisins B_1 and B_2 in corn by LC/MS with immunoaffinity column cleanup: interlaboratory study', *Journal of AOAC International*, **93**(2), 611–21.

Stroka, J., Anklam, E., Joerissen, U. and Gilbert, J. (2000), 'Immunoaffinity column cleanup with liquid chromatography using post-column bromination for determination of aflatoxins in peanut butter, pistachio paste, fig paste, and paprika powder: Collaborative study', *Journal of AOAC International*, **83**, 320–40.

Stroka, J., Anklam, E., Joerissen, U. and Gilbert, J. (2001), 'Determination of aflatoxin B_1 in baby food (infant formula) by immunoaffinity column cleanup liquid chromatography with postcolumn bromination: Collaborative study', *Journal of AOAC International*, **84**, 1116–23.

Stroka, J., Von Holst, C., Anklam, E. and Reutter, M. (2003), 'Immunoaffinity column cleanup with liquid chromatography using post-column bromination for determination of aflatoxin B_1 in cattle feed: Collaborative study', *Journal of AOAC International*, **86**, 1179–186.

Sulyok, M., Krska, R. and Schumacher, R. (2007), 'Application of a liquid chromatography–tandem mass spectrometric method to multi-mycotoxin determination in raw cereals and evaluation of matrix effects', *Food Additives and Contaminants*, **24**, 1184–95.

Thompson, M. (2000). 'Recent trends in inter-laboratory precision at ppb and sub-ppb concentrations in relation to fitness for purpose criteria in proficiency testing', *Analyst*, **125**, 385–6.

Thompson, M., Ellison, S. L.R. and Wood, R. (2002) 'Harmonized guidelines for single-laboratory validation of methods of analysis', *Pure and Applied Chemistry*, **74**, 835–55.

Trucksess, M. W., Weaver, C. M., Oles, C. J., Fry, J., Noonan, G. O., Betz, J. M. and Rader, J. I. (2008), 'Determination of aflatoxins B_1, B_2, G_1, and G_2 and ochratoxin A in ginseng and ginger by multitoxin immunoaffinity column cleanup and liquid chromatographic quantitation: Collaborative study', *Journal of AOAC International*, **91**, 511–23.

van Egmond, H.P. and Jonker, M.A. (2004), 'Worldwide regulations on aflatoxins – the situation in 2002', *Journal of Toxicology Toxins Reviews*, **23**, 273–93.

van Egmond, H.P., Schothorst, R.C. and Jonker, M.A. (2007), 'Regulations relating to mycotoxins in food', *Analytical and Bioanalytical Chemistry*, **389**, 147–57.

Vargas, E. A., Dos Santos, E. A. and Pittet, A. (2005), 'Determination of ochratoxin A in green coffee by immunoaffinity column cleanup and liquid chomatography: Collaborative study', *Journal of AOAC International*, **88**, 773–9.

Visconti, A., Pascale, M. and Centonze, G. (2001a), 'Determination of ochratoxin A in wine and beer by immunoaffinity column cleanup and liquid chromatographic analysis with fluorometric detection: Collaborative study', *Journal of AOAC International*, **84**, 1818–27.

Visconti, A., Solfrizzo, M. and De Girolamo, A. (2001b), 'Determination of fumonisins B_1 and B_2 in corn and corn flakes by liquid chromatography with immunoaffinity column cleanup, Collaborative study', *Journal of AOAC International*, **84**, 1828–37.

7

Ensuring the quality of results from food control laboratories: laboratory accreditation, method validation and measurement uncertainty

J. O. De Beer, Scientific Institute of Public Health, Brussels, Belgium and C. Van Poucke, Laboratory of Food Analysis, Faculty of Pharmaceutical Sciences, Ghent University, Belgium

Abstract: Within the framework of control activities by a competent food analysis laboratory, an important basic task is to develop and improve expertise and knowledge continuously in order to report unequivocally reliable results that guarantee safe food distribution and consumption. This scientific message is vital with respect to the whole concept of the general protection of public health. Internationally prescribed and adopted quality regulations are mandatory for quality assurance during production and in quality control before release for consumption. These quality regulatory demands are elaborated within internationally recognized organizations and directorates in order to develop specific aims and activities to assure efficient quality control of traded food. Suitable validated methods of analysis should be able to verify if these integral quality demands for food are fulfilled. In this way the exact composition, safety, falsifications and absence of harmful contaminants should be verified or detected. In this context it is of principal importance that a control laboratory has at its disposal selective, sensitive, rugged, accurate and precise methods of analysis providing highly reliable results with low uncertainty. It is a fundamental duty of a control laboratory to follow actual progress in analysis development and statistical method validation continuously within an accredited quality environment such as prescribed in the ISO 17025 norm and to apply relevant suitable and powerful statistical tools that ensure excellent quality results.

Key words: accepted reference value, accreditation, accuracy profile, quality assurance, ANOVA, calibration, competence, decision making tool, detection limit, interval hypothesis test, ISO 5725, ISO 17025, linearity, precision, quality assurance, quality management, response function, selectivity, specificity, total error, trueness, uncertainty, validation.

7.1 Introduction: why accreditation can be important for laboratories

Laboratory accreditation can be defined as a process which gives formal recognition to the technical competence of a laboratory to perform well-determined tests, types of tests or calibrations. This accreditation process requires the structured maintenance of well-documented quality management, the identification of personnel qualified and authorized to perform well determined tasks and analyses, related to the scope of accreditation, and the disposal of the necessary qualified and calibrated measurement equipment which enable the tests and analysis to be performed adequately within the scope of interest.

Accreditation involves a thorough evaluation of all structural parts of a laboratory which contribute to generating accurate and reliable test measurements results and data. The criteria to which auditors refer during quality assessments are based on an international standard (ISO/IEC 17025:2005) which is applicable to all laboratories involved in sampling, development of new methods, performing tests and calibrations. Official accreditation bodies use this international standard edited by the International Standards Organization (ISO) to assess the factors that influence a laboratory's ability to produce precise and accurate test and calibration results including suitable general quality assurance procedures, the qualification, training and experience of the laboratory staff and the appropriate calibrated and maintained equipment and apparatus. Obvious benefits affect accredited laboratories, their customers, the general public and the regulatory bodies. Lists of their accredited laboratories, contact details and scope of accreditation are published by the official accreditation bodies, which promote accredited activities to potential clients as a marketing tool.

Opportunities are provided to acquire better insight into current developments in the field of their work. Effective and efficient management of the quality system enhances and improves staff development, its competence and discipline. An accreditation allows the laboratory further to evaluate its analytical performance and provides it with a benchmark for maintaining its competence by regular assessments. Confidence in the technical capability of the accredited laboratory is consolidated, while reliable test results contribute significantly in the decision making process for customers and provide the confidence that the supplied products comply with the claimed specifications.

Appealing to and relying on accredited laboratories benefits governmental regulatory bodies by ensuring confidence in the data used to establish baselines for key decisions. Uncertainties linked to decisions to be taken that have an impact on the protection of human health and which are based on measurement results from accredited laboratories are reduced. This also leads to an increase in public confidence as an accreditation is recognized as proof of approval. False positive or negative results, which can directly affect compliance with legal regulations are reduced as well. From an economical point of view, considerable benefits resulting from laboratory accreditations are generated for the manufacturer, the supplier or the customer as the risk of producing, supplying or receiving an inferior product is limited. The chance of retesting of products is drastically reduced.

With respect to export of products to different continents and overseas markets, quality labels generated by an accredited laboratory lead to more ready acceptance. Most accreditation bodies worldwide have adopted the ISO/IEC 17025:2005 standard as the basis for accreditation, applying a uniform approach to assessing the laboratories' competence. This allows countries to establish multilateral agreements based on mutual recognition of their official accreditation systems. This system of international agreements provides accredited laboratories with international recognition, which allows their certificates to be more readily accepted in foreign markets. The main organization for laboratory accreditation bodies is the International Laboratory Accreditation Cooperation (ILAC, 2010), which has been established to contribute to the removal of technical barriers to trade.

7.2   Laboratory accreditation and ISO 17025

The first edition of ISO/IEC 17025 of the General Requirements for the Competence of Testing and Calibration Laboratories published in 1999 arose from wide and thorough experience of the implementation of previous standards (ISO/IEC Guide 25, 1990; EN 45001, 1989). It contained the necessary requirements for testing and calibration laboratories to prove that they operate a well-organized management system and to demonstrate their technical competence and ability to produce reliable and valid results. The management requirements of this first edition referred to other standards that were in effect at the time (ISO 9001:1994 and ISO 9002:1994). Both standards have been superseded by a new standard (ISO 9001:2000) and an alignment of the ISO/IEC 17025 was needed. In the second edition (ISO/IEC 17025:2005), articles were adapted or added to be in accordance with the new ISO 9001:2000. These included the principles of continuous improvement, process definition and mapping process metrics. Most of the new articles focus on the laboratory's effective implementation of core quality elements, for example audits, management review, corrective and preventive action and customer feedback. New terminology in the ISO/IEC 17025:2005 standard reflects the shift in thinking in ISO 9001:2000; quality system becomes management system, client becomes customer and non-conformance becomes non-conformity. Accreditation bodies that recognize the competence of testing and calibration laboratories refer to this ISO 17025 norm as the norm for accreditation.

7.2.1 Management requirements

Management requirements are mainly related to the operation and effectiveness of the quality management system within the laboratory. The part of ISO 17025 that deals with management requirements is divided into 15 chapters, of which those regulating quality assurance within the laboratory's analytical scope are reviewed below (ISO/IEC 17025 in analytical laboratories).

Organization (4.1)

The duties and responsibilities of the laboratory have to be defined unambiguously. The laboratory is required to have clear organizational structures with respect to its technical activities, support services as well as the organizational duties and tasks of its staff. As far as applicable all requirements in the ISO 17025 should be fulfilled. There should be no internal or external conflicts of interest, adversely influencing the required and guaranteed quality of the laboratory's delivered services. The laboratory and its personnel need to have sufficient independence and should be free from any commercial or financial pressure that may unfavourably affect the quality of the generated final results.

Policies and procedures should be developed, approved and standardized which warrant the confidentiality of the customers' interests. Adequate supervision of the laboratory staff by experienced, responsible and qualified persons, assessing the analysis results should be provided. A quality assurance manager has to be appointed who directly communicates with, and assesses on behalf of, the highest level of the laboratory's management.

Management system (4.2) and improvement (4.10)

The chapter on the management system in ISO 17025 (2005) indicates the conditions and the modalities necessary for the successful establishment, development and maintenance of a management system reflecting and referring to the scope of activities. Policies, standard operating procedures and instructions must be available to ensure the quality of the work performed. The general quality management system and its policy statements should be documented in a quality manual, distributed and put into practice by the top level management. These statements should be relevant and consistent with the scope of activities. Evidence of commitment by the laboratory's management should be demonstrated to develop and continually improve its effectiveness, meeting the customer's quality demands.

New requirements in the revised ISO document ISO/IEC 17025:2005 with respect to the edition of 1999, use a new language with the principle objective to align it closer with ISO 9001:2000. Highlights of these new requirements emphasize the importance of 'having managerial and technical personnel who, irrespective of other responsibilities, have the authority and resources needed to carry out their duties, including the implementation, maintenance and improvement of the management system *and* ensuring that its personnel are aware of the relevance and importance of their activities and how they contribute to the achievement of the objectives of the management system' (4.1.5a and k) (ISO/IEC, 2005).

Job descriptions should comply and the laboratory should implement relevant training. Continuous improvement in the effectiveness of the management system in the laboratory by applying the quality policy and objectives, audit results, analysis of data, corrective and preventive actions and management review must be pursued. Evidence of commitment to the development and implementation of the management system and improving its effectiveness should be provided by the laboratory management.

Another clause referring to ISO 9001:2000 is stated under 4.2.7: 'top management shall ensure the integrity of the management system is maintained when changes to the management system are planned or implemented' (ISO/IEC, 2005). Management of change is an important part of ISO 9001:2000. There might be no loss of quality over time by implementing changes in order to comply with this clause.

Document control (4.3)

Individual clauses in this section indicate how all documents and procedures relating to the management of the quality system are established, identified, approved and standardized, internally distributed and modified according to an all-encompassing system of document control. Internal documents, containing specifications for analytical results, standard operating procedures, instructions for maintenance and calibration, forms and tables, as well as external normative and regulatory documents, should be authorized and controlled. All documents should be reviewed periodically and adapted if necessary to ensure continuing suitability and compliance with applicable requirements. Specific demands are defined with respect to changing documents (4.3.3) (ISO/IEC, 2005). Changes to documents are reviewed and approved. The new or altered text shall be identified in the procedure or its attachments, where practicable. Revised documents are reissued as soon as practicable.

Control of non-conforming testing and/or calibration work (4.9)

This section requires the laboratory to introduce general procedures for dealing with delivered services that do not conform to its own procedures or the agreed requirements of the customer. The implemented policy and procedures should be sufficiently detailed to manage and apply remedial actions as well as recalling non-conforming work and notifying the customer. The necessary previously defined corrective actions should be provided and executed immediately to avoid reoccurrence. Corrective action procedures as stated under Section 4.11 should be considered.

Corrective action (4.11)

This section reveals the necessary measures to be taken in case of identification of non-conforming work or deviations from standardized policies and procedures in the management system or technical operations. Problems might be identified through a variety of activities such as control of non-conforming work, internal or external audits, management reviews, feedback from customers and from staff observations.

The root cause analysis of the non-conformity is the key and sometimes the most difficult part in the corrective action procedure and a careful analysis of all potential causes of the problem is required. The effectiveness of the corrective actions should be monitored and assessed. Changes resulting from corrective action investigations should be documented and implemented.

Preventive action (4.12)
Preventive actions are more proactive processes that identify opportunities for improvement rather than reaction measures to identified problems or complaints. Procedures should be developed and available to prevent potential sources of non-conformities as well as reoccurrence of established previous non-conformities. Preventive actions might also involve analysis of data, including trend and risk analyses and proficiency-testing results.

A new requirement in the revised ISO 17025 (2005) with respect to the 1999 edition is a clause that urges the identification of necessary improvements and potential sources of non-conformities. If such improvements are identified or if preventive actions are required, suitable action plans should be developed, implemented and monitored in the management system. The efficiency of the preventive measures should be evaluated. The new language in this clause emphasizes that improvements as well as potential non-conformity are both covered under preventive action.

Internal audits (4.14)
Accredited laboratories should organize internal audits, following a predetermined schedule and an internal procedure, to verify that they comply with the ISO 17025 for the requirements of the management system as well as to the technical competence for the testing and/or calibration activities in the scope. Internal audits are an excellent tool to prepare external audits and continuously to improve quality management. Internal audits should be carried out by independent and qualified assessors.

If the internal audit reveals failing and shortcomings of the reliability and validity of the laboratory's test results or the effectiveness of its quality management, the laboratory should include follow-up activities with corrective and preventive action plans. Customers should be informed in writing by the laboratory management if it is demonstrated that sent test results have been affected by the non-conformities found.

Management reviews (4.15)
In this section the requirements are stipulated which should guarantee the continuing suitability and effectiveness of the laboratory's management system and testing and/or calibrating activities. It is the responsibility of the laboratory's top management to conduct periodically a relevant review according to a predetermined schedule and standardized procedure.

Necessary improvements and their monitored effectiveness should be introduced by means of follow-up activities, carried out within a suitable timescale. The management review shall especially focus on the suitability of policies and procedures, reports from managerial and supervisory personnel, the outcome of recent internal audits, corrective and preventive actions, assessments by external bodies, the results of inter-laboratory comparisons or proficiency tests, changes in the volume and type of the work, customer feedback, complaints, recommendations for improvement and other relevant factors such as quality control activities, resources and staff training.

7.2.2   Technical requirements

The technical requirements of the ISO 17025 document mainly focus on the required technical competence of the staff and the personnel of the laboratory, the capability of the facility accommodation and environmental conditions, the sampling and handling of the tested objects together with the appropriate and efficient performance of the applied test and/or calibration methods, good operational conditions of the equipment involved, the traceability and quality assessment of the measurement raw data and the reporting of the experimental results. The whole part on the technical requirements is divided in ten chapters. Those chapters that refer to fundamental quality demands that ensure the reliability of the final analytical results are discussed below.

General (5.1)

This general clause deals with the several factors which determine the correctness and reliability of the measurement results as outlined in the following sections. The laboratory staff and personnel should be aware of the extent to which these different factors contribute to the measurement uncertainty of the developed test and calibration methods.

Personnel (5.2)

Emphasis on technical competence, education, training, experience and demonstration of skills is an important requirement in this clause as the personnel might significantly affect the quality of the test results. The implication is that unambiguous definitions of tasks and correct job descriptions covering the whole laboratory management activities should be worked out. Based on the required skills and the qualifications available, a training programme should be developed and supplied for each member of the staff.

A new requirement of ISO 17025 is that the effectiveness of the training action taken should be evaluated (5.2.2). Each feasible action taken to verify training effectiveness is acceptable. Amongst others, participating to proficiency tests or collaborative trials, analysing a sample already run by an experienced analyst, running a reference material or observing the test manipulations and the handling and processing of the measured results by a supervisor might be suitable evaluation criteria of the new analyst's technical competence. The supervisor's signature on the experimental results in the training records might provide a favourable assessment and serve as the evidence of the effectiveness of the training.

Test and calibration methods and method validation (5.4)
Validation can be defined as confirmation by experimental verification and by providing objective evidence that an analytical method fulfils the specific conditions and requirements for the applications it is intended for. This means that every method applied within the scope of activity of a laboratory should be validated. Another consequence for the laboratory is that modified or in-house developed methods are fully validated for new intended purposes. Reference or normalized methods might be considered as validated if approval statements are recognized and confirmed by competent authorities. In this case the laboratory should only demonstrate its competence in performing the test by using, for example, control samples, by fulfilling critical system suitability tests, controlling the experimental progress of the method or by running method specific proficiency test samples.

Analytical procedures for each validated method applied should be written and describe all means and operations required to perform the analysis correctly. The field of application, main principles of analysis, full equipment, reference standards and reagents, operational conditions and protocols, primary validation prescriptions, criteria and/or requirements for approval/rejection of the data recorded, system suitability demands, calculation and expression of results and their uncertainty estimation and the test reporting should be outlined.

The field of application might be understood as being the combination of different possible matrices and the actual concentration range within which the examined analyte might be present and has to be measured. The field of application also determines the right performance conditions which are adequate with respect to influencing matrix factors and warn of known interferences from other compounds present. It is more convenient to place a boundary on the field of application than to validate a too large procedure. In the ISO 17025 norm, different approaches for validation are recommended either alone or in combination, some of which are discussed in Section 7.3 later.

Equipment (5.5)
This section deals with the essential requirements for sampling and testing equipment installed in the laboratory, ensuring the correct performance of the tests and/or calibrations. Emphasis is put on the environmental conditions, the capacity and quality of the equipment, the maintenance and calibration procedures as basic prerequisites to produce precise and accurate measurement results. The equipment should fulfil the required specifications relevant to the tests performed, for example the detection and quantification limits, sensitivity, reproducibility or selectivity. This means that these specifications should first be described in detail so that the equipment purchased does not fail afterwards to produce the relevant required results. Relevant calibration programmes should be regularly executed to demonstrate that the significant equipment's effects on the measurement results are established and its specifications are maintained.

Records of the equipment's identification characteristics and documentation of the maintenance and calibration procedures performed should be registered as well as records kept of the individual modules and software that contributed to final measurement results. The calibration status of the equipment should be labelled as well as the last and next calibration dates.

Up-to-date operating instructions referring to the manuals provided by the manufacturer should be available to the authorized operating personnel. The whole equipment, including hardware and software, of which the function and calibration status are verified and shown to fulfil the required specifications, should be protected against adjustments that may invalide the measurement results (Huber, 1998, 1999).

Complete equipment qualification can be defined as the overall process of ensuring that the measuring equipment is appropriate for its intended use. Equipment qualification is often split into design qualification, installation qualification, operational qualification and performance qualification (verification).

The recommended steps to be considered for inclusion in the design qualification (DQ) are a clear description of the analytical problem, selection of the most suitable technique of analysis, a detailed description of the environmental conditions for punctual functioning of the equipment chosen, a preliminary selection of the functional and performance specifications, a preliminary selection of potential suppliers, testing the instrument, the final selection, full documentation of the functional and operational specifications and use of the supplier's guidelines.

The installation qualification (IQ) performs and documents the proper installation in the selected user environment, establishes that the equipment is received as designed and specified by the supplier, establishes and confirms that the environmental conditions are met for proper operation and use of the equipment.

The operational qualification (OQ) is a next step during the process of putting into use newly installed equipment. The operational qualification executes tests in the selected user environment to ensure that it meets the previously defined functional and performance specifications. Extensive testing is essential if all types of applications will be carried out on the equipment, some of which put high demands on the performance of the system. Convenient generic standards should be tested to check the equipment for its intended purpose. When the equipment is built up of several connected modules, tests verifying the system as a whole instead of individual modular testing are obvious (holistic testing). Individual modules are tested to diagnose deficiencies if the system fails.

The frequency of performing OQ depends on the type of instrument, the stability of the performance parameters and the acceptance criteria defined to ensure a high probability that critical parameters are within the operational specifications. If meeting this requirement becomes doubtful, the quality of the analytical results is questionable. As a consequence, the proper selection of suitable procedures to verify the presumed acceptance limits is critical. The frequency of the operational qualification also depends on the frequency of using the equipment. In many application bulletins on the qualification frequency of common laboratory equipment, a frequency of once a year is appropriate. However, if the equipment is operating continuously, the frequency of the operational qualification should be enhanced.

The performance qualification (PQ) (performance verification, PV) tests that the whole equipment consistently performs as intended for the selected field of applications according to the specifications appropriate for routine use. The test frequency is much higher than for OQ and should be performed under similar conditions of routine analysis. This means in practice that PQ happens daily or whenever the equipment is used.

Test criteria and frequency might be established during the development and validation of the analytical methods applied. In practice, PQ can mean system suitability testing, or comparing critical key system performance characteristics to previously documented preset limits. Suitable corrective measures should be defined and applied if obvious specification deviations are observed with the PQ tests.

A valuable recommended alternative for performance verification might be the analysis of quality control (QC) samples, enabling progressive construction of control charts. The known QC samples are spread between the actual analysed samples at intervals characterized by the total number of samples, the stability of the measurement system or the specified precision. In this way the measurement system's performance is continuously monitored under similar conditions as during the method application.

7.3 Statistical method validation approach for ensuring the quality of results from food control laboratories

7.3.1 Basic statistical concepts

Essential parameters to be considered when characterizing a fully validated in-house developed method are accuracy, detection limit, selectivity, linearity (and range), repeatability, reproducibility and robustness against external influences (ISO/IEC 17025:2005). The range and accuracy of the values for these parameters, as assessed for the intended use, should be relevant to the customer's needs.

A fundamental reference point of departure steering the whole validation process of an analytical method, is the concept of 'accepted reference value' for an analyte to be determined by the method to be validated (Hubert *et al.*, 2003, 2004). The accepted reference value is the 'conventionally true value' for a measured compound and might be defined as a generally agreed reference value to establish the essential validation parameters during the validation process. Its value might be derived from a theoretical or established value, defined from scientific evidence. It also could be an assigned or certified value, based on experimental data from an official national or international organization or a consensus value based on a collaborative study. In the case where the previous situations are not applicable, its value might be the mathematical expectation of the measurable quantity as the arithmetic mean of a specified population of repeated measurement values obtained by applying a reference procedure (ISO 5725-2:1994). If certified reference material is available in the field of application, the conventionally true value of the sample might be accepted as such without use of the reference procedure.

To establish the correct calibration so that the analytical process may be validated successfully, the disposal of suitable calibration and validation standards is essential. A calibration standard is a sample of an exactly known concentration of the analyte determined as such or in the expected samples' matrices, which allows the calibration curve to be performed within a previously fixed concentration range. A validation standard can be considered to be a reconstituted sample in the matrix with a known added analyte concentration and agreed to have a true value used to validate the analytical procedure.

Some of the ISO 17025 validation parameters mentioned might have several different and even inconsistent definitions according to the official organizations that proclaim official regulatory documents for their own fields of application. The terminology used varies between different official documents such as the Food and Drug Administration 'guide on validation of bio-analytical methods' (FDA, 2001), ICH Q2(R1) (ICH, 1995), ISO (ISO 5725-1 to 6, 1994), IUPAC (Thompson *et al.*, 2002) and AOAC (AOAC, 1990). This means that the subsequent statistical interpretation of the results obtained and the final decision about the validity of the analytical procedure depends on consistent and adequate definition of the criteria assessed (Rozet *et al.*, 2007a). This leads to highly critical consequences since the validated analytical process might be used daily in routine analysis to decide on economic and public health matters. In this chapter we give preference to the ISO definitions as they are totally compatible with the concept of the 'accuracy profile' which will be discussed further.

7.3.2   Definitions

Specificity and selectivity
The specificity of an analytical process can be defined as its ability to assess unequivocally the target analyte in the presence of components which may be expected to be present without interfering. A method which is perfectly selective for an analyte or group of analytes is said to be specific. This means that the method guarantees that the registered signal measured is directly related to the targeted compound and allows its identification and quantification. For chromatographic methods, the selectivity depends on the separation quality and on the selectivity of the detection method (e.g. liquid chromatography–mass spectrometry (LC–MS), liquid chromatography–mass spectrometry/mass spectrometry (LC–MS/MS), liquid chromatography–diode ray detection (LC–DAD), gas chromatography–mass spectrometry (GC–MS), gas chromatography–electron capture (GC–EC) etc).

Several general definitions of selectivity and specificity are given by different international organizations as, for example IUPAC (Vessman *et al.*, 2001), WELAC (WELAC, 1993), ICH (ICH Topic Q2 (R1), 1995) and AOAC (AOAC, 1990). ISO does not give a definition.

Response function and calibration
The response function for an analytical procedure describes the mathematical

relationship between a measurable response or signal (peak area, peak height, absorbance) and the concentration or quantity of the analyte in the sample within a suitable concentration range. This mathematical relationship is represented by the calibration curve resulting from the response function which is linear or non-linear, depending on the detection method or the concentration range covered. Common fitting methods allow the estimation of the mathematical equation of the response function that gives reliable measurements, for example the classical or weighed least-squared linear model. However, it is not required or even relevant systematically to force a linear function where the linear range is different from the working or dosing range. The inadequate choice of the statistical regression model for the calibration curve often provokes significant biases or imprecision in analytical measurements. In the same context, it is important to model the whole analytical process properly over a larger concentration range. In this way the complete analytical procedure should be modelled by an overall appropriate response function that allows accurate measurement (Rozet *et al.*, 2007a).

Linearity

The linearity of an analytical process is its ability to back-calculate by means of the response function resulting from calibration, quantitative test results which are directly proportional to the real quantities present in the sample.

As explained in a clarifying article about regulatory documents for analytical method validation (Rozet *et al.*, 2007a), linearity is often confused with the response function. These authors demonstrate that this confusion is maintained, for example in the ICH Topic Q2 (R1) document. In the terminology part, linearity is correctly defined as '… the ability (within a given range) to obtain test results which are directly proportional to the concentration (amount) of analyte in the sample'. However, the methodology section declares that '… linearity should be evaluated by visual inspection of a plot of signals as function of analyte concentration or content'.

It is obvious that here the signal and not the result is concerned and that the document confuses linearity and calibration curve (response function). The text in the same document continues 'if there is a linear relationship, test results should be evaluated by appropriate statistical methods, e.g. by calculation of a regression line by the method of least squares'. The 'test results' for an analyst are the back-calculated measurements evaluated by the 'regression line' which is in fact the suitable calibration curve, established by means of appropriate statistics. Paradoxically, the last sentence of this section states explicitly that no linearity is needed between the quantity and measured signal: 'In some cases, to obtain linearity between assays and sample concentrations, the test data may have to be subjected to a mathematical transformation prior to regression analysis'. It is further concluded that this section in the document intends to suggest that it might be convenient to apply an ordinary least squares (OLS) linear function by transforming the data if 'the visual plot' of signal versus concentration seems to be not 'straight'. However, it is emphasized that this rule should not be interpreted as there being a scientific necessity to have an implicit linear relationship between 'signal' and 'concentration' (Rozet *et al.*, 2007a,b).

In this context, the 'fit-for-purpose' principle might be introduced as a valuable alternative. It is the purpose of an analytical procedure to give accurate measurements in the future, so the standard calibration curve has to be evaluated on its ability to provide accurate measurements. As a consequence, a significant source of bias and imprecision in analytical procedures might be caused by inadequate modelling of the calibration curve. The precision of the results calculated from a chosen fitted regression model depends largely on the spread of the standard concentration values over the range of analysis of the target compounds. In this concept replicated standard calibration points are included at the extremes of the range as well as equally spread replicated standard points in between (Rozet *et al.*, 2007b).

Trueness

As in the ISO-5725 norm (ISO 5725-1:6, 1994), trueness expresses the closeness of agreement between the average value from a large series of test results and an accepted reference value (or a conventional true value). Trueness is a concept related to systematic error and is generally expressed in terms of bias. Bias expresses the difference between the expectation of the test results and an accepted reference value. Trueness is considered as a concept and refers to a characteristic or a quality of the measurement procedure and not to a result generated by this procedure. The trueness is expressed by measurement of the bias. Trueness is generally expressed in terms of recovery and of absolute or relative bias:

$$\text{Recovery} = \frac{\bar{x}_i}{\mu_\text{T}} \times 100 = 100 - \text{relative bias } (\%)$$

$$\text{Relative bias } (\%) = 100 \times \frac{\bar{x}_i - \mu_\text{T}}{\mu_\text{T}}$$

The ISO-5725 document (ISO 5725-2, 4 and 6, 1994) clearly explains and describes how to measure the trueness of an analytical procedure. Independent validation standards i with known true values of analyte concentrations or amounts μ_T are analysed several times to obtain their individual measured values. The mean value of these individual results $\bar{x}_i$ is calculated and compared to the known true values.

ISO considers 'bias' and 'trueness' essentially as the same, whereas IUPAC attributes the same meaning to bias as ISO but does not recognize the definition of 'trueness'. AOAC also accepts 'bias' in the same sense and defines it as 'long term' difference from the average of many groups of individual values from the 'true' or 'assigned' or 'accepted value'. Trueness is defined by AOAC as the difference between the single average of a group of individual values and the 'true' or 'assigned' or 'accepted value'. So AOAC considers a hierarchy in systematic errors and distinguishes between a single average (trueness) and bias (many averages). Accuracy is also defined by AOAC. In official regulatory documents the concept of trueness is not *per se* defined. Recoveries of methods used in the official control of aflatoxins, ochratoxin A, patulin, deoxynivalenol, zearalenone, fumonisin B_1 and B_2, T-2 or HT-2 toxin in foodstuffs should fall within the recommended range as defined in Commission Regulation 401/2006/EC (European Commission, 2006). This range is defined per toxin and for different concentrations.

It is essential to distinguish the difference between a result and an average value. It is the ultimate objective of an applied analytical procedure to deliver a final measurement result which decided the destiny of the verified entity as a whole. As a consequence, each measurement result obtained for a representative sample of the controlled product has to be determined adequately. Unlike a single measurement result, an average result represents only the central location of the distribution of all measurement results obtained for the same true result and not the position of each individual result. Similarly the bias, relative bias or recovery also position the distribution of the analytical results with respect to the accepted true value (Rozet *et al.*, 2007a). The usual statistical methodology applied to assess the fitness of the bias (relative bias, recovery) of an analytical procedure for a certain purpose is the Student *t*-test.

The significance level α is mostly set at 0.05, which means that the probability of wrongly rejecting the null hypothesis H_0 is 5% or that the bias is considered erroneously significant in five times out of 100. The only meaningful conclusion in accepting the null hypothesis is not that the test demonstrates the absence of a bias, but that it could not establish a bias that is different from zero. Moreover, the test might conclude that there is a significant bias, whereas it might be totally acceptable from an analytical point of view. So the question the analyst wants to answer is 'is the bias of my analytical procedure acceptable?' This question is answered by the concept of the 'interval hypothesis test', where acceptance limits for the bias are previously fixed. These limits enclose the true bias of the analytical procedure in such a way that the trueness of this procedure is acceptable, as a totally unbiased procedure does not exist (Boulanger *et al.*, 2007; Feinberg, 2007; Hubert *et al.*, 2008; Hartmann *et al.*, 1995).

Precision

In contrast to the definitions of trueness, consistent definitions of precision have been formulated by the FDA Bioanalytical Method Validation, ISO, Eurachem and IUPAC. Precision expresses the closeness of agreement (dispersion level, relative standard deviation) between a series of measurements of the same homogeneous sample (independent assays) under prescribed conditions. The precision is a measure of the size of random errors, irrespective of whether or not the mean of the measurements is a correct representation of the accepted true value. Precision is expressed as standard deviation s, variance s^2 or relative standard deviation (*rsd*). In the ISO 5725 document (ISO-5725:2, 1994) precision is distinguished at three levels: repeatability, intermediate precision (within laboratory) and reproducibility (between laboratories).

Repeatability is precision under conditions where the results of independent

Table 7.1   General table presenting total variance: the sum of the squares of the differences between each of the data x_{ij} and the grand mean $\bar{x}$, divided by $n{-}1$ degrees of freedom where n is the total number of data.

	Sample 1	Sample2	...Sample j...	...Sample k
	x_{11}	x_{12}	x_{1j}	x_{1k}
	x_{21}	x_{22}	x_{2j}	x_{2k}
	.	.	.	.
	x_{i1}	x_{i2}	x_{ij}	x_{ik}
	.	.	.	.
	x_{n1}	x_{n2}	x_{nj}	x_{nk}
Mean	$\bar{x}_1$	$\bar{x}_2$	$\bar{x}_j$	$\bar{x}_k$
Variance	s^2_1	s^2_2	s^2_j	s^2_k
Grand mean	$\bar{x}$			

assays are obtained by the same analytical procedure, on identical samples, in the same laboratory by the same operator, using the same equipment and during a short interval of time. Repeatability conditions involve the execution of the full procedure for the selection and preparation of the test sample and not only the replicate instrumental determinations on a single prepared sample.

Reproducibility is precision under conditions where results are obtained by the same analytical procedure, on an identical sample, in different laboratories, using different operators and different equipment. The reproducibility of an analytical procedure is established by an interlaboratory study and with standardization of the procedure.

For intermediate precision, ISO recognizes M-factor different intermediate conditions with $M = 1, 2$ or 3. For $M = 1$, only one of the three factors (operator, equipment, time) is different. For $M = 2$ or 3, two or all factors differ between the determinations.

For the official control of ochratoxin A, patulin, deoxynivalenol, zearalenone, fumonisin B_1 and B_2, T-2 and HT-2 toxin in foodstuffs, Commission Regulation 401/2006/EC (European Commission, 2006) defines maximum permitted values for the relative standard deviation calculated from results generated under reproducibility conditions (RSD_R) and calculated from results generated under repeatability conditions (RSD_r). For the aflatoxins the maximum permitted RSD_R value should not be greater than twice the value determined from the Horwitz equation. The maximum permitted value for the RSD_r for aflatoxins should not be greater than $0.66 \times RSD_R$.

$$\text{Horwitz equation} = 2^{(1-0.5\log C)}$$

The actual variance computations can be understood more easily by considering ANOVA as splitting up of the total variance in its components (Massart *et al.*, 1997a). The total variance is the sum of the squares of the differences between each of the data x_{ij} and the grand mean $\bar{x}_i$, divided by $n - 1$ degrees of freedom where n is the total number of data as presented in the general Table 7.1.

For reasons of computational convenience, the 'sums of squares' SS is used. SS_{Tot} is the sum of squared differences of each individual observation from the grand mean. SS_{Tot} also might be expressed (Massart *et al.*, 1997a) as:

$$SS_{Tot} = SS_{Res} + SS_A$$

where SS_{Res} is the residual sum of squares. SS_A is the sum of squares due to the effect of the studied factor, which is the composition heterogeneity among the samples. Variance estimates from the sum of squares are obtained by dividing with the number of degrees of freedom:

$$MS = SS/df$$

where *MS* or 'mean square' is a variance estimate and *df* is the number of degrees of freedom. Applied to SS_{Res} and SS_A, this yields:

$$MS_A = SS_A/(k - 1)$$

and

$$MS_{Res} = SS_{Res}/(n - k)$$

In practice the number of degrees of freedom can be derived by reasoning that the number of degrees of freedom for SS_{Tot} is $(n - 1)$, that $(k - 1)$ are those used by SS_A and that the rest $(n - 1)-(k - 1) = n - k$ is available for SS_{Res}. This helps to understand the reason for the term 'residual'. The residual sum of squares is the total sum of squares minus the sum of squares due to a specific factor ($SS_{Res} = SS_{Tot} - SS_A$) and the residual degrees of freedom are those that are not used up by this specific factor: $df_{Res} = df_{Tot} - df_A$.

This computational scheme can be summarized in a one way ANOVA table, Table 7.2.

Table 7.2 consists of up to five columns: the first column gives the source of the variation, the second and third the degrees of freedom and sums of squares, the fourth the mean square and the fifth the *F* values. Under the table, critical *F* values are often written that have to be compared with the experimental values in the fifth column and the conclusion about the significance of the effect at a certain level.

The concept of series and runs determines what makes the difference between repeatability and intermediate precision. The selection of different factors which will compose the runs or series, must simulate similar conditions during the routine

Table 7.2   General layout of a one-way ANOVA table

Source	Degrees of freedom	Sum of squares	Mean square	F
Between columns (*A*)	*k −1*	SS_A	$SS_A/(k - 1)$	MS_A/MS_R
Within columns (residual)	*n − k*	SS_R	$SS_R/(n - k)$	
Total	*n − 1*	SS_T		

$F_{0.05;k-1,n-k} = ...$, conclusion about significance of *A*.

use of the analytical procedures. As a consequence, including the variability, for example from one day to another, of the analytical procedure is mandatory. If during routine use, the analytical procedure will also be applied by more than one operator and performed on more than one instrument, these different factors should be introduced in the validation protocol, leading to a representative estimation of the overall variability of the analytical procedure.

Accuracy

According to ISO 5725, accuracy expresses the closeness of agreement between the test result and the value accepted as a conventional true value or as a reference value (ISO-5725:1-6, 1994). ISO defines 'test result' as a single result or the average of a set of results. The observed closeness of agreement in fact expresses the sum of the systematic (bias) and the random errors. From this definition it is clear that accuracy is a total error linked to the analytical result and not to the analytical method, the laboratory or the operators. As a consequence, the total error is expressed by the sum of trueness or bias and precision or standard deviation.

Bias and precision components for a single laboratory working under repeatability conditions should be distinguished from the interlaboratory situation (Massart *et al.*, 1997b). Two bias components are discerned: the method bias which is inherent in the method and the laboratory bias, which is considered as the bias introduced by the laboratory applying an unbiased method. ISO (ISO-5725:1994) states that the laboratory bias is the difference between the expectation of results, that is the mean of a sufficiently large number of results from a particular laboratory, and the accepted reference value.

The bias of a measurement method is defined as the difference between the expectation of the test results obtained from all laboratories using that method and the accepted reference value. The laboratory component of bias is the difference between the average of a large number of results in that laboratory and the overall average result for the measurement methods obtained by all laboratories. According to these definitions, the laboratory bias is the sum of the bias of the measurement method and the laboratory component of the bias. Depending on the situation, the laboratory component of the bias can be considered to be part of the systematic or of the random error. From the point of view of the individual laboratory, this component of bias is a systematic error. However, when carrying out method performance interlaboratory studies, the between laboratory component of reproducibility includes the laboratory component of the bias of the participating laboratories.

For most users it does not matter whether deviation from the true value is due to random error (lack of precision) or to systematic error (lack of trueness) as long as the total quantity of error remains acceptable. The concept of total analytical error or accuracy as a function of random and systematic error is essential. It is important that the total amount of error does not affect the interpretation of the test result and the subsequent decision to be taken.

Detection limit

The limit of detection (LOD) of a given method is used to discriminate between blank and contaminated samples. It is the smallest amount of analyte in the test sample which can be reliably distinguished from zero. The limit of quantification (LOQ) is the lowest concentration that can be quantified. Below this quantification limit a method cannot operate with acceptable precision (Thompson *et al.*, 2002).

Because mycotoxins belong to group B of annex I of Council Directive 96/23/EC (European Commission, 1996), official methods for the determination of mycotoxins in animal feed and animal products should, as long as no specific performance characteristics are officially prescribed (which currently is only the case for methods to determine mycotoxins in food (European Commission, 2006) be validated according to Commission Decision 2002/657/EC (European Commission, 2002). Consequently, for these methods it is not the LOD and LOQ but the decision limit (CCα) and detection capability (CCβ) that need to be determined. The decision limit (CCα) is defined as the limit at and above which it can be concluded with an error probability of α that a sample is non-compliant. The detection capability (CCβ) is the lowest concentration at which a method is able to detect truly contaminated samples with a statistical certainty of $1 - \beta$.

In practice this can cause some confusion as with a method for analysing for example maize, it is important to know whether the maize is intended for human consumption (LOD/LOQ) or will be used as animal feed (CCα and CCβ). This can be overcome by either defining the scope of the method well or by determining as well LOD and LOQ as CCα and CCβ during method validation.

7.3.3   Accuracy profile as a decision making tool

The objective of a quantitative analytical method is to quantify as accurately as possible the unknown quantities of measured compounds (Rozet *et al.*, 2007a). As a consequence, in an analytical procedure the difference between the measured result and the unknown true value μ_T in the test sample is as small as possible, which means within an acceptance limit λ which has been previously fixed:

$$-\lambda < X - \mu_T < \lambda \ \text{ or }\ |X - \mu_T| < \lambda$$

The acceptance limit λ depends on the required objectives of the analytical method and is linked to the predefined limits depending on the intented use of the results. During the validation phase, sufficient information should be gathered to guarantee that a large proportion of future results, approximate to the true value without being affected by matrix interferences. The difference between a measured result X and its true value μ_T is composed of a systematic error (bias) and a random error (precision). Their true values are unknown but might be estimated from adequate validation experiments. As a consequence, the objective of the validation phase is to demonstrate that for a certain experimentally estimated method bias $\hat{\mu}_M$ and precision, $\hat{\sigma}_M$ the expected proportion of measurements that will fall within the acceptance limits, is higher than a predefined proportion level β:

$$E_{\hat{\mu},\hat{\sigma}}\{P_X[\,|X - \mu_T| < \lambda\,]\,|\,\hat{\mu}_M, \hat{\sigma}_M|\} \geq \beta$$

There is no exact solution to calculate this proportion. To make a reliable decision, the β expectation tolerance interval is calculated (Mee, 1984, Lin and Liao, 2006):

$$E_{\hat{\mu}_M, \hat{\sigma}_M} \{ P_X [\hat{\mu}_M - k\hat{\sigma}_M < X < \hat{\mu}_M + k\hat{\sigma}_M | \hat{\mu}_M, \hat{\sigma}_M |]\} = \beta$$

The factor k is determined so that the expected proportion of the results falling within this calculated interval is equal to β. If this obtained β expectation tolerance interval is totally within the preset acceptance limits [−λ, +λ], the expected proportion of measurements within the same acceptance limits is greater than or equal to β. In practice the β-expectation tolerance interval might be expressed as follows:

$$\left[\bar{\bar{x}} \pm k_{tol} \times s_P \right]$$

This formula allows calculation of a functioning interval based on estimated statistical parameters obtained from experimental results. Here $\bar{\bar{x}}$ is the grand mean of all measured values for a certain content level which is an estimate of the reported value, k_{tol} is the 'coverage' coefficient of the tolerance interval and s_P is the standard deviation corresponding to the intermediate precision of the estimated result $\bar{\bar{x}}$. It is absolutely necessary to consider the intermediate precision standard deviation as defined in the ISO 5725 norm, as all possible variance sources have to be taken into account. Only in this way might it be assured that the analytical results are obtained and produced under identical conditions during future routine measurements. According to the ISO 5725 norm (ISO-5725:2, 1994), the intermediate precision standard deviation is calculated from the repeatability variance s^2_r and the between-days variance s^2_D:

$$s_P = \sqrt{s^2_D + s^2_r}$$

The exact calculation of k_{tol} is explained in different publications (Hubert *et al.*, 2003, 2004, 2007a,b; Boulanger *et al.*, 2003; Dewé *et al.*, 2007; Feinberg *et al.*, 2004; Gonzales and Herrador, 2006, 2007). This coefficient is directly proportional to the 1 + b/2 quantile of the Student-t distribution and also depends on the ratio s^2_D/s^2_r of the between-days variance and the repeatability variance.

The accuracy profile is a 'decision making' tool which might be graphically visualized (Fig. 7.1). The combination of the tolerance interval and the acceptation interval within the same graphic enables the analyst to decide if the analytical procedure is suitable at a certain concentration level. As an analytical procedure should quantify over a broader range of quantities, samples should be prepared during the validation phase within this range and the β-expectation tolerance interval is calculated at each level. In this way, the lowest and the highest compound quantity level, the tolerance interval of which still comprises the predetermined acceptation limits, might also be established. This is illustrated in Fig. 7.1 where both levels are appointed as the lower limit (LLQ) and upper limit (ULQ) of compound quantification in the sample.

In practice, well defined sequences of steps might be followed (Feinberg, 2007). The validation phase is the ultimate stage before the definite exploitation of

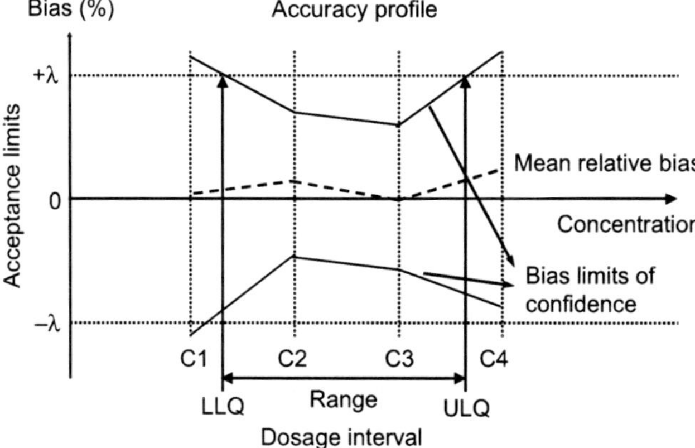

Fig. 7.1   Graphic representation of an accuracy (total error) profile, characterized by the fixed upper and lower bias acceptance limits (+/− λ), the examined quantity (concentration) range going from levels $C1$ to $C4$, the mean relative bias and the bias limits of confidence, composing the β-expectation tolerance interval at each quantity (concentration) level and the quantity (concentration) range with tolerance intervals not exceeding the acceptance limits. The lower and upper limits of quantification (LLQ and ULQ) are defined from this range.

the analytical procedure, allowing a reasonable estimate of its performances in the expected operational conditions as well as verification of its capability to quantify each unknown sample that has to be analysed (Hubert *et al.*, 2007a,b). Several calibration protocols with specified concentration levels for the determination of contaminants in food are described by the same authors (Hubert *et al.*, 2008). Using this practical decision tool as the accuracy profile, the analyst can choose the response function which is the most appropriate to answer the objectives of the analytical procedure. As such, the selected response function confirms the fitness for purpose which also validates the dosing range to be covered. Another benefit resulting from the accuracy profile is the possibility of estimating the overall accuracy of the analytical results produced by the calibration model by verifying the linearity of the relationship between the estimated and the known concentrations over the whole range.

In a recent article (Bouabidi *et al.*, 2010), a critical analysis of several analytical validation strategies is reported in the framework of the fit-for-purpose concept. They demonstrate that there is an obvious lack of clear guidance in methodologies for deciding adequately when an analytical method can be considered to be valid. In their article, Bouabidi *et al.* compare the classical decision processes applied to evaluate method validation such as the 'descriptive', the 'difference' and the 'equivalence' approaches with the validation approach developing the accuracy profile from the β-expectation tolerance (total error) interval . They conclude that these three classical validation methodologies give rise to inadequate and contradictory conclusions and do not allow them to answer adequately the objective of method validation, that is to give sufficient guarantee that each of the future results generated by the method during routine use will be close enough to the true value. The validation methodology which gives the best guarantee regarding the reliability or adequacy of the decision to consider methods as valid is the one based on the use of the accuracy profile (Boulanger *et al.*, 2007).

7.4   Comparison of a routine method with a reference method for validating the results from food and feed control laboratories

A procedure for comparing the performance precision and bias of an alternative measurement method and a reference method has been extensively developed. (Kuttatharmmakul *et al.*,1999, 2000). It is based on ISO 5725-6:1994 which has been adapted for the intralaboratory situation. This means that the proposed approach does not evaluate the reproducibility but considers the (operator + instrument + time)-different intermediate precision and/or the time-different intermediate precision. The calculation of different variance estimates from the experimental data is carried out by ANOVA. The Satterthwaite approximation (Satterthwaite, 1946) is included to determine the number of degrees of freedom associated with the compound variances. Taking into account the acceptable bias, the acceptable ratio between the precision parameters of the two methods, the significance level α and the probability β of wrongly accepting an alternative method with an unacceptable performance, the formulae for determining the number of measurements required for the comparison are given. To evaluate the bias, in addition to point hypothesis testing, interval hypothesis testing is also included as an alternative (Hartmann and Massart, 1994; Hartmann, 1995).

The ISO standard is meant to show that two methods have similar precision and/ or trueness whereas a laboratory that performs a method comparison study is rather more interested to evaluate whether the new method is at least as good as the reference method. This implies that some two-sided statistical tests in the ISO guidelines are not appropriate for comparison of the two methods, applied in a single laboratory, where one-sided tests have to be considered in the evaluation of the precision. In decision-making concerning the new alternative method, it is important not to reject an alternative method which is appropriate and not to accept an alternative method which is not appropriate. The former is related to the α-error of the statistical tests used in the comparison and is controlled through the selection of the significance level. The latter is related to the β-error and when it is considered it is generally taken into account by including sample size calculations. This approach is also included in the ISO guideline.

In their article Kuttatharmmakul *et al.* (2000) propose an adaptation of the ISO guidelines to the intralaboratory comparison of two methods. This new concept is also applicable when two laboratories of the same organization are involved, each laboratory specializing in one of the methods. To evaluate the bias, in addition to point hypothesis testing, interval hypothesis testing in which the probability of accepting a method that is too biased is controlled, is also included.

Owing to the specified acceptance criteria of the alternative method, the proposed approach might lead to a large number of measurements being performed. An alternative approach is to perform the methods comparison with a user-defined number of measurements and to evaluate the probability that a method with an unacceptable performance will be accepted. Indeed, the application of interval hypothesis testing in the evaluation of bias can lead to the false rejection of a method, which, in reality, has an acceptable bias. To limit the risk of false rejection, an appropriate number of measurements are required. However, the required number of measurements depends on the precision of the analytical methods, the bias that analysts are prepared to accept with a high probability and the risk that one is willing to take of incorrectly rejecting a method that has an acceptable bias.

The reliability of the formulae proposed depends on the quality of the precision estimates used in the formulae. When the precision estimates applied correspond well with the true precision parameters, the sample size determined assures that the risk of incorrectly rejecting an alternative method that has an acceptable bias does not exceed the specified level. In different case studies with a given repeatability and time different intermediate precision for the reference method and a given acceptable bias for the alternative method, the minimum number of days of analysis are calculated if the alternative method is not biased, if the bias of the alternative method is half the acceptable bias, if the bias of the alternative method is one quarter of the acceptable bias and if the bias of the alternative method is in between the acceptable bias and half the acceptable bias (Kuttatharmmakul *et al.*, 2000; Nguyen Minh Nguyet *et al.*, 2004). The probability might also be calculated where an alternative method that is too highly biased will be accepted, as well as the probability that the alternative method, which, in reality, is not biased will be rejected.

7.5 Measurement uncertainty in the results from food and feed control laboratories

'Uncertainty of the measurement is defined as a parameter, which is associated with the result of a measurement and characterizes the dispersion of the values that could reasonably be attributed to the measurand' (Hund *et al.*, 2001, 2003). The result of a measurement is considered to be the best estimate of the value of the measurand and all sources of uncertainty contribute to the spread of the results (Analytical Methods Committee, 1995). This means that the measurement result cannot be properly interpreted without knowledge of the uncertainty of the results. Several concepts are developed for the estimation of the uncertainty related to measurement result. One of the first approaches in analytical chemistry (Wernimont, 1985) used the precision estimates from interlaboratory method performance studies for uncertainty estimations. A completely different approach, referred to as the bottom-up, error budget or error propagation, is proposed in the 'guide to the expression of uncertainty in measurement' (GUM, 1995). This guideline was developed mainly from the physical-metrological field. It derives the uncertainty of the measurement result by combining the contributions of all uncertainty sources.

A requirement for accreditation according to the ISO 17025 standard is the expression of the measurement uncertainty associated with the result obtained from an analytical measurement. In the section about uncertainty, it states that 'reasonable estimation shall be based on knowledge of the performance of the method and on the measurement scope and to make use of, for example, previous experience and validation data'. For methods used in the official control of mycotoxin levels in foodstuffs Commission Regulation 401/2006/EC (European Commission, 2006) requires that results are reported as $x \pm U$, where x is the analytical result and U is the expanded measurement uncertainty. This expanded measurement uncertainty is calculated by multiplying the measurement uncertainty by a coverage factor of 2, resulting in a confidence interval of approximately 95%.

In a published guide (Eurachem, 2000, 2007) several practical examples of the basic principles of traditional metrology applied to chemical tests are presented. Two-thirds of this document contains examples. In this guide the strategy proposed in the ISO guide for the expression of uncertainty of measurement (GUM, 1995) is presented and illustrated in many examples which show how different uncertainty contributions can be combined. In this document the most applicable procedure for identifying uncertainty sources is by a cause and effect diagram, known as the Ishikawa or fishbone diagram.

In cases where the calculation of the measurement result does not completely refer to the analytical method, many important sources of uncertainty are not considered such as sampling, sample treatment or environmental sources, which are difficult to estimate. The traditional cause and effect diagram can be related to some important chapters of the ISO 17025 standard: personnel (5.2), environment (5.3), method (5.4), traceability (5.6), equipment (5.5), sampling (5.7) and sample handling (5.8). Most of the examples presented in the Eurachem guide emphasize the influence of the method or the equipment and underestimate the uncertainty (Feinberg $et\ al.$, 2004; Feinberg and Laurentie, 2006). As a consequence, there is a need to express more realistic uncertainty values which might be obtained from experimental data of precision studies. ISO (ISO/TS guide 21748, 2004) prescribes the use of repeatability, reproducibility and trueness (bias) estimates in measurement uncertainty estimation. As these three validation parameters are substantial and essential elements in the construction of an accuracy profile, they are available for estimating the measurement uncertainty of the validated method.

According to the recommendations of the ISO/TS 21748 guide (2004), a basic model for the uncertainty of the observed measurement result is expressed by:

$$u^2(y) = u^2(\hat{\delta}) + s_L^2 + \sum c_i^2 u^2(x_i) + s_r^2$$

where:

$s_r^2 + s_L^2 = s_R^2$ is the reproducibility variance,

$u(\hat{\delta})$ is the uncertainty associated with the bias δ and

$\sum c_i^2 u^2(x_i)$ is associated with the sum of all other deviation effects.

According to this same guide, the bias uncertainty is specified as:

$$u(\hat{\delta}) = s_{\hat{\delta}} = \sqrt{\frac{s_R^2 - (1 - 1/n)s_r^2}{p}}$$

where n is the number of replicates (within condition), p is the number of different conditions, s_r^2 is the estimate of the repeatability variance (within condition) and s_R^2 is the estimate of the reproducibility conditions (the sum of the repeatability and the between condition variance components).

The calculation of the β-expectation tolerance interval involves estimating the bias and the standard deviation of the intermediate precision of the analytical procedure of which the latter is denoted as $\hat{\sigma}$ (Feinberg *et al.*, 2004). Here the variance used to estimate the β-expectation tolerance interval is defined as:

$$\hat{\sigma}_{Tol}^2 = k_s^2 \hat{\sigma}_M^2$$

and this equation is developed further yielding:

$$\hat{\sigma}_{Tol}^2 = \hat{\sigma}_M^2 + \frac{ns_B^2 + s_W^2}{np}$$

In this equation the second term is an estimator of the uncertainty of the overall mean or bias for p conditions of experiments (between conditions) and n replicates within each condition (within conditions). This equation can be simplified as:

$$\hat{\sigma}_{Tol}^2 = \hat{\sigma}_M^2 + \hat{\sigma}_{\delta M}^2$$

where the second term represents the estimated uncertainty (variance) of the estimated bias. So this equation demonstrates that the variance used for calculation of β-expectation tolerance interval equals the sum of the total variance of the method and the variance of the bias. This means that it is possible to use the standard deviation of the β-expectation tolerance interval as an estimate of the standard uncertainty in the measurements. As a consequence, a practical and direct way to estimate the measurement uncertainty is available by using data collected from the validation step. As such, the measurement uncertainty might contribute to fulfilling customer satisfaction, which is an important aspect of laboratory activities demanded by ISO 17025. Other measurement uncertainty approaches in chromatographic analysis (Barwick, 1999a; Barwick and Ellison, 1999b; Barwick *et al.*, 2001) and accredited testing laboratories (Galappatti and De Silva, 2003; Van der Veen, 2001; Visser, 2004; Maroto *et al.*, 1999a,b, 2001; Meyer, 2007; Gonzales *et al.*, 2005) are described in literature.

7.6   Conclusions and future trends

Food control laboratories have a great interest in continuously installing and developing integral quality management, of which the ultimate aim is ensuring the production of reliable analytical measurement results. The importance of implementing and organizing fundamental management as well as technical requirements, resulting in an official accreditation according to the internationally approved and mandatory ISO 17025 norm is demonstrated.

Obvious aspects and examples of specific management requirements such as the laboratory organization structure, the laboratory policy of continuous improvement, efficient document control as well as control of non-conforming testing, the implementation of necessary corrective and preventive actions, the maintenance of internal quality assurance and quality control actions and the purpose of the management review are described.

The technical requirements, the competence of the personnel, the qualification of the measurement equipment and apparatus and the validation processes of the applied testing methods are emphasized as being basic elements of the quality assurance system. Special attention is paid to the definition of fundamental statistical concepts behind the required validation parameters, mentioned in the ISO 17025 norm, for example the response function used for the method calibration and the linked linearity range, the relationship between trueness, accuracy and precision according to the ISO 5725 documents and the limits of quantification and detection according to European Commission prescriptions.

The efficiency of the fit-for-purpose principle in constructing an accuracy profile as a decisive tool for assessing the method validation results with respect to the content range of analysis to be covered and the limits of acceptance, is demonstrated. According to predefined acceptance limits, this 'total error' concept which unifies trueness as well as intermediate precision of the applied analytical method, fixes the β expectation tolerance interval which predicts the percentage of future measurement results enclosed by the acceptance limits.

This integral concept of the β expectation tolerance interval, assessing method validation parameters with respect to predefined demands for a suitable analytical method, also enables prediction of the upper and lower limits of quantification as well as the measurement uncertainties for different levels within the examined content range. As an alternative to the validation strategy applied in the construction of an accuracy or 'total error' profile, an interval hypothesis test is referred to which compares the intralaboratory results from an alternative analytical method with those from a reference method at a single concentration level. This approach is based on sample size calculations. The performed measurements ensure a high probability that an alternative method with an unacceptable performance is rejected. At the same time the interval hypothesis test controls the probability that a method with unacceptable performance characteristics will be adopted.

Future guidelines for the validation of target analysis methods, combining simultaneously both precision and bias parameters obtained from relevant experimental designs and giving rise to the construction of β expectation tolerance intervals or reliable interval hypothesis tests should be propagated and stimulated for implementation in accredited laboratories, as the predefined risk of accepting a biased or rejecting a suitable analytical method is well controlled from a statistical point of view.

7.7   References

Analytical Methods Committee (1995), *Analyst*, **120**, 2303–8.

AOAC (Association of Official Analytical Chemists) (1990), *Official Methods of Analysis*, 15th edition, volume 1, AOAC, Arlington, VA.

Barwick V.J. (1999a), 'Review – Sources of uncertainty in gas chromatography and high-performance liquid chromatography', *J. Chromatogr. A*, **849**, 13–33.

Barwick V.J. and Ellison S.L.R. (1999b), 'Measurement uncertainty: approaches to the evaluation of uncertainties associated with recovery', *Analyst*, **124**, 981–90.

Barwick V.J., Ellison S.L.R., Lucking C.L. and Burn M.J. (2001), 'Experimental studies on uncertainties associated with chromatographic techniques, *J. Chromatogr. A*, **918**, 267–76.

Bouabidi A., Rozet E., Fillet M., Ziemons E., Chapuzet E., Mertens B., Klinkenberg R., Ceccato A., Talbi M., Streel B., Bouklouze A., Boulanger B. and Hubert Ph. (2010), 'Critical analysis of several analytical methods validation strategies in the framework of the fit for purpose concept', *J. Chromatogr. A*, **1217**, 3180–92.

Boulanger B., Chiap P., Dewé W., Crommen J. and Hubert Ph. (2003), 'An analysis of the SFSTP guide on validation of chromatographic bioanalytical methods: progress and limitations', *J. Pharm. Biomed. Anal.*, **32**, 753–65.

Boulanger B., Dewé W., Gilbert A., Govaerts B. and Maumy-Bertrand M. (2007), 'Risk management for analytical methods based on the total error concept: conciliating the objectives of the pre-study and in-study validation phases', *Chemometrics and Intelligent Laboratory Systems*, **86**, 198–207.

Dewé W., Govaerts B., Boulanger B., Rozet E., Chiap P. and Hubert Ph. (2007), 'Using total error as a decision criterium in analytical method transfer', *Chemometrics and Intelligent Laboratory Systems*, **85**, 262–8.

Eurachem (2000), *Eurachem/CITAC Guide CG4*: Quantifying uncertainty in analytical measurement (second edition 2000) http://www.eurachem.org/guidesanddocuments.htm.

Eurachem (2007), *Eurachem/CITAC Guide*: Use of uncertainty information in compliance assessment (first edition 2007) http://www.eurachem.org/guidesanddocuments.htm.

EN 45001 (1989), *General Criteria for the Operation of Testing Laboratories*.

European Commission (1996), 'Council Directive 96/23/EC of 29 April 1996 on measures to monitor certain substances and residues thereof in live animals and animal products and repealing Directives 85/358/EEC and 86/469/EEC and Decisions 89/187/EEC and 91/664/EEC'. *Official Journal of the European Union*, L 125/10.

European Commission (2002), 'Commission Decision 2002/657/EC of 12 August 2002 implementing Council Directive 96/23/EC concerning the performance of analytical methods and the interpretation of results'. *Official Journal of the European Union*, L 221/8.

European Commission (2006), 'Commission regulation 401/2006/EC of 23 February 2006 laying down the methods of sampling and analysis for the official control of the levels of mycotoxins in foodstuffs'. *Official Journal of the European Union*, L 70/12.

FDA (Food and Drug Administration) (2001), *Guidance for Industry: Bioanalytical Method Validation*, US Department of Health and Human Services, Center for Biologics Evaluation and Research.

Feinberg M. (2007), 'Validation of analytical methods based on accuracy profiles', *J. Chromatogr. A*, **1158**, 174–83.

Feinberg M. and Laurentie M. (2006), 'A global approach to method validation and measurement uncertainty', *Accred. Qual. Assur.*, **11**, 3–9.

Feinberg M., Boulanger B., Dewé W. and Hubert Ph. (2004), "New advances in the validation and measurement uncertainty aimed at improving the quality of chemical data", *Anal. Bioanal. Chem.*, **380**, 502–14.

Galappatti M.S. and De Silva (2003), 'Uncertainty of analytical determinations', *Journal of AOAC International*, **86**, 1077–83.

Gonzales A.G. and Herrador M.A. (2006), 'Accuracy profiles from uncertainty measurements', *Talanta*, **70**, 896–901.

Gonzales A.G. and Herrador M.A. (2007), 'A practical guide to analytical method validation, including measurement uncertainty and accuracy profiles', *Trends in Analytical Chemistry*, **26**, 227–38.

Gonzales A.G., Herrador M.A. and Asuero A.G. (2005), 'Practical digest for evaluating the uncertainty of analytical assays from validation data according to the LGC/VAM protocol', *Talanta*, **65**, 1022–30.

GUM (1995), *Guide to the expression of uncertainty in measurement*. BIPM, IEC, IFCC, ISO, IUPAC, IUPAP, OIML. International Organization for Standardization, First edition 1993. corrected and reprinted 1995.

Hartmann C. and Massart D.L. (1994), 'Display methods for visual comparison of the results of two measurement methods', *Journal of AOAC International*, **77**, 1318–25.

Hartmann C., Smeyers-Verbeke J., Penninckx W., Vander Heyden Y., Vankeerbergen P. and Massart D.L. (1995), 'Reappraisal of hypothesis testing for method validation: detection of systematic error by compairing the means of two methods or of two laboratories', *Anal. Chem.*, 67, 4491–4499.

Huber L. (1998), 'Equipment qualification in practice', *LC GC Magazine*, **16**, 148–56.

Huber L. (1999), 'How instrument manufactures help with accreditation and ISO certification', *Accred. Qual. Assur.*, **4**, 87–9.

Hubert Ph., Nguyen-Huu J.-J., Boulanger B., Chapuzet E., Chiap P., Cohen N., Compagnon P.-A., Dewé W., Feinberg M., Lallier M., Laurentie M., Mercier N., Muzard G., Nivet C. and Valat L. (2003), 'Validation of quantitative analytical procedures, harmonization of approaches', *STP Pharma Pratiques*, **13**, 101–38.

Hubert Ph., Nguyen-Huu J.-J., Boulanger B., Chapuzet E., Chiap P., Cohen N., Compagnon P.-A., Dewé W., Feinberg M., Lallier M., Laurentie M., Mercier N., Muzard G., Nivet C. and Valat L. (2004), 'Harmonization of strategies for the validation of quantitative analytical procedures A SFSTP proposal – Part I', *J. Pharm. Biomed. Anal.*, **36**, 579–86.

Hubert Ph., Nguyen-Huu J.-J., Boulanger B., Chapuzet E., Chiap P., Cohen N., Compagnon P.-A., Dewé W., Feinberg M., Lallier M., Laurentie M., Mercier N., Muzard G., Nivet C., Valat L. and Rozet E. (2007a), 'Harmonization of strategies for the validation of quantitative analytical procedures A SFSTP proposal – Part II', *J. Pharm. Biomed. Anal.*, **45**, 70–81.

Hubert Ph., Nguyen-Huu J.-J., Boulanger B., Chapuzet E., Cohen N., Compagnon P.-A., Dewé W., Feinberg M., Laurentie M., Mercier N., Muzard G., Valat L. and Rozet E. (2007b), 'Harmonization of strategies for the validation of quantitative analytical procedures A SFSTP proposal – Part III', *J. Pharm. Biomed. Anal.*, **45**, 82–96.

Hubert Ph., Nguyen-Huu J.-J., Boulanger B., Chapuzet E., Cohen N., Compagnon P.-A., Dewé W., Feinberg M., Laurentie M., Mercier N., Muzard G., Valat L. and Rozet E. (2008), 'Harmonization of strategies for the validation of quantitative analytical procedures A SFSTP proposal – Part IV. Examples of application', *J. Pharm. Biomed. Anal.*, **48**, 760–71.

Hund E., Massart D.L. and Smeyers-Verbeke J. (2001), 'Operational definitions of uncertainty', *Trends in Analytical Chemistry*, **20**, 394–406.

Hund E., Massart D.L. and Smeyers-Verbeke J. (2003), 'Comparison of different approaches to estimate the uncertainty of a liquid chromatographic assay', *Anal. Chim. Acta*, **480**, 39–52.

ICH Topic Q 2 (R1) (1995), *Note for Guidance on Validation of Analytical Methods: Text and Methodology (CPMP/ICH/381/95) – part I– I*, EMEA, International Conference on Harmonisation..

ILAC – (International Laboratory Accreditation Cooperation) (2010) http://www.ilac.org.

ISO (1994) 5725-1:1994. *Accuracy (Trueness and Precision) of Measurement Methods and Results – Part 1: General principles and definitions.*

ISO (1994) 5725-2:1994. *Accuracy (Trueness and Precision) of Measurement Methods and Results – Part 2: Basic method for the determination of repeatability and reproducibility of a standard measurement method.*

ISO (1994) 5725-3:1994. *Accuracy (Trueness and Precision) of Measurement Methods and Results – Part 3: Intermediate measures of the precision of a standard measurement method.*

ISO (1994) 5725-4:1994. *Accuracy (Trueness and Precision) of Measurement Methods and Results – Part 4: Basic methods for the determination of the trueness of a standard measurement method.*

ISO (1994) 5725-5:1994. *Accuracy (Trueness and Precision) of Measurement Methods and Results – Part 5: Alternative methods for the determination of the precision of a standard measurement method.*

ISO (1994) 5725-6:1994. *Accuracy (Trueness and Precision) of Measurement Methods and Results – Part 6: Use in practice of accuracy values.*

ISO (2000and 2008) 9001:2000 and ISO 9001:2008. *Quality Management Systems – Requirements.*

ISO/IEC (2005) 17025:2005. *General Requirements for the Competence of Testing and Calibration Laboratories.* http://www.labcompliance.com/tutorial/iso17025/.

ISO/TS (2004) 21748:2004. *Guidance for the Use of Repeatability, Reproducibility and Trueness Estimates in Measurement Uncertainty Estimation.*

ISO/IEC (1990) Guide 25:1990. *General Requirements for the Competence of Calibration and Testing Laboratories.*

Kuttatharmmakul S., Massart D.L. and Smeyers-Verbeke J. (1999), 'Comparison of alternative measurement methods', *Anal. Chim. Acta*, **391**, 203–25.

Kuttatharmmakul S., Massart D.L. and Smeyers-Verbeke J. (2000), 'Comparison of alternative measurement methods: determination of the minimal number of measurements required for the evaluation of the bias by means of interval hypothesis testing', *Chemometrics and Intelligent Laboratory Systems*, **52**, 61–73.

Lin T.-Y and Liao C.-T. (2006), 'A β-expectation tolerance interval for general balanced mixed linear models', *Computational Statistics and Data Analysis*, **50**, 911–25.

Maroto A., Boqué R., Riu J. and Rius F.X. (1999a), 'Evaluating uncertainty in routine analysis', *Trends in Analytical Chemistry*, **18**, 577–84.

Maroto A., Riu J., Boqué R. and Rius F.X. (1999b), 'Estimating uncertainties of analytical results using information from the validation process', *Anal. Chim. Acta*, **391**, 173–85.

Maroto A., Boqué R., Riu J. and Rius F.X. (2001), 'Measurement uncertainty in analytical methods in which trueness is assessed from recovery assays', *Anal. Chim. Acta*, **440**, 171–84.

Massart D.L., Vandeginste B.G.M., Buydens L.M.C., De Jong S., Lewi and Smeyers-Verbeke J. (1997a), *Data Handling in Science and Technology 20A – Handbook of Chemometrics and Qualimetrics: Part A*, Elsevier Science BV, Amsterdam, Netherlands, 121–50.

Massart D.L., Vandeginste B.G.M., Buydens L.M.C., De Jong S., Lewi and Smeyers-Verbeke J. (1997b), *Data Handling in Science and Technology 20A – Handbook of Chemometrics and Qualimetrics: Part A*, Elsevier Science BV, Amsterdam, Netherlands, 393–6.

Mee R.W. (1984), 'β-expectation and β-content tolerance limits for balanced one-way ANOVA random model', *Technometrics*, **26**, 251–4.

Meyer V.R. (2007), 'Review – Measurement uncertainty', *J. Chromatogr. A*, **1158**, 15–24.

Nguyen Minh Nguyet A., Van Nederkassel A.M., Tallieu L., Kuttatharmmakul S., Hund E.,

Hu Y., Smeyers-Verbeke J. and Vander Heyden Y. (2004), 'Statistical method comparison: short- and long-column liquid chromatography assays of ketoconazole and formaldehyde in shampoo', *Anal. Chim. Acta*, **516**, 87–106.

Rozet E., Ceccato A., Hubert C., Ziemons E., Oprean R., Rudaz S., Boulanger B. and Hubert Ph. (2007a), 'Analysis of recent pharmaceutical regulatory documents on analytical method validation', *J. Chromatogr. A*, **1158**, 111–25.

Rozet E., Hubert C., Ceccato A., Dewé W., Ziemons E., Moonen F., Michail K., Wintersteiger R., Streel B., Boulanger B. and Hubert Ph. (2007b), 'Using tolerance intervals in pre-study validation of analytical methods to predict in-study results. The fit-for-future purpose concept', *J. Chromatogr. A*, **1158**, 126–37.

Satterthwaite F.E. (1946), 'An Approximate Distribution of Estimates of Variance Components.', *Biometr. Bull.*, **2**, 110–4.

Thompson M., Ellison S.L.R. and Wood R. (2002), 'Harmonized guidelines for single-laboratory validation of methods of analysis', *Pure Appl. Chem.*, **74**, 835–55.

Van der Veen A.M.H. (2001), 'Uncertainty evaluation in proficiency testing: state-of-the-art, challenges, and perspectives', *Accred. Qual. Assur.*, **6**, 160–3.

Vessman J., Stefan R.I., Van Staden J.F., Danzer K., Lindner W., Burns D.T., Fajgeli A. and Müller H. (2001), 'Selectivity in analytical chemistry', *Pure Appl. Chem.*, **73**, 1381–16.

Visser R. (2004), 'Measurement uncertainty: practical problems encountered by accredited testing laboratories', *Accred. Qual. Assur.*, **9**, 717–23.

WELAC (1993) *Guidance Document No. WG D2*, 1st edition, EURACHEM/WELAC Chemistry, Teddington.

Wernimont G.T. (1985), *Use of Statistics to Develop and Evaluate Analytical Methods*, AOAC, Arlington VA.

Part III

Development and analysis of biomarkers for mycotoxins

8

Developing biomarkers of human exposure to mycotoxins

M. N. Routledge and Y. Y. Gong, University of Leeds, UK

Abstract: Exposure assessment is a critical part of epidemiological studies into the effect of mycotoxins on human health. Whilst exposure assessment can be made by estimating the quantity of ingested toxins from food analysis and questionnaire data, the use of biological markers (biomarkers) of exposure can provide a more accurate measure of individual level of exposure in reflecting the internal dose. Biomarkers of exposure can include the excreted toxin or its metabolites, as well as the products of interaction between the toxin and macromolecules such as protein and DNA. Samples in which biomarkers may be analysed include urine, blood, other body fluids and tissues, with urine and blood being the most accessible for human studies. Here we describe the development of biomarkers of exposure for the assessment of three important mycotoxins: aflatoxin, fumonisin and deoxynivalenol. A number of different biomarkers and methods have been developed that can be applied to human population studies and these approaches are reviewed in the context of their application to molecular epidemiology research.

Key words: aflatoxin albumin adducts, aflatoxin DNA adducts, biomarkers of mycotoxin exposure, urinary deoxynivalenol detection, urinary fumonisin detection.

8.1 Introduction to biomarkers of exposure

In order to understand possible links between mycotoxins and human disease it is necessary to measure the exposure of a population to the toxin in question. There are several approaches to estimating exposure to dietary environmental toxins, including making direct measurement of the toxin in a sample of the food or using a questionnaire to gather data on amounts of various foods that are eaten and relating this to health outcomes (Feskanich *et al.*, 1993). Measurement of toxins in a food sample may give a broad measure of exposure providing the toxin is uniformly distributed throughout the food. If it is not, individual exposure will not be adequately assessed by this method, which may confound the assessment of any association between exposure and effect. The quality of information gathered via questionnaire may also vary depending on how well the questionnaire is designed, who records the information and issues such as reliability of the information provided and accuracy of memory. It is also very difficult to obtain reliable and accurate dose estimations by this method (Kohlmeier and Bellach, 1995).

However, advances in analytical techniques in the fields of molecular biology and biochemistry have allowed the development and use of various biological markers (biomarkers) in human tissue or bodily fluids. This has led to the development of molecular epidemiology as a subdivision of epidemiology that uses biomarkers to study human disease (Wild *et al.*, 2008). Biomarkers of exposure include the parent compound, the metabolites of the toxin and products of the reaction of the toxin (or its metabolites) with molecules such as protein and DNA (Wild, 2009) in human blood, urine or tissue samples. Biomarkers allow more accurate and objective assessment of exposure at the individual level, avoiding the issues of confounding or bias that may be associated with measuring the toxin in food samples or estimating exposure dose by questionnaire (Wild *et al.*, 2002). Such specific biomarkers of exposure may, therefore, have advantages over traditional methods of exposure estimation through diet because factors such as the amount of absorption, metabolism and excretion of the toxin, which may vary between individuals, are integrated into the formation of the biomarker. For example, the mycotoxin aflatoxin is metabolised in the body by a variety of routes that can lead to either activation to the aflatoxin 8, 9-epoxide that binds to protein and DNA, or detoxification to metabolites that are more easily excreted (see Fig. 8.1). Enzymes mediate these metabolic reactions and interindividual variations in enzyme activity may influence the fate of aflatoxin in different individuals. It is the formation of mutagenic DNA adducts that is the key initiating step in the carcinogenesis of aflatoxin and it can be seen, therefore, that measurement of aflatoxin DNA adducts is particularly relevant to the carcinogenic pathway (Wild and Turner, 2002). However, biomarkers that are not part of the disease pathway, such as protein adducts can still be useful biomarkers of exposure, indicating the internal dose of toxin (Wild, 2009).

The choice of biomarker of exposure will depend on several factors, including the specific toxin being measured, the availability of certain tissue or fluids, the specific research question being asked and the methodology available. The choice of method is certainly a key one, with factors such as specificity, sensitivity, cost and throughput being important. Most methods for specific biomarkers need to be optimised and validated (Vineis and Garte, 2008). Although sensitivity is not usually an issue in experiments *in vitro* or in animal studies, in human population studies where carcinogen exposure is usually chronic and at low levels, sensitivity

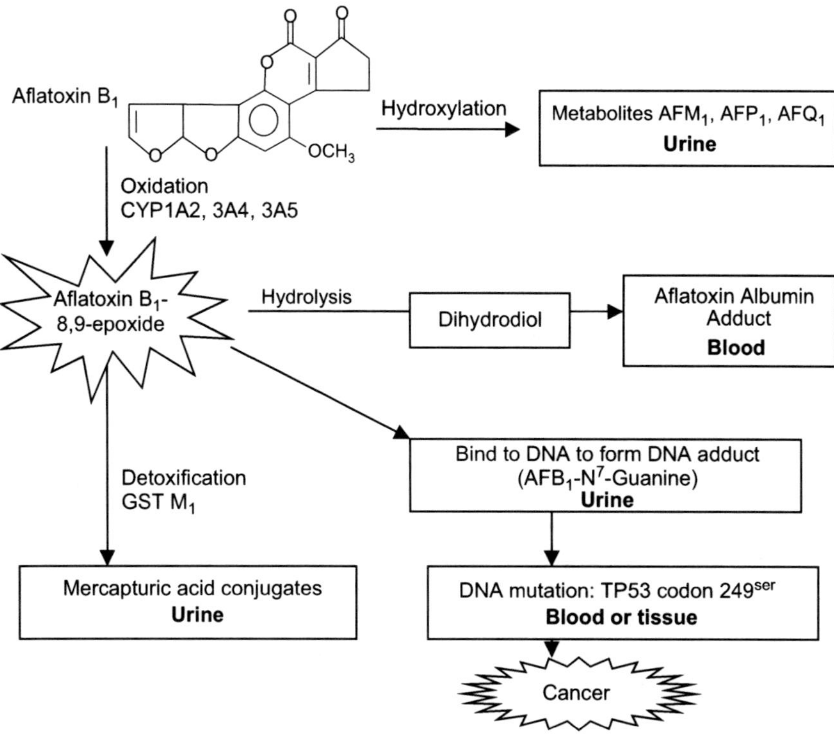

Fig. 8.1 Simplified outline of the metabolism of aflatoxin showing potential biomarkers. CYP, cytochrome P450; GST, glutathione S-transferase.

is a key factor. Whilst high sensitivity in an assay will help to avoid false negative results, high specificity will prevent false positive results. For example, in antibody-based methods the specificity of the antibody must be well established to avoid such false positive results. To be applied to human population studies the method should also be relatively high throughput, reliable and inexpensive and the biomarker should be available from non-invasive collection methods. In some cases, measuring the parent chemical in the blood or urine serves as an adequate marker of exposure but biomarkers such as DNA or protein adducts show that the toxin has also been metabolised and may have a biological effect. Furthermore, in the case of carcinogen DNA adducts, the adducts may be part of the carcinogenic process and, therefore, a better indicator between exposure and cancer outcome (Swenberg *et al.*, 2008). Protein adducts may also serve as good biomarkers of exposure; even for cancer, where the DNA adducts may be considered to be more mechanistically relevant, protein adducts may well have been formed from the same metabolites that form the DNA adducts.

Biomarkers of exposure have been applied to a number of human toxins and carcinogens in a variety of environmental, lifestyle and occupational settings. It is not surprising, therefore, that such biomarkers have been applied to the search for greater understanding of mycotoxin exposure. Indeed, the development and application of biomarkers of exposure to aflatoxin can be considered as a paradigm for the use of biomarkers in human population studies in general and not just for mycotoxin research (Groopman *et al.*, 2008). In this chapter we focus on the biomarkers of exposure that have been developed for three important mycotoxins: aflatoxin, fumonisins and deoxynivalenol.

8.2   Biomarkers of exposure for aflatoxin

There are few better examples of the application of biomarkers to the understanding of the aetiology of human disease than their use in unravelling the effects of aflatoxin in humans. A thorough understanding of the metabolism of aflatoxin in animals and later humans has allowed the development of a range of biomarkers for aflatoxin exposure, which have been used in population based studies in countries around the world. Table 8.1 shows some examples of health risk evaluation studies in which aflatoxin biomarkers of exposure have been used. These studies are further discussed below.

Aflatoxins are secondary metabolites produced by fungi of the *Aspergillus* species, which often contaminate crops such as groundnuts and maize (Wild and Hall, 2000). The toxin is produced by fungus that contaminates the crops in the field but higher levels of fungal growth and toxin production often occur during storage, particularly in warm humid conditions. There are several types of aflatoxin, with aflatoxin B_1 (AFB_1) being the most toxic and frequently occurring of the aflatoxins (Wild and Turner, 2002). The major product of aflatoxin binding to DNA is the mutagenic AFB_1-N^7-guanine adduct. When the aflatoxin epoxide reacts with proteins, the major adduct formed is with lysine residues in the protein (Sabbioni *et al.*, 1990). In addition, the AFB_1-epoxide is hydrolysed to AFB_1-dihydrodiol, which reacts with lysine in albumin to form aflatoxin albumin adducts in circulating plasma (Sabbioni *et al.*, 1990, Wild and Turner, 2002).

Since the recognition that aflatoxin exposure was a serious human health hazard, studies have been undertaken in exposed populations to obtain a fuller understanding of the contribution of aflatoxin exposure to human disease. Such studies have necessitated the estimation of exposure, either at a population or individual level. For a toxin such as aflatoxin, which in areas of high exposure often occurs in the staple food crop (usually either peanuts or maize or both), there are particular problems associated with assessing levels of exposure by food analysis. First, the distribution of mycotoxin contamination in crops can be very heterogeneous. Some maize kernels contain very high concentrations of aflatoxin, whilst in the same container or even the same cob, others may have undetectable levels of aflatoxin. This may lead to sampling error, in particular when the sample size is small. Accuracy may be further hindered by the limit of using food frequency questionnaires to estimate intake. As discussed above, biomarkers that indicate the internal dose may give a more accurate measurement of biologically relevant exposure, although individual variability could be a concern in some

Table 8.1   Examples of aflatoxin biomarker applications in human studies

Studies	Exposure categories based on biomarker	Disease risk		Reference
AFB exposure, HBV infection and human hepatocellular carcinoma in a cross-sectional study in Shanghai	HBV infection/ urinary aflatoxin biomarkers[a]	Human hepatocellular carcinoma relative risk (95% CI)		
	HBV neg./AFB absent	1		Qian *et al.*, 1994
	HBV neg./AFB present	3.4 (1.1–10.0)		
	HBV pos./AFB absent	7.3 (2.2–24.4)		
	HBV Pos./AFB present	59.4 (16.0–212.0)		
AFB exposure, child growth in a longitudinal study in Benin and Togo	Blood AFB–albumin adduct quartiled group	Body height increase cm over 8 months, mean (95% CI)		
	Low exposure	5.9 (5.2–6.6)		Gong *et al.*, 2004
	Mid–low exposure	5.3 (4.8–5.9)		
	Mid–upper exposure	4.8 (4.4–5.2)		
	Upper exposure	4.2 (3.9–4.6)		
AFB exposure, and immune function modulation in a cross-sectional study in Ghana	Blood AFB–albumin adduct dichotomised group	Immune profile[b] (mean ± SD)		
		CD3+/69+	CD19+/69+	
	Low exposure (<0.9 pmol mg^{-1})	3.9 ± 2.5	11.9 ± 7.1	Jiang *et al.*, 2005
	High exposure (>=0.9 pmol mg^{-1})	2.2 ± 1.2	6.9 ± 4.2	

CI = confidence interval
[a]Urinary aflatoxin B_1 metabolites (AFP_1, AFM_1) and DNA adduct (AFB_1-N^7-Guanine): one or more biomarkers present is defined as AFB present; HBV neg. refers to HBV surface antigen negative, HBV pos. refers to HBV surface antigen positive.
[b]Immune profile: T (CD3) and B (CD19) lymphocytes with CD69 activation marker are essential for a normal immune response against an infectious agent or a vaccine. Decreased number of CD3+/69+ and CD19+/69+ may indicate toxic effects on lymphocytes.

cases. Biomarkers of exposure may also be useful as short-term endpoints for intervention studies, in order to determine the effectiveness of the intervention (i.e. in reducing exposure) without waiting for the time that the disease takes to develop.

The demonstration that toxic groundmeal containing aflatoxin was carcinogenic in rats and other species in the early 1960s alerted scientists to the possibility that aflatoxin could be a human carcinogen and by the mid 1970s several studies in Africa showed an association between aflatoxin exposure and human liver cancer (Linsell and Peers, 1977). However, the presence of endemic infection with hepatitis B virus (HBV), a known human liver carcinogen, in the same geographical areas as where aflatoxin exposure was high, was a major confounding factor in the link between aflatoxin and human liver cancer. Confirmation of the role that aflatoxin played in this disease and subsequent findings that aflatoxin also has other human health effects depended on the development of a reliable biomarker for assessing human exposure.

Early reports of aflatoxin being detected in tissues of liver cancer patients exploited the fluorescent properties of aflatoxin (Stora, 1978). Fluorescence detection, following high-performance liquid chromatography (HPLC) separation of digested DNA, was also used to identify the major AFB_1-DNA adduct formed in rat liver *in vivo* and in human cells treated *in vitro* (Swenson *et al.*, 1977; Croy *et al.*, 1978; Autrup *et al.*, 1979). Experiments such as these helped to elucidate the structure and stability of the major AFB_1 DNA adduct, AFB_1-N^7-guanine (Groopman *et al.*, 1981). It was recognised that the ability to measure aflatoxin exposure accurately in human populations was essential to understanding the contribution of AFB_1 to human liver cancer and that the AFB_1 DNA adduct excreted into urine, which occurs when the aflatoxin base-adduct is excised from DNA during DNA repair (Bedard and Massey, 2006) or by spontaneous depurination, was a potential biomarker of exposure in human populations. To this end, Autrup and co-workers used HPLC separation with synchronous fluorescence spectroscopic detection to analyse human urine samples that had been collected from a district of Kenya in which foodborne exposure to aflatoxin was known to be high (Autrup *et al.*, 1983). This not only established that the AFB_1-N^7-guanine adduct in urine could be used as a biomarker of exposure, but also confirmed that AFB_1 was metabolised by humans *in vivo* to yield the same adduct as seen in rat liver, thus strengthening the case for the potential involvement of AFB_1 in human cancer.

However, the sensitivity of the fluorescence-based detection method was not high and a major step forward in the ability to measure aflatoxin adducts in human samples exposed to chronic low levels of AFB_1 was the development of high affinity monoclonal antibodies to the DNA adduct. In 1981, Haugen *et al.* (1981) reported the production of monoclonal antibodies raised against DNA containing the AFB_1-N^7-guanine and the ring opened AFB_1-formamidopyrimidine adducts. In 1984 Groopman *et al.* (1984) reported the development of monoclonal antibodies that recognised AFB_1, AFB_2, AFM_1, as well as the major AFB_1 adducts, enabling the development of more sensitive radio-immunoassays (RIA) and enzyme-linked immunosorbent assays (ELISA) for aflatoxin metabolites, including DNA adduct detection. More importantly these antibodies could be used to purify aflatoxin metabolites or adducts. Groopman *et al.* (1985) used aflatoxin antibodies bound to sepharose columns to isolate aflatoxin derivatives from the urine of exposed people in China. The aflatoxin adducts and metabolites purified by this immunoaffinity chromatography method were then further separated and identified by HPLC with UV detection. Without the antibody affinity purification step, this method alone would not have been sensitive enough to detect the adducts.

Another highly sensitive approach to measuring DNA adducts that was widely applied to human biomonitoring studies around this time is the ^{32}P-post-labelling method, in which radioactively labelled adducted nucleotides are separated by chromatographic methods and quantified by detection of radioactivity (Randerath *et al.*, 1981; Stiborova *et al.*, 2004). ^{32}P-post-labelling can be as sensitive as one adduct in 10^9 normal nucleotides using microgram quantities of DNA (Strickland *et al.*, 1993). However, although aflatoxin DNA adducts were detected using this method (Randerath *et al.*, 1985), the labelling efficiency (i.e. transfer of ^{32}P onto the adducted nucleotide) was subsequently shown to be very low (less than 1% of available adducts were labelled) for the aflatoxin DNA adduct (Routledge, 1991), thus limiting the application of the method for aflatoxin exposure.

Whilst DNA adducts may be the most relevant biomarker to measure for cancer-related studies, protein adducts have also been developed to serve as biomarkers of exposure. In 1986, Wild *et al.* (1986) showed that there was a correlation between DNA adducts and serum albumin adducts in rats dosed orally with environmentally relevant doses of AFB$_1$. Gan *et al.* (1988) used immunoaffinity purification coupled to a competitive RIA to measure aflatoxin albumin adducts in the serum of subjects from areas of China where AFB$_1$ exposure was known to be high. By extrapolating from their data and the known average food intake levels of AFB$_1$ in this region, the amount of adducts predicted from a chronic dose of AFB$_1$ was calculated and used to provide evidence that the adduct levels found in rats at carcinogenic doses of AFB$_1$ were relevant to human exposure levels. Hence, the use of suitable biomarkers of exposure in a human study provided evidence that the animal model for human liver carcinogenesis was indeed relevant. The use of immunoassay to quantify the aflatoxin albumin adduct as a biomarker of exposure has advantages, as the method allows high throughput at high sensitivity and serum albumin is isolated from blood, which is relatively easy to collect. Providing that the antibody has been thoroughly tested and validated, problems with low specificity caused by cross reactivity of the antibody can be avoided.

Hence, by the late 1980s, two major biomarkers of aflatoxin exposure had been developed: the urinary DNA adduct and the serum albumin adduct. The urinary DNA adduct, which can be measured in human urine using the immunoaffinity/ HPLC procedure (Groopman *et al.*, 1984, 1985, 1988), reflects recent exposure, as excretion occurs over 24–48 hours, whereas the serum albumin adduct reflects accumulated longer term exposure (2–3 months). This method for measuring serum albumin adducts utilises immunoaffinity or C18 Sep-Pak column purification and immunoassay to quantify the adduct levels (Wild *et al.*, 1986; Gan *et al.*, 1988). The previously used method of HPLC with fluorescence detection of DNA adducts (Autrup *et al.*, 1983, 1987; Tsuboi *et al.*, 1984) had now been superseded by immunoassay methods.

Both urinary DNA and serum albumin aflatoxin adducts were further validated through population studies in China and Africa (Groopman *et al.*, 1992a,b; Wild *et al.*, 1992; Groopman *et al.*, 1993). Studies that were carried out on samples from China using a RIA and from West Africa using immunoaffinity purification with HPLC-UV detection (Groopman *et al.*, 1992a; 1992b) showed a dose-dependent relationship between aflatoxin exposure and the excretion of the AFB$_1$-N^7-guanine adduct in urine in humans (Groopman *et al.*, 1992a). Again, calculations showed that the rat model was a good model for human carcinogenesis as the amount of aflatoxin excreted in the urine in the adducted form was similar in humans and rats, relative to dose and body weight (Groopman *et al.*, 1992a). The West African study also began to explore the link between HBV infection, aflatoxin exposure and liver cancer (Groopman *et al.*, 1992a). It had already been shown that there was a geographical variation in liver cancer amongst HBV positive subjects that correlated well with calculated aflatoxin dietary intake (Yeh *et al.*, 1989) and availability of a validated aflatoxin biomarker made it possible to explore this association more directly.

A key paper that illustrates the value of validated biomarkers of exposure followed in 1994 (Qian *et al.*, 1994, see Table 8.1), reporting the application of the urinary aflatoxin biomarkers, including the DNA adduct, with an HBV surface antigen marker to a cohort of middle-aged men from the Shanghai region of China. Using analysis of the biomarkers of aflatoxin and HBV exposure from a single sample collection at recruitment, and following the cohort until sufficient cases of human hepatocellular carcinoma (HCC) were diagnosed, it was shown that whilst being HBV positive in the absence of the aflatoxin marker incurred a 7.3-fold increased risk of HCC and being positive for the presence of the aflatoxin biomarker in urine incurred a 3.4 risk of HCC in HBV negative subjects, the combination of both HBV positive and aflatoxin biomarker positive status conferred a 59.4 fold increase in risk. Only the availability of well-validated biomarkers of exposure to HBV and aflatoxin allowed the role of both factors in human liver cancer to be identified in this way.

With ELISA becoming established as a suitable method for quantifying aflatoxin albumin adducts in human serum, several studies were carried out to validate this approach to human studies in populations from areas of Africa with high aflatoxin exposure (Wild *et al.*, 1990, 1992). Wild *et al.* (1990) found high levels of aflatoxin exposure in adults and children from Gambia, Senegal and Kenya, lower levels in adults from Thailand and the expected lack of aflatoxin exposure of adults and children from Poland and France.

The high throughput ELISA method of measuring aflatoxin albumin adduct in blood enables a large amount of exposure data to be collected and compared amongst various geographical areas, seasons and populations (Wild and Gong, 2010). Figure 8.2 shows examples of levels of aflatoxin albumin adduct in populations in different countries. For example, in some of the West African countries, aflatoxin albumin adduct was detected in over 90% of the population with the high values up to thousands or even tens of thousands of pictograms per milligram albumin. In contrast, the adduct levels in Europe and the USA are extremely low with the majority being undetectable. It has also been shown that 99% of the young children in Benin were exposed to aflatoxin as soon as they started weaning food (Gong *et al.*, 2002, 2003) and that aflatoxin exposure shows strong seasonal variations related to changes in aflatoxin levels in stored grains (Turner *et al.*, 2000, 2005; Gong *et al.*, 2004).

The availability of the biomarkers has also enabled two important advances in aflatoxin research. First, the validated biomarkers of exposure can serve as short term endpoints in intervention studies designed to identify ways of reducing

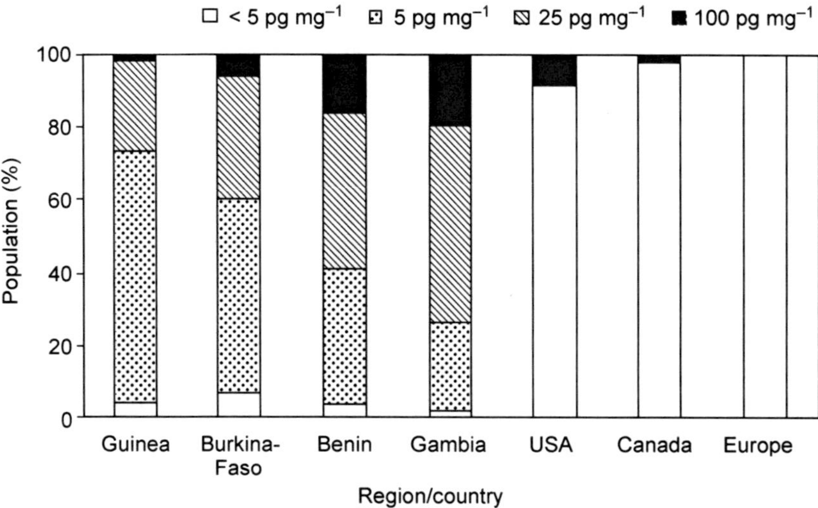

Fig. 8.2   Aflatoxin albumin adduct level (pg/mg albumin) in different countries/regions.

aflatoxin exposure. It is important to identify methods for reducing aflatoxin exposure and, therefore, liver cancer risk. In developing countries where fungal contamination of staple crops such as maize and peanuts is high and no regulation of food contamination exists, it may only be possible to modify farming or storage practices at a community level to lower aflatoxin contamination. Instead of waiting decades for the cancer endpoint, the biomarker of aflatoxin exposure can be measured before and after the intervention to determine the effectiveness of the intervention. In a post-harvest community intervention in West Africa, villagers were provided with a series of inexpensive measures to reduce post-harvest fungal contamination of ground nut crops (Turner *et al.*, 2005). These included mats to place the groundnuts on during sun drying (reducing moisture from the earth), natural fibre sacks to replace plastic sacks for storage and wooden pallets to store sacks above the ground, reducing the insect infestation that helps to spread the fungus. The aflatoxin albumin adduct was measured by ELISA in samples before and after intervention to show that there was about a 60% reduction in mean adduct level and an approximately ten-fold increase in the percentages of individuals with undetectable levels of aflatoxin adduct in the intervention villages compared to the control villages.

The other advance in understanding of aflatoxin health effects gained by the application of the aflatoxin albumin adduct biomarker was the demonstration that aflatoxin exposure in young children may contribute to impaired growth (Gong *et al.*, 2002, 2004; see Table 8.1) and impaired immune function (Jiang *et al.*, 2005; see Table 8.1). The findings of a cross-sectional study in West Africa showed that dietary aflatoxin exposure was associated with impaired child growth (Gong *et al.*, 2002). The level of serum aflatoxin albumin adduct in weaning age children in Benin and Togo was inversely correlated with height-for-age z score and weight-for-age z score. The Aflatoxin albumin adduct level was significantly higher in children with a z score of $<= -2$, a cut-off value for stunting and underweight according to WHO criteria, compared to those with normal growth. This finding was confirmed by a longitudinal study that related reduced child growth to aflatoxin levels measured over a eight-month period (Gong *et al.*, 2004).

Although considerable success has been gained by measuring aflatoxin DNA and protein adducts using immunoassay methods, method development continues in this field, with the development of isotope dilution tandem mass spectrometry (IDMS) methods for detection of both urinary aflatoxin DNA adducts and serum albumin adducts (Egner *et al.*, 2006; McCoy *et al.*, 2008). Mass spectrometry provides higher specificity than other methods because the detection focuses on the signature fragmentation pattern of the aflatoxin adduct (Egner *et al.*, 2006). With improvements in throughput for HPLC IDMS, this method provides an advantage of approximately ten-fold higher sensitivity compared to the ELISA method for quantification of adducts, following immunoaffinity or other solid phase extraction column purification.

8.3   Biomarkers of exposure for fumonisin

Fumonisins (FB), identified in 1988, are a family of mycotoxins produced primarily by the fungi *Fusarium verticillioides* and *F. proliferatum*, which contaminate food, primarily maize, world wide. Whilst a number of types of FB have been isolated, FB_1, FB_2 and FB_3 are most commonly detected in maize and FB_1 is recognised as the most toxic and possibly carcinogenic to humans. Levels of contamination of maize by FB vary from undetectable to 10 ppm in the USA to above 100 ppm in South Africa and even higher in parts of China (IARC, 2002; Shephard *et al.*, 2002; Chu and Li, 1994). First identified as being responsible for equine leukoencephalomalacia (ELEM) and porcine pulmonary oedema (PPE) (IARC, 2002), FB_1 has also been associated with neural tube defects in humans (Marasas *et al.*, 2004; Gelineau-van Waes *et al.*, 2005; Merrill *et al.*, 2004) and exposure to FB_1 has been shown to be high in certain areas with high incidence of oesophageal and liver cancer (Chu and Li, 1994; Rheeder *et al.*, 1992). FB_1 is a rat kidney carcinogen and mouse liver carcinogen (IARC, 2002; Voss *et al.*, 2002). It has been shown to have both promoting and weakly initiating properties (Gelderblom *et al.*, 1988, 2008). Although links between FB_1 intake and human cancer have been found in the above ecological studies, the lack of a reliable biomarker for FB_1 exposure has hindered attempts to show a causative association between FB_1 and human cancer (Wild and Gong, 2010).

The mechanism of FB_1 carcinogenesis is not yet clear. There is no evidence that FB_1 binds directly to DNA, although evidence from animal studies does show that indirect DNA damage may occur through increased oxidation and lipid peroxidation (Abel and Gelderblom, 1998; Domijan *et al.*, 2006, 2007; Gelderblom *et al.*, 2008).

Another possible mechanism of carcinogenesis is through inhibition of the enzyme ceramide synthase by FB_1, which results in disruption of sphingolipid metabolism resulting in changes in the sphinganine (Sa) to sphingosine (So) ratio (Voss *et al.*, 2002). This may contribute to carcinogenesis by disruption of various cell functions, including apoptosis and mitosis, that alter the balance of cell death and replication (Stockmann-Juvala and Savolainen, 2008). Because of this the Sa and So levels and the Sa:So ratio in serum and urine have received attention as surrogate biomarkers to reflect the internal mechanistic change induced by FB exposure. Whilst the Sa:So ratio has been reported for human populations from several countries this measurement is rarely sensitive enough to detect differences in exposure unless exposure was particularly high (Castegnaro *et al.*, 1998; Van der Westhuizen *et al.*,1999; Qiu and Liu, 2001; Solfrizzo *et al.*, 2004, see Table 8.2). For example, the finding that no differences in Sa:So ratio were seen between samples from people exposed to an estimated 3.8 μg kg^{-1} body weight per day and those exposed to 0.06 μg kg^{-1} bw/day (van der Westhuizen *et al.*, 1999), probably reflects the fact that the higher dose has not reached the possible threshold for effect of FB exposure on sphingolipid metabolism. In contrast, in a study of exposed people in Henan province China, in which some people were consuming up to 740 μg kg^{-1} bw/day, a difference between Sa:So ratio was observed between high and low exposure male groups, but the small magnitude of increase in the urinary Sa:So ratio before and after eating contaminated maize did not reflect the large dose increase in FB intake between subjects (Qiu and Liu, 2001).

In general, although a dose response trend was clearly demonstrated in animal studies, the Sa:So ratio is less effective in humans at natural exposure levels. Disruption of sphingolipid metabolism will only occur when a threshold for inhibition of ceramide synthesis has been reached and further work is required to determine the suitability of raised markers of disrupted sphingolipid metabolism for fumonisin exposure estimation in humans. Recently, other mechanism-based biomarkers, for example, the serum Sa and So 1-phosphates have been shown to increase in pigs and horses fed with FB (Constable *et al.*, 2005; Piva *et al.*, 2005) and it has been suggested that Sa 1-phosphate may be a more reliable biomarker than the Sa:So ratio (Riley and Voss, 2006).

As shown previously by the example of the aflatoxin albumin adduct in blood, a biomarker of exposure does not have to be involved in the mechanistic pathway which leads to disease. FB_1 has a short half-life in many species, with the majority of the compound being excreted in unmetabolised form via the faeces, and to a lesser extent (0.4–2%) the urine. Using HPLC with fluorescence detection of FB_1 following C18 cartridge extraction of FB_1 it was reported by Chelule *et al.* (2001) that 33% of faecal samples tested positive with levels up to 39 mg kg^{-1} in a population from a rural area of South Africa, compared to 6% positive with levels up to 16 mg kg^{-1} in the urban area samples. These results reflected the difference in the maize mean contamination level between rural (2.2 mg kg^{-1}) and urban (0.3 mg kg^{-1}) areas. However, although only a small proportion of FB_1 is excreted in the urine, as it is much easier to collect urine from human volunteers and then extract the FB_1 from urine, in recent work to develop a biomarker for FB_1 exposure we have focused on urinary FB_1. The detection method developed needed to be

Table 8.2 Plasma and urinary Sa:So ratio in representative human studies

Regions	N	Fumonisin mean PDI* (μg kg^{-1} bw/day)	Plasma Sa:So mean (range)	Urinary Sa:So mean (range) ±SE	References
Bomet, Kenya	29	0.06	0.28 (0.17–0.50)	0.34 (0.03–0.74)	Van der Westhuizen *et al.*,1999
KwaZulu-Natal, South Africa	27	nd**	0.43 (0.12–0.88)		
Centane, South Africa	154	3.8	0.34 (0.01–2.97)	0.41 (0.01 – 5.75)	
South Brazil	116	0.57		1.57 ± 0.49	Solfrizzo *et al.*, 2004
North Argentina	74	0.55		0.69 ± 0.12	
Italy and Central Argentina	86	0		0.36 ± 0.02	
Henan province, China (after intervention)	28	183.8		0.2 (0.04 – 0.87)	Qiu and Liu, 2001

*PDI: probable daily intake ; **nd: non-detectable.

Table 8.3   Validation of urinary FB$_1$ biomarkers of exposure

Studies	Intake category		Urinary FB$_1$ biomarker (pg/mg creatinine)* mean (range)	Reference
	Number of maize tortillas consumption			
Correlation study of	Low	2/day	35 (19–65)	Gong *et al.*, 2008
maize consumption	Medium	8/day	63 (37–108)	
and biomarker in a	High	25/day	147 (88–248)	
Mexican female				
cohort				

*Urinary FB$_1$ is adjusted for urine creatinine concentration in order to control for interindividual variation in urine concentration.

sufficiently sensitive to detect concentrations of FB$_1$ below 1 ng ml^{-1}. HPLC with tandem mass spectrometry (MS) had previously been developed to detect FB$_1$ in human hair samples in order to assess human exposure to FB (Sewram *et al.*, 2003). We therefore modified the LC–MS method to detect FB$_1$ in human urine. A deuterium labelled FB$_1$ (FBd6) serves as an internal standard to allow accurate quantification of FB$_1$. The method involves purification of the urine samples using a solid phase extraction cartridge. Extracted FB$_1$ is separated by HPLC followed by detection by electrospray ionisation tandem MS. Using this method, 90% recovery of FB$_1$ from urine with a detection limit of 20 pg ml^{-1} was achieved. When applied to human samples from Mexican women eating FB$_1$ contaminated tortillas in their diet both the frequency of positive samples and the levels of FB$_1$ were correlated with the number of tortillas consumed per day (Gong *et al.*, 2008; see Table 8.3).

Having established that the urinary FB$_1$ biomarker measured by the LC–MS detection method was sensitive enough to detect dietary levels of FB$_1$ exposure, we then applied the method to an intervention study aimed at reducing FB$_1$ contamination in a South African rural community in the former Transkei, an area of high FB$_1$ exposure that is also a high oesophageal cancer area. In the intervention, villagers were given training on how to hand sort and wash maize prior to using the maize for food preparation. FB levels from maize, maize meal and urinary FB$_1$ from morning urine were measured and the biomarker level was compared against the FB intake derived from a food frequency questionnaire and the maize FB contamination level. There was a clear correlation between the FB intake and urinary biomarker levels, with both being significantly reduced by the intervention (Van der Westhuizen *et al.*, submitted).

8.4   Biomarkers of exposure for deoxynivalenol

Deoxynivalenol (DON) is a trichothecene mycotoxin also produced by several *Fusarium* fungi that contaminate cereal crops throughout the world. It is toxic to animals and has been responsible for human toxicosis episodes in China and India (Bhat *et al.*, 1989; Luo, 1994). Long-term health effects of exposure to DON are not clear but there are indications that immune function could be affected in humans (Meky *et al.*, 2001). As with other mycotoxins, understanding the potential scope of the DON problem requires a validated human biomarker of exposure. Again, the development of a suitable human biomarker was based on studies of the metabolism and excretion of the mycotoxin in an animal model.

The availability of a monoclonal antibody immunoaffinity column (IAC) that did not cross react with related molecules was crucial for purification of DON and DON derivatives. Following analysis of DON metabolism and urinary excretion in rats, Meky *et al.* (2003) developed a biomarker method that included extraction of the DON with an anti-DON monoclonal antibody IAC. The eluate was treated with β-glucuronidase to release any DON conjugated to glucuronide, a pathway for excretion of DON metabolites. DON was then quantified by HPLC with UV detection and the presence of DON confirmed by electrospray ionisation MS. Although glucuronidase treatment of urine from DON treated rats increased the detection of DON 5–6-fold, the increase seen in enzyme-treated urine from exposed humans was lower (1.2–2.8-fold), suggesting inter species differences in metabolism. Nevertheless, the method was sensitive enough (detecting as little as 4 ng DON/per ml urine) to detect differences between DON levels in human urine from a high DON exposure region of China, Linxian County and a low exposure region, Gejiu, where DON levels were three-fold lower (Meky *et al.*, 2003).

It is important to recognise that the DON biomarker method used relies on three criteria to confirm that it is DON that is being quantified; binding to a selective monoclonal antibody on the IAC, presence of a single UV peak on HPLC that corresponds to the peak from a DON standard and structural identification of the DON by HPLC with mass spectrometry.

With the establishment of a reliable human DON biomarker of exposure and the recognition that DON contamination of cereal-based foods in the EU was frequent, further studies (see Table 8.4) have been carried out to assess human exposure to DON (Turner *et al.*, 2008a, 2008b). DON was found to be present in 296 of 300 UK subjects, with a correlation between cereal intake (from food frequency questionnaire data) and DON levels, with the possibility that some individuals could exceed the EU recommended maximum tolerable daily intake of 1000 ng DON/kg body weight (Turner *et al.*, 2008a). In an intervention study design, in which UK volunteers adopted a diet containing reduced cereal products (e.g. bread, pasta, breakfast cereal) the measurement of urinary DON levels measured by HPLC–MS after IAC purification showed that DON was present in the urine of all 25 volunteers prior to adopting the reduced cereal diet but that there was an 11-fold reduction in DON levels when cereal intake was reduced from a mean of 322 g per day to a mean of 26 g per day (Turner *et al.*, 2008b). Thus, the availability of a validated DON biomarker of exposure both revealed the presence of DON in human urine from UK subjects and showed that the DON was derived from ordinary dietary intake of cereal products.

Table 8.4   Validation of DON biomarkers of exposure

Studies	Dietary exposure category	Levels of biomarker	Reference
		Urinary DON level (µg/day)	
Correlation study: cereal consumption and biomarker in a UK population	Low (107 g/day) Medium (179 g/day) High (300 g/day)	6.6 (5.7–7.5) 9.6 (8.4–11.1) 13.2 (11.5–15.2)	Turner *et al.*, 2008 a,b
		Urinary DON level (ng mg^{-1} creatinine)*	
Intervention study: reduction of wheat intake in a UK based population	Before intervention 322 g/day After intervention 26 g/day	Before intervention 7.2 (4.9–10.5) After intervention 0.6 (0.4–0.9) 92% reduction	Turner *et al.*, 2008a,b

*Urinary DON level is adjusted for urine creatinine concentration in order to control for interindividual variation in urine concentration.

8.5   Summary

In summary, the aflatoxin biomarkers are well established and have contributed greatly to our understanding of the carcinogenesis and growth impairing effects of aflatoxin, as well as the intervention strategies for reducing aflatoxin exposure in human populations. The biomarker development for detecting exposure to fumonisin and DON, although in the early stages of application, has achieved significant success in recent years with the availability of advanced detection techniques and equipment. With recent advances in sensitivity strengthening the application of MS-based methods of detection, and particularly LC–MS, it seems likely that these methods, which provide accurate identification of the molecule being analysed and are both highly specific and sensitive enough to reflect human exposure levels, will become the most widely used methods for mycotoxin biomarker analysis in the future. These biomarker developments will provide key tools to promote research in evaluating the health risk of exposure and exploring effective intervention solutions.

8.6   References

Abel S and Gelderblom WCA (1998) 'Oxidative damage and fumonisin B$_1$ induced toxicity in primary rat hepatocytes and rat liver *in vivo*'. *Toxciology*, **131**, 121–31.
Autrup H, Essigman JM, Croy RG, Trump BF, Wogan GN and Harris CC (1979) 'Metabolism of aflatoxin-B$_1$ and identification of the major aflatoxin-B$_1$ DNA adducts formed in cultured human bronchus and colon', *Cancer Res.*, **39**, 694–68.

Autrup H, Bradley KA, Shamsuddin AKM, Wakhisi J and Wasunna A (1983) 'Detection of putative adduct with fluorescence characteristics identical to 2,3-dihydro-2-(7'-guanyl)-3-hydroxyaflatoxin B_1 in human urine collected in Murang'a district, Kenya', *Carcinogenesis*, **4**, 1193–5.

Autrup H, Seremet T, Wakhisi J and Wasunna A (1987) 'Aflatoxin exposure measured by urinary excretion of aflatoxin B_1-guanine adduct and hepatitis B virus infection in areas with different liver cancer incidence in Kenya', *Cancer Res.*, **47**, 3430–33.

Bedard LL and Massey TE (2006) 'Aflatoxin B_1-induced DNA damage and its repair'. *Cancer Letters*, **241**, 174–83.

Bhat RV, Beedu SR, Ramakrishna Y and Munshi KL (1989) 'Outbreak of trichothecenes mycotoxicosis associated with consumption of mould-damaged wheat production in Kashmir Valley, India'. *Lancet*, **1**, 35–37

Castegnaro M, Garren L, Galendo D, Gelderblom WC, Chelule P, Dutton MF and Wild CP (1998) 'Analytical method for the determination of sphinganine and sphingosine in serum as a potential biomarker for fumonisin exposure'. *J. Chromatogr. B Biomed. Sci. Appl.*, **720**, 15–24.

Chelule PK, Gqaleni N, Dutton MF and Chuturgoon AA (2001) 'Exposure of rural and urban populations in KwaZulu Natal, South Africa, to fumonisin B_1 in maize'. *Environ. Health Perspect.*, **109**, 253–6.

Chu FS and Li GY (1994) 'Simultaneous occurrence of fumonisin B_1 and other mycotoxins in moldy corn collected from the People's Republic of China in regions with high incidences of esophageal cancer'. *Appl. Environ. Microbiol.*, **60**, 847–52.

Constable PD, Riley RT, Waggoner AL, Hsiao S-H, Foreman JH, Tumbleson ME and Haschek WM (2005) *Serum Sphingosine-1-Phosphate and Sphinganine-1-phosphate are Elevated in Horses Exposed to Fumonisin B*. AOAC International Midwest Section Final Program, 63–4 AOAC International, Gaithersburg, MD (Abstract).

Croy RG, Essigman JM, Reinhold VN and Wogan GN (1978) 'Identification of the principal aflatoxin B_1-DNA adduct formed *in vivo* in rat liver'. *Proc. Natl Acad. Sci. USA*, **75**, 1745–9.

Domijan AM, Zeljezic D, Kopjar N and Peraica M (2006) 'Standard and Fpg modified comet assay in kidney cells of ochratoxin A and fumonisin B_1 treated rats'. *Toxicology*, **222**, 53–9.

Domijan AM, Zeljezic D, Milic M and Peraica M (2007) 'Fumonisin B_1: oxidative status and DNA damage in rats'. *Toxicology*, **232**, 163–9.

Egner PA, Groopman JD, Wang J-S, Kensler TW and Friesen MD (2006) 'Quantification of aflatoxinB_1-*N*7-guanine in human urine by high-performance liquid chromatography and isotope dilution tandem mass spectrometry', *Chem. Res. Toxicol.*, **19**, 1191–5.

Feskanich D, Rimm EB, Giavannucci EL, Colditz, GA, Stampfer MJ, Liton LB and Willett WC (1993) 'Reproducibility and validity of food-intake measurements from a semi quantitative food frequency questionnaire'. *J. Am. Dietetic Assoc.*, **93**, 790–6.

Gan LS, Skipper PL, Peng XC, Groopman JD, Chen JS, Wogan GN and Tannenbaum SR (1988) 'Serum albumin adducts in the molecular epidemiology of aflatoxin carcinogenesis – correlation with aflatoxin B_1 intake and urinary excretion of aflatoxin M_1'. *Carcinogenesis*, **9**, 1323–5.

Gelderblom WC, Jaskiewicz K, Marasas WF, Thiel PG, Horak RM, Vleggaar R and Kriek NP (1988) 'Fumonisins – novel mycotoxins with cancer-promoting activity produced by *Fusarium moniliforme*. *Appl. Environ. Microbiol.*, **54**, 1806–11.

Gelderblom WCA, Marasas WFO, Lebepe-Mazur S, Swanevelder S and Abel S (2008) 'Cancer initiating properties of fumonisin B_1 in a short term rat liver carcinogenesis assay'. *Toxicology*, **250**, 89–95.

Gelineau-van Waes J, Starr L, Maddox J, Aleman F, Voss KA, Wilberding J and Riley RT (2005) 'Maternal fumonisin exposure and risk for neural tube defects: Mechanisms in an *in vivo* mouse model'. *Birth Defects Res. Part A, Clinical Mole. Teratology*, **73**, 487–97.

Gong YY, Cardwell K, Hounsa A, Egal S, Turner PC, Hall AJ and Wild CP (2002) 'Dietary aflatoxin exposure and impaired growth in young children from Benin and Togo: cross sectional study, *Brit. Med. J.*, **325**, 20–1.

Gong YY, Egal S, Hounsa A, Turner PC, Hall AJ, Cardwell KF, and Wild CP (2003) 'Determinants of aflatoxin exposure in young children from Benin and Togo, West Africa: the critical role of weaning'. *Int. J. Epidemiol.*, **32**(4), 556–62.

Gong YY, Hounsa A, Egal S, Turner PC, Sutcliffe AE, Hall AJ, Cardwell K and Wild CP (2004) 'Postweaning exposure to aflatoxin results in impaired child growth: a longitudinal study in Benin, West Africa', *Environ. Health Perspect.*, **112**, 1334–8.

Gong YY, Torres-Sanchez L, Lopez-Carrillo L, Peng JH, Sutcliffe AE White KL, Humpf HU, Turner PC and Wild CP (2008) 'Association between tortilla consumption and human urinary fumonisin B_1 levels in a Mexican population'. *Cancer Epidemiol. Biomarkers Prevention*, **17**, 688–94.

Groopman JD, Croy RG and Wogan GN (1981) '*In vitro* reactions of aflatoxin B_1-adducted DNA. *Proc. Natl Acad. Sci. USA*, **78**, 5445–59.

Groopman JD, Trudel LJ, Donahue PR, Marshakrothstein A and Wogan GN (1984) 'High affinity monoclonal antibodies for aflatoxins and their application to solid phase immunoassays'. *Proc. Natl Acad. Sci. USA*, **81**, 7728–31.

Groopman JD, Donahue PR, Zhu J, Chen J and Wogan GN (1985) 'Aflatoxin metabolism in humans: detection of metabolites and nucleic acid adducts in urine by affinity chromatography', *Proc. Natl Acad. Sci. USA*, **82**, 6492–6.

Groopman JD, Cain LG and Kensler TW (1988) 'Aflatoxin exposure in human-populations – measurements and relationship to cancer', *CRC Crit. Rev. Toxicol.*, **19**, 113–45.

Groopman JD, Hall AJ, Whittle H, Hudson GJ, Wogan GN, Montesano R and Wild CP (1992a) 'Molecular dosimetry of aflatoxin-N7-guanine in human urine obtained in the Gambia, West Africa', *Cancer Epidemiol. Biomarkers Prevention*, **1**, 221–7.

Groopman JD, Zhu J, Donahue PR, Pikul A, Zhang L, Chen J and Wogan GN (1992b) 'Molecular dosimetry of urinary aflatoxin DNA adducts in people living in Guangxi autonomous region, Peoples Republic of China', *Cancer Res.*, **52**, 45–52.

Groopman JD, Wild CP, Hasler H, Chen J, Wogan GN and Kensler TW (1993) 'Molecular epidemiology of aflatoxin exposures – validation of aflatoxin-N7-guanine levels in urine as a biomarker in experimental rat models and humans', *Environ. Health Perspect.*, **99**, 107–13.

Groopman JD, Kensler KW and Wild CP (2008) 'Aflatoxin, hepatitis B virus and liver cancer: a paradigm for molecular epidemiology'. In: *Molecular Epidemiology of Chronic Diseases*, Wild CP, Vineis P and Garte S (eds), John Wiley & Sons, Chichester, 323–42.

Haugen A, Groopman JD, Hsu IC, Goodrich GR, Wogan GN and Harris CC (1981) 'Monoclonal antibody to aflatoxin B_1-modified DNA detected by enzyme immunoassay'. *Proc. Natl. Acad. Sci. USA*, **78**, 4124–4127.

International Agency for Research on Cancer (IARC) (2002) Fumonisin B_1. *Some Traditional Herbal Medicines, Some Mycotoxins, Naphthalene and Styrene*. IARC Monographs on the Evaluation of Carcinogenic Risks to Humans. Vol 82, IARC, Lyon, 301–66.

Jiang Y, Jolly PE, Ellis WO, Wang JS, Phillips TD and Williams JH (2005) 'Aflatoxin B_1-albumin adduct levels and cellular immune status in Ghanaians'. *Int. Immunol.*, **17**, 807–14.

Kohlmeier L and Bellach B (1995) Exposure assessment error and its handling in nutritional epidemiology. *Annual Review of Public Health*, **16**, 43–59.

Linsell CA and Peers FG (1977) 'Aflatoxins and liver-cell cancer'. *Trans. Royal Soc. Tropical Medicine and Hygiene*, **71**, 471–3.

Luo X (1994) 'Food poisoning caused by Fusarium toxins'. *Proceedings of the Second Asian Conference on Food Safety*. International Life Sciences Institute, Chatuchak, Thailand, 129–36.

Marasas WFO, Riley RT, Hendricks KA, Stevens VL, Sadler TW, Gelineau-van Waes J, Missmer SA, Cabrera J, Torres O, Gelderblom WCA, Allegood J, Martinez C, Maddox

J, Miller JD, Starr L, Sullards MC, Roman AV, Voss KA, Wang E and Merrill AH (2004) 'Fumonisins disrupt sphingolid metabolism, folate transport and neural tube development in embryo culture and *in vivo*: a potential risk factor for human neural tube defects among populations consuming fumonisin-contaminated maize'. *J. Nutrition*, **134**, 711–6.

McCoy LF, Scholl PF, Sutcliffe AE, Kieszak SM, Powers CD, Rogers HS, Gong YY and Groopman JD (2008) 'Human aflatoxin albumin adducts quantitatively compared by ELISA, HPLC with fluorescence detection, and HPLC with isotope dilution mass spectrometry'. *Cancer Epidemiol. Biomarkers Prevention*, **17**, 1653–7.

Meky FA, Hardie LJ, Evans SW and Wild CP (2001) Deoxynivalenol induced immunomodulation of human lymphocyte proliferation and cytokine production. *Food and Chemical Toxicology*, 39, 827–836.

Meky FA, Turner PC, Ashcroft AE, Miller JD, Qiao YL, Roth MJ and Wild CP (2003) 'Development of a urinary biomarker of human exposure to deoxynivalenol', *Food Chem.Toxicol.*, **41**, 265–73.

Piva A, Casadei G, Pagliuca G, Cabassi E, Galvano F, Solfrizzo M, Riley RT and Diaz DE (2005) 'Activated carbon does not prevent the toxicity of culture material containing fumonisin B_1 when fed to weanling piglets'. *J. Animal Sci.*, **83**, 1939–47.

Qian GS, Ross RK, Yu MC, Yuan JM, Gao YT, Henderson BE, Wogan GN and Groopman JD (1994) 'A follow-up study of urinary markers of aflatoxin exposure and liver cancer risk in Shanghai, Peoples Republic of China', *Cancer Epidemiol. Biomarkers Prevention*, **3**, 3–10 .

Qiu MF and Liu XF (2001) 'Determination of sphinganine, sphingosine and Sa/So ratio in urine of humans exposed to dietary fumonisin B_1.. *Food Addit. Contam.*, **18**, 263–9.

Randerath K, Reddy MV and Gupta RC (1981) '^{32}P-labeling test for DNA damage'. *Proc. Natl Acad. Sci. USA*, **78**, 6126–9.

Randerath K, Randerath E, Agrawal HP, Gupta RC, Schurdak ME and Reddy MV (1985) 'Postlabeling methods for carcinogen–DNA adduct analysis'. *Environ. Health Perspect.*, **62**, 57–5.

Rheeder JP, Marasas WFO, Thiel PG, Sydenham EW, Shephard GS and Van Schalkwyk DJ (1992) 'Fusarium moniliforme and fumonisins in corn in relation to human esophageal cancer in Transkei'. *Phytopathology*, **82**, 353–37.

Riley RT and Voss KA (2006) 'Differential sensitivity of rat kidney and liver to fumonisin toxicity: Organ-specific differences in toxin accumulation and sphingoid base metabolism'. *Toxicol. Sci.*, **92**, 335–45.

Routledge MN (1991) *Analysis of Carcinogen-DNA Adducts by ^{32}P-Postlabelling Assay*, PhD Thesis, University of York, UK.

Sabbioni G, Ambs S, Wogan GN and Groopman JD (1990) 'The aflatoxin-lysine adduct quantified by high-performance liquid chromatography from human serum albumin samples', *Carcinogenesis*, **11**, 2063–6

Sewram V, Mshicileli N, Shephard GS, Marasas WFO (2003) 'Fumonisin mycotoxins in human hair'. *Biomarkers*, **8**(2),110–18.

Shephard GS, Leggott NL, Stockenstrom S, Somdyala NIM and Marasas WFO (2002) 'Preparation of South African maize porridge: effect on fumonisin mycotoxin levels'. *South African J. Sci.*, **98**, 393–6.

Solfrizzo M, Chulz SN, Malmann C, Visconti A, De Girolamo A, Rojo F and Torres A (2004) 'Comparison of urinary sphingolipids in human populations with high and low maize consumption as a possible biomarker of fumonisin dietary exposure'. *Food Addit. Contam. Part A – Chem. Anal. Control Exposure and Risk Assess.*, **21**, 1090–5.

Stiborova M, Rupertova M, Hodek P, Frei E and Schmeiser HH (2004) 'Monitoring of DNA adducts in humans and ^{32}P-postlabelling methods. A review', *Collection of Czechoslovak Chem. Commun.*, **69**, 476–98.

Stockmann-Juvala H and Savolainen K (2008) 'A review of the toxic effects and mechanisms of action of fumonisin B-1'. *Human Experiment. Toxicol.*, **27**, 799–809.

Stora C (1978) 'L'aflatoxine est presente dans des cancers primitifs du foie developpés chez les habitants du Zaire', *Compte. R. Hebd. Seances Acad. Sci. Ser. D*, **286**, 917–20.

Strickland PT, Routledge MN and Dipple A (1993) 'Methodologies for measuring carcinogen adducts in humans'. *Cancer Epidemiol., Biomarkers Prevention*, **2**, 607–19.

Swenberg, JA, Fryar-Tita, E; Jeong, YC, Boysen G, Starr T, Walker VE and Albertini RJ (2008) 'Biomarkers in toxicology and risk assessment: Informing critical dose-response relationships'. *Chem. Res. Toxicol.*, **21**, 253–65.

Swenson DH, Lin JK, Miller EC and Miller JA (1977) 'Aflatoxin B_1-2,3-oxide as a probable intermediate in the covalent binding of aflatoxins B_1 and B_2 to rat liver DNA and ribosomal RNA *in vivo*', *Cancer Res.*, **37**, 172–81.

Tsuboi S, Nakagawa T, Tornita M, Seo T, Ono H, Kawamura K and Iwamura N (1984) 'Detection of aflatoxin B_1 in serum samples of male Japanese subjects by RIA and high performance liquid chromatography'. *Cancer Res.*, **44**, 1231–4.

Turner PC, Mendy M, Whittle H, Fortuin M, Hall AJ, and Wild CP (2000) 'Hepatitis B infection and aflatoxin biomarker levels in Gambian children'. *Tropical Medicine Int. Health*, **5**, 837–41.

Turner PC, Sylla A, Gong YY, Diallo MS, Sutcliffe AE, Hall AJ and Wild CP (2005) 'Reduction in exposure to carcinogenic aflatoxins by postharvest intervention measures in west africa: a community-based intervention study'. *Lancet*, **365**, 1950–6.

Turner PC, Rothwell JA, White KLM, Gong YY, Cade JE and Wild CP (2008a) 'Urinary deoxynivalenol is correlated with cereal intake in individuals from the United Kingdom', *Environ. Health Perspect.*, **116**, 21–5.

Turner PC, Burley VJ, Rothwell JA, White KLM, Cade JE and Wild CP (2008b) 'Dietary wheat reduction decreases the level of urinary deoxynivalenol in UK adults', *J. Exposure Sci. Environ. Epidemiol.*, **18**, 392–99.

Van der Westhuizen L, Brown NL, Marasas WF, Swanevelder S and Shephard GS (1999) 'Sphinganine/sphingosine ratio in plasma and urine as a possible biomarker for fumonisin exposure in humans in rural areas of Africa'. *Food Chem. Toxicol.*, **37**, 1153–8.

Van der Westhuizen L, Shephard GS, Burger HM, Rheeder JP, Gelderblom WCA, Wild CP and Gong YY (20xx) 'Fumonisin B_1 as a urinary biomarker of exposure in a subsistence maize intervention study', submitted.

Vineis P and Garte S (2008) Biomarker validation In: *Molecular Epidemiology of Chronic Diseases*, Wild CP, Vineis P and Garte S (eds), John Wiley & Sons, Chichester, 71–82.

Voss KA, Howard PC, Riley RT, Sharma RP, Bucci TJ and Lorentzen RJ (2002) 'Carcinogenicity and mechanism of action of fumonisin B_1: a mycotoxin produced by *Fusarium moniliforme (=verticillioides)*'. *Cancer Detect. Prevention*, **26**, 1–9.

Wild CP (2009) 'Environmental exposure measurement in cancer epidemiology'. *Mutagenesis*, **24**, 117–25.

Wild CP and Gong YY (2010) 'Mycotoxins and human disease: a largely ignored global health issue'. *Carcinogenesis*, **31**, 71–82.

Wild CP and Hall AJ (2000) 'Primary prevention of hepatocellular carcinoma in developing countries'. *Mutation Res.*, **462**, 381–93.

Wild CP and Turner PC (2002) 'The toxicology of aflatoxins as a basis for public health decisions'. *Mutagenesis*, **17**, 471–81.

Wild CP, Garner RC, Montesano R and Tursi F (1986) 'Aflatoxin B_1 binding to plasma albumin and liver DNA upon chronic administration to rats'. *Carcinogenesis*, **7**, 853–8.

Wild CP, Jiang YZ, Allen SJ, Jansen LAM, Hall AJ and Montesano R (1990) 'Aflatoxin albumin adducts in human sera from different regions of the world'. *Carcinogenesis*, **11**, 2271–4.

Wild CP, Hudson GJ, Sabbioni R, Chapot B, Hall AJ, Wogan GN, Whittle H, Montesano R and Groopman JD (1992) 'Dietary-intake of aflatoxins and the level of albumin bound aflatoxin in peripheral blood in the Gambia, West Africa', *Cancer Epidemiol. Biomarkers Prevention*, **1**, 229–34.

Wild CP, Law GR and Roman E (2002) 'Molecular epidemiology and cancer: promising areas for future research in the post-genomic era'. *Mutation Res.*, **499**, 3–12.
Wild CP, Vineis P and Garte S (eds) (2008) *Molecular Epidemiology of Chronic Diseases*, John Wiley & Sons, Chicester.
Yeh F-S, Yu MC, Mo C-C, Luo S, Tong M-J and Henderson BE (1989) 'Hepatitis B virus, aflatoxins and hepatocellular carcinoma in southern Guangxi, China'. *Cancer Res.*, **49**, 2506–9.

9

Developing mechanism-based and exposure biomarkers for mycotoxins in animals

R. T. Riley and K. A. Voss, United States Department of Agriculture, Agricultural Research Service, USA, R. A. Coulombe, Department of Veterinary Sciences, Utah State University, USA, J. J. Pestka, Michigan State University, USA and D. E. Williams, Oregon State University, USA

Abstract: The purpose of this review is to summarize briefly the toxicology and current state of biomarker development for commercially important mycotoxins. Combining information about known exposure, clinical indicators and biomarkers will provide a potential bioinformatic 'silver bullet' for identifying disease causation in the most economical and definitive manner. There is no single diagnostic approach that can identify/pinpoint when a disease outbreak is due to exposure to a mycotoxin or even when a mycotoxin could be a possible contributing factor to a disease outbreak of unknown etiology. One major problem is that the dose-response studies necessary to reveal the threshold for changes in mechanism-specific biochemical alterations (mechanism-based biomarkers) often have not been statistically correlated with thresholds for disease progression and exposure biomarkers (parent compound or a metabolite in tissues/fluids). The key is to better define the underlying biochemical changes and thresholds that ultimately lead to adverse effects. To accomplish this, the first site of action or more precisely, the proximate cause for all the downstream effects must be identified. The development of validated biomarkers is critical to the effort to reduce the existing uncertainty in the risk assessment of most mycotoxins and to be able to predict with some degree of certainty when a specific mycotoxin is a contributing factor in a disease outbreak.

Key words: aflatoxin, *Aspergillus*, biomarkers, deoxynivalenol, fumonisin, *Fusarium*, mycotoxins, *Penicillium*, ochratoxin, trichothecene, zearalenone.

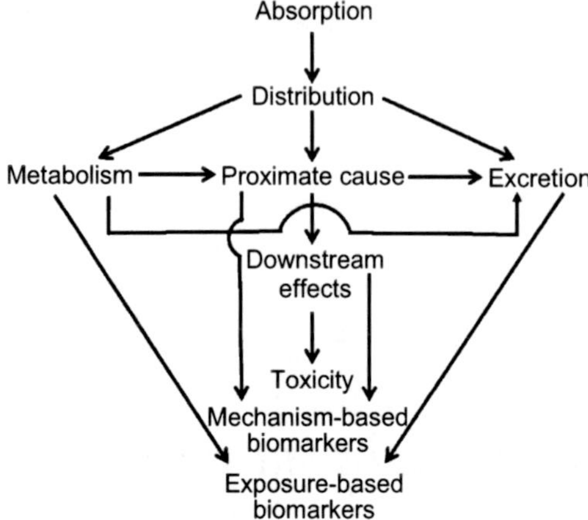

Fig. 9.1   Components of the toxicology knowledge base needed for developing biomarkers of exposure and effect in animals. A proximate cause is that event which sets everything downstream into motion. The simplest to understand mechanism of action is a classical agonist binding to an extracellular receptor. The act of binding sets in motion a series of downstream effects ending ultimately in some biological/physiological responses. The action which constitutes the proximate cause and the downstream effects which can be directly linked to the proximate cause are those biochemical or molecular events best suited for use as mechanism-based biomarkers. In order for the downstream effects to be predictive they should be specific in their linkage to the proximate cause. Exposure-based biomarkers are most often the parent compound or a metabolite of the parent compound that can be detected in easily accessed biological fluids or tissues.

9.1   Background

The purpose of this review is to summarize briefly the toxicology and current state of biomarker development of commercially important mycotoxins in animals. In addition some approaches and concepts will be presented that could prove useful in revealing the pathogenesis of disease outbreaks in which mycotoxins could be a contributing factor. Because mycotoxins are chemically diverse and exhibit a wide spectrum of biochemical mechanisms, there is no single diagnostic approach that can identify/pinpoint when a disease outbreak is due to exposure to a mycotoxin or even when a mycotoxin could be a possible contributing factor to a disease outbreak of unknown etiology. Currently, combining information about clinical indicators, suspected or confirmed exposure, and mechanism-based biomarkers is the best approach for identifying disease causation in the most economical and definitive manner. In the best of all worlds we will have specific biochemical markers that are easily quantified and definitively link adverse effects with proximate mechanisms; these include the proximate cause and down stream effects (Fig. 9.1).

Since the discovery of aflatoxins (AFBs) (Blount, 1961; Hartley *et al.*, 1963) there have been many recent excellent reviews of the toxicology and mechanisms of action of mycotoxins and unless noted, the following brief summary will draw from all of these reviews (Coulombe, 1993; Bolger *et al.*, 2001; Canady *et al.*, 2001; CAST, 2003; Cousin *et al.*, 2005; Diaz, 2005; Dietrich *et al.*, 2005; Eaton *et al.*, 2010; Fink-Gremmels and Malekinejad, 2007; Haschek *et al.*, 2002; Jouany and Diaz, 2005; O'Brien and Dietrich, 2005; Marin-Kuan *et al.*, 2008; Mally and Dekant, 2009; Pfohl-Leszkowicz and Manderville, 2007; Shephard *et al.*, 2007; Voss *et al.*, 2007; Zinedine *et al.*, 2007).

The mycotoxins that present the greatest risk to farm animals are those that occur in commodities consumed in large amounts and include aflatoxin B_1 (AFB_1), deoxynivalenol (DON), fumonisin B_1 (FB_1), ochratoxin A (OTA) and zearalenone (ZEA). There are also many other mycotoxins that are known to cause or contribute to animal disease, such as ergot alkaloids, macrocyclic trichothecenes and T-2 toxin, but these will not be covered in this review. For each of the selected mycotoxins there will be a brief summary of clinical signs and other empirical information suggesting involvement of a particular mycotoxin and also a brief description of the toxicokinetics, metabolism, proposed mechanisms and potential or existing biomarkers of exposure and effect.

9.2   Aflatoxin B_1

Aflatoxins are produced primarily by *Aspergillus flavus* and *Aspergillus parasiticus* and are often found in maize, peanuts, wheat, rice, fermented soy beans, cottonseed and some tree nuts and spices (reviewed in IARC, 2002). AFM_1, a metabolite of AFB_1, can be found in the milk of animals that eat contaminated food crops (IARC, 2002). In maize, aflatoxins often co-occur with fumonisins and when the producing fungus is *A. flavus*, the mycotoxin cyclopiazonic acid can also be a common co-contaminant (Riley and Pestka, 2005). Notably, there have been recent outbreaks of aflatoxicosis in Kenya and elsewhere in which hundreds of people have died and thousands sickened with a variety of hepatic diseases (Probst *et al.*, 2007). The outbreak in humans was significantly correlated with reported deaths of dogs and livestock (MMWR, 2004; http://www.cdc.gov/mmwr/preview/mmwrhtml/mm5334a4.htm), supporting the notion that domestic animals can serve as sentinels for mycotoxin-contaminated commodities.

9.2.1   Animal diseases

In animals, the overt symptoms of AFB poisoning are not definitive. Nonetheless, AFBs are hepatotoxic in all species. Typical signs in liver include fatty changes, hepatocyte degeneration, biliary hyperplasia, necrosis and altered function. A common clinical sign is jaundice. Grossly, the liver appears pale and swollen or fatty with variable texture. In poultry, which are among the most susceptible animals to aflatoxicosis, AFB is believed to cause 'fatty liver syndrome', although OTA can also cause fatty liver in poultry (Trenholm *et al.*, 1988). Liver damage ultimately can lead to coagulopathy as evidenced by anemia and hemorrhaging, which in poultry and pigs, contributes to internal bruising during handling. This can occur at levels as low as 150 ng g^{-1} feed in pigs (Edds, 1979). In a disease outbreak, the combination of hepatotoxicity and the presence of aflatoxigenic *Aspergilli* in the feed are supportive of AFB as a contributing factor.

9.2.2   Toxicokinetics

One of the major problems with implicating mycotoxins in a disease outbreak is that the feed is often destroyed before it can be analyzed and mycotoxins are often rapidly cleared from the body. Thus, understanding the toxicokinetics is critical for weighing the importance of biomarkers in tissues and body fluids. AFB_1 absorption, distribution and elimination is rapid. AFB_1 is well absorbed and accumulates in liver where it is extensively metabolized. It is also metabolized in other tissues but to a lesser degree. The binding of AFB_1 metabolites to proteins and nucleic acids occurs soon after absorption. The parent compound is excreted in feces and unbound water-soluble metabolites are excreted in urine and other fluids; however, some metabolites bound to nucleic acids can persist in tissues for relatively long periods of time.

The metabolism of AFB_1 has been extensively reviewed (IARC, 1993, 2002; Eaton *et al.*, 2010). In most animals, AFB_1 in liver and other tissues is metabolized by microsomal cytochromes P450 (CYPs) 3A4 and 1A2 (Guengerich, 2003; Kamdem *et al.*, 2006) to AFP_1, AFM_1, AFQ_1 and AFB_1 -8,9-epoxide (AFBO). CYPs are responsible for activation of AFB_1, AFM_1 and AFP_1, which can form nucleic acid adducts or undergo conjugation to glutathione (GSH), conversion to dihydrodiols, or binding to serum proteins or other macromolecules (Fig. 9.2). In cows, pigs and sheep, AFM_1 is the main unconjugated metabolite in urine. AFB_1 nucleic acid adducts are also found in urine of rodents and presumably farm animals, with 80% of the depurinated adducts being excreted within 48 hours of dosing. The correlation between dietary intake and adducts in urine and serum is more variable (IARC, 2002) compared to the close correlation between the levels of adducts in urine and levels in liver (IARC, 1993).

9.2.3   Mechanisms and biomarkers

The relative sensitivity of animals/individuals to the toxic effects of AFB_1 is closely linked to differences in metabolism between species/individuals (IARC, 2002). Risk factors contributing to an individual's sensitivity to liver tumors and hepatotoxicity include level of exposure to AFB, expression of AFB activation/ detoxification pathways, nutritional status and, in humans, chronic infection with hepatitis B or C virus (Henry *et al.*, 1999). In farm animals chronic microbial infection could sensitize animals to the adverse effects of AFB. AFB_1 induction of liver tumors in laboratory animals is closely correlated with mutations to specific genes that are known to control tumorigenicity (Shen and Ong, 1996) and this also is likely to be true in farm animals.

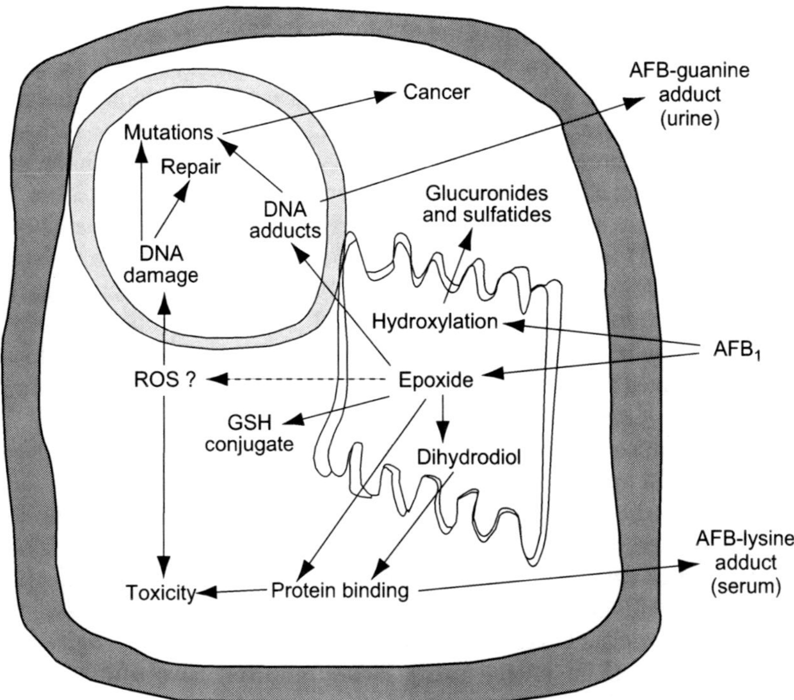

Fig. 9.2 Mechanism of action of aflatoxins (AFB) (after Eaton *et al.*, 2010) and the pathways of AFB_1 metabolism leading to cancer and toxicity. Briefly, AFB_1 enters the cell and is either metabolized via mono-oxygenases in the endoplasmic reticulum to hydroxylated metabolites which are further metabolized to glucuronide and sulfate conjugates or oxidized to the reactive epoxide (AFB_1-8,9-epoxide; the *exo* enantiomer is much more reactive with DNA than the *endo*) which undergoes spontaneous decomposition to the AFB_1-8,9-dihydrodiol (there is little evidence that epoxide hydrolase plays a significant role (Johnson *et al.*, 1997)). Both the epoxide and the dihydrodiol can bind to proteins resulting in toxicity. The epoxide can also react with DNA or be detoxified through conjugation by an inducible GST which is constitutively expressed in resistant species such as mouse. Both the DNA adduct and the protein adducts have proven useful as both exposure and mechanism-based biomarkers in humans and animals. There is also some evidence for AFB_1-mediated production of reactive oxygen species (ROS) resulting in oxidation of DNA bases such as in the formation of 8-hydroxy-2'-deoxyguanosine (Guindon *et al.*, 2007).

The metabolism to reactive intermediates and subsequent binding to nucleic acids and proteins is the proximate cause of AFB_1 toxicity and carcinogenicity. The critical downstream effects are a consequence of the subsequent disruption of transcriptional, translational processes and the binding to proteins can lead to numerous non-specific changes in pathways (metabolic pathways, signaling pathways, etc.) critical to cellular viability (Fig. 9.2). The susceptibility of an animal to AFB toxicity and carcinogenicity is to a large extent dependent on the rate of metabolism and the type of metabolites that are produced.

In most animals studied thus far, it appears that principal route of AFB_1 detoxification is by conjugation of the AFBO with GSH by the enzyme glutathione S-transferase (GST). Evidence suggests that the critical factor determining species sensitivity is the rate at which AFBO can be conjugated by glutathione (Eaton and Gallagher, 1994; Eaton *et al.*, 2010; Hayes *et al.*, 1992). As examples, both quail and rat are much more sensitive to AFB toxicity/carcinogenicity than the more resistant mouse, yet all three species exhibit high rates of epoxide formation. The relatively high rate of glutathione conjugation by a constitutive glutathione-S-transferase (GST) in the alpha class, GSTA3-3, provides mice with resistance (Hayes *et al.*, 1992). Resistance in the rat to AFB_1-induced hepatocarcinogenesis can by enhanced by induction of a similar GST in liver which is constitutively expressed at low levels (Stresser *et al.*, 1994). In non-human primates, evidence has been presented that GSTs in the mu-class, rather than alpha, are most efficient in conjugation of AFBO (Wang *et al.*, 2000).

Modern commercial turkeys are one of the animals most susceptible to aflatoxicosis, a condition associated with a high rate of hepatic AFB_1 epoxidation, mediated by CYPs 1A5 and 3A37, homologues of the human 1A2, and 3A37, respectively (Yip and Coulombe, 2006; Rawal *et al.*, 2009). In addition modern commercial turkeys are deficient in GST-mediated detoxification of AFBO (Klein *et al.*, 2000). Constitutively, turkey liver expresses high amounts of alpha-class GSTs, one of which is the homologue of the high affinity mouse GSTA3-3 (Kim *et al.*, 2010). None, however, have affinity toward AFBO.

In laboratory animal models of AFB_1-induced hepatocarcinogenesis, such as rat, rainbow trout and poultry, liver AFB_1-DNA adducts have proven to be a useful biomarker of risk (Bailey, 1994; Guarisco *et al.*, 2008). In humans urinary levels of AFB_1-N7-guanine have proven to be a useful biomarker of exposure (Egner *et al.*, 2006). Other predictive markers for AFB_1 exposure and effect that have been used in clinical trials include AFB adducts to lysine and albumin (Turner *et al.*, 2005). In AFB_1-exposed people, in Africa and China, the 'molecular fingerprint' commonly found in liver tumor tissue is the G→T transversion mutation in codon 249 in the *p53* tumor suppressor gene (Groopman and Kensler, 2005). AFB adducts in urine and blood of farm animals could be very useful as a biomarker of exposure during suspected field outbreaks; however, the current cost of the analyses would make such a practice impractical as a routine procedure. Conversely, if predictive thresholds for relating levels of biomarkers to performance were available then the analysis might be cost effective in a large animal operation. In agricultural settings, measuring AFB_1 concentrations in feed remains the most economically viable approach to diagnosing problems associated with AFB_1 diseases.

9.3   Deoxynivalenol and other trichothecenes

The trichothecenes are a group of over 200 structurally related sesquiterpenoid metabolites produced by *Fusarium, Stachybotrys* and other molds during growth in food and the environment (Grove, 1988, 1993, 2000). These low molecular weight compounds ($\approx$200–500 D) interact with the eukaryotic ribosomes and prevent polypeptide chain initiation or elongation (Ueno, 1984). All trichothecenes have in common a 9, 10-double bond and a 12, 13-epoxide group, but extensive variation exists relative to ring oxygenation patterns. There have been several recent in-depth reviews of deoxynivalenol (Canady *et al.*, 2001; Pestka *et al.*, 2004; Cousin *et al.*, 2005; Pestka and Smolinski, 2005; Pestka, 2007, 2008).

9.3.1 Animal diseases

Deoxynivalenol (DON), known colloquially as 'vomitoxin', is the most commonly detected trichothecene in cereal grains used for human and animal foods (Canady *et al.*, 2001; Pestka and Smolinski, 2005). Experimentally, low dose chronic exposure to DON causes reduced weight gain, anorexia and decreased feed conversion efficiency whereas acute high exposure causes vomiting, diarrhea and gastroenteritis. Typically, DON concentrations > 2 to 5 ppm are required for decreased feed intake and reduced weight gain and > 20 ppm for vomiting and feed refusal (Trenholm *et al.*, 1984). However, concentrations as low as 1 ppm have been associated with feed refusal in pigs. Impaired hematopoiesis is one major hallmark of trichothecene intoxication (Josse *et al.*, 2006; Parent-Massin, 2004). DON can also disrupt gut epithelial function by decreasing expression of claudins thereby increasing permeability to toxins and intestinal bacteria (Pinton *et al.*, 2009). Extremely high doses can evoke severe damage to the lymphoid tissues and epithelial cells of the gastrointestinal mucosa resulting in hemorrhage, endotoxemia and shock-like death (Ueno, 1984).

Numerous studies have determined that the immune system is exquisitely sensitive to trichothecenes and can be either suppressed or stimulated depending on dose, exposure frequency and timing. Leukocytes, most notably mononuclear phagocytes, play a central role in the acute and chronic toxicity from this trichothecene (Pestka *et al.*, 2004). High concentrations of DON and other trichothecenes promote rapid onset of leukocyte apoptosis *in vitro*. Similarly administration of trichothecenes to rodents causes apoptosis in Peyer's patches, thymus and bone marrow and resultant lymphoid tissue depletion is very likely to contribute to immunosuppression (Pestka, 2008).

9.3.2 Toxicokinetics

Deoxynivalenol absorption, distribution and elimination is rapid following oral or parenteral dosing (Amuzie *et al.*, 2008; Pestka *et al.*, 2008). There is no evidence for accumulation in tissues or transmission to eggs or milk. There is the potential for extensive metabolism in the gastrointestinal tract via de-epoxidation reactions. There is no evidence that DON is metabolized by CYPs; however, there is

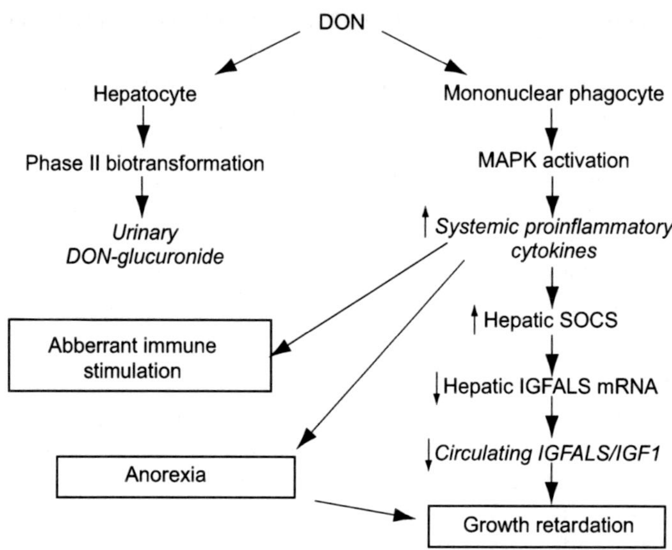

Fig. 9.3   Metabolism and aspects of the mode of action of deoxynivalenol (DON) and possibly other trichothecenes and potential biomarkers of exposure and effect. Urinary DON-glucuronide has been used as a biomarker of DON exposure. DON induces proinflammatory cytokines and suppressors of cytokine signaling (SOCS) expression in organs such as the spleen and liver. The latter can impair hepatic insulin-like growth factor acid labile subunit (IGFALS) mRNA expression and reduce circulating insulin-like growth factor 1 (IGF1). Thus plasma IGFALS and IGF1 might serve as biomarkers of effect (Amuzie and Pestka, 2010).

evidence for glucuronide conjugation in sheep and pigs (Canady *et al.*, 2001; Eriksen *et al.*, 2003). In sheep, the glucuronic acid-conjugate of DON can account for a large percentage of the total plasma DON. In pigs fed diets containing 3-acetyl deoxynivalenol, 42% and 33% of the deoxynivalenol in plasma and urine, respectively, is conjugated to the glucuronide (Eriksen *et al.*, 2003). In this same study, a significant portion of the deoxynivalenol in feces was the de-epoxide.

9.3.3   Mechanisms and biomarkers

Given the capacity of DON to evoke toxicity in animals, accurate exposure assessment is necessary to understand the potential for adverse health effects in both farm animals and humans. Based on the observation that DON is conjugated with glucuronic acid in liver (Fig. 9.3) and the resultant metabolite found in animal tissue and excreta, urinary DON-glucuronide can be used as a biomarker of exposure in human populations (Meky *et al.*, 2001, 2003; Turner *et al.*, 2008). Development of a routine method for measuring urinary DON in farm animals, similar to that used in human studies, will prove useful in suspected field disease outbreaks in farm animals and also to establish thresholds for effects when coupled with the use of mechanism-based biomarkers.

Proinflammatory gene expression
Elucidation of the subtler stimulatory effects of low DON concentrations on leukocytes has provided general insight into the molecular basis of trichothecene toxicity. A summary of the main biochemical and molecular events underlying the toxicity of DON that have shed light on possible biomarkers is illustrated in Fig. 9.3. Of critical interest, is DON's capacity to induce protein kinase-mediated proinflammatory gene expression as well as downstream alteration of endocrine function related to growth.

Low level trichothecene exposure modulates, in a hormetic-like fashion, the expression of numerous cytokines and chemokines *in vitro* and *in vivo* (Pestka, 2008). DON-mediated elevations in cytokines, chemokines and other immune-related proteins are immediately preceded by up-regulation of the respective mRNAs for these genes; both transcriptional and post-transcriptional effects underlie the increased mRNA expression (Chung *et al.*, 2003; Moon and Pestka, 2002, 2003a,b). DON exposure also activates several transcription factors including NF-κB, CREB, AP-1 and C/EBP which have binding sites in the promoters of many immune- and inflammation-related genes. Measurements of promoter reporter construct expression and transcription factor binding suggest that trichothecenes can induce gene transcription.

DON-induced mRNA stabilization is also often observed and can be explained by the presence of multiple copies of the AUUUA motifs in the 3'-untranslated region (3'-UTR) of mRNAs which target a transcript for rapid degradation (Moon *et al.*, 2003; Jia *et al.*, 2004; Chung *et al.*, 2003; Li *et al.*, 1997). Conversely, mRNA stabilization does not play a role in DON-induced IL-8 in monocytes (Gray and Pestka, 2007).

Protein kinase activation
Mitogen-activated protein kinases (MAPKs) modulate numerous physiological processes including cell growth, differentiation and apoptosis and are crucial for signal transduction in the immune response. DON and other trichothecenes activate three major MAPK families including (i) extracellular signal regulated protein kinase 1 and 2 (ERK1 and 2), (ii) p54 and p46 c-Jun N-terminal kinase 1 and 2 (JNK 1/2), and (iii) p38 *in vitro* and *in vivo* (Pestka, 2008). Accordingly, MAPKs are very likely to contribute to DON-induced immune dysregulation and other toxic manifestations.

Using chemical inhibitors of potential upstream kinases of the MAPKs, both double-stranded RNA-activated protein kinase (PKR) and hematopoietic cell kinase (Hck) have been identified as upstream transducers of MAPKs. Thus, the potential exists for PKR to mediate early events leading to immunotoxicity associated with leukocyte exposure to DON and other trichothecenes and, furthermore, to be an early step in the ribotoxic stress response. Hck belongs to the highly conserved Src-family of cytoplasmic protein tyrosine kinases and is specifically expressed in myelomonocytic cell lineages (Tsygankov and Shore, 2004). Hck transduces extracellular signals that modulate cellular processes involving proliferation, differentiation and migration (Ernst *et al.*, 2002) including up regulation of cytokines (English *et al.*, 1993).

Ribotoxic stress response

Trichothecenes and other translational inhibitors appear to modulate protein kinase activity via a process referred to as 'ribotoxic stress response' (RSR) (Iordanov *et al.*, 1997, 1998). This activity has been related to interference in the functioning of the 3'-end of the large 28S ribosomal RNA which functions in aminoacyl-tRNA binding, peptidyl transferase activity and ribosomal translocation which can lead to activation of p38, JNK or ERK depending on type of cell and toxin. Common among ribotoxic agents is their capacity to bind to or damage a specific region at the 3'-end of the 28S ribosomal RNA.

A fundamental question regarding the DON-induced RSR relates to how ribosomal damage is linked to MAPK and other signaling cascades. One possibility is that DON induces interaction of p38 and other MAPKs with the ribosome. Two *in vitro* models, U937 human monocytes and RAW 264.7 murine macrophages, were used to test this hypothesis based on their capacity to evoke rapid and robust p38 phosphorylation responses to DON (Bae and Pestka, 2008). The results showed that in mononuclear phagocytes, DON induces MAPK mobilization to the ribosome and their subsequent activation. The ribosome might thus play a central role as a scaffold in the RSR.

Recently, Bae *et al.* (2010) elucidated critical linkages that exist among p38, PKR, Hck and the ribosome in mononuclear phagocytes during DON-induced ribotoxic stress. Using PKR and Hck inhibitors it was shown that PKR and Hck appear to be critical for DON-induced ribosomal recruitment of p38, its subsequent phosphorylation and, ultimately, p38-driven proinflammatory cytokine expression.

Biomarkers of effect

Relative to biomarkers of effect, studies on DON toxicity suggest that several approaches are possible (Fig. 9.3). Proinflammatory cytokines are one obvious choice; however, there are many stressors and chemical agents which could induce these genes, making such assays non-selective relative to DON. However, proinflammatory cytokines are known to induce several suppressors of cytokine signaling (SOCS), some of which impair growth hormone (GH) signaling (Fig. 9.3). Indeed, DON induces up-regulation of mRNAs for four well-characterized SOCS including cytokine-inducible SH2 domain protein (CIS), SOCS1, SOCS2, and SOCS3 concurrently with or after cytokine up-regulation (Amuzie *et al.*, 2009). Notably, DON induces SOCS3 mRNAs in muscle, spleen and liver, with CIS1, SOCS1 and SOCS2 occurring to a lesser extent. Hepatic SOCS3 mRNA is a particularly sensitive indicator of DON exposure with SOCS3 protein being detectable in the liver well after the onset of cytokine decline (5 hours). Measurement of SOCS3 and other SOCS might therefore be indicative of DON effect.

However, there are issues of selectivity and the possible requirement for an invasive procedure such as liver biopsy.

Since SOCs can interfere with growth hormone (GH) signaling (Pass *et al.*, 2009), it is possible that GH axis impairment precedes DON-induced growth retardation in the mouse. Subchronic dietary exposure of young (4-week old) mice to DON (20 ppm) over a period of 2–8 weeks was found to: (1) impair weight gain; (2) result in a steady-state plasma DON concentration (40–60 ng ml); (3) downregulate hepatic insulin-like growth factor acid labile subunit (IGFALS) mRNA expression and (4) reduce circulating insulin-like growth factor 1 (IGF1) and IGFALS levels (Amuzie and Pestka, 2010). Acute oral exposure to DON at 0.5–12.5 mg kg^{-1} bw markedly suppressed hepatic IGFALS mRNA levels within 2 hours in a dose-dependent fashion whereas 0.1 mg kg^{-1} bw was without effect. DON-induced IGFALS mRNA up-regulation occurred both with and without exogenous GH treatment. These latter effects co-occurred with robust hepatic SOCS3 up-regulation. Thus, oral DON exposure perturbs the GH axis by suppressing two clinically relevant growth-related proteins, IGFALS and IGF1. Both have potential to serve as biomarkers of effect in populations exposed to this common foodborne mycotoxin.

9.4   Fumonisin

There are many naturally produced forms of fumonisins (FB), of which FB$_1$ is the most significant based upon its prevalence in the environment (Rheeder *et al.*, 2002) and toxicity (Howard *et al.*, 2002). One of the properties which all the FBs have in common is that their backbone is structurally similar to free sphingoid bases, most notably, 1-deoxysphinganine (Zitomer *et al.*, 2009). While there are many FBs (Bartok *et al.*, 2006), those most commonly found in maize are FB$_1$, B$_2$ and B$_3$ (Rheeder *et al.*, 2002). Structurally, the biological activity and the toxicity of the fumonisins *in vivo* and *in vitro* require the presence of the free primary amino group (Norred *et al.*, 1997, 2001) although the tricarballylic acid side chains also are important, as demonstrated by the lack of toxicity of hydrolyzed FB$_1$ (Howard *et al.*, 2002; Collins *et al.*, 2006; Voss *et al.*, 2009). The targets of FB toxicity in farm and laboratory animals are species, sex and strain specific, although, in all species FB$_1$ is hepatotoxic and in many nephrotoxic. In horses and pigs the unique clinical signs of these diseases coupled with the presence of FB or FB-producing *Fusaria* in a maize-based feed are highly suggestive of FB as a contributing factor in outbreaks of equine leukoencephalomalacia or porcine pulmonary edema (see below). There have been several comprehensive reviews of the animal diseases, toxicokinetics and mechanism of action of FB (Bolger *et al.*, 2001; WHO, 2000; Voss *et al.*, 2007; Eaton *et al.*, 2010).

9.4.1   Animal diseases

Equine leukoencephalomalacia (ELEM) syndrome is a fatal disease that apparently occurs only in horses and related species (reviewed in Voss *et al.*, 2007). The first symptoms are lethargy, head pressing and decreased feed intake, followed by convulsions and death after several days. Early clinical signs also include mild proprioceptive dysfunction (Foreman *et al.*, 2004). The characteristic finding at necropsy is one or more foci of liquifactive necrosis with hemorrhage that most often occur in the white matter.

The effects of FB in *Equidae* are not confined to the central nervous system. The liver can also be involved and, in this case, icterus and hyperbilirubinemia usually occur early. Elevated serum enzyme levels indicative of liver damage are preceded by elevations in free sphingoid bases (Wang *et al.*, 1992; Riley *et al.*, 1997) and sphingoid base 1-phosphates in serum (unpublished data). Serum enzyme levels often return to near normal concentrations but usually increase markedly immediately prior to or at the onset of behavioral changes (Wang *et al.*, 1992; Riley *et al.*, 1997). Cardiotoxicity similar to that found in swine has been induced in horses by intravenous administration of FB_1 (Smith *et al.*, 2002).

Clinical signs indicative of porcine pulmonary edema (PPE) typically occur soon (2–7 days) after pigs consume diets containing large amounts of FBs over a short period of time (reviewed in Voss *et al.*, 2007). Clinical signs usually include decreased feed consumption, dyspnea, weakness, cyanosis and death. At necropsy, the lungs of the animals exhibit varying degrees of interstitial and interlobular edema, as well as hydrothorax. The pulmonary edema can be attributed to cardiac dysfunction including decreased cardiac contractility, decreased heart rate and increased pulmonary artery pressure (Constable *et al.*, 2000). Toxic hepatosis is also induced in swine that consume large amounts of FBs and can either occur concurrently with PPE or in the absence of pulmonary effects.

Several reports have been published associating *F. verticillioides* and by implication FB contamination of feed in diseases of poultry including ducks, chickens and turkeys. The clinical features of the disease often include diarrhea, weight loss, increased liver weight and poor performance (reviewed in Voss *et al.*, 2007).

The effects of pure FBs, contaminated maize screenings or maize culture materials of *F. verticillioides* have been studied in other farm animals including catfish, trout, mink, rabbits, sheep and cattle. Like Sprague-Dawley, F344N and Wister rats, rabbits are especially sensitive to renal toxicity. In all animals where toxicity is evident it involves liver and/or kidney and where examined there is evidence of disruption of sphingolipid metabolism (reviewed in Bolger *et al.*, 2001; WHO, 2000; Voss *et al.*, 2007; Eaton *et al.*, 2010).

9.4.2   Toxicokinetics

In most animals, FB_1 absorption, distribution and elimination are rapid (reviewed in Shephard *et al.*, 2007; Voss *et al.* 2007). FBs are poorly absorbed and while there is some evidence that FBs can be metabolized by gut flora, there is currently no convincing evidence of metabolism *in vitro* or *in vivo* by animals. Absorbed FBs are clearly excreted in bile and unabsorbed FBs are eliminated primarily in feces either unchanged or with loss of one or both of the tricarballylic acid side changes, presumably due to microbial metabolism. Low levels are retained in liver and kidney as unmetabolized FB_1 (reviewed in Voss *et al.*, 2007). FB_1 persists in kidney much longer than in plasma or liver, and in male Sprague-Dawley (Riley and Voss, 2006) and Wister rats (Martinez-Larranaga *et al.*, 1999), the levels of FB_1 in kidney can be 10 times the amount in liver. FB_1 can be detected in the urine of animals exposed experimentally to dietary FB including rabbits (Orsi *et al.*, 2009), rats (Cai *et al.*, 2007), pigs (Fodor *et al.*, 2008) and vervet monkeys (Shephard *et al.*, 2007). There is evidence from feeding studies that FB_1 is preferentially accumulated compared to FB_2 and FB_3 (Riley and Voss, 2006; Fodor *et al.*, 2008); however, diets containing predominantly FB_2 from culture material can induce liver toxicity in both rats and horses (Riley *et al.*, 1997; Voss *et al.*, 1998) although pure FB_2 did not induce any toxicity in mice (Howard *et al.*, 2002). While FB_1 is rapidly eliminated in rats, mice and non-human primates, in pigs exposure to dietary FB_1 in feed required a withdrawal period as long as two weeks to be completely eliminated (Prelusky *et al.* 1996a, 1996b).

9.4.3   Mechanisms and biomarkers

There is considerable evidence that the underlying mechanism by which FBs cause toxicity to animals is disruption of sphingolipid metabolism (reviewed in WHO, 2000; Bolger *et al.*, 2001; Voss *et al.*, 2007; Eaton *et al.*, 2010). FBs are specific inhibitors of certain ceramide synthases (sphinganine and sphingosine N-acyltransferase) (Wang *et al.*, 1991), key enzymes in the pathway leading to formation of ceramide and more complex sphingolipids (Fig. 9.4). There are currently six isoforms of ceramide synthase (CerS1-6). Each has a unique fatty acyl CoA preference (Pankova-Kholmyansky and Futerman, 2006) and at least one of the isoforms may be insensitive to FB_1 when over-expressed. One of the most easily detected consequences of CerS inhibition is the accumulation of sphingoid bases and sphingoid base 1-phosphates. The elevation in serum and tissue of free sphingoid bases and sphinganine 1-phosphate are biomarkers for both exposure and toxicity of FBs and have been used in studies in horses, pigs, mink, rabbits, poultry, trout, catfish and other farm and laboratory animals (WHO, 2000; Bolger *et al.*, 2001; Piva *et al.*, 2005; unpublished data).

ELEM, PPE and the hepato- and nephrotoxicity seen in many species (including rats and mice) that result from FB exposure are closely correlated to the degree of disruption of the sphingolipid metabolism (Riley *et al.*, 1993, 2001; Wang *et al.*, 1992; Delongchamp and Young, 2001; NTP, 2002; Riley and Voss, 2006; Burns *et al.*, 2008; Voss *et al.*, 2009). For ELEM, the downstream effects associated with sphingolipid-induced alterations in cardiovascular function include the deregulation of cerebral arteries responsible for autoregulation of blood flow to the horse's brain (Haschek *et al.*, 2002). PPE is hypothesized to be a result of acute left-sided heart failure as a consequence of sphingoid base-induced inhibition of L-type calcium channels (Haschek *et al.*, 2002). In both horses and pigs FB causes large increases in sphinganine 1-phosphate in blood (Piva *et al.*, 2005; unpublished

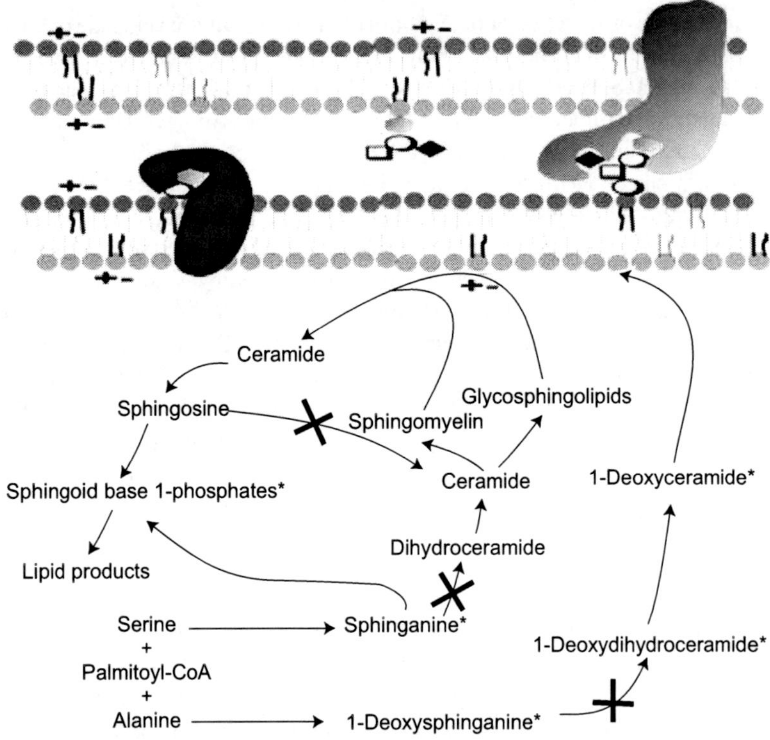

Fig. 9.4   The proximate cause/key event and mechanism of action of fumonisins (FBs). The initial site of action of FB is inhibition of ceramide synthase (sphingoid base N-acyltransferase); marked with an 'X'. At the top of the figure is the cell membrane showing the lipid bilayer (polar head groups (circles) and fatty acid 'tails') of two adjoining cells. Embedded in the membranes are globular membrane proteins. The association of two membrane proteins (large irregular shapes embedded in and spanning the membrane) with glycosphingolipids and cell–cell interactions between membrane proteins and glycosphingo-lipids which contain various carbohydrates (ovals, rectangles, diamonds, etc.) linked to ceramide (indicated by the bold hydrophobic lipid tails) and the charged phosphocholine headgroup (+/–) of sphingomyelin are also depicted. The accumulation of sphinganine (and sometimes sphingosine), sphinganine 1-phosphate and 1-deoxysphinganine in tissues (primarily liver and kidney), serum/plasma, blood spots and urine have been used as mechanism-based biomarkers in farm and laboratory animals. Depletion of more complex sphingolipids and the sphingoid base backbone composition of complex sphingolipids may also prove useful as a biomarker of effect. The major biochemical and cellular consequences subsequent to blockage of ceramide biosynthesis are inset. For additional details see Eaton *et al.* (2010). FBs are not metabolized but the parent compound in tissues, serum and urine of animals has been used as an exposure marker.

data). For both ELEM and PPE it is likely that FB exposure results in disruption of signaling pathways associated with the extracellular G-protein coupled sphingosine 1-phosphate receptors, $S1P_{1-5}$ (formerly endothelial differentiation gene (EDG) receptors). Typically, the $S1P_1$ (and possibly other S1PR) receptor functions to maintain endothelial barrier homeostasis (reviewed in Maceyka *et al.*, 2009). FB-induced deregulation of sphinganine 1-phosphate production could disrupt normal $S1P_1$ regulation of vascular permeability and other aspects of vascular physiology (reviewed in Gelineau van Waes *et al.*, 2009). The elevation of free sphingoid bases (Riley *et al.*, 1996) in serum and urine and sphingoid base 1-phosphates (Riley and Pestka, 2005) has been proposed as a functional biomarker for exposure to FBs. These biomarkers of effect work well in farm animals and in mechanism-based bioassays to assess the efficacy of alkali processing of maize (nixtamalization) using both *in vitro* and *in vivo* models (for example, Palencia *et al.*, 2003; Burns *et al.*, 2008; Voss *et al.*, 2009).

Newly discovered sphingoid bases may also prove useful as biomarkers of effect. For example, in mouse liver 1-deoxysphinganine accumulates to levels similar to sphinganine but because it is lacking a hydroxyl on carbon-1 it cannot be phosphorylated but can be acetylated to 1-deoxydihydroceramide (Zitomer *et al.*, 2009). Complex sphingolipids that contain 1-deoxysphinganine as their sphingoid base backbone have been detected in mouse liver and other tissues (Zitomer *et al.*, 2009; Voss *et al.*, 2009) including blood (unpublished data). It is likely that 1-deoxysphingoid bases will be found in many other tissues in many species.

One major problem with urinary FB as an exposure marker and urinary sphingoid bases and sphingoid base 1-phosphates as indicators of disrupted sphingolipid metabolism is that both are reversible. Nonetheless the elevation in sphingoid bases can persist in tissues (especially kidney) much longer than FB_1 (Enongene *et al.*, 2002; Garren *et al.*, 2001; Wang *et al.*, 1999; Cai *et al.*, 2007). In urine from rats fed FB_1, > 95% of the free sphinganine was recovered in dead cells (Riley *et al.*, 1994). FB_1-induced elevation of free sphingoid bases and toxicity are both reversible, although the elimination of free sphinganine from the liver is more rapid than from the kidney (Enongene *et al.*, 2002; Garren *et al.*, 2001). However, the study in mice (Enongene *et al.*, 2002) was done before the discovery of 1-deoxysphinganine which can accumulate to high levels in mouse liver (Zitomer *et al.*, 2009; Voss *et al.*, 2009).

While the inhibition of ceramide synthase is the proximate cause for the disruption of sphingolipid and glycerophospholipid metabolism and is closely correlated with the extent and severity of the animal disease, the linkages between various aspects of disrupted sphingolipid metabolism and the onset of disease is not well understood. One reason for the confusion is that there are several hundred distinct sphingolipids and each is likely to have specific functions (Hirabayashi *et al.*, 2006). To appreciate the complexity of sphingolipid metabolism it is worthwhile visiting the sphingolipid maps website (http://www.sphingomap.org).

The specific biochemical aspects of FB-disruption of sphingolipid metabolism that are most likely to alter cell regulation leading to increased cells death and altered cell proliferation are (1) increased free sphingoid bases and their 1-

phosphates, (2) alterations in complex sphingolipids and (3) decreased ceramide and ceramide 1-phosphate biosynthesis. Free sphingoid bases and ceramide can induce cell death; thus, FB inhibition of ceramide synthase can inhibit cell death induced by ceramide-mediated processes but can promote free sphingoid base-induced cell death. In addition, FB-induced elevation in sphingoid base 1-phosphates in tissues and blood can disrupt signaling associated with extracellular receptors (S1P receptors) that regulate vascular physiology and other physiological processes (Chalfant and Spiegel, 2005). Decrease in pools of more complex sphingolipids will alter vitamin uptake via glycosylphosphatidylinosital (GPI)- anchored proteins associated with lipid rafts (Stevens and Tang, 1997; Gelineau van Waes *et al.*, 2005; Nour-Abedel *et al.*, 2007). The kinetics of the biosynthesis and degradation of the various bioactive and structurally critical sphingolipid pools in tissues will be important factors in the observed downstream effects and ultimately the toxicity and carcinogenicity of fumonisins. Recent studies have shown that mice deficient in CerS2 spontaneously develop liver tumors (Imgrund *et al.*, 2009) providing additional indirect evidence for the key role that disrupted sphingolipid metabolism plays in FB tumorigenicity.

9.5   Ochratoxin A

Ochratoxins (OT) are a family of mycotoxins (OTA, OTB, OTC) that contain a phenylalanine and dihydroisocumarin moiety joined by a peptide bond. OTA is the most prevalent and toxic. OTA is produced by both *Aspergillus* and *Penicillium* species and is found on numerous small grains and maize used in animal feeds. There have been several comprehensive reviews of OTA-induced animal diseases, toxicokinetics and mechanisms of action (Benford *et al.*, 2001; Dietrich *et al.*, 2005; O'Brien and Dietrich, 2005; Pfohl-Leszkowicz and Manderville, 2007; Marin-Kuan *et al.*, 2008; Mally and Dekant, 2009).

9.5.1   Animal diseases

The primary effect of OTA in all farm animals is nephrotoxicity. In pigs and poultry the proximal tubules are mainly affected and the kidney is pale and grossly enlarged. Fatty liver can occur in poultry. The most sensitive indicator of acute ochratoxicosis in chickens is a reduction in total serum proteins and albumin while in pigs a decrease in phosphoenolpyruvate carboxykinase in kidney is a sensitive and specific indicator (Marquardt and Frohlich, 1992; Krogh, 1992). In pigs, large increases in proteins excreted in urine are indicative of glomerular proteinuria and were correlated with histological observations of renal damage.

Exposure to lower levels of OTA in poultry and pigs can result in altered performance including reduced feed consumption and weight gain and at higher levels delayed response to immunization and increased susceptibility to infection (Stoev *et al.*, 2000a,b). Other effects in poultry include decreased egg production,

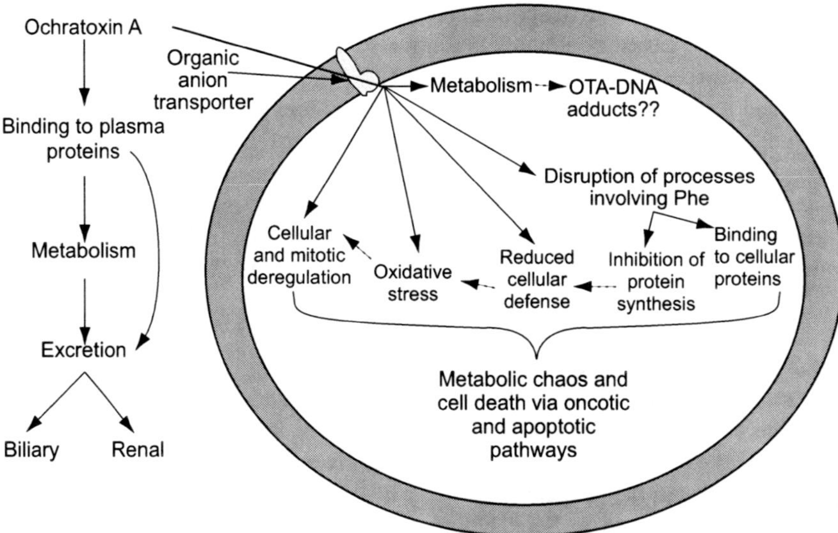

Fig. 9.5 Metabolism and hypothesized mechanisms involved in ochratoxin A (OTA) toxicity. OTA can be transported into cells via a multispecific organic anion transporter (reviewed in O'Brien and Dietrich, 2005). Studies have shown that OTA can alter processes that require phenylalanine and some of the biological effects of OTA can be partially prevented by supplementation with phenylalanine or phenylalanine analogs (for example Zanic-Grubisic *et al.*, 2000; Schwerdt *et al.*, 1999; Baudrimont *et al.*, 2001). OTA in serum/ plasma is potentially a useful marker for OTA exposure (reviewed in Scott, 2005; Dietrich *et al.*, 2005; O'Brien and Dietrich, 2005). Potential OTA specific indicators of effect in target tissues include the development of unique gene expression profiles specific to OTA-induced alterations of genes involved in cellular defense, cell proliferation (reviewed in Marin-Kuan *et al.*, 2008; Mally and Dekant, 2009, Adler *et al.*, 2009) and oxidative stress (Cavin *et al.*, 2009; Arbillaga *et al.*, 2008). Expression profiling could also help to identify specific proteins that could be used as biomarkers of effect. Dotted lines indicate a high degree of uncertainty with regard either to the relationship between the mechanisms or to the contribution of the mechanism to the toxicity of OTA in animals.

coagulopathy (increased susceptibility to bruising during processing), decreased force to break bones, decreased tensile strength of the large intestines, under-pigmentation and glycogen accumulation in liver.

9.5.2   Toxicokinetics

Comprehensive reviews of the toxicokinetics of OTA are available (Marquardt and Frohlich, 1992; Pfohl-Leszkowicz *et al.*, 1999; Benford *et al.*, 2001; Dietrich *et al.*, 2005). OTA is rapidly and well absorbed and the kinetics is best described by a three- or four-compartment model (Dietrich *et al.*, 2005). The half-life in plasma is dependent on the extent of binding to plasma proteins. The protein-binding capacity correlates well with the biological half life. OTA is widely distributed and is accumulated in kidney and other tissues in pigs. The concentration of OTA specific transporters in tissues has been proposed (Fig. 9.5) as an explanation for the relative species and sex sensitivities (reviewed in Dietrich *et al.*, 2005). The extent of albumin binding also markedly decreases the uptake of OTA by transporters (Bow *et al.*, 2005). OTA can occur in edible tissues of pig. OTA and its metabolites are reabsorbed by the kidney and excreted in urine and also undergo enterohepatic circulation and excretion in feces. There is the potential for extensive metabolism to less toxic metabolites in the gastrointestinal tract by microbes. In addition to gastrointestinal degradation by microbes, OTA can be oxidized by CYPs to their less toxic hydroxyochratoxin A metabolites using microsomal preparations from rabbits and pigs (reviewed in Benford *et al.*, 2001).

9.5.3   Mechanisms and biomarkers

Because OTA binds tightly to albumin and serum/plasma proteins, OTA in blood or urine could be a useful biomarker for exposure in pigs and probably other animals. However, in humans the urinary levels are poorly correlated with the plasma levels and the presence of urinary metabolites of OTA could complicate quantitation (reviewed in Scott, 2005). Nonetheless, the mechanism of action (proximate cause) of OTA is unclear (Fig. 9.5). The structural similarity to phenylalanine and the fact that it inhibits many enzymes and processes that are dependent on phenylalanine makes it likely that at least some of the effects of OTA are due to the disruption of phenylalanine metabolism (CAST, 2003) including inhibition of protein synthesis. Several studies have shown that supplementation of feed with L-phenylalanine or proteins protects against the toxic effects (including mortality) of OTA (reviewed in Benford *et al.*, 2001; Marquardt and Frohlich, 1992). The evidence for OTA DNA adducts (reviewed in Pfohl-Leszkowicz and Manderville, 2007) is controversial (reviewed in Mally and Dekant, 2009; Mantle *et al.*, 2010). Potential biomarkers of effect in target tissues include the development of unique gene expression profiles specific to OTA-induced alterations of genes involved in cellular defense, cell proliferation (reviewed in Marin-Kuan *et al.*, 2008; Mally and Dekant 2009; Adler *et al.*, 2009) and oxidative stress (Cavin *et al.*, 2009; Arbillaga *et al.*, 2008). A metabonomic approach using gas chromatography–mass spectrometry (GC–MS), liquid chromatography–mass spectrometry (LC–MS) and 1H nuclear magnetic resonance (NMR) assessed the changes in urinary metabolite profiles as a tool for developing a predictive model for OTA toxicity (Sieber *et al.*, 2009). The results, while not specific for OTA, were indicative of kidney damage and general toxicity. Nonetheless, this approach could prove useful for elucidating more specific mechanisms that are unique to OTA.

9.6   Zearalenone

Zearalenone (ZEA) is an estrogenic mycotoxin that often co-occurs with DON on scabby wheat and on maize. It is produced by many of the same *Fusarium* species that produce trichothecenes, most notably *F. culmorum and F. graminearum.*

Several recent reviews are available describing the occurrence and toxicity of ZEA (CAST 2003; Fink-Gremmels and Malekinejad, 2007; Tiemann and Dänicke, 2007; Minervini and Dell'Aquila, 2008; Zinedine *et al.*, 2007).

9.6.1   Animal diseases

There is no doubt that ZEA is a cause of farm animal performance problems (reviewed in Fink-Gremmels and Malekinejad, 2007; Zinedine *et al.*, 2007). In animals the primary effect of ZEA is estrogenic and prepubertal female pigs are clearly the most affected farm animals. Clinical signs of estrus can be induced in ovariectomized sows with doses as low as 1–5 ppm inducing vulvovaginitis, tenesmus, vaginal and rectal prolapse (Osweiler, 1986). Reduced libido, plasma testosterone and other effects have been reported in prepubertal boars (Osweiler, 1986). Dietary levels of 3–10 ppm ZEA can induce anestrus in sows, reduced litter size, fetal resorption and implantation failure. While ruminants are more resistant to the estrogenic effects, reduced conception rates have been reported in cattle. Other effects in ruminants include enlarged mammary glands, sterility and infertility (reviewed in Fink-Gremmels and Malekinejad, 2007).

9.6.2   Toxicokinetics

ZEA is well absorbed and in pigs dosed via either the intravenous or oral route the calculated plasma half life was 87 hours. ZEA is metabolized in liver and excreted in urine and feces as the parent compound, or its metabolites (zearalenol (ZOL) and/or zearalanol (ZAL)) or as their respective glucuronide conjugates after considerable enterohepatic recirculation (Fig. 9.6) (reviewed in Fink-Gremmels and Malekinejad, 2007; Zinedine *et al.*, 2007). There is some evidence that the intestinal mucosa is active in reducing ZEA to alpha-ZOL and conjugating with glucuronic acid (Biehl *et al.*, 1993). Accumulation in tissues is minimal. ZEA is also metabolized by rumen microbes. The rapid hepatic conversion of ZEA to the more easily excreted alpha and beta ZOL derivatives in cattle, along with microbial metabolism in the rumen, could explain the resistance of cattle to the reproductive effects of ZEA relative to pigs (Raisbeck *et al.*, 1991). The alpha ZOL metabolite is more frequently detected than the beta ZOL metabolite in liver and trace amounts of ZEA and its metabolites can be detected in muscle tissue of pigs fed ZEA-contaminated oats (Zollner *et al.*, 2002). In pig serum and urine, the alpha-ZOL glucuronide appears to be the major metabolite detected (Dänicke *et al.*, 2005) after prolonged exposure.

The enzymes responsible for the conversion of ZEA to ZOL are 3-alpha and 3-beta hydroxysteroid dehydrogenase (3 alpha- and 3 beta-HSD) which are important enzymes in steroid metabolism. The fact that ZEA is a substrate for these enzymes gives it the potential to disrupt steroid metabolism since the substrates for these enzymes include natural steroid hormones (reviewed in Fink-Gremmels and Malekinejad, 2007).

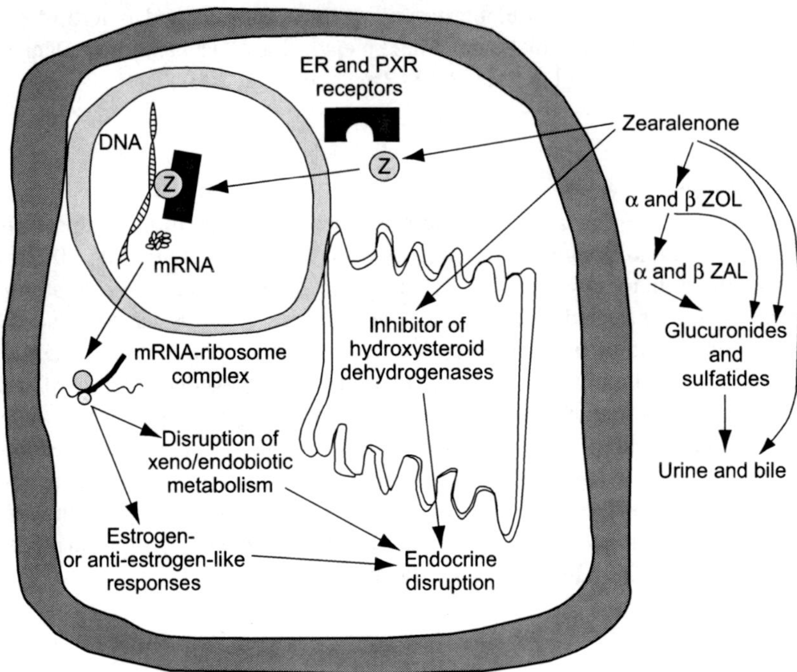

Fig. 9.6   Metabolism and mechanism of action of zearalenone (ZEA). ZEA (Z) passively crosses the cell membrane and binds to the cytosolic estrogen receptor (ER) and more recently the pregnane X receptor (PXR) which regulates the expression of genes involved in metabolism of xenobiotics and endobiotics (reviewed in Fink-Gremmels and Malekinejad, 2007). Briefly, the receptor–Z complex is transferred into the nucleus where it binds to specific nuclear receptors and generates responses via gene activation or suppression resulting in the up-regulation or down regulation of mRNAs that code for proteins that are normally expressed by ZEA-ER- or ZEA-PXR-complex binding (for a more complete description of ER and PXR activation see Parkinson and Ogilvie, 2008). Studies of the estrogen mimicry have shown distinct dose-dependent expression patterns indicative of both estrogen-like and anti-estrogen-like responses (Boehme *et al.*, 2009; Parveen *et al.*, 2009) which could serve as the basis for developing specific biomarkers of effect in animals. The presence of the parent compound (ZEA) or its metabolites (ZOL, ZAL) or conjugates in urine and feces has been used as an exposure marker in animal studies (reviewed in Zinedine *et al.*, 2007).

9.6.3   Mechanisms and biomarkers

The presence of the parent compound (ZEA) or its metabolites (ZOL, ZAL) or their conjugates in urine and feces have been used as exposure markers in animal studies (reviewed in Zinedine *et al.*, 2007). The glucuronic acid conjugates have potential as a marker of exposure in suspected field outbreaks. The probable co-occurrence of DON during an outbreak would necessitate the DON conjugate also being measured and there is some *in vitro* data suggesting that co-exposure could exacerbate reproductive performance in pigs including steroidogenesis (Ranzenigo *et al.*, 2008).

The basis for the estrogenic effect is well established and is due to a close structural similarity between ZEA (and many of its metabolites) and estradiol (Osweiler, 2000). Alpha ZOL is three times more potent in its estrogenic activity compared to ZEA and the relative binding affinity for estrogen receptors (ER) was greater in pig than in other species, which may explain the interspecies differences in sensitivity to the estrogenic effects (Fitzpatrick *et al.*, 1989). In addition, in pigs, the most sensitive species, the predominant form of hydroxysteroid dehydrogenase is the one that yields the alpha isomer of ZOL. With regards to the estrogen mimicry, *in vitro* studies have shown distinct dose-dependent expression patterns indicative of both estrogen-like (high dose) and anti-estrogen-like (low dose) responses (Fig. 9.6) (Boehme *et al.*, 2009; Parveen *et al.*, 2009), a finding that is consistent with the fact that ZEA can interfere with the metabolism of steroid hormones.

ZEA also has been shown to activate the human xenobiotic receptor pregnane X (PXR) which is known to regulate the biotransformation of endobiotics and xenobiotics (Fig. 9.6) (Ding *et al.*, 2006). The differential expression of genes in the pathways controlled by these two receptors (ER and PXR) could be fruitful in identifying expression profiles or in the development of proteomic approaches that could be used as specific indicators/biomarkers for ZEA-induced effects in animals.

9.7   Future trends

In the field (farm animals) one must consider all of the forensic information that is available in order to come up with a best guess about causation of disease. Even in laboratory animal studies where exposure is carefully controlled, there are a diversity of factors that can modify the extent and severity of disease expression including genetic, nutritional, age, sex and other environmental factors and pre-existing conditions that can modify disease expression in ways that are not easily identified and sometimes overlooked. For example, in horses consuming maize-based diets, leucoencephalomalacia is pathognomonic for what we now know to be a disease caused by fumonisin. Nonetheless, just observing behavioral changes (head pressing, circling, tongue lolling) does not prove that FB exposure is the cause. If the feed is available and it is found to be heavily contaminated with *F. verticillioides* or other FB producing *Fusaria* and if FB is detected in the feed at greater than 10 ppm then it can be stated that the likely cause of the horse's condition is/was consumption of FB (Ross *et al.*, 1991). Additional presumptive evidence that the animals were exposed to FBs would be the demonstration of elevated levels of free sphingoid bases or the sphingoid base metabolite sphinganine 1-phosphate in liver, kidney or serum samples and at concentrations that are likely to correlate experimentally with the onset of ELEM. The findings are described as presumptive (likely) because the dose–response relationships in the tissues and in feeding studies have not been done so it is impossible to state with absolute certainty that the cause of ELEM in the horse is FB; as described in this example, however, the weight of the evidence makes a compelling case for FB as the cause of the horse's brain disease (and probable liver disease).

This example is provided to underscore the fact that even with a pathognomonic disease correlating exposure, to biochemical changes that are specific to the accepted mechanism of action does not provide a definitive diagnosis. The reason for this is that the dose–response studies necessary to reveal the threshold for mechanism-specific biochemical and molecular changes (mechanism-based biomarkers) often have not been statistically correlated with the thresholds for disease progression and the levels of exposure biomarkers (parent compound or a metabolite in tissues/fluids). The concept of dose–response is critical to developing diagnostic mechanism-based biomarkers that can predict increased risk of disease. For example, AFB_1-DNA adduct levels in liver of some animal models have proven to be useful biomarkers for subsequent risk of hepatocarcinogenesis. However, one should not assume that such a relationship will hold for all hepatocarcinogens. In a recent ED_{001} study, almost 40 000 trout were administered dibenzo(*def,p*)chrysene (DBC) (formally called dibenzo[a,l]pyrene) in the diet to determine the dose that statistically would give one additional cancer in 1 000 animals (ED_{001}) (Bailey *et al.*, 2009). Although DBC–DNA adducts in liver were linear at the lowest dose administered, hepatocarcinogenesis statistically became sub-linear at low dose. Therefore, the use of this biomarker would have significantly overestimated the cancer risk at low (environmentally relevant) doses.

The key to developing better predictive tools for diagnosing mycotoxin-induced diseases in farm animals is to define the underlying biochemical changes better and the thresholds that ultimately lead to undesired consequences (adverse effects). But to do this we must first identify the initial site of action or more precisely, the proximate cause/key event for the downstream effects (Fig. 9.1). Proximate cause is a legal term that equates to 'mechanism of action' in toxicology. Proximate cause is that key event which sets everything downstream into motion. Effects are everything that happens after the proximate cause and can be linked back to it. For example, for FB the proximate cause is ceramide synthase inhibition (Wang *et al.*, 1991). However, the proximate cause at high doses may be quite different than the proximate cause at low doses. One reason for this is that at high doses the effects are often acute effects such as oxidative stress and oncotic cell death, whereas at low doses the effects are often more subtle and slower to develop such as cancer, diabetes, and so on. Thus there is a need for research to validate both disease-specific mechanism-based and exposure biomarkers for several of the most important mycotoxins including FB, OTA, DON and other trichothecenes and ZEA. The one possible exception is AFB where dose–response data is available that equates exposure to both long-term and acute effects and levels of serum and urinary biomarkers. Nonetheless, our understanding of the mechanistic basis for chronic effects of AFB in humans is probably much better than our understanding of the mechanistic basis for the acute effects, effects that are equally relevant to both humans and farm animals. The development of validated biomarkers is critical to the effort to reduce the existing uncertainty in the risk assessment of most mycotoxins in humans but also to develop a better understanding of the impact of mycotoxin exposure on farm animal productivity.

9.8   Acknowledgements

This work was supported in part by the USDA-ARS (RTR, KAV), USDA, under a cooperative project with US Wheat and Barley Scab Initiative (JJP) and in part by Public Health Service Grant ES 3358 (JJP) from the National Institute for Environmental Health Sciences and NRI competitive grant 2007-35205-17880 from the USDA Cooperative State Research, Education, and Extension Service (CSREES) Animal Genome Program (RAC). Support was also provided by the USDA-CSREES Project W-2122: Beneficial and Adverse Effects of Natural, Bioactive Dietary Chemicals on Human Health and Food Safety. Any findings, opinions, conclusions or recommendations expressed in this publication are those of the authors and do not necessarily reflect the views of USDA.

9.9   References

Adler M, Müller K, Rached E, Dekant W and Mally A (2009), 'Modulation of key regulators of mitosis linked to chromosomal instability is an early event in ochratoxin A carcinogenicity', *Carcinogenesis*, **30**, 711–19.

Amuzie C J and Pestka J J (2010), 'Suppression of insulin-like growth factor acid-labile subunit (IGFALS) expression – a novel mechanism for deoxynivalenol-induced growth retardation', *Toxicol Sci*, **113**, 412–21.

Amuzie C J, Harkema, J R and Pestka J J (2008), 'Tissue distribution and proinflammatory cytokine induction by the trichothecene deoxynivalenol in the mouse: comparison of nasal vs. oral exposure', *Toxicology*, **248**, 39–44.

Amuzie C J, Shinozuka J and Pestka J J (2009), 'Induction of suppressors of cytokine signaling by the trichothecene deoxynivalenol in the mouse', *Toxicol Sci*, **111**, 277–87.

Arbillaga L, Vettorazzi A, Gil A G, van Delft J H, García-Jalón J A and López de Cerain A (2008), 'Gene expression changes induced by ochratoxin A in renal and hepatic tissues of male F344 rat after oral repeated administration', *Toxicol Appl Pharmacol*, **230**, 197–207.

Bae H K and Pestka J J (2008), 'Deoxynivalenol induces p38 interaction with the ribosome in monocytes and macrophages', *Toxicol Sci*, **105**, 59–66.

Bae H K, Gray JS, Li M, Vines L, Kim J and Pestka JJ. (2010), 'Double-stranded RNA-activated protein kinase and hematopoetic cell kinase associate with the 40S ribosomal subunit and mediate the ribotoxic stress response to deoxynivalenol in mononuclear phagocytes.' *Toxicol Sci*, **115**, 444–52

Bailey G S (1994), 'Role of aflatoxin-DNA adducts in the cancer process', *The Toxicology of Aflatoxins: Human Health, Veterinary, and Agricultural Significance*, Academic Press, New York, 137–48.

Bailey G S, Reddy A P, Pereira C B, Harttig U, Baird W M, Spitsbergen J M, Hendricks J D, Orner G A, Williams D E, and Swenberg J S (2009), 'Non-linear cancer response at ultra-low dose: a 40,800-animal ED_{001} tumor and biomarker study, *Chem Res Toxicol*, **22**, 1264–76.

Bartok T, Szecsi A, Szekeres A, Mesterhazy A and Bartok M (2006), 'Detection of new fumonisin mycotoxins and fumonisin-like compounds by reversed-phase high-performance liquid chromatography/electrospray ionization ion trap mass spectrometry', *Rapid Comm Mass Spectrom*, **20**, 2447–62.

Baudrimont I, Sostaric B, Yenot C, Betbeder A M, Dano-Djedje S, Sanni A, Steyn P S and Creppy E E (2001), 'Aspartame prevents the karyomegaly induced by ochratoxin A in rat kidney', *Arch Toxicol*, **75**, 176–83.

Benford D, Boyle C, Dekant W, Fuchs R, Gaylor D, Hard G, McGregor D, Pitt J, Plestina

R, Shepard G, Solfrizzo M, Verger P and Walker R (2001), 'Ochratoxin A', in Food and Agriculture Organization of the United Nations, *World Health Organization Food Additives*, Paper 74, Series 47, 281–387.

Biehl M L, Prelusky D B, Koritz G D, Hartin K E, Buck, W B and Trenholm H L (1993), 'Biliary excretion and enterohepatic cycling of zearalenone in immature pigs', *Toxicol Appl Pharmacol*, **121**, 152–9.

Blount W P (1961), 'Turkey "X" disease', *Turkeys*, **9**, 52–77.

Boehme K, Simon S and Mueller S O (2009), 'Gene expression profiling in Ishikawa cells: a fingerprint for estrogen active compounds', *Toxicol Appl Pharmacol*, **236**, 85–96.

Bolger M, Coker R D, Dinovi M, Gaylor D, Gelderblom M O, Paster N, Riley R T, Shephard G and Speijers J A (2001), 'Fumonisins', in *Safety Evaluation of Certain Mycotoxins in Food*. Food and Agriculture Organization of the United Nations,*World Health Organization Food Additives*, Paper 74, Series 47, 103–279.

Bow D A, Perry J L, Simon J D and Pritchard J B (2005), 'The impact of plasma protein binding on the renal transport of organic anions', *J Pharmacol Exp Ther*, **316**, 349–55.

Burns T D, Snook M E, Riley R T and Voss K A (2008), 'Fumonisin concentrations and *in vivo* toxicity of nixtamalized *Fusarium verticillioides* culture material: Evidence for fumonisin-matrix interactions', *Food Chem Toxicol*, **46**, 2841–8.

Cai Q, Tang L and Wang J S (2007), 'Validation of fumonisin biomarkers in F344 rats', *Toxicol Appl Pharmacol*, **225**, 28–39.

Canady R A, Coker R D, Rgan S K, Krska R, Kuiper-Goodman T, Olsen M, Pestka J J, Resnik S and Schlatter J (2001), 'Deoxynivalenol. Safety evaluation of certain mycotoxins in food', *Fifty-sixth report of the Joint FAO/WHO Expert Committee on Food Additives*, *International Programme on Chemical Safety*, World Health Organization, Geneva.

CAST (2003), *Mycotoxins: Risks in Plant and Animal Systems*. Report no. 139 of the Council for Agricultural Science and Technology, Ames, Iowa.

Cavin C, Delatour T, Marin-Kuan M, Fenaille F, Holzhäuser D, Guignard G, Bezençon C, Piguet D, Parisod V, Richoz-Payot J, and Schilter B (2009), 'Ochratoxin A-mediated DNA and protein damage: roles of nitrosative and oxidative stresses', *Toxicol Sci*, **110**, 84–94.

Chalfant C E and Spiegel S (2005), 'Sphingosine 1-phosphate and ceramide 1-phosphate: expanding roles in cell signaling', *J Cell Sci*, **118**, 4605–12.

Chung Y J, Zhou H R and Pestka J J (2003), 'Transcriptional and posttranscriptional roles for p38 mitogen-activated protein kinase in upregulation of TNF-alpha expression by deoxynivalenol (vomitoxin)', *Toxicol Appl Pharmacol*, **193**, 188–201.

Collins T F, Sprando R L, Black T N, Olejnik N, Eppley R M, Shackelford M E, Howard P C, Rorie J I, Bryant M and Ruggles D I (2006), 'Effects of aminopentol on *in utero* development in rats', *Food Chem Toxicol*, **44**, 161–9.

Constable P D, Smith G W, Rottinghaus G E, and Haschek W M (2000), 'Ingestion of fumonisin B_1-containing culture material decreases cardiac contractility and mechanical efficiency in swine', *Toxicol Appl Pharmacol*, **162**, 151–60.

Coulombe R A, Jr (1993), 'Biological action of mycotoxins', *J Dairy Sci*, **76**, 880–91.

Cousin M A, Riley R T and Pestka J J (2005), 'Foodborne mycotoxins: chemistry, biology, ecology, and toxicology', in *Foodborne Pathogens: Microbiology and Molecular Biology*, Fratamico P M and Bhunia A K (eds), Horizon Scientific Press, Norfolk, UK, 164–226.

Dänicke, S, Swiech E, Buraczewska L and Überschär K H (2005), 'Kinetics and metabolism of zearalenone in young female pigs', *J Anim Physiol Anim Nutr Berlin*, **89**, 268–76.

Delongchamp R R and Young J F (2001), 'Tissue sphinganine as a biomarker of fumonisin-induced apoptosis', *Food Addit Contam*, **18**, 255–61.

Diaz D (2005), *The Mycotoxin Blue Book*, Nottingham University Press, Nottingham,UK.

Dietrich D R, Heussner A H, and O'Brien E (2005), 'Ochratoxin A: comparative pharmacokinetics and toxicological implications (experimental and domestic animals and humans)', *Food Addit Contam*, **22**, 45–52.

Ding X, Lichti K and Staudinger J L (2006) 'The mycoestrogen zearalenone induces CYP3A through activation of the pregnane X receptor', *Toxicol Sci*, **91**, 448–55.

Eaton D L and Gallagher E P (1994), 'Mechanisms of aflatoxin carcinogenesis', *Ann Rev Pharmacol Toxicol*, **34**, 135–72.

Eaton D L, Beima K M, Bammler T K, Riley R T and Voss K A (2010), 'Hepatotoxic mycotoxins', in *Comprehensive Toxicology*, Roth R A and Ganey P E (eds), Elsevier, Amsterdam, in press.

Edds G T (1979), 'Biological effects of aflatoxins in swine', in *Interactions of Mycotoxins in Animal Production*, National Academy of Sciences, Washington, 67–76.

Egner P A, Groopman J D, Wang J-S, Kensler T W and Friesen M D (2006), 'Quantification of aflatoxin-B_1-N^7-guanine in human urine by high-performance liquid chromatography and isotope dilution tandem mass spectrometry, *Chem Res Toxicol*, **19**, 1191–5.

English B K, Ihle J N, Myracle A and Yi T (1993), 'Hck tyrosine kinase activity modulates tumor necrosis factor production by murine macrophages', *J Exp Med*, **178**, 1017–22.

Enongene E N, Sharma R P, Bhandari N, Meredith F I, Voss K A and Riley R T (2002), 'Persistence and reversibility of the elevation in free sphingoid bases induced by fumonisin inhibition of ceramide synthase', *Toxicol Sci*, **67**, 173–81.

Eriksen G S, Pettersson H and Lindberg J E (2003), 'Absorption, metabolism and excretion of 3-acetyl DON in pigs', *Arch Animal Nutrition-Archiv Fur Tierernahrung*, **57**, 335–45.

Ernst M, Inglese M, Scholz G M, Harder K W, Clay F J, Bozinovski S, Waring P, Darwiche R, Kay T, Sly P, Collins R, Turner D, Hibbs M L, Anderson G P and Dunn A R (2002), 'Constitutive activation of the SRC family kinase Hck results in spontaneous pulmonary inflammation and an enhanced innate immune response', *J Exp Med*, **196**, 589–604.

Fink-Gremmels J and Malekinejad H (2007), 'Clinical effects and biochemical mechanisms associated with exposure to the mycoestrogen zearalenone', *Anim Feed Sci Technol*, **137**, 326–41.

Fitzpatrick D W, Picken C A, Murphy L C, and Buhr M M (1989), 'Measurement of the relative binding affinity of zearalenone, alpha-zearalenol and beta-zearalenol for uterine and oviduct estrogen receptors in swine, rats and chickens: an indicator of estrogenic potencies', *Comp Biochem Physiol*, **C94**, 691–4.

Fodor J, Balogh K, Weber M, Miklós M, Kametler L, Pósa R, Mamet R, Bauer J, Horn P, Kovács F and Kovács M (2008), 'Absorption, distribution and elimination of fumonisin B(1) metabolites in weaned piglets', *Food Addit Contam Part A, Chem Anal Control Expo Risk Assess*, **25**, 88–96.

Foreman J H, Constable P D, Waggoner A L, Levy M, Eppley R M, Smith G W, Tumbleson M E and Haschek W M (2004), 'Neurological abnormalities and cerebrospinal fluid changes in horses administered fumonisin B_1 intravenously', *J Vet Intern Med*, **18**, 223–30.

Garren L, Galendo D, Wild C P and Castegnaro M (2001), 'The induction and persistence of altered sphingolipid biosynthesis in rats treated with fumonisin B_1', *Food Addit Contam*, **18**, 850–6.

Gelineau-van Waes J, Starr L, Maddox J R, Aleman F, Voss K A, Wilberding J and Riley R T (2005), 'Maternal fumonisin exposure and risk for neural tube defects: Disruption of sphingolipid metabolism and folate transport in an *in vivo* mouse model', *Birth Defects Res*, **73**, 487–97.

Gelineau-van Waes J, Voss K A, Stevens V L, Speer M C and Riley R T (2009), 'Maternal fumonisin exposure as a risk factor for neural tube defects' *Adv Food Nutr Res*, **56**, 145–81.

Gray J S and Pestka J J (2007), 'Transcriptional regulation of deoxynivalenol-induced IL-8 expression in human monocytes', *Toxicol Sci*, **99**, 502–11.

Groopman J D and Kensler T W (2005), 'Role of metabolism and viruses in aflatoxin-induced liver cancer', *Toxicol Appl Pharmacol*, **206**, 131–7.

Grove J F (1988), 'Non-macrocyclic trichothecenes', *Nat Prod Rep*, **5**, 187–209.

Grove J F (1993), 'Macrocyclic trichothecenes', *Nat Prod Rep*, **10**, 429–48.

Grove J F (2000), 'Non-macrocyclic trichothecenes. Part 2', *Prog Chem Org Nat Prod*, **69**, 1–70.

Guarisco J A, Hall J O, and Coulombe R A Jr (2008), 'Butylated hydroxytoluene chemoprevention of aflatoxicosis – effects on aflatoxin B(1) bioavailability, hepatic DNA adduct formation, and biliary excretion', *Food Chem Toxicol*, **46**, 3727–31.

Guengerich F P (2003), 'Cytochrome P450 oxidations in the generation of reactive electrophiles: epoxidation and related reactions', *Arch Biochem Biophys*, **409**, 59–71.

Guindon K A, Bedard L L and Massey T E (2007), 'Elevation of 8-hydroxydeoxyguanosine in DNA from isolated mouse lung cells following *in vivo* treatment with aflatoxin B(1)', *Toxicol Sci*, **98**, 57–62.

Hartley R D, Nesbitt B and O'Kelley J (1963), 'Toxic metabolites of *Aspergillus flavus*', *Nature*, **198**, 1056.

Haschek W M, Voss K A and Beasley V R (2002), 'Selected mycotoxins affecting animal and human health', in *Handbook of Toxicologic Pathology*, Academic Press, San Diego 645–99.

Hayes J D, Judah D J, Neal G E, and Nguyen T (1992), 'Molecular cloning and heterologous expression of a cDNA encoding a mouse glutathione S-transferase Yc subunit possessing high catalytic activity for aflatoxin B_1-8,9-epoxide', *Biochem J*, **285**, 173–80.

Henry S, Bosch F X, Troxell T C and Bolger P M (1999), 'Reducing liver cancer – global control of aflatoxin', *Science*, **286,** 2453–4.

Hirabayashi Y, Igarashi Y and Merrill A H Jr (2006) 'Sphingolipid synthesis, transport and cellular signaling', in *Sphingolipid Biology*, Toyko, Springer Verlag, 3–22.

Howard P C, Couch L H, Patton R E, Eppley R M, Doerge D M, Churchwell M I, Marques M M and Okerberg C V (2002), 'Comparison of the toxicity of several fumonisin derivatives in a 28-day feeding study with female $B6C3F_1$ mice', *Toxicol Appl Pharmacol*, **185**, 153–65.

IARC (1993), IARC *Monographs on the Evaluation of Carcinogenic Risks of Chemicals to Humans, Vol. 56, Toxins derived from* Fusarium moniliforme: *Fumonisins B_1 and B_2 and fusarin C.* International Agency for Research on Cancer, Lyon, France.

IARC (2002), *IARC Monographs on the Evaluation of Carcinogenic Risks of Chemicals to Humans, Vol. 82, Some Traditional Herbal Medicines, Some Mycotoxins, Naphthalene and Styrene*, International Agency for Research on Cancer, Lyon, France.

Imgrund S, Hartmann D, Farwanah H, Eckhardt M, Sandhoff R, Degen J, Gieselmann V, Sandhoff K and Willecke K (2009), 'Adult ceramide synthase 2 (CERS2)-deficient mice exhibit myelin sheath defects, cerebellar degeneration, and hepatocarcinomas', *J Biol Chem*, **284**, 33549–60.

Iordanov M S, Pribnow D, Magun J L, Dinh T H, Pearson J A, Chen S L and Magun B E (1997), 'Ribotoxic stress response: activation of the stress-activated protein kinase JNK1 by inhibitors of the peptidyl transferase reaction and by sequence-specific RNA damage to the alpha-sarcin/ricin loop in the 28S rRNA', *Mol Cell Biol*, **17**, 3373–81.

Iordanov M S, Pribnow D, Magun J L, Dinh T H, Pearson J A and Magun B E (1998), 'Ultraviolet radiation triggers the ribotoxic stress response in mammalian cells', *J Biol Chem*, **273**, 15794–803.

Jia Q, Zhou H R, Bennink M and Pestka J J (2004), 'Docosahexaenoic acid attenuates mycotoxin-induced immunoglobulin a A nephropathy, interleukin-6 transcription, and mitogen-activated protein kinase phosphorylation in mice', *J Nutr*, **134**, 3343–49.

Johnson W W, Yamazaki H, Shimade T, Ueng Y-F, and Guengerich F P (1997), 'Aflatoxin B_1 8,9-epoxide hydrolysis in the presence of rat and human epoxide hydrolase', *Chem Res Toxicol*, **10**, 672–6.

Josse R, Hymery N, Sibiril Y, Batina P and Parent-Massin D (2006), 'In vitro effects of trichothecene metabolites on human hematopoietic progenitors', *Toxicol Lett*, **164**, S221–2.

Jouany J P and Diaz D E (2005), 'Effects of mycotoxins in ruminants', *The Mycotoxin Blue Book*, in Diaz D (ed), Nottingham University Press, Nottingham,UK, 295–321.

Kamdem L K, Meineke I., Godtel-Armbrust U, Brockmoller J, and Wojnowski L (2006), 'Dominant contribution of P450 3A4 to the hepatic carcinogenic activation of aflatoxin B_1', *Chem Res Toxicol*, **19**, 577–86.

Kim J E, Bauer M M, Mendoza K M, Reed K M and Coulombe R A Jr (2010), 'Comparative genomics identifies new alpha class genes within the avian glutathione S-transferase gene cluster', *Gene*, **452**, 45–53.

Klein P J, Buckner R, Kelly J and Coulombe R A Jr (2000), 'Biochemical basis for the extreme sensitivity of turkeys to aflatoxin B(1)', *Toxicol Appl Pharmacol*, **165**, 45–52.

Krogh P (1992), 'Role of ochratoxin A in disease causation', *Food Chem Toxicol*, **30**, 213–24.

Li S, Ouyang Y L, Dong W and Pestka J J (1997), 'Superinduction of IL-2 gene expression by vomitoxin (deoxynivalenol) involves increased mRNA stability', *Toxicol Appl Pharmacol*, **147**, 331–42.

Maceyka M, Milstien S and Spiegel S (2009), 'Sphingosine-1-phosphate: the Swiss army knife of sphingolipid signaling', *J Lipid Res*, **50**, S272–6.

Mally A and Dekant W (2009), 'Mycotoxins and the kidney: modes of action for renal tumor formation by ochratoxin A in rodents', *Mol Nutr Food Res*, **53**, 467–78.

Mantle P G, Faucet-Marquis V, Manderville R A, Squillaci B and Pfohl-Leszkowicz A (2010), 'Structures of covalent adducts between DNA and ochratoxin A:A new factor in debate about genotoxicity and human risk assessment', *Chem Res Toxicol*, **23**, 89–98.

Marin-Kuan M, Cavin C, Delatour T and Schilter B (2008), 'Ochratoxin A carcinogenicity involves a complex network of epigenetic mechanisms', *Toxicon*, **52**, 195–202.

Marquardt R R and Frohlich A A (1992), 'A review of recent advances in understanding ochratoxicosis', *J Anim Sci*, **70**, 3968–88.

Martinez-Larranaga M R, Anadon A, Diaz M J, Fernandez-Cruz M L, Martinez M A, Frejo M T, Martinez M, Fernandez R, Anton R M, Morales M E and Tafur M (1999), 'Toxicokinetic and oral bioavailability of fumonisin B_1', *Vet Human Toxicol*, **41**, 357–62.

Meky F A, Hardie L J, Evans S W and Wild C P (2001), 'Deoxynivalenol-induced immunomodulation of human lymphocyte proliferation and cytokine production', *Food Chem Toxicol*, **39**, 827–36.

Meky F A, Turner P C, Ashcroft A E, Miller J D, Qiao Y L, Roth M J and Wild C P (2003), 'Development of a urinary biomarker of human exposure to deoxynivalenol', *Food Chem Toxicol*, **41**, 265–73.

Minervini F and Dell'Aquila M E (2008), 'Zearalenone and Reproductive Function in Farm Animals', *Int J Mol Sci*, **9**, 2570–84.

MMWR (Mortality and Morbibity Weekly Report) (2004), 'Outbreak of aflatoxin poisoning – Eastern and Central Provinces, Kenya, January–July 2004', *MMWR*, **53**(34), 790–3.

Moon Y and Pestka J J (2002), 'Vomitoxin-induced cyclooxygenase-2 gene expression in macrophages mediated by activation of ERK and p38 but not JNK mitogen-activated protein kinases', *Toxicol Sci*, **69**, 373–82.

Moon Y and Pestka J J (2003a), 'Cyclooxygenase-2 mediates interleukin-6 upregulation by vomitoxin (deoxynivalenol) *in vitro* and *in vivo*', *Toxicol Appl Pharmacol*, **187**, 80–8.

Moon Y and Pestka J J (2003b), 'Deoxynivalenol-induced mitogen-activated protein kinase phosphorylation and IL-6 expression in mice suppressed by fish oil', *J Nutr Biochem*, **14**, 717–26.

Moon Y, Uzarski R and Pestka J J (2003), 'Relationship of trichothecene structure to COX-2 induction in the macrophage: selective action of type B (8-keto) trichothecenes', *J Toxicol Environ Health A*, **66**, 1967–83.

Norred W P, Plattner R D, Meredith F I and Riley R T (1997), 'Mycotoxin-induced elevation of free sphingoid bases in precision-cut rat liver slices: Specificity of the response and structure-activity relationships', *Tox Appl Pharmacol*, **147**, 63–70.

Norred W P, Riley R T, Meredith F I, Poling S M and Plattner R D (2001), 'Instability of N-acetylated fumonisin B_1 (FA1) and the impact on inhibition of ceramide synthase in rat liver slices', *Food Chem Toxicol*, **39**, 1071–8.

Nour-Abdel M A, Ringot D, Gueant J L and Chango A (2007), 'Folate receptor and human reduced folate carrier expression in HepG2 cell line exposed to fumonisin B_1 and folate deficiency', *Carcinogenesis*, **28**, 2291–7

NTP Technical Report (2002), *Toxicology and Carcinogenesis Studies of Fumonisin B_1 in F344/N Rats and B6C3F1 Mice*, National Institution Health, Publication no. 99–3955.

O'Brien E and Dietrich D R (2005), 'Ochratoxin A: the continuing enigma', *Crit Rev Toxicol*, **35**, 33–60.

Orsi RB, Dilkin P, Xavier JG, Aquino S, Rocha LO and Corrêa B (2009), 'Acute toxicity of a single gavage dose of fumonisin B_1 in rabbits', *Chem Chem-Biol Interact*, **179**, 351–5.

Osweiler G D (1986), 'Occurrence and clinical manifestations of trichothecene toxicoses and zearalenone toxicoses', in *Diagnosis of Mycotoxicoses*, Richard J L and Thurston J R(eds), Martinus Nijhoff, Dordrecht, 31–42.

Osweiler, G D (2000), 'Mycotoxins – contemporary issues of food animal health and productivity', *Vet Clin North Am Food Anim Pract*, **16**, 511–30.

Palencia E, Torres O, Hagler W, Meredith F I, Williams L D and Riley R T (2003), 'Total fumonisins are reduced in tortillas using the traditional nixtamalization method of Mayan communities', *J Nutr*, **133**, 3200–3.

Pankova-Kholmyansky I and Futerman A H (2006), 'Ceramide synthase', *Sphingolipid Biology*, in Hirabayashi Y, Igarashi Y and Merrill A H Jr (eds), Springer Verlag, Toyko, 49–56.

Parent-Massin D (2004), 'Haematotoxicity of trichothecenes', *Toxicol Lett*, 153, 75–81.

Parkinson A and Ogilvie B W (2008), 'Biotransformation of Xenobiotics', in *Casarett and Doull's Toxicology: the Basic Science of Poisons*, Klaassen C D (ed.), McGraw Hill, New York, 161–304.

Parveen M, Zhu Y and Kiyama R (2009), 'Expression profiling of the genes responding to zearalenone and its analogues using estrogen-responsive genes', *FEBS Lett*, **583**, 2377–84.

Pass C, Macrae V E, Ahmed S F and Farquharson C (2009), 'Inflammatory cytokines and the GH/IGF-I axis: novel actions on bone growth', *Cell Biochem Funct*, **27**, 119–27.

Pestka J J (2007), 'Deoxynivalenol: toxicity, mechanisms and animal health risks', *Anim Feed Sci Technol*, **137**, 283–98.

Pestka J J (2008), 'Mechanisms of deoxynivalenol-induced gene expression and apoptosis', *Food Addit Contam Part A Chem Anal Control Expo Risk Assess,* **25**, 1128–40.

Pestka J J and Smolinski A T (2005), 'Deoxynivalenol: toxicology and potential effects on humans', *J Toxicol Environ Health B Crit Rev*, **8**, 39–69.

Pestka J J, Zhou H R, Moon Y and Chung Y J (2004), 'Cellular and molecular mechanisms for immune modulation by deoxynivalenol and other trichothecenes: unraveling a paradox', *Toxicol Lett*, **153**, 61–73.

Pestka J J, Islam Z and Amuzie C J (2008), 'Immunochemical assessment of deoxynivalenol tissue distribution following oral exposure in the mouse', *Toxicol Lett*, **178**, 83–7.

Pfohl-Leszkowicz A and Manderville R A (2007), 'Ochratoxin A: An overview on toxicity and carcinogenicity in animals and humans', *Mol Nutr Food Res*, **51**, 61–99.

Pfohl-Leszkowicz A, Guerre P and Galtier P (1999), 'Metabolisation des mycotoxins', in *Les Mycotoxines dans l'alimentation: Evaluation et gestion di risqué*, Technique & Documentation, Paris, 37–68.

Pinton P, Nougayrede J P, Del Rio J C, Moreno C, Marin D E, Ferrier L, Bracarense A P, Kolf-Clauw M and Oswald I P (2009), 'The food contaminant deoxynivalenol, decreases intestinal barrier permeability and reduces claudin expression', *Toxicol Appl Pharmacol*, **237**, 41–8.

Piva A, Casadei G, Pagliuca G, Cabassi E, Galvano F, Solfrizzo M, Riley R T and Diaz D E (2005), 'Inability of activated carbon to prevent the toxicity of culture material containing fumonisin B_1 when fed to weaned piglets', *J Anim Sci*, **83**, 1939–47.

Prelusky D B, Trenholm H L, Rotter B A, Miller J D, Savard M E, Yeung J M and Scott P M (1996a), 'Biological fate of fumonisin B_1 in food-producing animals', *Adv Exp Med Biol*, **392**, 265–78.

Prelusky D B, Miller J D, Trenholm H L (1996b), 'Disposition of ^{14}C-derived residues in tissues of pigs fed radiolabelled fumonisin B$_1$' *Food Addit Contam*, **13**, 155–62.

Probst C, Njapau H, and Cotty P J (2007), 'Outbreak of an acute aflatoxicosis in Kenya in 2004: identification of the causal agent', *Appl Environ Microbiol*, **73**, 2762–4.

Raisbeck M F, Rottinghaus G E and Kendall J D (1991), 'Effects of naturally occurring mycotoxins on ruminants', in *Mycotoxins and Animal Foods*, Smith J E and Henderson R S (eds), CRC Press, Boca Raton, 647–77.

Ranzenigo G, Caloni F, Cremonesi F, Aad P Y and Spicer L J (2008), 'Effects of Fusarium mycotoxins on steroid production by porcine granulosa cells', *Anim Reprod Sci*, **107**, 115–30.

Rawal S, Mendoza K M, Reed K M and Coulombe R A Jr (2009), 'Structure, genetic mapping, and function of the cytochrome P450 3A37 gene in the turkey (Meleagris gallopavo)', *Cytogenet Genome Res*, **125**, 67–73.

Rheeder J P, Marasas W F O and Vismer H F (2002), 'Production of fumonisin analogs by *Fusarium* species', *Appl Environ Microbiol*, **68**, 2101–05.

Riley R T and Pestka J (2005), 'Mycotoxins: metabolism, mechanisms and biochemical markers', in *The Mycotoxin Blue Book*, Diaz D (ed.), Nottingham University Press, UK, 295–321.

Riley R T and Voss K A (2006), 'Differential sensitivity of rat kidney and liver to fumonisin toxicity: organ-specific differences in toxin accumulation and sphingoid base metabolism', *Toxicol Sci*, **92**, 335–45

Riley R T, An N -H, Showker J L, Yoo H S, Norred W P, Chamberlain W J, Wang E, Merrill, A H, Motelin G, Beasely V R and Haschek W M (1993), 'Alteration of tissue and serum sphinganine to sphingosine ratio: An early biomarker in pigs of exposure to fumonisin containing feeds', *Toxicol App Pharmacol*, **118**, 105–12.

Riley R T, Hinton D M, Chamberlain W J, Bacon C W, Wang E, Merrill A H Jr and Voss K A (1994), 'Dietary fumonisin B$_1$ induces disruption of sphingolipid metabolism in Sprague Dawley rats: A new mechanism of nephrotoxicity', *J Nutr*, **124**, 594–603.

Riley R T, Wang E, Schroeder, J J, Smith, E R, Plattner, R D, Abbas, H, Yoo, H S and Merrill, A H Jr (1996), 'Evidence for disruption of sphingolipid metabolism as a contributing factor in the toxicity and carcinogenicity of fumonisins', *Natural Toxins*, **4**, 3–15.

Riley R T, Showker J L, Owens, D L and Ross P F (1997), 'Disruption of sphingolipid metabolism and induction of equine leukoencephalomalacia by *Fusarium proliferatum* culture material containing fumonisin B$_2$ or B$_3$', *Environ Toxicol Pharmacol*, **3**, 221–8.

Riley R T, Enongene E, Voss K A, Norred W P, Meredith F I, Sharma R P, Williams L D E, Carlson D B, Spitsbergen J and Merrill A H Jr (2001), 'Sphingolipid perturbations as mechanisms for fumonisin carcinogenesis', *Environ Health Perspect*, **109**, 301–8.

Ross P F, Rice L G, Plattner R D, Osweiler G D, Wilson T M, Owens D L, Nelson H A and Richard J L (1991), 'Concentrations of fumonisin B$_1$ in feeds associated with animal health problems', *Mycopathologia*, **114**, 129–135.

Schwerdt G, Freudinger R, Silbernagl S and Gekle M (1999), 'Ochratoxin A – binding proteins in rat organs and plasma and in different cell lines of the kidney', *Toxicology*, **135**, 1–10.

Scott P M (2005), 'Biomarkers of human exposure to ochratoxin A', *Food Addit Contam*, Suppl **1**, 99–107.

Shen H-M and Ong C-N (1996), 'Mutations of the *p53* tumor suppressor gene and *ras* oncogenes in aflatoxin hepatocarcinogenesis', *Mut Res*, **366**, 23–44.

Shephard G S, Van Der Westhuizen L and Sewram V (2007), 'Biomarkers of exposure to fumonisin mycotoxins: a review', *Food Addit Contam*, **24**, 1196–201.

Sieber M, Wagner S, Rached E, Amberg A, Mally A and Dekant W (2009), 'Metabonomic study of ochratoxin a toxicity in rats after repeated administration: phenotypic anchoring enhances the ability for biomarker discovery', *Chem Res Toxicol*, **22**, 1221–31.

Smith G W, Constable P D, Foreman J H, Eppley R M, Waggoner A L, Tumbleson M E and Haschek W M (2002), 'Cardiovascular changes associated with intravenous administration of fumonisin B_1 in horses', *Am J Vet Res*, **63**, 538–45.

Stevens V L and Tang J (1997), 'Fumonisin B_1-induced sphingolipid depletion inhibits vitamin uptake via the glycosylphosphatidylinositol-anchored folate receptor', *J Biol Chem*, **272**, 18020–25.

Stoev S D, Anguelov G, Ivanov I and Pavlov D (2000b), 'Influence of ochratoxin A and an extract of artichoke on the vaccinal immunity and health in broiler chicks', *Exp Toxicologic Pathol*, **52**, 43–55.

Stoev S D, Goundasheva D, Mirtcheva T and Mantle P G (2000a), 'Susceptibility to secondary bacterial infections in growing pigs as an early response in ochratoxicosis', *Exp Toxicologic Patholy*, **52**, 287–96.

Stresser D M, Williams D E, McLellan L I, Harris T M and Bailey G S (1994), 'Indole-3-carbinol induces a rat liver glutathione transferase subunit (Yc2) with high activity toward aflatoxin B_1 *exo*-epoxide. Association with reduced levels of hepatic aflatoxin-DNA adduct *in vivo*', *Drug Metabol Dispos*, **22**, 392–9.

Tiemann U and Dänicke S (2007), '*In vivo* and *in vitro* effects of the mycotoxins zearalenone and deoxynivalenol on different non-reproductive and reproductive organs in female pigs: A review' *Food Addit Contam, Part A*, **24**, 306–14.

Trenholm H L, Hamilton R M, Friend D W, Thompson B K and Hartin K E (1984), 'Feeding trials with vomitoxin (deoxynivalenol)-contaminated wheat: effects on swine, poultry, and dairy cattle', *J Am Vet Med Assoc*, **185**, 527–31.

Trenholm H L, Prelusky D B, Young J C and Miller J D (1988), 'Reducing mycotoxins in animal feeds', in *Agriculture Canada Publication 1827E*, Ottawa, Agriculture Canada, Communications Branch.

Tsygankov A Y and Shore S K (2004), 'SRC: Regulation, role in human carcinogenesis and pharmacological inhibitors', *Curr Pharmaceut Design*, **10**, 1745–56.

Turner P C, Sylla A, Gong Y Y, Diallo M S, Sutcliffe A E, Hall A J and Wild C P (2005), 'Reduction in exposure to carcinogenic aflatoxins by postharvest intervention measures in west Africa: a community-based intervention study', *Lancet*, **365**, 1950–6.

Turner P C, Rothwell J A, White K L, Gong Y, Cade J E and Wild C P (2008), 'Urinary deoxynivalenol is correlated with cereal intake in individuals from the United Kingdom', *Environ Health Perspect*, **116**, 21–5.

Ueno Y (1984), 'Toxicological features of T-2 toxin and related trichothecenes', *Fundam Appl Toxicol*, **4**, S124–S132.

Voss K A, Riley R T, Bacon C W, Meredith F I and Norred W P (1998), 'Toxicity and sphinganine levels are correlated in rats fed fumonisin B_1 or hydrolyzed FB$_1$', *Environ Toxicol Pharmacol*, **5**, 101–04.

Voss K A, Smith G W and Haschek W M (2007), 'Fumonisins: toxicokinetics, mechanism of action and toxicity', *Anim Feed Sci Technol*, **137**, 299–325.

Voss K A, Riley R T, Snook M E and Gelineau-van Waes J (2009), 'Comparing the reproductive and sphingolipid metabolic effects of fumonisin B_1 and its alkaline hydrolysis product in LM/Bc mice: hydrolyzed fumonisin B_1 did not cause neural tube defects', *Toxicol Sci*, **112**, 459–67.

Wang C, Bammler T K, Guo Y, Kelly E J and Eaton D L (2000) '*Mu*-Class GSTs are responsible for aflatoxin B_1-8,9-epoxide-conjugating activity in the nonhuman primate *Macaca fascicularis* liver', *Toxicol Sci*, **56**, 26–36.

Wang, E, Norred, W P, Bacon, C W, Riley, R T and Merrill, A H Jr. (1991), 'Inhibition of sphingolipid biosynthesis by fumonisins: implications for diseases associated with Fusarium moniliforme', *J Biol Chem*, **266**, 14486–90.

Wang E, Ross P F, Wilson T M, Riley R T and Merrill A H Jr. (1992), 'Alteration of serum sphingolipids upon exposure of ponies to feed containing fumonisins, mycotoxins produced by *Fusarium moniliforme*', *J Nutr*, **122**, 706–16.

Wang E, Riley R T, Meredith F I and Merrill A H Jr. (1999), 'Fumonisin B_1 consumption by rats causes reversible dose-dependent, increases in urinary sphinganine and sphingo-sine', *J Nutr*, **129**, 214–20.

WHO (2000), 'Environmental Health Criteria 219:Fumonisin B$_1$', in *International Programme on Chemical Safety,* Marasas W H O, Miller J D, Riley R T and Visconti A (eds), United Nations Environmental Programme, The International Labour Organization, World Health Organization, Geneva, Switzerland.

Williams B R (2001), 'Signal integration via PKR', *Sci. STKE* 2001, (89)re2. http://stke.sciencemag.org/cgi/content/full/OC_sigtrans;2001/89/re2

Yip S S, and Coulombe R A Jr (2006), 'Molecular cloning and expression of a novel cytochrome p450 from turkey liver with aflatoxin B$_1$ oxidizing activity', *Chem Res Toxicol*, **19**, 30–7.

Zanic-Grubisic T, Zrinski R, Cepelak I, Petrik J, Radic B and Pepeljnjak S (2000), 'Studies of ochratoxin A-induced inhibition of phenylalanine hydroxylase and its reversal by phenylalanine', *Toxicol Appl Pharmacol*, **167**, 132–9.

Zinedine A, Soriano J M, Moltó J C and Mañes J (2007), 'Review on the toxicity, occurrence, metabolism, detoxification, regulations and intake of zearalenone: an oestrogenic mycotoxin', *Food Chem Toxicol*, **45**, 1–18.

Zitomer N C, Mitchell T, Voss K A, Bondy G S, Pruett S T, Garnier-Amblard E C, Liebeskind L S, Park H, Wang E, Sullards M C, Merrill AH Jr and Riley R T (2009), 'Ceramide synthase inhibition by fumonisin B$_1$ causes accumulation of 1-deoxy-sphinganine: A novel category of bioactive 1-deoxy-sphingoid bases and 1-deoxy-dihydroceramides biosynthesized by mammalian cell lines and animals', *J Biol Chem*, **284**, 4786–95.

Zollner P, Jodlbauer J, Kleinova M, Kahlbacher H, Kuhn T, Hochsteiner W and Lindner W (2002), 'Concentration levels of zearalenone and its metabolites in urine, muscle tissue, and liver samples of pigs fed with mycotoxin-contaminated oats', *J Agric Food Chem*, **50**, 2494–501.

Part IV

Determining mycotoxigenic fungi in food and feed

10

Rationale for a polyphasic approach in the identification of mycotoxigenic fungi

J. C. Frisvad, Technical University of Denmark, Denmark

Abstract: The taxonomy of mycotoxigenic fungi is complicated, especially within the genera *Penicillium, Aspergillus, Alternaria* and *Fusarium*, and classification is still in a state of flux. Furthermore new species are still being described. Since a polyphasic classification of filamentous fungi is generally recommended, a polyphasic identification is also recommended, even though keys based on such an approach are not yet fully developed. For practical reasons the following practice is recommended. First, fungal isolates involved in mycotoxicoses are identified to genus level using morphological criteria. Depending on the genus, isolates are then grown on a battery of indicative media and identified to species level based on a combination of morphological, physiological, nutritional and chemical data. The identification is then validated by a blast search of the DNA sequence of the β-tubulin and calmodulin genes (or other house-keeping genes with a good resolution at the species level), but avoiding ribosomal DNA genes that often have a poor resolution at the species level. Several books published in 2009 and 2010 will assist in a proper identification and established connections between species and mycotoxins will also help in avoiding misidentification. Unusual records should be retested using more taxonomic features or sequencing of more genes in order to be sure that the identification is correct.

Key words: *Aspergillus, Fusarium*, mycotoxins, *Penicillium*, polyphasic identification.

10.1 Introduction

Mycotoxins are fungal specific (secondary) metabolites that are toxic to verte-brates when introduced in small amounts via a natural route. The first mycotoxins known include the ergot alkaloids from *Claviceps purpurea* and *C. paspali*, but since the outbreaks of 'St. Anthony's fire' in the middle ages many other mycotoxins were discovered, for example T-2 toxin during World War II and in 1960, when mycotoxins became famous by the outbreak of the aflatoxin induced 'turkey X-disease' (Purchase, 1971, 1974). Many more mycotoxicoses and mycotoxins have been discovered in the last 50 years (Forgacs and Carll, 1962; Wogan 1965; Rodricks, 1976; Rodricks *et al.*, 1977; Jemmali, 1977; Wyllie and Morehouse, 1978; Moreau, 1979; Steyn, 1980; Cole and Cox, 1981; Reiss, 1981; Shank, 1981; Keller and Tu, 1983; Ueno, 1983; Betina, 1984; Kurata and Ueno, 1984; Marasas *et al.*, 1984; Lacey, 1985; Smith and Moss, 1985; Watson, 1985; Cole, 1986; Richard and Thurston, 1986; Steyn and Vleggaar, 1986; Krogh, 1987; Marasas and Nelson, 1987; Frisvad, 1988; Eaton and Groopman, 1989; Betina, 1989; Chelkowski, 1989; Egmond, 1989; Natori *et al.*, 1989; Roth *et al.*, 1990; Bhatnagar *et al.*, 1991; Champ *et al.*, 1991; Chelkowski, 1991; Frisvad and Samson, 1991; Sharma and Salunkhe, 1991; Chelkowski and Visconti, 1992; Betina, 1993; Pohland, 1993; Scudamore, 1993; Miller and Trenholm, 1994; Smith and Solomons, 1994; Egmond, 1996; Miraglia *et al.*, 1998; Reiss, 1998; Sinha and Bhatnagar, 1998; Watson, 1998; Weidenbörner, 2001; DeVries *et al.*, 2002; Frisvad and Thrane, 2004; Magan and Olsen, 2004; Diaz, 2005; Barug *et al.*, 2006; Mokiuddin, 2007; Weidenbörner, 2007; Barkai-Golan and Paster, 2008; Leslie *et al.*, 2008; Njapau et al., 2008; Rai and Varma, 2009; Anyanwu, 2010; Gonzales *et al.*, 2010). In order to determine mycotoxins in foods and feedstuffs and prevent mycotoxin formation, it is necessary to identify the mycotoxin producer(s). Identification of mycotoxin producers has always been difficult for several reasons. Taxonomic concepts have changed over the last centuries, sometimes species appear to be closely related, but have quite different mycotoxin profiles, and the methods used for identification have rapidly evolved while the equipment need may not be available in all laboratories working with mycotoxins. These difficulties have resulted in a large number of misidentifications and incorrect connections between fungal species and mycotoxin production (Frisvad, 1988, 1989; Frisvad *et al.*, 2004, 2006a,b).

A recent development in classification and identification has been a polyphasic approach, where several kinds of taxonomic features are used in conjunction, for example morphology, physiology, chemistry and gene sequencing (Frisvad and Samson, 2004). The polyphasic approach was proposed by Vandamme *et al.* (1996) in bacteriology, but often this approach has been neglected, even in bacteriology, where only 16S rDNA has been used for identification. In mycology, sequence-based identification, barcoding and microcoding have been suggested for identification of common filamentous fungi, but there are still many technical, economical and bioinformative problems to be solved before this may eventually become a common practice (Summerbell *et al.*, 2005; Seifert *et al.*, 2007; Seifert, 2009), so for both scientific and practical reasons, polyphasic identification will probably be the way to identify fungi correctly. For example in *Penicillium*, a known genus with many species producing mycotoxins, COX1 and ITS gene sequences will help in identification, but are not sufficiently informative to allow correct identification of all *Penicillium* species (Seifert *et al.*, 2007; Chen *et al.*, 2009), while β-tubulin sequences are more effective, but still not entirely sufficient to allow species level identification in all cases (Samson *et al.*, 2004). Identification should thus be polyphasic and an identified organism should be compared to a control, a well-known authenticated isolate from a culture collection, in order to be sure it is correctly identified.

10.2   Mycotoxigenic fungi

A large number of genera have been claimed to contain mycotoxigenic species. Usually isolates in the same species are chemoconsistent, that is if one isolate produces a mycotoxin other isolates in the same species will do the same (Frisvad *et al.*, 2004; Larsen *et al.*, 2005). However, in many cases, especially in indirectly domesticated species, such as *Aspergillus flavus* and *Penicillum verrucosum*, isolates of the former species will not always produce aflatoxins and isolates of the latter will not always produce ochratoxins. There is a high probability that this is caused by silenced gene clusters caused by epigenetic factors (Cichewicz, 2010). A list of correctly identified species producing the major mycotoxins is given in Table 10.1. This list may help in deciding whether a fungal species – mycotoxin connection is correct, and if it is deviating from the list, a more thorough validation may be needed.

In analytical chemistry at least three different kinds of data to proof identity of the mycotoxin are recommended (Frisvad *et al.*, 2006a). For example if the identification of a particular compound is based on an accurate mass spectrum, a high performance liquid chromatography (HPLC) retention time and/or a circular dichroism (CD) spectrum compared to an authentic standard, it will probably be correctly identified, but if it is only based on TLC retardation factor and a fluorescing spot it may be several compounds and may in several cases be misidentified. Standards may be difficult to obtain and in these cases four pieces of evidence are needed for confirmation of identify. There are no minimum standards for the identification of fungi, but Frisvad *et al.* (2006a) recommended a series of tests to assure that the fungus was correctly identified. Many genera of fungi contain mycotoxigenic fungi, but most mycotoxins are produced by species in the genera *Fusarium, Penicillium* and *Aspergillus*.

Most mycotoxins are produced by several species in one genus, but some mycotoxins are produced by species in several phylogenetically different genera. For example fumonisins B_2, B_4 and B_6 are produced by *Aspergillus niger* (Frisvad *et al.*, 2007; Sørensen *et al.*, 2009; Mogensen *et al.*, 2009, 2010; Månsson *et al.*, 2010), while fumonisins B_1, B_2 and B_3 are produced by a series of *Fusarium* species from four sections: *Arthrosporiella, Dlaminia, Elegans* and *Liseola* (Fotso *et al.*, 2002; Rheeder *et al.*, 2002; Sewram *et al.*, 2005). It is not yet known whether the production of similar mycotoxins in widely different genera is caused by horizontal gene transfer (for example mobile chromosomes) or by parallel evolution.

A large number of isolates of mycotoxigenic fungi have been misidentified over

Table 10.1   Producers of the known mycotoxins (based on Cole and Cox, 1981; Hermann *et al.*, 1996; Frisvad and Thrane, 2004; Frisvad and Samson, 2004; Thrane *et al.*, 2004; Varga *et al.*, 2009). Most species are listed under their correct holomorphic name, but *Aspergillus fumigatus* has been used instead of *Neosartorya fumigata*, *Fusarium* species have been listed rather than under their *Gibberella* state and *Isaria* has been used rather than its teleomorphic state *Cordyceps* (Luangsa-ard *et al.*, 2009)

Mycotoxin	Producing species	Common in foods?
AAL-toxins (alperisins)	*Alternaria arborescens*	Yes
Aflatoxins B$_1$ and B$_2$	*Aspergillus flavus*	Yes
	Aspergillus parasiticus	Yes
	Aspergillus nomius	Yes
	Aspergillus arachidicola	Yes
	Aspergillus minisclerotigenes	Yes
	Aspergillus pseudotamarii	No
	Aspergillus parvisclerotigenus	No
	Aspergillus bombycis	No
	Aspergillus ochraceoroseus	No
	Aspergillus rambellii	No
	Emericella venezuelensis	No
	Emericella olivicola	No
	Emericella astellata	No
Aflatoxins G$_1$ and G$_2$	*Aspergillus parasiticus*	Yes
	Aspergillus nomius	Yes
	Aspergillus arachidicola	Yes
	Aspergillus minisclerotigenes	Yes
	Aspergillus parvisclerotigenus	No
	Aspergillus bombycis	No
Aflatrems	*Aspergillus flavus*	Yes
	Aspergillus minisclerotigenes	Yes
	Aspergillus parvisclerotigenus	No
Alternariols	*Alternaria tenuissima*	Yes
	Alternaria alternata	No
	Alternaria spp.	Yes
	Botrytis aclada	Yes
	Penicillium coprophilum	No
	Penicillium diversum	No
	Penicillium verruculosum	No
Aranotins	*Arachnoitus aureus*	No
	Aspergillus terreus	Yes
Asteltoxin	*Aspergillus flocculosus*	No
	Aspergillus insulicola	No
	Emericella variecolor	No
	Penicillium concentricum	No
	Penicillium confertum	No
	Penicillium formosanum	No
	Penicillium tricolor	No (but do occur)
	Ponchonia bulbilosa	No
	Ponchonia suclasporia 'var. *catenata*'	No
Austocystins	*Aspergillus puniceus*	No
	Aspergillus ustus	Yes
Beauvericin	*Beauveria bassiana*	No
	Fusarium langsethiae	Yes

Table 10.1 *cont.*

Mycotoxin	Producing species	Common in foods?
	Fusarium poae	Yes
	Fusarium sporotrichioides	Yes
	Isaria cicadae	No
	Isaria fumosorosea	No
	Isaria japonica	No
	Isaria tenuipes	No
Botryodiploidin	*Botrysphaeria rhodina*	Yes
	Macrophomina phaseolina	Yes
	Penicillium brevicompactum	Yes
	Penicillium paneum	Yes
Butenolide	*Fusarium crookwellense*	Yes
	Fusarium equiseti	Yes
	Fusarium graminearum	Yes
	Fusarium tricinctum	Yes
Byssotoxin*	*Byssochlamys fulva*	Yes
Byssochlamic acid	*Byssochlamys fulva*	Yes
	Byssochlamys nivea	Yes
Chaetocin, chaetomin	*Chaetomium cochlioides*	No
	Chaetomium globosum	No
Chaetoglobosins A-J	*Calonectria morganii*	No
	Chaetomium cochlioides	No
	Chaetomium globosum	No
	Chaetomium mollipileum	No
	Chaetomium rectum	No
	Cylindrocladium moridanum	No
	Diplodia macrospora	No
	Discosia sp.	No
	Penicillium discolor	Yes
	Penicillium expansum	Yes
	Penicillium marinum	No
Chaetoglobosin K	*Diplodia macrospora*	No
Citreoviridin	*Aspergillus terreus*	Yes
	Eupenicillium ochrosalmoneum	Yes
	Penicillium citreonigrum	Yes
	Penicillium manginii	No
	Penicillium miczynskii	No
	Penicillium smithii	No
	Ponchonia suclasporia 'var. *catenata*'	No
Citrinin	*Aspergillus alabamensis*	No
	Aspergillus carneus	No
	Aspergillus hortai	No
	Aspergillis niveus	No
	Blennoria sp.**	No
	*Clavariopsis aquatica***	No
	Monascus purpureus	Yes
	Monascus ruber	Yes
	Monascus spp.	Yes
	Penicillium chrzaszczii	No
	Penicillium citrinum	Yes
	Penicillium decaturense	No

Table 10.1   *cont.*

Mycotoxin	Producing species	Common in foods?
	Penicillium expansum	Yes
	Penicillium gorlenkoanum	No
	Penicillium manginii	No
	Penicillium odoratum	No
	Penicillium radicicola	Yes
	Penicillium verrucosum	Yes
	Penicillium westlingii	No
Citromycetin	*Penicillium glabrum*	Yes
	Penicillium vinaceum	No
Communesins	*Penicillium atrovenetum*	No
	Penicillium expansum	Yes
	Penicillium marinum	No
	Penicillium rivulorum	No
Cyclochlorotine, islanditoxin	*Penicillium islandicum*	Yes
Cyclopiazonic acid	*Aspergillus flavus*	Yes
	Aspergillus lentulus	No
	Aspergillus minisclerotigenes	Yes
	Aspergillus oryzae	Yes
	Aspergillus pseudotamarii	No
	Aspergillus tamarii	Yes
	Penicillium camemberti	Yes
	Penicillium commune	Yes
	Penicillum dipodomyicola	No
	Penicillium griseofulvum	Yes
	Penicillium palitans	Yes
Cytochalasins A, B and F	*Drechslera dematioidea*	No
	Phoma sp.	No
Cytochalasins C and D	*Metarrhizium anisopliae*	No
	Zygosporium masonii	No
Cytochalasins E and K	*Aspergillus clavatus*	Yes
	Rosellinia necatrix	No
Cylochalasin H	*Phomposis paspali*	No
Cylochalasin G	*Nigrosabulum* sp.	No
Deoxaphomin, proxiphomin, protophomin	*Phoma* sp.	No
Duclauxin	*Talaromyces macrosporus*	Yes
Emodin	*Aspergillus wentii*	Yes
	Cladosporium fulvum	Yes
	Penicilliopsis clavariaformis	No
	Penicillium brunneum	No
	Penicillium islandicum	Yes
	Phoma foveata	No
	Talaromyces avellaneus (Hamigera avellanea)	No
Enniatins	*Fusarium acuminatum*	Yes
	Fusarium arthrosporioides	Yes
	Fusarium avenaceum	Yes
	Fusarium compactum	Yes
	Fusarium dimerum	Yes
	Fusarium kyushuense	No

Table 10.1   *cont.*

Mycotoxin	Producing species	Common in foods?
	Fusarium langsethiae	Yes
	Fusarium lateritium	Yes
	Fusarium merismoides	No
	Fusarium oxysporum f.sp. *batatas, melonis, lupini, pisi*	Yes
	Fusarium sambucinum	Yes
	Fusarium scirpi	No
	Fsuarium torulosum	Yes
	Fusarium tricinctum	Yes
	Fusarium venenatum	Yes
	Verticillium hemiptigerum	No
Ergot alkaloids	*Claviceps paspali*	Yes
	Claviceps purpurea	Yes
Erythroskyrin	*Penicillium islandicum*	Yes
Fumigaclavines A, B and C	*Aspergillus fumigatus*	Yes
Fumonisins B_1, B_2 and B_3	*Fusarium acutatum*	No
	*Fusarium andiyazi***	No
	Fusarium anthophilum	No
	Fusarium begoniae	No
	Fusarium brevicatenulatum	No
	Fusarium dlamini	No
	Fusarium fujikuroi	Yes
	Fusarium globosum	No
	Fusarium napiforme	No
	Fusarium nygamai	No
	Fusarium oxysporum	Yes
	Fusarium phyllophilum	No
	Fusarium polyphialidicum	No
	Fusarium proliferatum	Yes
	Fusarium pseudocircinatum	No
	*Fusarium pseudonygamai***	No
	Fusarium sacchari	No
	Fusarium subglutinans	Yes
	Fusarium thapsinum	No
	Fusarium verticillioides	Yes
Fumonisins B_2, B_4 and B_6	*Aspergillus niger*	Yes
Gliotoxin	*Aspergillus fumigatus*	Yes
	Dichotomomyces cejpii	No
	Dichotomomyces spinosus	No
	Penicillium lilacinoechinulatum	No
	Trichoderma virens	No
Hyalodendrin	*Hyalodendron* sp.	No
Isofumigaclavins	*Penicillium palitans*	Yes
	Penicillium roqueforti	Yes
Luteoskyrin	*Penicillum islandicum*	Yes
Moniliformin	*Fusarium acuminatum*	Yes
	Fusarium avenaceum	Yes
	Fusarium oxysporum	Yes
	Fusarium subglutinans	Yes
	Fusarium verticillioides	Yes

Table 10.1 *cont.*

Mycotoxin	Producing species	Common in foods?
Mycophenolic acid	*Penicillium bialowiezense*	Yes
	Penicillium brevicompactum	Yes
	Penicillium carneum	Yes
	Penicillium fagi	No
	Penicillium roqueforti	Yes
	*Phaerosphaeria nodorum***	No
Nephrotoxic glycopeptides	*Penicillium aurantiogriseum*	Yes
	Penicillium polonicum	Yes
3-Nitropropionic acid	*Arthrinium aureum*	No
	Arthrinium phaerospermum	Yes
	Arthrinium sacchari	Yes
	Arthrinium saccharicola	Yes
	Arthrinium sereanis	No
	Arthrinium terminalis	No
	Aspergillus flavus	Yes
	Aspergillus oryzae	Yes
	Aspergillus sojae	Yes
	Mucor circinelloides	Yes
	Penicillium atrovenetum	No
Ochratoxin A	*Aspergillus carbonarius*	Yes
	Aspergillus cretensis	No
	Aspergillus flocculosus	No
	Aspergillus lacticoffeatus	No
	*Aspergillus melleus****	Yes
	Aspergillus niger	Yes
	Aspergillus ochraceus	Yes
	*Aspergillus ostianus****	Yes
	Aspergillus persii	No
	*Aspergillus petrakii****	No
	Aspergillus pseudoelegans	No
	Aspergillus roseoglobulosus	No
	Aspergillus sclerotioniger	No
	Aspergillus sclerotiorum	Yes
	Aspergillus steynii	Yes
	Aspergillus sulphureus	No
	Aspergillus westerdijkiae	Yes
	*Neopetromyces muricatus*****	Yes
	*Petromyces albertensis*****	No
	*Petromyces alliaceus*****	Yes
	Penicillium nordicum	Yes
	Penicillium verrucosum	Yes
Paspaline, paspalicine, paspalinine	*Aspergillus flavus*	Yes
	Aspergillus minisclerotigenes	Yes
	Aspergillus parviscletigenus	No
	Claviceps paspali	Yes
Paspalitrems A and B	*Claviceps paspali*	Yes
Patulin	*Aspergillus clavatus*	Yes
	Aspergillus giganteus	No
	Aspergillus longivesica	No

Table 10.1   *cont.*

Mycotoxin	Producing species	Common in foods?
	Byssochlamys nivea	Yes
	Eupenicillium lapidosum	No (but do occur)
	Paecilomyces saturatus	No
	Penicillium antarcticum	No
	Penicillium atrovenetum	No
	Penicillium carneum	Yes
	Penicillium clavigerum	No
	Penicillium concentricum	No
	Penicillium coprobium	No
	Penicillium dipodomyicola	No (but do occur)
	Penicillium estinogenum	No
	Penicillium expansum	Yes
	Penicillium formosanum	No
	Penicillium gladioli	No
	Penicillium glandicola	No
	Penicillium griseofulvum	Yes
	Penicillium marinum	No
	Penicillium paneum	Yes
	Penicillium sclerotigenum	Yes
	Penicillium vulpinum	No
Paxillin	*Emericella foveolata*	No
	Emericellla similis	No
	Eupenicillium shearii	No
	Penicillium paxilli	No
	Penicillium thiersii	No
Penicillic acid	*Aspergillus auricomus*	No
	Aspergillus bridgeri	No
	Aspergillus cretensis	No
	Aspergillus flocculosus	No
	Aspergillus insulicola	No
	Aspergillus melleus	Yes
	Aspergillus neobridgeri	No
	Aspergillus ochraceus	Yes
	Aspergillus ostianus	Yes
	Aspergillus persii	No
	Aspergillus petrakii	No
	Aspergillus pseudoelegans	No
	Aspergillus roseoglobulosus	No
	Aspergillus sclerotiorum	Yes
	Aspergillus sulpureus	No
	Aspergillus westerdijkiae	Yes
	Eupenicillium baarnense	No
	Eupenicillium bovifimosum	No
	Eupenicillium egyptiacum	No
	Eupenicillium molle	No
	Malbranchea aurantiaca	No
	Neopetromyces muricatus	Yes
	Penicillium aurantiogriseum	Yes
	Penicillium brasilianum	No (but do occur)
	Penicillium carneum	Yes

Table 10.1   *cont.*

Mycotoxin	Producing species	Common in foods?
	Penicillium cyclopium	Yes
	Penicillium fennelliae	No
	Penicillium flavidostipitatum	No
	Penicillium freii	Yes
	Penicillium herquei	No
	Penicillium jamesonlandense	No
	Penicillium matriti	No
	Penicilllium megasporum	No
	Penicillium melanoconidium	Yes
	Penicillium neoechinulatum	No
	Penicillium ochrochloron	No
	Penicillium persicinum	No
	Penicillium polonicum	Yes
	Penicillium radiatolobatum	No
	Penicillium raistrickii	No (but do occur)
	Penicillium rolfsii	No
	Penicillium scabrosum	Yes
	Penicillium viridicatum	Yes
Penitrem A	*Penicillium antarcticum*	No
	Penicillium crustosum	Yes
	Penicillium flavigenum	Yes
	Penicillium glandicola	No
	Penicillium janczewskii	No (but do occur)
	Penicilllium melanoconidium	Yes
	Penicillium ochrochloron	No
	Penicillium radiatolobatum	No
	Penicillium tulipae	No
PR-toxin	*Penicillium chrysogenum*	Yes
	Penicillium roqueforti	Yes
Roquefortine C	*Penicillium albocoremium*	Yes
	Penicillium allii	Yes
	Penicillium atramentosum	Yes
	Penicillium carneum	Yes
	Penicillium chrysogenum	Yes
	Penicillium concentricum	No
	Penicillium coprobium	No
	Penicillium coprophilum	No
	Penicillium crustosum	Yes
	Penicillium expansum	Yes
	Penicillium flavigenum	Yes
	Penicillium glandicola	No
	Penicillium griseofulvum	Yes
	Penicillium hirsutum	Yes
	Penicillium hordei	Yes
	Penicillium marinum	No
	Penicillium melanoconidium	Yes
	Penicillium paneum	Yes
	Penicillium persicinum	No
	Penicillium radicicola	Yes
	Penicillium roqueforti	Yes

Table 10.1   *cont.*

Mycotoxin	Producing species	Common in foods?
	Penicillium sclerotigenum	Yes
	Penicillium tulipae	No
	Penicillium venetum	Yes
	Penicillium vulpinum	No
Roridins and verrucarins	*Myrothecium roridum*	No
	Myrothecium verrucaria	No
Roseotoxin B	*Trichothecium roseum*	Yes
Rubratoxins A and B	*Penicillium crateriforme******	Yes
Rugulosin	*Aschersonia calendulina*	No
	Aschersonia samoensis	No
	Endothia fluens	No
	Endothia gyrosa	No
	Hypocrella discoidea	No
	Myrothecium verrucaria (antipode produced)	No
	Penicillium allahabadense	No
	Penicillium concavorugulosum	Yes
	Penicillium brunneum	No
	Penicillium islandicum	Yes
	Penicillium radicum	No
	Penicillium rugulosum	Yes
	Penicillium tardum	Yes
	Penicillium variabile	Yes
	Sepedonium ampullosporum	No
	Talaromyces wortmannii	Yes
Rugulovasins	*Penicillium atramentosum*	Yes
	Penicillium commune	Yes
	Penicillium concavorugulosum	Yes
Satratoxins	*Stachybotrys chartarum*	No
Secalonic acid A, B, C, G	*Claviceps purpurea*	Yes
	Phoma terrestris	No
	*Parmelia entotheichroa***	No, is a lichen
Secalonic acid D and F	*Aspergillus aculeatinus*	Yes
	Aspergillus aculeatus	Yes
	Aspergillus uvarum	Yes
	Claviceps purpurea	Yes
	*Eupenicillium egyptiacum***	No
	Penicillium chrysogenum	Yes
	Penicillium confertum	No
	Penicillium dendriticum	No
	Penicillium isariiforme	No
	Penicillium krugeri	No
	Penicillium oxalicum	Yes
Slaframin	*Rhizoctonia solani*	Yes
Sporidesmins A-J	*Pithomyces chartarum*	Yes
Sterigmatocystin******	*Aspergillus aureolatus*	No
	Aspergillus ochraceoroseus	No
	Aspergillus rambellii	No
	Aspergillus togoensis	No
	Aspergillus versicolor	Yes
	Chaetomium cellulolyticum	No

Table 10.1   *cont.*

Mycotoxin	Producing species	Common in foods?
	Chaetomium longicolleum	No
	Chaetomium malaysiense	No
	Chaetomium udagawae	No
	Chaetomium virescens	No
	Bipolaris sorokiniana	Yes
	Emerciella acristata	No
	Emericella aurantiobrunnea	No
	Emericella bicolor	No
	Emericella cleistominuta	No
	Emericella dentata	No
	Emericella discophora	No
	Emericella echinulata	No
	Emericella falconensis	No
	Emericella foeniculicola	No
	Emericella foveolata	No
	Emericella fructiculosa	No
	Emericella heterothallica	No
	Emericella navahoensis	No
	Emericella nidulans	Yes
	Emericella olivicola	No (but may occur)
	Emericella parvathecia	No
	Emericella rugulosa	No
	Emericella stella-maris	No
	Emericella striata	No
	Emericella venezuelensis	No
	Humicola fuscoatra	No
Tenuazonic acid	*Alternaria kikuchiana*	Yes
	Alternaria longipes	Yes
	Alternaria tenuissima	Yes
	Aspergillus nomius	Yes
	Phoma sorghina	Yes
	Pyricularia oryzae	Yes
Territrems	*Aspergillus terreus*	Yes
	Penicillium echinulatum	Yes
	Penicillium cavernicola	No (but do occur)
Trichodermin	*Trichoderma brevicompactum*	No
Trichothecenes (TCTC)	*Fusarium acuminatum* (type A TCTC)	Yes
	Fusarium crookwellense (type B TCTC)	Yes
	Fusarium culmorum (type B TCTC)	Yes
	Fusarium graminearum (type B TCTC)	Yes
	Fusarium equiseti (type A and B TCTC)	Yes
	Fusarium langsethiae (type A TCTC)	Yes
	Fusarium poae (type A and B TCTC)	Yes
	Fusarum sambucinum (type A TCTC)	Yes
	Fusarium sporotrichioides (type A TCTC)	Yes
	Fusarium venenatum (type A TCTC)	Yes
Trichothecin	*Trichothecium roseum*	Yes
Tryptoquivalins and tryptoquivalons	*Aspergillus clavatus*	Yes
	Aspergillus fumigatus	Yes

Table 10.1   *cont.*

Mycotoxin	Producing species	Common in foods?
Verrucosidin	*Penicillium aurantiogriseum*	Yes
	Penicillium melanoconidium	Yes
	Penicillium polonicum	Yes
Verruculogen and fumitremorgins	*Aspergillus caespitosus*	No
	Aspergillus fumigatus	Yes
	Eupenicillium crustaceum	No
	Neosartorya fischeri	Yes
	Penicillium brasilianum	Yes
	Penicillium mononematosum	No
Verruculotoxin	*Penicillium verruculosum*	No
Viridic acid	*Penicillium nordicum*	Yes
	Penicillium viridicatum	Yes
Verticillins	*Verticillium* sp.	No
Viridicatumtoxin	*Penicillium aethiopicum*	Yes
	Penicillium brasilianum	No (but do occur)
Viriditoxin	*Aspergillus viridinutans*	No
	Byssochlamys spectabilis (anamorph is *Paecilomyces variotii*)	Yes
	Penicillium mononematosum	No
Xanthomegnin, viomellein, vioxanthin	*Aspergillus auricomus*	No
	Aspergillus bridgeri	No
	Aspergillus elegans	No
	Aspergillus flocculosus	No
	Aspegillus insulicola	No
	Aspergillus melleus	Yes
	Aspergillus neobridgeri	No
	Aspergillus ochraceus	Yes
	Aspergillus ostianus	Yes
	Aspergillus persii	No
	Aspergillus petrakii	No
	Aspergillus roseoglobulosus	No
	Aspergillus sclerotiorum	Yes
	Aspergillus steynii	Yes
	Aspergillus sulphureus	No
	Aspergillus westerdijkiae	Yes
	Eupenicillium javanicum	
	Microsporum cookei	No
	*Neopetromyces muricatus*****	Yes
	Penicillium cyclopium	Yes
	Penicillium freii	Yes
	Penicillium janthinellum	No
	Penicillium mariae-crucis	No
	Penicillium melanoconidium	Yes
	Penicillium tricolor	No (but do occur)
	Penicillium viridicatum	Yes
	Trichophyton megninii	No
	Trichophyton rubrum	No
	Trichophyton violaceum	No

Table 10.1   *cont.*

Mycotoxin	Producing species	Common in foods?
Zearalenone	*Fusarium avenaceum*	Yes
	Fusarium crookwellense	Yes
	Fusarium culmorum	Yes
	Fusarium equiseti	Yes
	Fusarium graminearum	Yes
	Fusarium sambucinum	Yes
	Fusarium semitectum	Yes
Zygosporins D-G	*Zygosporium masonii*	No

*Structure not known. **Not confirmed since its original publication. ***Only trace production. ****Also known under the same name in *Aspergillus*. *****The taxonomy of this species is not yet clear, and the species has been called *P. rubrum* and/or *P. purpurogenum* earlier. ******Also produced by the producers of aflatoxins, although often transiently.

the years and furthermore sometimes the mycotoxin is not correctly identified (Frisvad *et al.*, 2006a). Examples are given in Varga *et al.* (2008, 2009), where authors of various papers have stated that a number of penicillia and other fungi were producers of aflatoxins. These statements were based on misidentification of the fungi or secondary metabolites found, but may also have been caused by less efficient separation and detection techniques or even contamination by mycotoxins in the media used for mycotoxin production. For example, corn is often used as a substrate for mycotoxin production, but may already contain mycotoxins. Also occasionally a mycotoxin is claimed to be produced by a fungal species, but that species may have been contaminated by another fungus. In one example Maskey *et al.* (2003) identified sterigmatocystin in *Penicillium chrysogenum*, but we later examined the strain and discovered it was contaminated with *Aspergillus versicolor*, a known producer of sterigmastocystin. The strain of *P. chrysogenum* did not produce any sterigmatocystin.

10.3   Identification methods

For filamentous fungi, polyphasic identification is strongly recommended. The identification procedure should start with identification of the fungi to genus level morphologically, for example using the books by Samson *et al.* (2010) and Domsch *et al.* (2007). When the fungus is identified to genus level, different identification procedures may be used. For example *Penicillium* and *Aspergillus* isolates are usually inoculated on the media Czapek yeast extract agar (CYA), malt extract agar (MEA), yeast extract sucrose agar (YES), creatine sucrose agar (CREA) and oat meal agar (OAT), while *Fusarium* isolates are identified on Specifikke nutrient-arme agar (SNA), potato dextrose agar (PDA) and YES agar. *Alternaria* species are identified via dichloran rose Bengal yeast extract sucrose (DRYES) agar and potato carrot agar (PCA) (Simmons, 2007). *Penicillium* isolates can be identified to species level according to Samson *et al.* (2010), Pitt and Hocking (2009), Frisvad and Samson (2004) and Pitt (1979). *Aspergillus* species can be identified to species level using Samson *et al.* (2010), Pitt and Hocking (2009), Samson and Varga (2007) and Raper and Fennell (1965). *Fusarium* species can be identified according to Samson *et al.* (2010) and Pitt and Hocking (2009). The identifications can be validated using authenticated cultures of each species and by comparison to the descriptions in the taxonomic works listed. However in order to be polyphasic, it is also recommended the isolates be identified using molecular methods.

10.4   Molecular methods for identification

Molecular methods are based on sequencing of house-keeping genes such as β-tubulin, calmodulin and actin genes. In general it has been shown that ribosomal and mitochondrial genes are not sufficiently specific for identification at the species level (Skouboe *et al.*, 1999, 2000; Seifert *et al.*, 2007). The methods are described in detail in the next chapter. Fingerprinting methods are more difficult to use, for example AFLP, as they require comparison with authenticated isolates of several species as controls, and may include both species and isolate specific information. Bar-coding methods are being developed, and may be used much more in the future, as mentioned in the introduction (Summerbell *et al.*, 2005; Seifert *et al.*, 2007; Seifert, 2009).

10.5   Conclusions

Filamentous fungi should be identified using a polyphasic approach in order to avoid mistakes and the mycotoxins they may produce should be identified using proper chemical/physical techniques, in order to validate the findings. This is even more important for discoveries of new connections between fungal species and mycotoxins.

10.6   Acknowledgements

The EEC project MycoRed (KBBE-2007-222690-2) is acknowledged for financial support.

10.7   References

Anyanwu EC (2010), *Advances in Environmental Health Effects of Toxigenic Mold and Mycotoxins*, NOVA Science Publishing, Hauppauge.
Barkai-Golan R and Paster N (2008), *Mycotoxins in Fruits and Vegetables*, Amsterdam, Elsevier.

Barug D, Bhatrnagar D, van Egmond HP, van der Karp JW, van Osenbruggen NJ and Visconti A (2006), *The Mycotoxin Fact Book. Food and Feed Topics*, Wageningen Academic Publishers, Wageningen.

Betina V (1984), *Mycotoxins – Production, Isolation, Separation and Purification*, Elsevier, Amsterdam, .

Betina V (1989), *Mycotoxins – Chemical, Biological and Environmental Aspects*, Elsevier, Amsterdam.

Betina V (1993), *Chromatography of Mycotoxins. Techniques and Applications*, Elsevier, Amsterdam.

Bhatnagar D, Lillehoj EB and Arora DK (1991), *Mycotoxins in Ecological Systems*, Marcel Dekker, New York.

Champ BR, Highly E, Hocking AD and Pitt JI (1991), *Fungi and Mycotoxins in Stored Products*, ACIAR Proceedings 36, CSIRO, Sydney.

Chelkowski J (1989), Fusarium. *Mycotoxins, Taxonomy and Pathogenicity*, Elsevier, Amsterdam.

Chelkowski J (1991), *Cereal Grain. Mycotoxins, Fungi and Quality in Drying and Storage*, Elsevier, Amsterdam.

Chelkowski J and Visconti A (1992), Alternaria – *Biology, Plant Diseases and Metabolites*, Elsevier, Amsterdam.

Chen W, Seifert KA and Lévesque CA (2009), 'A high density COX1 barcode oligonucleotide array for identification and detection of species of *Penicillium* subgenus *Penicillium*', *Mol Ecol Resour*, **9**, 114–29.

Cichewicz RH (2010), 'Epigenome manipulation as a pathway to new natural product scaffolds and their congeners', *Nat Prod Rep*, **27**, 11–22.

Cole RJ (1986), *Modern Methods in the Analysis and Structural Elucidation of Mycotoxins*, Academic Press, Orlando.

Cole RJ and Cox RH (1981), *Handbook of Toxic Fungal Metabolites*, Academic Press, New York.

Devries JW, Trucksess MW and Jackson LS (2002), *Mycotoxins and Food Safety*, Springer, New York.

Diaz D (2005), *Mycotoxin Blue Book*, Nottingham University Press, Nottingham.

Domsch KH, Gams W and Anderson T-H (2007) *Compendium of Soil Fungi*, IHW-Verlag, Eching.

Eaton DL and Groopman JD (1989), *The Toxicology of Aflatoxins*, Academic Press, New York.

van Egmond HP (1989), *Mycotoxins in Dairy Products*, Elsevier, London.

van Egmond HP (1996), *Mycotoxins and Toxic Plant Components*, Wiley Liss, London.

Forgacs J and Carll WT (1962), 'Mycotoxicoses', *Adv Vet Sci,* **7**, 273–382.

Fotso J, Leslie JF and Smith JS (2002), 'Production of beauvericin, moniliformin, fusaproliferin, and fumonisins B_1, B_2, and B_3 by fifteen ex-type strains of *Fusarium* species', *Appl Environ Microbiol*, **68**, 5195–57.

Frisvad JC (1988), 'Fungal species and their specific production of mycotoxins', in *Introduction to Food-borne Fungi*, Samson RA and van Reenen-Hoekstra ES (eds), 3rd ed.n, Centraalbureau voor Schimmelcultures, Baarn, 239–49.

Frisvad JC (1989), 'The connection between the penicillia and aspergilli and mycotoxins with special emphasis on misidentified isolates', *Arch Environ Contam Toxicol*, **18**, 452–467.

Frisvad JC and Samson RA (1991), 'Filamentous fungi in foods and feeds: ecology, spoilage and mycotoxin production', in *Handbook of Applied Mycology*, Arora DK, Mukerji KG and Marth, EH (eds), Marcel Dekker, New York, 31–68.

Frisvad JC and Samson RA (2004), 'Polyphasic taxonomy of *Penicillium* subgenus *Penicillium*. A guide to identification of the food and air-borne terverticillate Penicillia and their mycotoxins', *Stud Mycol*, **49**, 1–173.

Frisvad JC and Thrane U (2004), 'Mycotoxin production by common filamentous fungi', in

Introduction to Food- and Airborne Fungi, Samson RA, Hoekstra ES and Frisvad JC (eds), 7th edn, Centraalbureau voor Schimmelcultures, Utrecht, 321–30.

Frisvad JC, Smedsgaard J, Larsen TO and Samson RA (2004), 'Mycotoxins, drugs and other extrolites produced by species in *Penicillium* subgenus *Penicillium*', *Stud Mycol*, **49**, 201–41.

Frisvad JC, Nielsen KF and Samson RA (2006a), 'Recommendations concerning the chronic problem of misidentification of mycotoxigenic fungi associated with foods and feeds', *Adv Exp Med Biol*, **571**, 33–46.

Frisvad JC, Thrane U, Samson RA and Pitt JI (2006b), 'Important mycotoxins and the fungi which produce them', *Adv Exp Med Biol*, **571**, 1–31.

Frisvad JC, Smedsgaard J, Samson RA, Larsen TO and Thrane U (2007), 'Fumonisin B_2 production by *Aspergillus niger*', *J Agric Food Chem*, **55**, 9727–32.

Gonzales RR, Martinez Vidal JL and Frenich AG (2010), *Liquid Chromatography for the Detection of Mycotoxins in Food*, NOVA Science Publishing, Hauppauge.

Hermann M, Zocher R and Haese A (1996), 'Enniatin production by *Fusarium* strains and its effect on potato tuber tissue', *Appl Environ Microbiol*, **62**, 393–8.

Jemmali M (1977), *Mycotoxins in Foodstuffs*, Elsevier, Amsterdam.

Keller RF and Tu AT (1983), *Handbook of Natural Toxins. Vol. 1. Plant and Fungal Toxins*, Marcel Dekker, New York.

Krogh P (1987), *Mycotoxins in Food*, Academic Press, London.

Kurata H and Ueno Y (1984) *Toxigenic Fungi – Their Toxins and Health Hazard*, Elsevier, Amsterdam.

Lacey J (1985), *Trichothecenes and Other Mycotoxins*, John Wiley and Sons, New York.

Larsen TO, Smedsgaard J, Nielsen KF, Hansen ME and Frisvad JC (2005), 'Phenotypic taxonomy and metabolite profiling in microbial drug discovery'. *Nat Prod Rep*, 22, 672–695.

Leslie JF, Bandypadhyay R and Visconti A (2008), '*Mycotoxins: detection methods, management, public health and agricultural trade*', Wallingford, CABI.

Luangsa-ard JJ, Berkaew P, Ridkaew R, Hywel-Jones NL and Isaka M (2009), 'A beauvericin hot spot in the genus *Isaria*', *Mycol Res*, **113**, 1389–95.

Magan N and Olsen M (2004), *Mycotoxins in Food: Detection and Control*, Woodhead, Abington, Cambridge.

Månsson M, Klejnstrup ML, Phipps RK, Nielsen KF, Frisvad JC, Gotfredsen CH and Larsen TO (2010), 'Isolation and NMR characterization of fumonisin B_2 and B_6, a new fumonisin from *Aspergillus niger*', *J Agric Food Chem*, **58**, 949–53.

Marasas WFO and Nelson PE (1987), *Mycotoxicology: Introduction to the Mycology, Plants Pathology, Chemistry, Toxicology, and Pathology of Naturally Occurring Mycotoxicosis in Animals and Man*, The Pennsylvania State University Press, University Park.

Marasas WFO, Nelson PE and Tousson TA (1984), *Toxigenic* Fusarium *Species, Identity and Mycotoxicology*, The Pennsylvania State University Press, University Park.

Maskey RP, Gren-Wollny I and Laatsch H (2003), 'Isolation, structure elucidation and biological activity of 8-O-methylaverufin and 1,8-O-dimethylaverantin as new antifungal agents from *Penicillium chrysogenum*', *J Antibiot*, **56**, 459–63.

Miller JD and Trenholm HL (1994), *Mycotoxins in Grain. Compounds other Than Aflatoxins*, Eagan, St. Paul.

Miraglia M, van Egmond HP, Brera C and Gilbert J (1998), *Mycotoxins and Phycotoxins – Developments in Chemistry, Toxicology and Food Safety*, Alaken Press, Denver.

Mogensen JM, Nielsen KF, Frisvad JC, Samson RA and Thrane U (2009), 'Effect of temperature and water activity on the production of fumonisin B_2 by *Aspergillus niger* and *Fusarium* species', *BMC Microbiol*, **9**, 281.

Mogensen JM, Frisvad JC, Thrane U and Nielsen KF. (2010), 'Production of fumonisin B_2 and B_4 by *Aspergillus niger* on grapes and raisins', *J Agric Food Chem*, **58**, 954–8.

Mokiuddin SM (2007), *Mould and Mycotoxin in Poultry Diseases*, International Book Distribution Co, Lucknow.

Moreau C (translated by Moss M) (1979), *Moulds, Toxins and Foods*, John Wiley & Sons, New York.

Natori S, Hashimoto K and Ueno Y (1989), *Mycotoxins and Phycotoxins 1988*, Elsevier, Amsterdam.

Njapau H, Trujillo S, Pohland AE and Park DL (2008), *Mycotoxin Contamination and Control*, AuthorHouse, Bloomington.

Pitt JI (1979), *The genus* Penicillium *and its Teleomorphic States* Eupenicillium *and* Talaromyces, Academic Press, London.

Pitt JI and Hocking AD (2009), *Fungi and Food Spoilage*, Springer, Dordrecht.

Pohland AE (1993), 'Mycotoxins in review', *Food Addit Contam*, **10**, 17–28.

Purchase IHF (1971), *Mycotoxins in Human Health*, MacMillan, London.

Purchase IHF (1974), '*Mycotoxins*', Amsterdam, Elsevier.

Rai M and Varma A (2009), '*Mycotoxins in food, feed and bioweapons*', Berlin, Springer.

Raper KB and Fennell DI (1965), *The Genus* Aspergillus, Williams and Wilkins, Baltimore.

Reiss J (1981), *Mykotoxine in Lebensmitteln*, Gustav Fisher, Stuttgart.

Reiss J (1998), *Schimmelpilze, Lebensweise, Nutzen, Schaden, Bekämpfung*, 2nd edn, Springer, Berlin.

Rheeder JP, Marasas WFO and Vismer HF (2002), 'Production of fumonisin analogs by *Fusarium* species', *Appl Environ Microbiol*, **68**, 2101–05.

Richard JL and Thurston JR (1986), *Diagnosis of Mycotoxicoses*, John Wiley and Sons, Chichester.

Rodricks JC (1976), *Mycotoxins and Other Fungal Related Food Problems*, American Chemical Society, Washington DC.

Rodricks JV, Hesseltine CW and Mehlmann MA (1977), *Mycotoxins in Human and Animal Health*, Pathotox, Park Forest South.

Roth L, Frank H and Kromann K (1990), *Giftpilze – Pilzgifte*, Ecomed, Landsberg/Lech.

Samson RA and Varga J (2007), '*Aspergillus* systematics in the genomic era', *Stud Mycol*, **59**, 1–206.

Samson RA, Seifert KA, Kuijpers AFA, Houbraken JAMP and Frisvad JC (2004), 'Phylogenetic analysis of *Penicillium* subgenus *Penicillium* using partial β-tubulin sequences', *Stud Mycol*, **49**, 175–200.

Samson RA, Houbraken J, Thrane U, Frisvad JC and Andersen B (2010), *Food and Indoor Fungi*, CBS/KNAW/Fungal Biodiversity Centre, Utrecht.

Scudamore KA (1993), *Occurrence and Significance of Mycotoxins*, Central Science Laboratory, Slough.

Seifert KA (2009), 'Progress towards DNA barcoding of fungi', *Mol Ecol Resour*, **9**, 83–9.

Seifert KA, Samson RA, deWaard JR, Houbraken J, Lévesque A, Moncalvo J-M, Louis-Seize G and Hebert PDN (2007), 'Prospects for fungus identification using CO1 DNA barcodes, with *Penicillium* as a test case', *Proc Natl Acad Sci USA*, **104**, 3901–6.

Sewram V, Mshicileli N, Shephard GS, Vismer HF, Rheeder JP, Lee Y-W, Leslie JF and Marasas WFO (2005), 'Production of fumonisin B and C analogues by several *Fusarium* species', *J Agric Food Chem*, **53**, 4861–4866.

Shank RC (1981), *Mycotoxins and N-nitroso Compounds: Environmental Risks*, CRC Press, Boca Raton.

Sharma RP and Salunkhe DK (1991), *Mycotoxins and Phytoalexins*, CRC, Boca Raton.

Simmons EG (2007), Alternaria. *An Identification Manual*, CBS Fungal Biodiversity Center, Utrecht.

Sinha KK and Bhatnagar D (1998), *Mycotoxins in Agriculture and Food Safety*, Marcel Dekker, New York.

Skouboe P, Frisvad JC, Lauritsen D, Boysen M, Taylor JW and Rossen L (1999), 'Nucleotide sequences from the ITS region of *Penicillium* species', *Mycol Res*, **103**, 873–81.

Skouboe P, Taylor JW, Frisvad JC and Rossen L (2000), 'Molecular methods for differentiation of closely related *Penicillium* species', in *Integration of Modern Taxonomic*

Methods for Aspergillus *and* Penicillium *Classification*, Samson RA and Pitt JI (eds), Harwood Scientific Publ., Reading, 179–88.

Smith J and Moss M (1985), *Mycotoxins. Formation, Analysis and Significance*, John Wiley & Sons, New York.

Smith J and Solomons GL (1994), *Mycotoxins in Human Nutrition and Health*, European Commission, Bruxelles.

Sørensen LM, Lametsch R, Andersen MR, Nielsen PV and Frisvad JC (2009), 'Proteome analysis of *Aspergillus niger*: Lactate added in starch-containing medium can increase production of the mycotoxin fumonisin B_2 by modifying acetyl-CoA metabolism', *BMC Microbiol*, **9**, 255.

Steyn PS (1980), *The Biosynthesis of Mycotoxins. A Study in Secondary Metabolism*, Academic Press, New York.

Steyn PS and Vleggaar R (1986), *Phycotoxins and Mycotoxins*, Elsevier, Amsterdam.

Summerbell RC, Lévesque CA, Seifert KA, Bovers M, Fell JW, Diaz MR, Boekhout T, de Hoog GS, Stalpers J and Crous PW (2005), 'Microcoding: the second step in DNA barcoding', *Phil Trans R Soc B*, **360**, 1897–903.

Thrane U, Adler A, Clasen P-E, Galvano F, Lengseth W, Lew H, Logrieco, A, Nielsen KF and Ritieni A (2004), 'Diversity in metabolite production by *Fusarium langsethiae, Fusarium poae*, and *Fusarium sporotrichioides*', *Int J Food Microbiol*, **95**, 257–66.

Ueno Y (1983), *Trichothecenes – Chemical, Biological and Toxicological aspects*, Kodansha/ Elsevier, Tokyo.

Vandamme P, Pot B, Gillis M, de Vos P, Kersters K and Swings J (1996), 'Polyphasic taxonomy, a consensus approach to bacterial systematics', *Microbiol Rev*, **60**, 407–38.

Varga J, Houbraken J, Samson RA and Frisvad JC (2008), 'Molecular diversity of *Aspergillus* and *Penicillium* on fruits and vegetables', in *Mycotoxins in Fruits and Vegetables*, Barkai-Golan R and Paster N (eds), Academic Press, Amsterdam, 205–23.

Varga J, Frisvad JC and Samson RA (2009), 'A reappraisal of fungi producing aflatoxin', *World Mycotoxin J*, **2**, 263–77.

Watson DH (1985), 'Toxic fungal metabolites', *CRC Crit Rev Food Sci Nutr*, **22**, 177–98.

Watson DH (1998), *Natural Toxicants in Food*, Sheffield Academic, Sheffield.

Weidenbörner M (2001), *Encyclopedia of Food Mycotoxins*, Springer, Berlin.

Weidenbörner M (2007), *Mycotoxins in Feedstuffs*, Springer, Berlin.

Wogan GN (1965), *Mycotoxins in Foodstuff*, MIT Press, Boston.

Wyllie TD and Morehouse LG (1978), *Mycotoxic Fungi, Mycotoxins, Mycotoxicoses. An Encyclopedic Handbook'*, Marcel Dekker (3 vol.), New York.

11

Molecular identification of mycotoxigenic fungi in food and feed

F. Munaut and F. Van Hove, Université catholique de Louvain, Belgium and A. Moretti, National Research Council (CNR), Italy

Abstract: Classical culture-based methods for detection and identification of the main toxigenic fungi occurring on food and feedstuffs (belonging to *Aspergillus*, *Fusarium* and *Penicillium* genera, as well as to *Alternaria*, *Claviceps*, *Monascus*, *Phoma*, *Phomopsis*, *Pithomyces*, *Stachybotrys* genera) present numerous limitations, such as their time-consuming and labour-intensive aspects. Therefore, most detection procedures are now DNA-based and, by using polymerase chain reaction (PCR) based-methods, range from protocols allowing identification of a single species, or of multiple species belonging to the same genus or to mixed populations of different genera. In the present context, the PCR approach has been considered under two main complementary approaches: by targeting conserved functional genes or regions of taxonomical interest, or by focusing on the mycotoxigenic genes. The use of real-time PCR technology, which monitors the DNA (or RNA) amplification products in real time and allows fast detection and identification of the fungal content of food and feedstuffs in a single assay, will be described. Moreover, although less practicable for quantitative purposes, some applications that combine different technologies from 'basic' PCR, such as PCR-restriction fragment length polymorphism and PCR and enzyme immunoassay, denaturing gradient gel electrophoresis PCR-based, PCR and single strand conformational polymorphism, microsatellite length polymorphism, loop-mediated-isothermal amplification assay, will also be shown for their applicability to toxigenic fungi detection. Finally, since novel molecular technologies for sequencing the whole genome provide a mine of important information for either detection or characterization of fungal species, their benefit for qualitative and quantitative molecular technologies will be illustrated.

Key words: *Aspergillus*, *Fusarium*, genome, mycotoxin, *Penicillium*, PCR, real-time PCR methods.

11.1   Introduction

Classical culture-based methods for detection and identification of fungal toxin producing fungi present numerous limitations, such as their time-consuming and labour-intensive aspects, their environment-dependent character, the difficulties of standardization and poor reliability of results. Therefore, culture-independent techniques are increasingly used in feed and food microbiology, as has been the case for a long time in medical applications. Most detection procedures are now DNA-based and range from protocols allowing identification of a single species, or of multiple species belonging to the same genera or to mixed populations of different species.

Polymerase chain reaction (PCR)-based methods have been extensively applied since, for two decades for the most widely spread mycotoxigenic fungal contaminants belonging to the genera *Aspergillus*, *Fusarium* and *Penicillium*, as well as for *Alternaria*, *Claviceps*, *Monascus*, *Phoma*, *Phomopsis*, *Pithomyces*, *Stachybotrys* genera (Frisvad *et al.*, 2006). In the present context, the PCR strategy has been considered under two main complementary approaches: by targeting conserved functional genes or regions of taxonomical interest or by focusing on the mycotoxigenic genes.

Real-time PCR technology, which monitors both DNA and RNA amplification products in real time, is most probably one of the major technological advances in food security. Indeed, it allows the detection, identification and determination of fungal content of food and feedstuffs in a single assay. In addition to the extreme sensitivity and the high specificity of the technique, its rapidity con-stitutes a major advantage for quality control of food and feedstuffs. Furthermore, multiplex real-time PCR reactions allow specific and sensitive detection, identification and quantification of different DNA (or RNA) targets in a single reaction.

Although less practicable for quantitative purposes, some applications combined technologies that are different from 'basic' PCR, such as PCR–restriction fragment length polymorphism (RFLP), PCR and enzyme immunoassay (PCR-EIA), denaturing gradient gel electrophoresis (DGGE) PCR-based, PCR and single strand conformational polymorphism (SSCP), microsatellite length polymorphism (MLP) and loop-mediated isothermal amplification (LAMP) assay.

Nowadays, novel molecular technologies for sequencing the whole genome provide a mine of information which is important, or will exponentially gain importance for either detection or characterization of fungal species. Qualitative and quantitative molecular technologies are already benefiting from these progresses, such as DNA microarray or DNA barcoding. Increasingly or even widely used in medical applications, their use in food microbiology is only at the early stages.

11.2   Polymerase chain reaction (PCR) detection and quantification using conserved genes

Species-specific PCR methodologies developed on the basis of conserved genes originated mainly for taxonomic and phylogenetic purposes. They were first designed to analyse the possible sequence polymorphism between species within the same genus and for subsequent studies of their phylogenetic relationships. Nevertheless, the level of conservation of some of the genes described hereunder was sometimes too high for full resolution within a genus. When discriminating, conserved genes are the ideal template for the development of real-time PCR primers and probes, allowing simultaneously the detection, identification and quantification of the target fungal species. Hundreds of applications are reported for phytopathological diagnostic assays, as well as for fungal contaminants detection and quantification in food processes, either for raw materials or end-products analyses.

Although the sensitivity of the method proposed is crucial, there are unfortunately no standard rules for reporting it. Therefore, some authors refer to the minimum quantity of DNA, to the minimum number of infected grains required for detection, to the minimum number of conidia or more recently to the number of haploid genomes detected.

11.2.1 Conserved genes

Ribosomal gene family
The ribosomal genes are represented by the highly conserved 18S, 5.8S and 28S units, on which universal primers are easily developed. They are separated from each other by two internal transcribed spacers regions (ITS1 and 2) and repeated modules are linked by the intergenic spacer (IGS). Owing to their high variability, these last three sequences were thoroughly exploited for discrimination between taxa (Richard *et al.*, 2008). Furthermore, their multi-copy feature is an indubitable advantage in detection and quantification of very small quantities of DNA, compared to single gene copy. Indeed, the sensitivity can be more than 100 times more important than for assays using single gene copy (Edwards *et al.*, 2002).

Calmodulin
Calmodulin (CaM, for CALcium MODULated proteIN) is a calcium-binding protein expressed in all eukaryotic cells where it participates in signalling pathways that regulate many crucial processes such as growth, proliferation and movement. Belonging to the E–F-hand family of Ca^{2+}-sensors, it is most probably the most intensively studied. It is relatively small and evolutionarily highly conserved (Chin and Means, 2000), what makes its coding gene a good candidate for taxonomic and phylogenetic studies, and then for subsequent PCR specific assays.

β-tubulin
Microtubules, the major cytoskeletal elements found in all eukaryotic cells, are composed of equimolar amounts of two 55 kDa subunits, α- and β-tubulin, and play key roles in several vital cellular mechanisms, such as chromosome segregation in mitosis, intracellular transport, ciliary and flagellar bending, and structural support of the cytoskeleton. Nevertheless, the α- and β-tubulin coding genes

present some evolutionary sequence differences between genera or species (Raff, 1984), differences that were also pointed out for taxonomical purposes and further molecular detection protocols.

Elongation factor

Translation elongation factors are responsible for two main processes during protein synthesis on the ribosome (Andersen and Nyborg, 2001; Andersen *et al.*, 2003; Nilsson and Nissen, 2005). TEF1-α (or EF-Tu) is responsible for the selection and binding of the cognate aminoacyl-tRNA to the A-site (acceptor site) of the ribosome. TEF2 (or EF-G) is responsible for the translocation of the peptidyl-tRNA from the A-site to the P-site (peptidyl-tRNA site) of the ribosome, thereby freeing the A-site for the next aminoacyl-tRNA to bind. Elongation factors are responsible for achieving accuracy of translation and both TEF1-α and TEF2 are remarkably conserved throughout evolution, which makes their coding regions interesting in detection.

11.2.2 *Fusarium* spp.

In addition to being among the most problematic mycotoxigenic genera, numerous *Fusarium* spp. also present a high level of phytopathogenicity. Therefore, the first published PCR primers were often designed on conserved or functional genes instead of on mycotoxigenic genes.

rDNA primers

Within the genus, ITS regions do not allow a full taxonomic resolution and are only usable for discrimination of several similar species. These primers were often developed for practical uses, for example to check if any pathogenic *Fusarium* species were present simultaneously to other genera on a plant host.

A lot of ITS-based primers were developed by authors, then used or modified by others, and finally incorporated into real-time PCR protocols. In addition to the development of sequence characterized amplified region-random amplified polymorphic DNA (SCAR-RAPD)-based primers, Schilling *et al.* (1996) designed an ITS-based primer pair specific to *F. avenaceum*. On the base of primers designed into the ITS regions of five important pathogenic Fusaria (*F. avenaceum, F. culmorum, F. equiseti, F. oxysporum* and *F. sambucinum*), Mishra *et al.* (2003) conjugated the upstream primers to various fluorescent dyes that allowed a rapid visualization of the PCR-products directly into the test tubes, without the need for gel electrophoresis. In a large study that aimed to characterize the major species causing head blight in Japan, Chung *et al.* (2008) developed ITS specific primers for *F. graminearum* and *F. culmorum* based only on two fixed nucleotides differences. Still for detection and differentiation purposes, Kulik *et al.* (2004) designed a forward ITS2-based primer that was used with a reverser primer previously published by Hue *et al.* (1999). The primer set amplified specifically *F. sporotrichioides* and was successfully tested on infected wheat samples. The primers will be used later in a multiplex PCR (see hereunder).

Numerous IGS-based primers sets are available for PCR detection of myco-toxigenic *Fusarium* species. Jurado *et al.* (2005) designed a set of specific primers based on the IGS sequence, for detection of the most important trichothecene producers of wheat: *F. culmorum, F. equiseti, F. poae, F. sporotrichioides* and *F. graminearum*. Primers were tested successfully on wheat samples. In 2006, Jurado *et al.* designed an additional primer set for *F. proliferatum* (sensitivity >5.5 pg of total DNA) and included it in an improved protocol combining a *FUM1*-PCR based step (sensitivity >87.5 pg) (Jurado *et al.*, 2006). Tested against 21 *Fusarium* species, the primers designed by Kulik (2008b) amplified the targeted *F. tricinctum*, but also *F. acuminatum* and *F. nurragi*, hampering further quantitative detection; a taxonomic issue was however pointed out. Patiño *et al.* (2004) designed two pairs of IGS-based primers. The first one was *F. verticillioides* species-specific, the second one amplified the *F. verticillioides* strains that were fumonisin producers, but not the *F. verticillioides* strains isolated from banana. These latter did not produce fumonisins and were formally described later on as the novel species *F. musae* (Van Hove *et al.*, in press).

Calmodulin

For a phytopathological purpose, Mulé *et al.* (2004a,b), became interested in developing PCR specific-primer for *F. oxysporum* and *F. proliferatum* strains pathogenic on asparagus. Both species were successfully detected from plant material, but not quantified. Mulé *et al.* (2004a) also designed calmodulin specific primers for *F. oxysporum, F. proliferatum* and *F. verticillioides* for detection and identification in maize kernels. The sensitivity of the primers was 12.5 pg of total genomic DNA in pure cultures and was qualitatively validated on infected kernels.

β-tubulin

Reischer *et al.* (2004) proposed a quantification protocol of *F. graminearum* in plants, by using primers and TaqMan probes targeting the β-tubulin gene. With a sensitivity down to five gene copies, the method proved to be a valuable tool for routine analysis or monitoring of the pathogen.

TEF1-α

Up to now, the TEF1-α seem to be the most discriminating gene available for the *Fusarium* spp. In 2004 Geiser *et al.* (2004) provided to the scientific community with a well-documented TEF1-α database (FUSARIUM-ID), containing sequences obtained from unequivocally identified species. At present, sequences data are available for 77 species and seven species complexes.

Nicolaisen *et al.* (2009) put considerable effort into designing species-specific primers on the TEF1-α sequence of nine *Fusarium* species: *F. avenaceum, F. equiseti, F. graminearum, F. langsethiae, F. poae, F. proliferatum, F. sporo-trichioides, F. subglutinans, F. tricinctum* and *F. verticillioides*. Only *F. culmorum* and *F. cerealis* could not be distinguished. Quantitative real-time PCRs were conducted using SYBR Green and validated on artificially and naturally infected grain samples. Simultaneously to the formal description of *F. pseudograminearum* as a new species, Aoki and O'Donnell (1999) provided a specific primer pair for it, encouraging the scientific community to use it for PCR diagnostic assays.

11.2.3 *Aspergillus* spp.

rDNA
A PCR detection assay was set up by Patiño *et al.* (2005) for early detection of *A. carbonarius* and *A. ochraceus*, two important OTA contaminants of grapes. The sensitivity calculated for pure cultures ranged between 1 and 10 ng. González-Salgado *et al.* (2009) set up an assay for quantification of *A. carbonarius* and compared the TaqMan and the SYBR Green results. Both methods reached a sensitivity of 0.6 genome equivalent. Gil-Serna *et al.* (2009a) proposed a quantitative PCR method based on the ITS1 region that allowed a SYBR Green quantification of *A. ochraceus* and *A. westerdijkiae* simultaneously, with the lowest detection limit achieved at 2.5 pg per reaction. The authors also proposed a modified protocol (still ITS1 based) that discriminated between *A. ochraceus* and *A. westerdijkiae*, as well as *A. steynii*, the third most important ochratoxigenic *Aspergillus* species of this section *Circumdati* (Gil-Serna *et al.*, 2009b). Although not really linked to food microbiology but of great interest when considering the tool that was developed, the study of Luo *et al.* (2009) proposed an efficient quantitative method for detection of *A. flavus* and *A. parasiticus* in soil from almond orchards. Two specific primer pairs were designed on the ITS regions and quantification was performed using tSYBR Green technology. Taking into account the DNA extraction yield efficiency, the method can detect 0 to 23 conidia per gram of soil.

β-tubulin
Morello *et al.* (2007) proposed an efficient real-time PCR assay for detection and quantification of *A. westerdijkiae* in coffee beans. Quantification by the SYBR Green technology allowed sensitivity between 1 and 10 haploid genomes.

Calmodulin
Perrone *et al.* (2004) focused on the calmodulin gene for design of specific primers for *A. carbonarius* and *A. japonicus*, two species of the section *Nigri*. From the calmodulin sequences obtained by Perrone *et al.* (2004), Mulè *et al.* (2006) re-designed specific primers and TaqMan probes for developing a quantitative assay providing a sensitivity of approximately 5×10^{-4} ng of total DNA. Abdin *et al.* (2010) and Niessen *et al.* (2005) have also reviewed some interesting medical applications for different *Aspergillus* species.

11.2.4 *Penicillium* spp.
In order to assess the growth dynamics of two species, *P. roqueforti* and *P. camemberti*, during the ripening of cheese, Le Dréan *et al.* (2010) quantified their

growth by real-time PCR using ITS and β-tubulin genes, respectively. The method allowed a sensitivity of 0.25 μg and 4 μg mycelium per gram of curd, respectively.

11.3   Polymerase chain reaction detection and quantification using anonymous DNA sequences

As reviewed in Sartori *et al.* (2010), the target specific sequence obtained from genomic DNA may or may not contain functional conserved genes. Those so-called anonymous DNA sequences (Carter and Vetrie, 2004) can be generated by the random amplified polymorphic DNA (RAPD) (Williams *et al.*, 1990) or the amplified fragment length polymorphism (AFLP) methodologies (Vos *et al.*, 1995). Although used for years and nowadays considered as a pioneer technique in molecular biology, and despite inherent problems of reproducibility, RAPD has found a new youth as source of species-specific DNA fragments that are useful for detection, identification and further quantification of fungal contaminants. Numerous RAPD or AFLP markers were converted into PCR markers and take part in robust detection, identification and quantification protocols.

11.3.1 *Fusarium* spp.
Wilson *et al.* (2004) designed RAPD-based primers specific to *F. langsethiae* and *F. sporotrichioides*, with a detection sensitivity as low as 100 fg of total DNA. Specific RAPD primers are also available for *F. subglutinans* (Möller *et al.*, 1999) and for *F. verticillioides* (Möller *et al.*, 1999; Murillo *et al.*, 1998). For the protocol developed by Möller *et al.* (1999), 5–10 pg of total fungal DNA was sufficient. Several RAPD-based primers which were developed some years ago were used afterwards as templates for development of real-time PCR protocols. For example, several authors developed specific RAPD-based PCR primers for *F. culmorum* (Nicholson *et al.*, 1998; Schilling *et al.*, 1996), *F. graminearum* (Nicholson *et al.*, 1998; Schilling *et al.*, 1996) and *F. poae* (Parry and Nicholson, 1996). From these efficient primers, Waalwijk *et al.* (2004b) designed TaqMan probes that were usable for real-time PCR. These authors also proposed primers and probes for *F. avenaceum*. The detection limits ranged from 9000 pg down to 0.09 pg, this last value corresponding approximately to five genome equivalents of *F. graminearum*.

11.3.2 *Aspergillus* spp.
Schmidt *et al.* (2004a,b, 2003) designed AFLP-based primers for a detection and quantification of *A. ochraceus* in green coffee. In pure cultures, the sensitivity was evaluated at 4–7 pg of total DNA.

11.3.3 Other mycotoxigenic fungal species
Ding *et al.* (2008) designed a pair of sequences characterized amplified region

(SCAR) primers based on RAPD markers for rapid and accurate identification of *Monascus* sp. Later on, Shinzato *et al.* (2009) distinguished *Monascus* rubber, *M. pilosus*, *M. purpureus* and *M. kaoliang* on the basis of RAPD profiles. The authors encouraged the construction of a fingerprint database that would be useful for quality control of food products.

11.4 Polymerase chain reaction detection and quantification using mycotoxin biosynthesis pathway genes

Correct as well as fast identification of fungal contaminants is essential for the food and feed industries and is of course of primary importance in enabling detection of mycotoxigenic fungi. In this respect, soon after the discovery of the fungal toxin/ biosynthetic genes, several studies were carried out to develop sensitive, robust and specific detection methods by means of PCR amplification of one or more genes coding for key enzymes and the regulatory factor involved in mycotoxin biosynthesis (Paterson, 2006a). This part of the chapter discusses the results obtained on some of the most important food and feed mycotoxigenic contaminants, that is producers of aflatoxins/sterigmatocystin, trichothecenes, ochratoxin/ citrinin, fumonisins, and other mycotoxins.

11.4.1 Aflatoxin and sterigmatocystin genes

The genetic deciphering of the aflatoxin and sterigmatocystin biosynthetic pathways was published in the early 1990s (Yu *et al.*, 1993). The aflatoxin biosynthesis in *Aspergillus* is based on a complex pathway involving at least 25 structural and two regulatory genes (Yu *et al.*, 2004). The sterigmatocystin biosynthesis pathway shares homologous genes except for the final step which converts sterigmatocystin to aflatoxin (Brown *et al.*, 1996). Using the gene sequences, early studies showed that the use of PCR primers designed on *aflD* (*nor*-1), *aflM* (*ver*-1), *aflP* (*omt*-1, *omtA*) and/or *aflR* (*afl*-2, *apa*-2) genes of the aflatoxin biosynthesis pathway was efficient in distinguishing aflatoxin producer *A. flavus* and *A. parasiticus* from other *Aspergillus*, *Fusarium* and *Penicillium* contaminants of stored or artificially inoculated grains (Chen *et al.*, 2002; Färber *et al.*, 1997; Geisen, 1996; Shapira *et al.*, 1996). Using primers designed on *aflD* (*nor*-1), *aflM* (*ver*-1) and *aflP* (*omt*-1, *omtA*) genes, Geisen developed a triplex PCR that gave a triplet banding pattern with aflatoxin producing strains of *A. flavus*, *A. parasiticus* and also with sterigmatocystin-producing strains of *A. versicolor* (Geisen, 1996). Similarly, in a quadruplex-PCR with four sets of primers for *aflD* (*nor*-1), *aflM* (*ver*-1), *aflP* (*omt*-1, *omtA*) and *aflR* genes, Criseo and collaborators (2001) showed that aflatoxigenic strains gave a quadruplet pattern. In both studies, non-aflatoxigenic strains gave varying results with one, two, three and/or four banding patterns. Because the banding pattern in non-aflatoxigenic strains resulted in non-differentiation between these and aflatoxigenic strains, it was concluded that further studies were needed to develop a technique that allows complete differentiation between aflatoxin-producing and non-producing strains of the *A. flavus* group (Criseo *et al.*, 2001; Geisen, 1996).

A real-time PCR system has been developed to detect and quantify the copy number of *aflD* (*nor*-1) gene from *A. flavus* and *A. parasiticus* in various foods (Mayer *et al.*, 2003). A multiplex PCR was successfully applied for the detection of potential aflatoxigenic fungal species using primers designed on three aflatoxin biosynthesis genes, *aflI* (*avfA*), *aflM* (*ver*-1) and *aflP* (*omt*-1, *omtA*), and on the ITS region as positive PCR control (Yang *et al.*, 2004). The four DNA fragments were amplified in all aflatoxigenic *A. flavus* and *A. parasiticus* strains. However, this study also revealed that none of the non-aflatoxigenic *A. flavus* and *A. oryzae* strains possessed all three aflatoxin gene fragments, indicating that different types of mutations have inactivated the aflatoxin biosynthetic pathway (Yang *et al.*, 2004).

The first detection system based on reverse transcription PCR (RT-PCR) has been developed with primers specific for *aflR* and *aflQ* (*ord1*) to monitor aflatoxin gene expression in *A. parasiticus* and with primers specific for β-tubulin as internal control (Sweeney *et al.*, 2000). Testing a RT-PCR with specific primers designed on nine aflatoxin structural genes (*aflD*, *aflG*, *aflH*, *aflI*, *aflK*, *aflM*, *aflO*, *aflP*, *aflQ*), two regulatory genes (*aflR*, *aflS*) and the β-tubulin gene as positive PCR and expression control, Scherm and collaborators have shown that the expression of *aflD*, *aflO* and *aflP* was consistently correlated with the ability and inability of *A. flavus* and *A. parasiticus* strains to produce aflatoxins in yeast extract sucrose (YES) and yeast extract peptone (YEP) media, respectively, while the other aflatoxin analysed genes of all strains varied in relation to the aflatoxin-producing ability and the growth conditions (Scherm *et al.*, 2005). Data from Degola and collaborators obtained with a multiplex RT-PCR including a set of five primers for *aflR*, *aflS*, *aflD*, *aflO* and *aflQ*, showed that the sampling time after inoculum was a crucial factor because gene expression may vary according to physiological and environmental conditions (Degola *et al.*, 2007). More recent results on expression of four aflatoxin genes (*aflD*, *aflP*, *aflQ*, *aflS*) and β-tubulin, as internal amplification control, indicated a perfect correlation between gene expression and aflatoxin production for 27 *A. flavus* strains isolated from corn grains or flour. However, five strains transcribed the four aflatoxin genes but apparently did not release aflatoxin in the medium after five days of growth on coconut agar medium (CAM) but were finally aflatoxin positive after ten days in the same culture conditions (Degola *et al.*, 2009). Since these slow aflatoxin-producing strains may display a different behaviour when grown in field conditions, these results raise concerns about the possible underestimation of the aflatoxin risk under environmental conditions (Degola *et al.*, 2009).

11.4.2 Trichothecene genes

Trichothecenes are a group of numerous metabolites sharing common structural characteristics and at least the initial steps of their biosynthetic pathway. They are produced by several fungal species belonging to different genera, that is *Cryptomela*,

Fusarium, Myrothecium, Stachybotrys, Trichoderma/Hypocrea, Trichothecium and *Verticimonosporium* (Niessen, 2007). Owing to the frequent occurrence of trichothecene producing *Fusarium* contaminant in important cereal crops and the large variety and high toxicity of trichothecene derivatives they produce, most studies have been conducted to develop specific and sensitive PCR methods rapidly to discriminate *Fusarium* species and/or trichothecene chemotypes. The analysis of the trichothecene biosynthesis (*TRI*) genes in *F. sporotrichioides* and *F. graminearum* revealed that 12 *TRI* genes form a core *TRI* cluster in which genes are in the same order and orientation while two genes (*TRI1* and *TRI16*) are at another locus and one gene (*TRI101*) is at a third locus (Brown *et al.*, 2004). In *F. equiseti*, the *TRI* cluster differs from *F. graminearum* and *F. sporotrichioides* clusters as (i) *TRI1* and *TRI101* are part of the cluster, (ii) *TRI12* is lacking and (iii) the order and orientation of *TRI3*, *TRI7* and *TRI8* are different (Proctor *et al.*, 2009). Sequences from *TRI4*, *TRI5* and *TRI6* genes have been exploited in the development of generic PCR assays for the detection of trichothecene producers, as these genes are present in all trichothecene producers (Mulè *et al.*, 2005; Niessen *et al.*, 2004).

More recently, to improve the determination of the trichothecene producing *Fusarium* content of a contaminated plant sample, the codetermination of a plant gene together with the *TRI5* gene was used to compensate for unequal DNA-extraction efficiencies (Brunner *et al.*, 2009). It has been demonstrated that differences in the trichothecene production pattern among *Fusarium* species and/or among strains of the same *Fusarium* species are due to variations in the gene clusters. On the basis of this information, several studies have demonstrated the possibility of discriminating between different trichothecene chemotypes. For example, *TRI7* and *TRI13* PCR assays have been developed to determine the nivalenol (NIV) and deoxynivalenol (DON) chemotype of *F. cerealis*, *F. culmorum* and *F. graminearum* (Chandler *et al.*, 2003). Indeed, *TRI7* and *TRI13* genes are responsible for conversion of DON to NIV and acetylation of NIV to 4-acetyl-nivalenol (4-ANIV), respectively, and they are both disrupted in all DON chemotype isolates of *F. graminearum* studied (Brown *et al.*, 2002; Lee *et al.*, 2001).

Using primer pairs designed on *TRI3*, *TRI5* and *TRI7* genes, a multiplex PCR was successfully used to identify the NIV, 3-acetyl-DON and 15-acetyl-DON chemo-types among *F. cerealis*, *F. culmorum* and *F. graminearum* contaminating wheat kernels (Quarta *et al.*, 2006). Recently, another multiplex PCR has been developed for accurate simultaneous identification of NIV, 3-acetyl-DON and 15-acetyl-DON chemotypes of *F. asiaticum* and *F. graminearum sensu stricto* by means of three pairs of primers designed for the *TRI6* gene and one pair for the *TRI3* gene (Suzuki *et al.*, 2010). Besides PCR methods developed for *Fusarium*, PCR primers have been designed on a putative *TRI5* gene from *Trichoderma harzianum* and allowed the detection of this *TRI5* gene in *T. harzianum* and homologous *TRI5* in *T. atroviride*, *T. viride*, *T. koningii* and *T. parceramosum* (Gallo *et al.*, 2004).

11.4.3 Ochratoxin and citrinin genes

Although ochratoxin A (OTA) is a very important mycotoxin, its biosynthesis pathway has not yet been fully elucidated. A putative OTA biosynthetic gene cluster including complete gene coding for an alkaline serine protease (ASP), an OTA non-ribosomal peptide synthetase (NRPS) and one partial gene coding for an OTA polyketide synthase (PKS) have been identified in *P. nordicum* (Geisen *et al.*, 2006; Karolewiez and Geisen, 2005). In addition a PKS gene which is involved in OTA biosynthesis in *A. ochraceus* is expressed together with two putative P450 type monooxygenase genes, suggesting their involvement together with the PKS in OTA biosynthesis (O'Callaghan *et al.*, 2006).

Interestingly, the first description of a gene implicated in OTA biosynthesis was performed with a quantitative reverse transcription real-time PCR method generating a 500 bp fragment of the OTA PKS of *P. nordicum* (Geisen *et al.*, 2004). Moreover, the real-time PCR method with the same primers allows specific detection of and monitoring of OTA production by *P. nordicum*, while giving negative results with other *Penicillium* and *Aspergillus* species, including OTA producing *P. verrucosum*, *A. carbonarius*, *A. niger* and *A. ochraceus* (Geisen *et al.*, 2004). Dao and co-workers (2005) have designed two sets of specific primers on the DNA sequence of the PKS gene from *A. ochraceus* that amplified a unique band from either OTA (*A. carbonarius*, *A. melleus*, *A. ochraceus*, *A. sulfureus* and *P. verrucosum*) or citrinin (*P. citrinum*, and *Monascus ruber*) producers or only from *A. ochraceus*, respectively (Dao *et al.*, 2005).

To study the natural occurrence of *P. nordicum* in cured meat and ripening rooms, a PCR method based upon the OTA PKS and the OTA NPS genes has been developed (Bogs *et al.*, 2006). The NPS specific primer pairs allowed amplification of the two important OTA producing *Penicillium* species, *P. nordicum* and *P. verrucosum*, while the PKS primer pair was specific only for *P. nordicum*. The main source of OTA contamination in grapes and wine is *A. carbonarius*. Therefore, primer pairs specific for the acyl transferase domain (Atoui *et al.*, 2007) or the β-ketosynthase domain (Selma *et al.*, 2008) of the *A. carbonarius* OTA PKS gene have been designed and used in PCR and quantitative real-time PCR for direct detection and quantification in grape samples. By comparing PKS acyl transferase (AT) domain sequences obtained from 11 *Aspergillus* and four *Penicillium* OTA-producing species using degenerate primers, one PCR primer pair has been designed specifically to amplify sequences of PKS AT domains from 13 *Aspergillus* species belonging to the section *Nigri* (Martínez-Culebras *et al.*, 2009).

11.4.4 Fumonisin genes

Fumonisins are polyketide derivatives produced by at least 11 *Fusarium* species, that is *F. anthophilum*, *F. bulbicola*, *F. fujikuroi*, *F. globosum*, *F. nygamai*, *F. oxysporum* (only a few singular strains), *F. phyllophilum*, *F. proliferatum*, *F. ramigenum*, *F. verticillioides* (Desjardins, 2006; Proctor *et al.*, 2004) and at least one *Aspergillus* species, *A. niger* (Frisvad *et al.*, 2007; Mogensen *et al.*, 2010; Noonim *et al.*, 2009). The biosynthetic genetic pathway of fumonisins has been quite extensively elucidated in *F. verticillioides* and it was revealed that 17 fumonisin biosynthetic (*FUM*) genes are clustered and coregulated (Brown *et al.*,

2007; Proctor *et al.*, 1999, 2003). A *FUM* gene cluster (FGC) has also been described in *A. niger*, *F. oxysporum* and *F. proliferatum* (Baker, 2006; Pel *et al.*, 2007; Proctor *et al.*, 2008; Waalwijk *et al.*, 2004a). Primer pairs designed on the *FUM1* PKS gene, responsible for the initial step of fumonisin biosynthesis, and *FUM6* and *FUM8* genes, were used to detect *Fusarium* fumonisin-producing strains by PCR. The lack of *FUM1*, *FUM6* and *FUM8* PCR products when testing fumonisin non-producing strains of *F. verticillioides* suggests they have lost the *FUM* cluster (González-Jaén *et al.*, 2004). These primer pairs have also been used in reverse transcription PCR assays. For the PCR detection of fumonisin – producing strains of different *Fusarium* species isolated from maize samples in Mexico, one primer pair was designed on the exon 3 of the *FUM1* gene (Sánchez-Rangel *et al.*, 2005). As expected, the *FUM1* gene was positive in *F. verticillioides* strains only, given that the other isolated *Fusarium* species were non-fumonisin producers, that is *F. avenaceum*, *F. chlamydosporum*, *F. oxysporum*, *F. semitectum*, *F. sporotrichioides* and *F. subglutinans*. Interestingly, among the 54 *F. verticillioides* strains recovered, 35 were fumonisin producers and *FUM1* positive, but 19 were fumonisin non-producers and *FUM1* negative. Either these strains were misidentified or they have a mutated *FUM1* gene.

In another study, López-Errasquín and co-workers effectively developed primers for a specific real-time RT-PCR assay to detect and quantify the expression level of *FUM1* and *FUM19* (López-Errasquín *et al.*, 2007). With this method, they have demonstrated that a good correspondence exists between the *FUM1* and *FUM19* expression levels and the fumonisin production in *F. verticillioides* cultures. Moreover, they have designed these primers on regions of *FUM1* and *FUM19* that differ in *F. verticillioides* and *F. proliferatum* to permit their discrimination. These primers amplified the expected specific product in two fumonisin-producing *Fusarium* species, *F. verticillioides* and *F. nygamai*, whereas no product was observed in other fumonisin producers, *F. fujikuroi* and *F. proliferatum*, and non-fumonisin producers, *i.e. F. circinatum*, *F. graminearum F. oxysporum*, *F. sacchari*, *F. subglutinans*, *F. thapsinum* and *F. tricinctum*.

Testing four different primer pairs designed on the *FUM1* gene, Baird and co-workers (2008) found one primer pair that consistently amplified the expected DNA fragment size from all 24 tested *F. verticillioides* strains except one, and from all 12 *F. proliferatum* strains except two, while other 10 tested *Fusarium* strains and 13 non-*Fusarium* strains were negative. However, PCR assay with this primer pair detected all the 22 *F. verticillioides* strains that were producing fumonisin in culture but only four of the five fumonisin producing *F. proliferatum* strains.

Another real-time RT-PCR assay based on the *FUM1* gene has been developed for *F. proliferatum*, which allowed discrimination from *F. verticillioides* (Jurado *et al.*, 2010). Moreover, *FUM1* gene expression detected with this assay was significantly correlated (0.77) with fumonisin production. To determine simultaneously the identity and the fumonisin biosynthetic ability of *Fusarium* contaminants of asparagus, a multiplex PCR has been developed based on four primer pairs designed on the internal transcribed spacer sequence (ITS) and *FUM1* and *FUM8* genes (Wang *et al.*, 2010).

To distinguish the fumonisin-producing *F. verticillioides* species from its closely related species, *F. musae*, which is morphologically very difficult to discriminate but has lost the *FGC* and the capacity to produce fumonisin, PCR assays have been developed (1) based on a primer pair with one forward primer (fvh59) designed in the first FGC gene (*FUM21*) and one reverse primer (fvh55) in the last FGC gene (*FUM19*), respectively, to detect *F. musae* strains, and (2) based on primer pairs designed on *FUM3*, *FUM6*, *FUM7* and *FUM8*, to detect *F. verticillioides* strains (Glenn *et al.*, 2008; Van Hove *et al.*, 2011; Van Hove *et al.*, 2008). Indeed, on one hand, PCR with primers fvh59 and fvh55 generated a 1178-bp amplicon from *F. musae* strains as they have lost a major internal part of the FGC bringing the priming sites closer, while this primer pair did not amplify a corresponding fragment in *F. verticillioides* strains due to the presence of the full FGC (44.7 kb) between the priming sites. On the other hand, PCR with *FUM3*, *FUM6*, *FUM7* or *FUM8* primers generated amplicons only from *F. verticillioides* and not from *F. musae* because *F. musae* strains have lost the FGC.

11.4.5 Example of other mycotoxin genes

Besides detection and quantification of mycotoxigenic fungi via different PCR methods designed on well-documented biosynthesis pathways, some publications deal with the use of newly or less described biosynthesis pathways. Hereunder we give three additional examples linked to patulin, AM-toxin and zearalenone biosynthetic genes.

Primers and PCR assays have been designed on the two known genes of the patulin biosynthesis pathway, that is the 6-methylsalicylic acid synthase (6msas) and the isoepoxydon dehydrogenase (*idh*) genes, which are associated with patulin production (Paterson, 2006b; Paterson *et al.*, 2003; Puel *et al.*, 2007). Using these methods, it has been possible to obtain profiles of DNA to identify and classify fungi and to obtain insight into patulin production.

After the *AMT* gene involved in the biosynthesis of AM-toxin has been cloned and characterized (Johnson *et al.*, 2000), specific *AMT* gene primers were designed to develop PCR and real-time PCR methods to detect and quantify AM-toxin-producing isolates of *Alternaria alternata* apple pathotype (Andersen *et al.*, 2006; Johnson *et al.*, 2000).

A recent paper reports the first method to detect and quantify the zearalenone producing *Fusarium* species by real-time PCR assay based on the zearalenone synthase gene *PKS4* (Meng *et al.*, 2010). This method is rapid, sensitive and specific for detection of zearalenone-producing *F. crookwellense*, *F. culmorum* and *F. graminearum* in maize flour.

11.5 Multistep strategies

Sreenivasa *et al.* (2008a) adopted a mycotoxin-risk oriented approach for a two-step detection of *Fusarium* species producing fumonisin in sorghum grains. The authors amplified their ITS regions for assessment of their membership to the *Fusarium* genus. Then, they tested a primer set amplifying the *FUM1* gene for evaluation of their fumonisin potential.

In another study, Sreenivasa *et al.* (2008b) developed a three-step strategy to control sorghum grains as follows: a first genus-specific primer pair detected the *Fusarium* spp. strains among the whole set of isolates; a second ITS-based primer pair identified the *F. verticillioides* among the other *Fusarium* spp.; the third *FUM1* gene primer pair amplified all the potentially fumonisin producer strains among the *Fusarium* strains detected in the first step. The PCR reactions were performed as single reaction, not in a multiplex.

Jurado *et al.* (2006) proposed a flowchart PCR strategy to detect the presence of the most common mycotoxigenic species of *Fusarium* in cereals. The method combined the use of IGS-based primers followed by amplification with trichothecene or fumonisin gene-based primers for detection of potentially trichothecene or fumonisin producers, respectively.

A six-step molecular strategy was recently developed for identification of the *Aspergillus* section *Flavi* species (Godet and Munaut, 2010). The strategy was based mainly on real-time PCR targeting the *Aflt* and *AflR* genes as well as the ITS1-5.8S region, combined with RAPD and SmaI digestion protocols developed previously (Yuan *et al.*, 1995). The resulting decision-making tree allows an accurate identification of nine of the 11 species of section *Flavi*, which is particularly useful in addressing toxigenic problems in the food industry.

11.6 Multiplex detection

11.6.1 Multiplex PCR

Fusarium spp.
Bluhm *et al.* (2002, 2004) combined the use of genus-specific primers based on the ITS1-2 regions with trichothecene and fumonisin group-specific primers designed from *TRI6* and *FUM1* (*FUM5*) genes. This allowed the differentiation of *F. graminearum* from *F. verticillioides* directly from cornmeal. Demeke *et al.* (2005) proposed a multiplex method able to identify the three main species pathogenic on wheat in Canada: *F. culmorum* and *F. graminearum* were amplified using RAPD primers while a *TRI13* gene was used to detect *F. sporotrichioides*. Brandfass and Karlovsky (2006) described a RAPD-primer-based duplex PCR for simultaneous detection of *F. culmorum* and *F. graminearum* in plant material. Yli-Mattila *et al.* (2008) developed specific primer TEF1-α based, as well a TaqMan probes for a multiplex quantitative PCR of *Fusarium* species from cereal grains (*F. poae, F. graminearum, F. langsethiae/F. sporotrichioides*). A duplex PCR detection was developed for simultaneous detection of *F. poae* and *F. sporotrichioides* (Kulik, 2008a). The authors designed a new primer set based on the ITS2 region for *F. poae* and used the primers pairs developed previously for the single detection reaction of *F. sporotrichioides* (Kulik *et al.*, 2004).

In some cases, the multiplex protocol is not conceived for bringing accurate information on the exact species present, but is more mycotoxin-risk oriented, such as the one reported by Sreenivasa *et al.* (2008a) . For example, Bezuidenhout *et al.* (2006) proposed a multiplex detection of different fumonisin-producing species simultaneously (species were not identified individually) in order to prevent any consumption of potentially mycotoxin-contaminated traditional vegetables by African people.

Aspergillus spp.
A multiplex PCR assay was successfully developed by Sartori *et al.* (2006) for detection of *A. carbonarius*, *A. ochraceus* and *A. niger* simultaneously from coffee beans. Therefore, a RAPD primer-pair was designed for *A. niger* and used in combination with RAPD-based primers previously developed by Pelegrinelli-Fungaro *et al.* (2004) for *A. carbonarius* and by Schmidt *et al.* (2003) for *A. ochraceus*. To distinguish the two very closely related species *A. niger* and *A. tubingensis*, Susca *et al.* (2007a) designed species-specific primers that were successfully tested in a duplex assay.

Alternaria spp.
Zur *et al.* (2002) proposed a multiplex reaction for simultaneous detection and quantification of *A. alternata* and *A. solani* in different cereal grains, providing a rapid practical tool for control before grain shipment.

Multispecies multiplex
Suanthie *et al.* (2009) designed genus-specific TaqMan probes from the ITS sequences of the most important mycotoxigenic species of *Fusarium*, *Penicillium* and *Aspergillus*. A multiplex assay (range of detection from 1 pg to 10 ng of DNA) was validated by analysing fungal growth in distiller's grain, an animal feedstock that is a by-product when ethanol is produced from corn. This assay was proposed as an initial step to evaluate the mycotoxigenic potential of various agricultural commodities.

Although developed for indoor environmental control and despite of a lack of specificity that hampers a full taxonomic resolution, the primers and TaqMan probes developed by Haugland *et al.* (2004) may be considered as another mycotoxin-risk oriented methodology. The multiplex assay, targeting the ITS regions of *Aspergillus* spp., *Penicillium* spp. and *Paecylomyces* spp., provided an indubitable practical tool for preventing health problems generated in buildings, and could provide inspiration to food microbiologists.

11.6.2 Combined multiplex methods

The two following studies concern the detection of fungal species related to the medical field. Nevertheless, some of them are also mycotoxigenic and food contaminants (like *A. flavus*), and the technological approaches as well as the tools developed seem of interest and transferable to the food microbiology field.

Padlock probes, generic reverse transcription polymerase chain reaction (RT-PCR) and array

Eriksson *et al.* (2009) present more details of the padlock probe which is a long oligonucleotide designed to have the 5' end meet the 3' end when hybridizing to a target sequence. The hybridized padlock probe becomes circularized via a ligation reaction. Internal parts of the probe contain two general PCR primer sequences common for all the different padlock probes and a tag sequence unique for each padlock probe. The circularized probes are amplified via rolling circle amplification (RCA) and PCR using a single set of primers and for subsequent capture to specific anti-tag sequences on a microarray (Banér *et al.*, 2007).

In order to speed up the time required for identification of fungal invasive infection in immuno-compromized patients, Eriksson *et al.* improved the technique by using first a SYBR Green real-time PCR for signal amplification and an additive quantification of all circularized padlock probes and second, a suspension array using Luminex™ technology. Thanks to these developments, the authors were able to detect simultaneously ten different pathogenic fungi, from which there are two important mycotoxin-producers: *A. flavus* and *A. fumigatus*.

Multiplex tandem PCR (MT-PCR)

This type of assay incorporates two separate steps for multiplexing and quantification: a short-cycle multiplex amplification followed by 10–12 simultaneous PCRs. MT-PCR is not considered to be a nested PCR assay since the first-step multiplex PCR is not run to completion and serves only as a preamplification process.

Lau *et al.* (2008) proposed targeting the ITS1-2 region, the TEF1-α and the β-tubulin genes to identify 11 fungal human pathogens: seven *Candida* spp., the *Cryptococcus neoformans* complex, *F. solani*, *Fusarium* sp. and *Scedosporium prolificans*.

11.7   Polymerase chain reaction-based methods

Some applications combined PCR-based and other molecular approaches. It is noticeable that important progresses in detection and identification of fungal species originated, or have been obtained, in the medical field. Nevertheless, the practical aspect of very early detection of fungal infection, of a pathogenic group species or at the genus level, often overcomes sharp taxonomical considerations.

Most of these techniques, like PCR-RFLP or DGGE, were initially applied for the qualitative identification of targets, but only the real-time PCR tended to be practicable for quantitative detection methods (Brunner and Mach, 2010).

11.7.1   Polymerase chain reaction–restriction fragment length polymorphisms (PCR–RFLP)

Fusarium spp.

Several papers reported the combination of PCR and RFLP for identification and discrimination of various species, but more often without development of specific primers that could have been useful for subsequent detection and quantification purposes.

For example, an important diversity was demonstrated within *F. proliferatum* (Láday *et al.*, 2004a) and *F. graminearum* (Láday *et al.*, 2004a,b) based on mitochondrial DNA PCR–RFLP products analyses. A similar approach based on the TEF1-α gene allowed the differentiation of eleven species isolated from sugar beet, but again, without a subsequent species-specific primer (Nitschke *et al.*, 2009). Whilst Llorens *et al.* (2006) confirmed the interest of PCR–RFLP using the IGS region as an additional tool for delineation of six *Fusarium* species (*F. culmorum, F. graminearum, F. cerealis, F. poae, F. oxysporum* and *Gibberella fujikuroi* species complex), Konstantinova and Yli-Mattila (2004) focused on the *F. poae, F. langsethiae, F. sporotrichioides* and *F. kuyushuense* and finally developed species-specific primers for *F. poae* and *F. langsethiae/F. sporotrichioides*, at least in pure cultures. In view of discriminating between *F. verticillioides* and *F. proliferatum*, two species morphologically very similar and closely related phylogenetically, Visentin *et al.* (2009) amplified the ITS region, then proceeded to a restriction of the PCR products. The protocol is presented as an additional tool usable for distinction of these two species and not to solve the taxonomic identification within the *G. fujikuroi* complex species.

Aspergillus spp.
A. tubingensis was clearly distinguished from other *Aspergillus* species of the section *Nigri* using the PCR–RFLP technique based on rDNA, but without species-specific primer development (Medina *et al.*, 2005). In the same section, Gonzales-Salgado *et al.* (2005) have discriminated *A. niger* and other *Aspergillus* species, using some species-specific primers designed on the ITS1-5.8S-ITS2 regions. Bau *et al.* (2006) separated the ochratoxin producers from the non-producers using a simple RFLP restriction of the ITS1-5.8S-ITS2, using the universal primers ITS1 and ITS4 (White *et al.*, 1990). Martínez-Culebras and Ramón (2007) developed a more complex protocol with three restriction enzymes, enabling the identification of *A. aculeatus, A. carbonarius, A. niger* and *A. tubingensis*. Zanzotto *et al.* (2006) used PCR–RFLP analysis of the ITS, IGS and β-tubulin genes to distinguish between OTA-producing and non-producing isolates of the *A. niger* aggregate.

Penicillium spp.
A simple identification protocol was set up for *P. aurantiogriseum*, a common and widely spread cereal-borne *Penicillium* species. The methods consisted of a PCR amplification of the ITS region followed by enzymatic restriction with BgII enzyme (Colombo *et al.*, 2003). Garcia *et al.* (2006) proposed a cleaved amplified polymorphic sequence (CAPS) test allowing a rapid identification of *P. expansum* among fungal strains collected on grape berries. The test was based on the sequence of the mitochondrial SSU-rDNA and was validated against 17 other fungal species isolated from grapes.

11.7.2 Real-time reverse transcriptase PCR

Instead of targeting DNA, Bleve *et al.* (2003) developed a high-throughput assay combining reverse transcriptase PCR and real-time PCR (SYBR Green detection) that targeted the actin mRNA of yeasts and moulds contaminating yogurts and pasteurized food products. mRNA detection is considered to be a better indicator of cell viability than DNA-based detection methods. Furthermore, actin is absent in the prokaryotic cell, what is a major advantage when testing dairy products.

11.7.3 Polymerase chain reaction–enzyme immunoassay (PCR–EIA)

Polymerase chain reaction–enzyme immunoassay (PCR–EIA) uses enzyme-bound antibody to detect the PCR amplicons. This method proved to be specific and sensitive (pg DNA) for six medically important *Aspergillus* species: *A. flavus*, *A. fumigatus*, *A. nidulans*, *A. niger*, *A. terreus* and *A. versicolor* (Hinrikson *et al.*, 2005). In this study, biotinylated and digoxigenin-labelled probes targeting the ITS2 rDNA region were hybridized directly onto the ITS2 amplicons produced after PCR amplification. After hybridization, the biotin probe was captured onto streptavidin-coated microtiter plate wells and the digoxigenin probe, hybridized adjacent to the biotin probe, was detected spectrophotometrically (A650 nm) after addition of horseradish peroxidase-conjugated anti-digoxigenin antibodies, H_2O_2, and a colorimetric substrate.

A reliable PCR-enzyme linked immunosorbent assay (ELISA) diagnostic assay was developed to investigate *Fusarium* dry rot in potato stock and in soil (Cullen *et al.*, 2005), since several *Fusarium* species seem to be involved (*F. avenaceum*, *F. culmorum*, *F. coeruleum* and *F. sulphureum*). In addition to the new RAPD-based primers developed for *F. coeruleum* and *F. sulphureum*, the authors used the primers designed previously for *F. avenaceum* (Turner *et al.*, 1998) and *F. culmorum* (Nicholson *et al.*, 1998). TaqMan probes were designed for all four species quantification. The method allowed a quantification of 50–100 fg of genomic DNA.

11.7.4 Polymerase chain reaction–single strand conformational polymorphism (PCR–SSCP)

Single-strand conformational polymorphism (SSCP) is based on electrophoretic detection of conformational changes in single-stranded DNA molecules resulting from base difference. The PCR-based–SSCP assay developed by Susca *et al.* (2007b) provided a rapid, computer-assisted and relatively low-cost tool for the identification of 11 *Aspergillus* species belonging to the *Nigri* section. It consists of an amplification of a species-specific region of the calmodulin gene, with a single pair of fluorescent primers and PCR conditions. Again in the medical field, Rath and Ansorg (2000) developed a fast diagnosis method for five pathogenic *Aspergillus* species, by amplification of the ITS1-5.8S-ITS2 region by SSCP.

11.7.5 Denaturing gradient gel electrophoresis (DGGE)–amplification refractory mutation system (ARMS)

Denaturing gradient gel electrophoresis (DGGE) is one of the most commonly used methods among the culture-independent fingerprinting techniques. Briefly, amplified DNA fragments of the same length but different sequences are separated on a denaturating electrophoresis gel, according to their melting temperature. This technique has been extensively used for characterization and identification in bacterial communities from the natural environment or food matrices (Ercolini, 2004), but very poorly applied for fungal food controls.

Most of the protocols are rDNA based. Mach *et al.* (2004) focused on the β-tubulin gene in proposing a DGGE protocol that detected and discriminated *F. kyushuense, F. robustum, F. langsethiae, F. poae, F. sambucinum, F. sporotrichioides* and *F. tumidum* from different infected cereal crops. In the same paper, the authors proposed an amplification refractory mutation system (ARMS)-PCR that used different primers for different regions of the β-tubulin and generated three distinct patterns allowing the distinction of *F. langsethiae, F. sporotrichioides* and *F. kyushuense/poae*.

11.7.6 PCR-microsatellite

Microsatellites (MST) are composed of simple sequence repeats that are found in all eukaryotic organisms (Li *et al.*, 2002). For the development of MST markers, specific primers are designed on flanking regions of the repeats, identified in genome databases or repeat-enriched clone libraries. Many MST contain a variable number of repeats in different individuals, resulting in length polymorphism of the amplified fragment commonly used in population genetic studies (Tenzer *et al.*, 1999). MST methods are mostly used for typing strains, below the species level. In a few papers, the MST were used at the genus/species level.

For example, Naef *et al.* (2006) developed a competitive microsatellite PCR assay allowing reliable biomass quantification of *T. atroviride* and *F. graminearum* in different samples which contained DNA and PCR inhibitors, such as air-dried maize leaves and field-overwintered maize residues.

11.7.7 Single-nucleotide polymorphism (SNP)

A single-nucleotide polymorphism (SNP) is a DNA sequence variation occurring when a single nucleotide in the genome (or other shared sequence) differs between members of a species or between paired chromosomes in an individual. Insertion or deletion SNPs may shift the translational frame. A SNP may fall within coding or non-coding sequences, or in the intergenic regions between genes. SNPs within a coding sequence will not necessarily change the amino acid sequence of the protein that is produced, owing to degeneracy of the genetic code.

Although SNPs look like infinitesimal differences in a whole genome, it has been proved that they can be very useful for discrimination within species. For example, a SNP in the IGS of *F. oxysporum* allows the development of PCR specific primers for detection, identification and quantification of the *F. oxysporum* f.sp. *vasinfectum* specifically (Zambounis *et al.*, 2007). Kristensen *et al.* (2007a) presented a SNP assay simultaneously to detect 16 trichothecene and moniliformin-producing *Fusarium* species (*F. avenaceum*, *F. cerealis*, *F. culmorum*, *F. equiseti*, *F. flocciferum*, *F. graminearum*, *F. kyushuense*, *F. langsethiae*, *F. lunulosporum*, *F. poae*, *F. graminearum*, *F. sambucinum*, *F. sporotrichioides*, *F. torulosum*, *F. tricinctum* and *F. venetatum*). Primers were derived from the elongation factor TEF1-α, and the SNP assay was validated against cereal samples. The authors proposed a *Fusarium* chip prototype for these 16 species (Kristensen *et al.*, 2007b).

11.7.8. Loop-mediated isothermal amplification (LAMP) assay

Loop-mediated isothermal amplification (LAMP) of DNA is a quick, simple and relatively cheap method for the specific detection of genomic DNA. It consists of using a set of oligonucleotide primers with binding sites hybridizing specifically to different regions of a target gene and a thermophilic DNA polymerase from *Geobacillus stearothermophilus* for DNA amplification (Notomi *et al.*, 2000). Up to now, the method has been applied mainly in the medical and veterinary fields, for various diagnostic assays of viral (Imai *et al.*, 2007; Nagamine *et al.*, 2001; Notomi *et al.*, 2000) and bacterial (Iwamoto *et al.*, 2003; Lucas *et al.*, 2009) infections. Only recently, first applications for fungal organisms were published such as for the pathogenic yeast *Candida* (Inacio *et al.*, 2008), the *Brettanomyces/ Dekkera* yeasts in brewing process (Hayashi *et al.*, 2007), the human pathogens *Paracoccidioides brasiliensis* and *Ochroconis gallopava* (Endo *et al.*, 2004; Ohori *et al.*, 2006; Sano and Itano, 2010), the phytopathogenic *Pythophtora ramorum* (Tomlinson *et al.*, 2007) and several arbuscular mycorrhizal fungi (Gadkar and Rillig, 2008).

Recently, Niessen and Vogel (2010) were the first authors to describe successful LAMP assay for the detection and identification of *F. graminearum*. The assay was based on the gaoA gene (galactose oxidase) of the fungus and the amplification product was indirectly detected *in situ* by using calcein fluorescence as a marker without the electrophoretic analysis. Only *F. graminearum* was recognized on the 132 fungal species tested, at less than 2 pg of purified DNA per reaction within 30 min. The method was validated on wheat and barley bulk seed, providing a promising tool for detection and identification of this important mycotoxin-producing species.

11.8   Novel technologies

Recently, novel molecular technologies have gained importance for either detection or characterization of fungal species. The use of integrated platforms combining detection, identification, typing and quantification facilitates the control of myco-toxin-producing fungi in food commodities.

The considerable efforts and resources allocated in the last few years to sequencing the entire genome of (mycotoxigenic) fungi will help to find species-specific sequences usable in detection, identification and quantification throughout the different methods presented in this paper and future novel ones. Various platforms were recently developed for high-throughput sequencing using different technologies such as the nanopore patented technology (Illumina Solexa 1G Genome Analyzer), pyrosequencing (Roche 454 GS-FLX platform), the use of self-assembling array (Complete Genomics) or single-molecule real-time SMRT sequencing technology (Pacific Bioscience) and the ABI SOLID sequencing system.

11.8.1 DNA microarray

A DNA (or RNA) microarray consists of thousands of spots of DNA (or RNA) oligonucleotides fixed to a small solid surface (chip), each spot containing pmoles of one specific sequence (probe). Labelled targets from the sample hybridize fully or partially to the complementary probes by forming hydrogen bonds. After washing, the non-specific bonding sequences are discarded and a more or less intense fluorescent signal indicates the probe-target sites. Results can be either qualitatively or quantitatively analysed. The probe can be designed from conserved or anonymous genes, but also from genes linked to host specificity and pathogenicity or to mycotoxin pathways. Finally, metagenomic microarrays are also designed that contain DNA fragments from environmental DNA.

Güldener *et al.* (2006) reported the design and validation of the first Affymetrix GeneChip microarray based on the entire genome of *F. graminearum*, aiming to profile fungal gene expression *in vitro* and *in planta*. Nicolaisen *et al.* (2005) developed a qualitative oligonucleotide array for the differentiation of toxigenic and non-toxigenic *Fusarium* isolates. Although based on the ITS2 regions, the array could distinguish type-B trichothecene producers (*F. graminearum*/*F. culmorum* and *F. pseudograminearum*), type-A trichothecene producers (*F. sporotrichioides* and *F. langsethiae*), type-A and -B producers (*F. equiseti* and *F. poae*) and non-trichothecene producers (*F. avenaceum* and *F. tricinctum*). A single array-based method is also available for the detection, identification and (semi) quantification of *Fusarium* species from cereals (Kristensen *et al.*, 2007b). Probes were based on the TEF-1α sequence and 15 species can be quantified in a single run, from which several trichothecene or moniliformin producers are found. The authors demonstrated a limit of quantification below 16 haploid *Fusarium* genomes.

Bufflier *et al.* (2004) set up a low-complexity oligonucleotide microarray (OLISA) based on DNA probes obtained from sequences of the calmodulin gene, in order to detect *A. carbonarius*, *A. japonicus*/*A. aculeatus* and *A. ibericus* isolated from grape. The designed microarray distinguished all *Aspergillus* species, with a detection limit of 3.2 pg of DNA for *A. carbonarius*.

Detection (and identification and quantification) of a mycotoxigenic fungus in food is not always correlated with its mycotoxin production. Indeed, production of secondary metabolites is environment dependent and very variable. Therefore, some authors developed a microarray assay targeting the expression of genes involved in the biosynthetic pathway of several mycotoxins. For example, Schmidt-Heydt and Geisen (2007) developed a microarray carrying oligonucleotides designed from the fumonisin (*Gibberella moniliformis*), the aflatoxin (*A. flavus*), the ochratoxin (*Penicillia*), the trichothecenes (*F. graminearum* and *F. sporotrichioides*) and the patulin (*P. expansum*) biosynthetic pathways. The array was used for gene expression-kinetics studies.

Recently, Lezar and Barros (2010) developed a microarray chip for identification of several potential mycotoxigenic species from the genera *Aspergillus*, *Fusarium* and *Penicillium*. A set of probes were designed based on the ITS and on the TEF-1α regions, while a second set targeted biosynthetic genes of the most important mycotoxin threatening the safety and quality of food products derived from maize and produced by the species studied (aflatoxin, deoxynivalenol, fumonisin, nivalenol, trichothecene). The diagnostic microarray allowed the identification of 32 single pure strains as well as their potential to produce mycotoxigenic fungi. Because the method is DNA-based and not based on the gene expression, it did not predict whether the mycotoxins were produced or not.

11.8.2 DNA barcoding

DNA barcoding is a taxonomic method that uses a short genetic marker in an organism's DNA to identify at the species level. Several loci have been suggested, but the mitochondrial cytochrome oxidase 1 (COX1) gene proposed by Hebert *et al.* (2003) is the most commonly used marker. For toxigenic fungi, other markers have been also proposed. Druzhinina *et al.* (2005) presented a DNA oligonucleotide barcode method for the species identification of *Hypocrea* and *Trichoderma*, based on the ITS1-2 regions. Geiser *et al.* (2007) have thoroughly studied the potential of COX1 gene as candidate marker for DNA barcoding of *Aspergilli*. Owing to an important occurrence of introns, they concluded that either β-tubulin or calmodulin sequences would be more informative and more accurate for species identification within this genus. In contrast, the COX1 gene seems to be a promising marker for *Penicillia*, as the number of introns was very low for the strains studied (in two of the 270 stains tested) (Seifert *et al.*, 2007). The authors developed COX1 gene-based primers for 58 *Penicillium* species, as well as for 12 allied taxa. Gilmore *et al.* (2009) evaluated the utility of COX1 as a DNA barcode for identifying species of *Fusarium* using newly designed primers. Nevertheless, multiple copies of COX1 confound the interpretation of the barcode tree of *Fusarium*.

11.8.3 Metagenomic

The ultimate facility would be to detect the microbial food contaminants directly in their environment, without isolation and time-consuming subsequent processes. A large-scale parallel pyrosequencing system, such as the 454 Genome Sequencher FLX system will help to achieve this goal.

11.9 Conclusion and future prospects

Increasing concern regarding food safety worldwide has led to the control of contamination of food and feed with mycotoxins and related producing species to be considered a major issue. Considerable progress in the taxonomy of the contaminating fungi and also in the biosynthetic pathways elucidation of various mycotoxins has contributed to enlarging the range of target sequences that are useful for accurate molecular detection, identification and quantification. Technical developments combining molecular, nanotechnology and physical aspects have brought novel technologies that solve most of the problems related to classical methods, raise specificity and sensitivity and speed up mycotoxin-producing fungi detection to less than one hour.

With increased interest in research dealing with molecular methods, biosensors, automation and miniaturization, there are good prospects for development of fast, accurate, sensitive, user friendly and cost-effective methods in food mycology in the near future, including the detection of toxigenic species.

Finally, further challenges in the detection of toxigenic fungal species include the possibility of combining the great potential of microarray technology with the current availability of many probes for biosynthetic genes involved in mycotoxin pathways. This would not only allow the detection of the main toxigenic species that occur directly in field samples by implanting them in DNA biochips, but also investigation of mycotoxin gene expression. Specific expression profiles of whole mycotoxin biosynthetic pathways can be generated by an RNA microarray that allows the exact prediction and control of food contamination. This kind of microarray could be used in the future to study the effects of different relevant parameters on mycotoxin gene expression. In this respect, the use of DNA-based assay to estimate the risk of contamination is severely problematic because (1) mycotoxins could be present without the fungi, (2) the fungi could be present without mycotoxins and (3) DNA quantification is not 'one gene/one spore' since the spores of many fungal species can have multiple nuclei. Thus, it will be important to use an innovative approach for diagnosis of toxigenic fungi allowing the PCR assay to detect, in field samples, the main occurring toxigenic species, their relevant mycotoxin genes and to determine whether or not such genes are expressed, in order to assess reliably the real toxicological risks related to contamination of food and feedstuffs by toxigenic fungi.

11.10 References

Abdin M., Ahmad M. and Javed S. (2010). 'Advances in molecular detection of *Aspergillus*: an update'. *Archives of Microbiology*, **192**, 409–25.

Andersen B., Smedsgaard J., Jorring I., Skouboe P. and Pedersen L.H. (2006). 'Real-time PCR quantification of the AM-toxin gene and HPLC qualification of toxigenic metabolites from *Alternaria* species from apples'. *Int J Food Microbiol*, **111**, 105–11.

Andersen G.R. and Nyborg J. (2001). 'Structural studies of eukaryotic elongation factors'. *Cold Spring Harbor Symposia on Quantitative Biology: The Ribosome*, May 30–June 4, Cold Spring Harbor Laboratory, Cold Spring Harbor, New York, USA, Volume 66, 425–37.

Andersen G.R., Nissen P. and Nyborg J. (2003). 'Elongation factors in protein biosynthesis'. *Trends Biochem Sci*, **28**, 434–41.

Aoki, T. and O'Donnell K. (1999). 'Morphological and molecular characterization of *Fusarium pseudograminearum* sp. nov., formerly recognized as the group 1 population of *F. graminearum*'. *Mycologia*, **91**, 597–609.

Atoui A., Mathieu F. and Lebrihi A. (2007). 'Targeting a polyketide synthase gene for *Aspergillus carbonarius* quantification and ochratoxin A assessment in grapes using real-time PCR'. *Int J Food Microbiol*, **115**, 313–8.

Baird R., Abbas H.K. , Windham G., Williams P., Baird S., Ma P., Kelley R., Hawkins L. and Scruggs M. (2008). 'Identification of select fumonisin forming *Fusarium* species using PCR applications of the polyketide synthase gene and its relationship to fumonisin production *in vitro*'. *Int J Molecul Sci*, **9**, 554–70.

Baker S.E. (2006). '*Aspergillus niger* genomics: past, present and into the future'. *Medical Mycology*, **44**, S17–S21.

Banér J., Gyarmati P., Yacoub A., Hakhverdyan M., Stenberg J., Ericsson O., Nilsson M., Landegren U. and Belák S. (2007). 'Microarray-based molecular detection of foot-and-mouth disease, vesicular stomatitis and swine vesicular disease viruses, using padlock probes'. *J Virological Methods*, **143**, 200–6.

Bau M., Castella G., Bragulat U.R. and Cabañes F.J. (2006). 'RFLP characterization of *Aspergillus niger* agregate species from grapes from Europe and Israel'. *Int J Food Microbiol*, **115**, 313–8.

Bezuidenhout C.C., Prinsloo M. and Van de Walt A.M. (2006). 'Multiplex PCR-based detection of potential fumonisin-producing *Fusarium* in traditional African vegetables'. *Environ Toxicol*, **21**, 360–6.

Bleve G., Rizzotti L., Dellaglio F. and Torriani S. (2003). 'Development of reverse transcription (RT)-PCR and real-time RT-PCR assays for rapid detection and quantification of viable yeasts and molds contaminating yogurts and pasteurized food products'. *Appl Environ Microbiol*, **69**, 4116–22.

Bluhm B.H., Cousin M.A. and Woloshuk C.P. (2004). 'Multiplex real-time PCR detection of fumonisin-producing and trichothecene-producing groups of *Fusarium* species'. *J Food Protect*, **67**, 536–43.

Bluhm B.H., Flaherty J.E., Cousin M.A. and Woloshuk C.P. (2002). 'Multiplex polymerase chain reaction assay for the differential detection of trichothecene- and fumonisin-producing species of *Fusarium* in cornmeal'. *J Food Protect*, **65**, 1955–61.

Bogs C., Battilani P. and Geisen R. (2006). 'Development of a molecular detection and differentiation system for ochratoxin A producing *Penicillium* species and its application to analyse the occurrence of *Penicillium nordicum* in cured meats'. *Int J Food Microbiol*, **107**, 39–47.

Brandfass C. and Karlovsky P. (2006). 'Simultaneous detection of *Fusarium culmorum* and *F. graminearum* in plant material by duplex PCR with melting curve analysis'. *BMC Microbiol*, **6**, 4.

Brown D.W., Yu J.H., Kelkar H.S., Fernandes M., Nesbitt T.C., Keller N.P., Adams T.H. and Leonard T.J. (1996). 'Twenty-five coregulated transcripts define a sterigmatocystin gene cluster in *Aspergillus nidulans*'. *Proc Natl Acad Sci USA*, **93**, 1418–22.

Brown D.W., McCormick S.P., Alexander N.J., Proctor R.H. and Desjardins A.E. (2002). 'Inactivation of a cytochrome P-450 is a determinant of trichothecene diversity in *Fusarium* species. *Fungal Genetics Biol*, **36**, 224–33.

Brown D.W., Dyer R.B., McCormick S.P., Kendra D.F. and Plattner R.D. (2004). 'Functional demarcation of the *Fusarium* core trichothecene gene cluster'. *Fungal Genetics Biol*, **41**, 454–62.

Brown D.W., Butchko R.A.E., Busman M. and Proctor R.H.(2007). 'The *Fusarium verticillioides FUM* gene cluster encodes a Zn(II)2Cys6 protein that affects *FUM* gene expression and fumonisin production'. *Eukaryotic Cell*, **6**, 1210–8.

Brunner K. and Mach R.L. (2010). 'Quantitative detection of fungi by molecular methods:

a case study on *Fusarium*', in *Molecular Identification of Fungi*. Gherbawy Y. and Voigt K. (eds). Springer Berlin Heidelberg, 93–105.

Brunner K., Paris M.P.K., Paolino G., Bürstmayr H., Lemmens M., Berthiller F., Schuhmacher R., Krska R. and Mach R.L. (2009). 'A reference-gene-based quantitative PCR method as a tool to determine *Fusarium* resistance in wheat'. *Anal Bioanal Chem* , **395**, 1385–94.

Bufflier E., Susca A., Baud M., Mulè G., Brengel K. and Logrieco A. (2004). 'Detection of *Aspergillus carbonarius* and other black aspergilli from grapes by DNA OLISA microarray'. *Food Addit Contam*, **24**, 1138–47.

Carter N.P. and Vetrie D. (2004). 'Applications of genomic microarrays to explore human chromosome structure and function'. *Human Molec Genetics*, **13**, 297–302.

Chandler E.A., Simpson D.R., Thomsett M.A. and Nicholson P. (2003). 'Development of PCR assays to *TRI7* and *TRI13* trichothecene biosynthetic genes, and characterisation of chemotypes of *Fusarium graminearum*, *Fusarium culmorum* and *Fusarium cerealis*'. *Physiol Molec Plant Pathol*, **62**, 355–67.

Chen R.S., Tsay J.G. , Huang Y.F. and Chiou R.Y.Y. (2002). 'Polymerase chain reaction-mediated characterization of molds belonging to the *Aspergillus flavus* group and detection of *Aspergillus parasiticus* in peanut kernels by a multiplex polymerase chain reaction'. *J Food Protect*, **65**, 840–844.

Chin D. and Means A.R. (2000) 'Calmodulin: a prototypical calcium sensor'. *Trends Cell Biol*, **10**, 322–8.

Chung W.-H., Ishii H., Nishimura K., Ohshima O., Iwama T. and Yoshimatsu H. (2008). 'Genetic analysis and PCR-based identification of major *Fusarium* species causing head blight on wheat in Japan'. *J General Plant Pathol*, **74**, 364–74.

Colombo F., Vallone L., Giaretti M. and Dragoni I. (2003). 'Identification of *Penicillium aurantiogriseum* species with a method of polymerase chain reaction-restriction fragment length polymorphism'. *Food Control*, **14**, 137–40.

Criseo G., Bagnara A. and Bisignano G. (2001). 'Differentiation of aflatoxin-producing and non-producing strains of *Aspergillus flavus* group'. *Lett Appl Microbiol*, **33**, 291–5.

Cullen D.W., Toth I.W., Pitkin Y., Boonham N., Walsh K., Barker I. and Lees A.K. (2005). 'Use of quantitative molecular diagnostic assays to investigate *Fusarium* dry rot in potato stocks and soil'. *Phytopathology*, **95**, 1462–71.

Dao H.P., Mathieu F. and Lebrihi A.(2005). 'Two primer pairs to detect OTA producers by PCR method'. *Int J Food Microbiol*, **104**, 61–7.

Degola F., Berni E., Dall'Asta C., Spotti E., Marchelli R., Ferrero I. and Restivo F.M. (2007). 'A multiplex RT-PCR approach to detect aflatoxigenic strains of *Aspergillus flavus*'. *J Appl Microbiol*, **103**, 409–17.

Degola F., Berni E., Spotti E., Ferrero I. and Restivo F.M. (2009). 'Facing the problem of "false positives": Re-assessment and improvement of a multiplex RT-PCR procedure for the diagnosis of *A. flavus* mycotoxin producers'. *Int J Food Microbiol*, **129**, 300–5.

Demeke T., Clear R.M., Patrick S.K. and Gaba D. (2005). 'Species-specific PCR-based assays for the detection of *Fusarium* species and a comparison with the whole seed agar plate method and trichothecene analysis'. *Int J Food Microbiol*, **103**, 271–84.

Desjardins A.E. (2006). *Fusarium Mycotoxins: Chemistry, Genetics and Biology*, APS Press, American Phytopathological Society, St. Paul, MN.

Ding H.M., Ding Z.S., Li H.N. and Chen N.P. (2008). 'Application of SCAR molecular marker technology in identification of *Monascus* strains'. *Zhongguo Zhong Yao Za Zhi*, **33**, 359–62.

Druzhinina I.S., Kopchinskiy A.G., Komon M., Bissett J., Szakacs G. and Kubicek C.P. (2005). 'An oligonucleotide barcode for species identification in *Trichoderma* and *Hypocrea*'. *Fungal Genetics Biol*, **42**, 813–28.

Edwards S.G., O'Callaghan J. and Dobson A.D.W. (2002). 'PCR-based detection and quantification of mycotoxigenic fungi'. *Mycological Res*, **106**, 1005–25.

Endo S., Komori T., Ricci G., Sano A., Yokoyama K., Ohori A., Kamei K., Franco M., Miyaji

M. and Nishimura K. (2004). 'Detection of gp43 of *Paracoccidioides brasiliensis* by the loop-mediated isothermal amplification (LAMP) method'. *FEMS Microbiol Lett*, **234**, 93–7.

Ercolini D. (2004). 'PCR-DGGE fingerprinting: novel strategies for detection of microbes in food'. *J Microbiol Methods*, **56**, 297–314.

Eriksson R., Jobs M., Ekstrand C., Ullberg M., Herrmann B., Landegren U., Nilsson M. and Blomberg J. (2009). 'Multiplex and quantifiable detection of nucleic acid from pathogenic fungi using padlock probes, generic real time PCR and specific suspension array readout'. *J Microbiol Methods*, **78**, 195–202.

Färber P., Geisen R. and Holzapfel W.H. (1997). 'Detection of aflatoxinogenic fungi in figs by a PCR reaction'. *Int J Food Microbiol*, **36**, 215–20.

Frisvad J.C., Thrane U., Samson R.A. and Pitt J.I. (2006). 'Important mycotoxins and the fungi which produce them', in *Advances in Food Mycology*. Hocking A.D., Pitt J.I., Samson R.A. and Thrane U. (eds). Springer USA,Volume 571, 3–31.

Frisvad J.C., Smedsgaard J., Samson R.A., Larsen T.O. and Thrane U. (2007). 'Fumonisin B-2 production by *Aspergillus niger*'. *J Agric Food Chem*, **55**, 9727–32.

Gadkar V. and Rillig M.C. (2008). 'Evaluation of loop-mediated isothermal amplification (LAMP) to rapidly detect arbuscular mycorrhizal fungi'. *Soil Biol Biochem*, **40**, 540–3.

Gallo A., Mulè G., Favilla M. and Altomare C. (2004). 'Isolation and characterisation of a trichodiene synthase homologous gene in *Trichoderma harzianum*'. *Physiol Molec Plant Pathol*, **65**, 11–20.

Garcia C., La Guerche S., Mouhamadou M., Férandon C., Labarère J., Blancard D., Darriet P. and Barroso G. (2006). 'A CAPS test allowing a rapid distinction of *Penicillium expansum* among fungal species collected on grape berries, inferred from the sequence and secondary structure of the mitochondrial SSU-rRNA'. *Int J Food Microbiol*, **111**, 183–90.

Geisen R. (1996). 'Multiplex polymerase chain reaction for the detection of potential aflatoxin and sterigmatocystin producing fungi'. *Systematic Appl Microbiol*, **19**, 388–92.

Geisen R., Mayer Z., Karolewiez A. and Färber P.F. (2004). 'Development of a real time PCR system for detection of *Penicillium nordicum* and for monitoring ochratoxin A production in foods by targeting the ochratoxin polyketide synthase gene'. *Systematic Appl Microbiol*, **27**, 501–7.

Geisen R., Schmidt-Heydt M. and Karolewiez A. (2006). 'A gene cluster of the ochratoxin A biosynthetic genes in *Penicillium*'. *Mycotoxin Res*, **22**, 134–41.

Geiser D., del Mar Jiménez-Gasco M., Kang S., Makalowska I., Veeraraghavan N., Ward T., Zhang N., Kuldau G. and O'Donnell K. (2004). 'Fusarium-ID v. 1.0: a DNA sequence database for identifying *Fusarium*'. *Eur J Plant Pathol*, **110**, 473–9.

Geiser D.M., Klich M.A., Frisvad J.C., Peterson S.W., Varga J. and R.A S. (2007). 'The current status of species recognition and identification in *Aspergillus*'. *Studies Mycol*, **59**, 1–10.

Gil-Serna J., Vázquez C., Sardiñas N., González-Jaén M.T. and Patiño B. (2009a). 'Discrimination of the main Ochratoxin A-producing species in *Aspergillus* section *Circumdati* by specific PCR assays'. *Int J Food Microbiol*, **136**, 83–7.

Gil-Serna J., González-Salgado A., González-Jaén M.T., Vázquez C. and Patiño B. (2009b). 'ITS-based detection and quantification of *Aspergillus ochraceus* and *Aspergillus westerdijkiae* in grapes and green coffee beans by real-time quantitative PCR'. *Int J Food Microbiol*, **131**, 162–7.

Gilmore S.R., Gräfenham T., Louis-Seize G. and Seifert K.A. (2009). 'Multiple copies of cytochrome oxidase 1 in species of the fungal genus *Fusarium*'. *Molec Ecol Resources*, **9**, 90–8.

Glenn A.E., Zitomer N.C., Zimeri A.M., Williams L.D., Riley R.T. and Proctor R.H. (2008). 'Transformation-mediated complementation of a *FUM* gene cluster deletion in *Fusarium verticillioides* restores both fumonisin production and pathogenicity on maize seedlings'. *Molec Plant-Microbe Interactions*, **21**, 87–97.

Godet M. and Munaut F. (2010) 'Molecular strategy for identification in *Aspergillus* section *Flavi'*. *FEMS Microbiol Lett*, **304**, 157–68.

González-Jaén M.T., Mirete S., Patiño B., López-Errasquín E. and Vázquez C. (2004). 'Genetic markers for the analysis of variability and for production of specific diagnostic sequences in fumonisin-producing strains of *Fusarium verticillioides'*. *Eur J Plant Pathol*, **110**, 525–32.

González-Salgado A., Patiño B., Vázquez C. and González-Jaén M.T. (2005). 'Discrimination of *Aspergillus niger* and other *Aspergillus* species belonging to section Nigri by PCR assays'. *FEMS Microbiol Lett*, **245**, 353–61.

González-Salgado A., Patiño B., Gil-Serna J., Vázquez C. and González-Jaén M.T. (2009). 'Specific detection of *Aspergillus carbonarius* by SYBR Green and TaqMan quantitative PCR assays based on the multicopy ITS2 region of the rRNA gene. *FEMS Microbiol Lett*, **295**, 57–66.

Güldener U., Seong K.-Y., Boddu J, Cho S., Trail F., Xu J.-R., Adam G., Mewes H.-W., Muehlbauer G.J. and Kistler H.C. (2006). 'Development of a *Fusarium graminearum* Affymetrix GeneChip for profiling fungal gene expression *in vitro* and *in planta'*. *Fungal Genetics Biol*, **43**, 316–25.

Haugland R.A., Varma M., Wymer L.J. and Vesper S.J. (2004). 'Quantitative PCR analysis of selected *Aspergillus*, *Penicillium* and *Paecilomyces* species. *Systematic Appl Microbiol*, **27**, 198–210.

Hayashi N., Arai R., Tada S., Taguchi H. and Ogawa Y. (2007). 'Detection and identification of *Brettanomyces/Dekkera* sp. yeasts with a loop-mediated isothermal amplification method'. *Food Microbiol*, **24**, 778–85.

Hebert P.D.N., Cywinska A., Ball S.L. and de Waard J.R. (2003). 'Biological identifications through DNA barcodes'. *Proc R Soc Lond B*, **270**, 313–21.

Hinrikson H.P., Hurst S.F., de Aguirre L. and Morrison C.J. (2005). 'Molecular methods for the identification of *Aspergillus* species. *Medical Mycol*, **43**, 129–37.

Hue F.-X., Huerre M., Rouffault M.A. and de Bievre C. (1999). 'Specific detection of *Fusarium* species in blood and tissues by a PCR technique'. *J Clin Microbiol*, **37**, 2434–8.

Imai M., Ninomiya A., Minekawa H., Notomi T., Ishizaki T., Van Tu P., Tien N.T.K., Tashiro Mand Odagiri T. (2007). 'Rapid diagnosis of H5N1 avian influenza virus infection by newly developed influenza H5 hemagglutinin gene-specific loop-mediated isothermal amplification method'. *J Virological Methods*, **141**, 173–80.

Inacio J., Flores O. and Spencer-Martins I. (2008). 'Efficient identification of clinically relevant *Candida* yeast species by use of an assay combining panfungal loop-mediated isothermal DNA amplification with hybridization to species-specific oligonucleotide probes. *J Clin Microbiol*, **46**, 713–20.

Iwamoto T., Sonobe T. and Hayashi K. (2003). 'Loop-mediated isothermal amplification for direct detection of *Mycobacterium tuberculosis* complex, *M. avium*, and *M. intracellulare* in sputum samples'. *J Clin Microbiol*, **41**, 2616–22.

Johnson R.D., Johnson L., Kohmoto K., Otani H., Lane C.R. and Kodama M. (2000) 'A polymerase chain reaction-based method to specifically detect *Alternaria alternata* apple pathotype (*A. mali*), the causal agent of *Alternaria* blotch of apple'. *Phytopathology*, **90**, 973–6.

Jurado M., Vázquez C., Patiño B. and González-Jaén M.T.(2005). 'PCR detection assays for the trichothecene-producing species *Fusarium graminearum*, *Fusarium culmorum*, *Fusarium poae*, *Fusarium equiseti* and *Fusarium sporotrichioides'*. *Systematic Appl Microbiol*, **28**, 562–8.

Jurado M., Vázquez C., Marín S., Sanchis V. and González-Jaén M.T. (2006). 'PCR-based strategy to detect contamination with mycotoxigenic *Fusarium* species in maize'. *Systematic Appl Microbiol*, **29**, 681–9.

Jurado M., Marin P., Callejas C., Moretti A., Vazquez C. and Gonzalez-Jaen M.T. (2010) 'Genetic variability and fumonisin production by *Fusarium proliferatum'*. *Food Microbiol*, **27**, 50–7.

Karolewiez A. and Geisen R. (2005). 'Cloning a part of the ochratoxin A biosynthetic gene cluster of *Penicillium nordicum* and characterization of the ochratoxin polyketide synthase gene'. *Systematic Appl Microbiol*, **28**, 588–95.

Konstantinova P. and Yli-Mattila T. (2004). 'IGS-RFLP analysis and development of molecular markers for identification of *Fusarium poae, Fusarium langsethiae, Fusarium sporotrichioides* and *Fusarium kyushuense*'. *Int J Food Microbiol*, **95**, 321–31.

Kristensen R., Berdal K.G. and Holst-Jensen A. (2007a). 'Simultaneous detection and identification of trichothecene- and moniliformin-producing *Fusarium* species based on multiplex SNP analysis'. *J Appl Microbiol*, **102**, 1071–81.

Kristensen R., Gauthier G., Berdal K.G, Hamels S., Remacle J. and Holst-Jensen A. (2007b). 'DNA microarray to detect and identify trichothecene- and moniliformin-producing *Fusarium* species'. *J Appl Microbiol*, **102**, 1060-70.

Kulik T. (2008a). 'Development of a duplex PCR assay for the simultaneous detection of *Fusarium poae* and *Fusarium sporotrichioides* from wheat'. *J. Plant Pathol*, **90**, 441–7.

Kulik T. (2008b). 'Detection of *Fusarium tricinctum* from cereal grain using PCR assay'. *J Appl Genetics*, **49**, 305–11.

Kulik T., Fordonski G., Pszczólkowska A., Plodzien K. and Lapinski M. (2004). 'Development of PCR assay based on ITS2 rDNA polymorphism for the detection and differentiation of *Fusarium sporotrichioides*'. *FEMS Microbiol Lett*, **239**, 181–6.

Láday M., Juhász Á., Mulè G., Moretti A, Szécsi Á. and Logrieco A. (2004a). 'Mitochondrial DNA diversity and lineage determination of European isolates of *Fusarium graminearum (Gibberella zeae)*'. *Eur J Plant Pathol*, **110**, 545–50.

Láday M., Mulè G., Moretti A., Hamari Z., Juhász Á., Szécsi Á. and Logrieco A. (2004b). 'Mitochondrial DNA variability in *Fusarium proliferatum (Gibberella Intermedia)*'. *Eur J Plant Pathol*, **110**, 563–71.

Lau A., Sorrell T.C., Chen S., Stanley K., Iredell J. and Halliday C. (2008). 'Multiplex tandem PCR: a novel platform for rapid detection and identification of fungal pathogens from blood culture specimens'. *J Clin Microbiol*, **46**, 3021–7.

Le Dréan G., Mounier J., Vasseur V., Arzur D., Habrylo O. and Barbier G. (2010). 'Quantification of *Penicillium camemberti* and *P. roqueforti* mycelium by real-time PCR to assess their growth dynamics during ripening cheese'. *Int J Food Microbiol*, **138**, 100–7.

Lee T., Oh D.-W., Kim H.-S., Lee J., Kim Y.-H., Yun S.-H. and Lee Y.-W. (2001). 'Identification of deoxynivalenol- and nivalenol-producing chemotypes of *Gibberella zeae* by using PCR'. *Appl Environ Microbiol*, **67**, 2966–72.

Lezar S. and Barros E. (2010) 'Oligonucleotide microarray for the identification of potential mycotoxigenic fungi'. *BMC Microbiol*, **10**, 87.

Li Y.C., Korol A.B., Fahima T., Beiles A. and Nevo E. (2002). 'Microsatellites: genomic distribution, putative functions and mutational mechanisms: a review'. *Mol Ecol*, **11**, 2453–65.

Llorens A., Hinojo M.J., Mateo R., González-Jaén M.T., Valle-Algarra F.M., Logrieco A. and Jiménez M. (2006). 'Characterization of *Fusarium* spp. isolates by PCR-RFLP analysis of the intergenic spacer region of the rRNA gene (rDNA)'. *Int J Food Microbiol*, **106**, 297–306.

López-Errasquín E., Vázquez C., Jiménez M. and González-Jaén M.T. (2007). 'Real-time RT-PCR assay to quantify the expression of *FUM1* and *FUM19* genes from the fumonisin-producing *Fusarium verticillioides*'. *J Microbiol Methods*, **68**, 312–7.

Lucas S., Martins M.D.L., Flores O., Meyer W., Spencer-Martins I. and Inácio J. (2009). 'Differentiation of *Cryptococcus neoformans* varieties and *Cryptococcus gattii* using CAP59-based loop-mediated isothermal DNA amplification'. *Clin Microbiol Infection*, **16**(6), 711–14.

Luo Y., Gao W., Doster M. and Michailides T.J. (2009). 'Quantification of conidial density of *Aspergillus flavus* and *A. parasiticus* in soil from almond orchards using real-time PCR'. *J Appl Microbiol*, **106**, 1649–60.

Mach R.L., Kullnig-Gradinger C.M., Farnleitner A.H., Reischer G., Adler A. and Kubicek C.P. (2004). 'Specific detection of *Fusarium langsethiae* and related species by DGGE and ARMS-PCR of a [beta]-tubulin (tub1) gene fragment'. *In J Food Microbiol*, **95**, 333–9.

Martínez-Culebras P.V. and Ramón D. (2007). 'An ITS-RFLP method to identify black *Aspergillus* isolates responsible for OTA contamination in grapes and wine'. *Int J Food Microbiol*, **113**, 147–53.

Martínez-Culebras P.V., Crespo-Sempere A., Gil J.V. and Ramón D. (2009). 'Acyl transferase domains of putative polyketide synthase (PKS) genes in *Aspergillus* and *Penicillium* producers of ochratoxin A and the evaluation of PCR primers to amplify PKS sequences in black *Aspergillus* species'. *Food Sci Technol Int*, **15**, 97–105.

Mayer Z., Bagnara A., Färber P. and Geisen R. (2003). 'Quantification of the copy number of *nor-1*, a gene of the aflatoxin biosynthetic pathway by real-time PCR, and its correlation to the cfu of *Aspergillus flavus* in foods'. *Int J Food Microbiol*, **82**, 143–51.

Medina A., Mateo R., Lopez-Ocana L., Valle-Algarra F.M. and Jimenez M. (2005). 'Study of Spanish grape mycobiota and ochratoxin A production by isolates of *Aspergillus tubingensis* and other members of *Aspergillus* section *Nigri*'. *Appl Environ Microbiol*, **71**, 4696–702.

Meng K., Wang Y., Yang P., Luo H., Bai Y., Shi P., Yuan T., Ma R. and Yao B. (2010) 'Rapid detection and quantification of zearalenone-producing *Fusarium* species by targeting the zearalenone synthase gene *PKS4*'. *Food Control*, **21**, 207–11.

Mishra P.K., Fox R.T.V. and Culham A. (2003). 'Development of a PCR-based assay for rapid and reliable identification of pathogenic Fusaria'. *FEMS Microbiol Lett*, **218**, 329–32.

Mogensen J.M., Frisvad J.C., Thrane U. and Nielsen K.F. (2010) 'Production of fumonisin B-2 and B-4 by *Aspergillus niger* on grapes and raisins'. *J Agric Food Chem*, **58**, 954–8.

Möller E.M., Chekowski J. and Geiger H.H. (1999). 'Species-specific PCR assays for the fungal pathogens *Fusarium moniliforme* and *Fusarium subglutinans* and their application to diagnose maize ear rot disease'. *J Phytopathol*, **147**, 497–508.

Morello L.G., Sartori D., de Oliveira Martinez A.J., Vieira M.L.C., Taniwaki M.H. and Fungaro M.H.P. (2007). 'Detection and quantification of *Aspergillus westerdijkiae* in coffee beans based on selective amplification of β-tubulin gene by using real-time PCR'. *Int J Food Microbiol*, **119**, 270–6.

Mulè G., Susca A., Stea G. and Moretti A. (2004a). 'Specific detection of the toxigenic species *Fusarium proliferatum* and *F. oxysporum* from asparagus plants using primers based on calmodulin gene sequences'. *FEMS Microbiol Lett*, **230**, 235–40.

Mulè G., Susca A., Stea G. and Moretti A. (2004b). 'A species-specific PCR assay based on the calmodulin partial gene for identification of *Fusarium verticillioides*, *F. proliferatum* and *F. subglutinans*'. *Eur J Plant Pathol*, **110**, 495–502.

Mulè G., González-Jaén M.T., Hornok L., Nicholson P. and Waalwijk C. (2005). 'Advances in molecular diagnosis of toxigenic *Fusarium* species: A review. *Food Addit Contam*, **22**, 316–23.

Mulè G., Susca A., Logrieco A., Stea G. and Visconti A. (2006). 'Development of a quantitative real-time PCR assay for the detection of *Aspergillus carbonarius* in grapes'. *Int J Food Microbiol*, **111**, S28–S34.

Murillo I., Cavallarin L. and San Segundo B. (1998). 'The development of a rapid PCR assay for detection of *Fusarium moniliforme*'. *Eur J Plant Pathol*, **104**, 301–11.

Naef A., Senatore M. and Défago G. (2006). 'A microsatellite based method for quantification of fungi in decomposing plant material elucidates the role of *Fusarium graminearum* DON production in the saprophytic competition with *Trichoderma atroviride* in maize tissue microcosms'. *FEMS Microbiol Ecol*, **55**, 211–20.

Nagamine K., Watanabe K., Ohtsuka K., Hase T. and Notomi T. (2001). 'Loop-mediated isothermal amplification reaction using a nondenatured template'. *Clin Chem*, **47**, 1742–3.

Nicholson P., Simpson D.R., Weston G., Rezanoor H.N., Lees A.K., Parry D.W. and Joyce D. (1998). 'Detection and quantification of *Fusarium culmorum* and *Fusarium graminearum* in cereals using PCR assays'. *Physiol Molec Plant Pathol*, **53**, 17–37.

Nicolaisen M., Justesen A.F., Thrane U., Skouboe P. and Holmstrøm K. (2005). 'An oligonucleotide microarray for the identification and differentiation of trichothecene producing and non-producing *Fusarium* species occurring on cereal grain'. *J Microbiol Methods*, **62**, 57–69.

Nicolaisen M., Suproniene S., Nielsen L.K., Lazzaro I., Spliid N.H. and Justesen A.F. (2009). 'Real-time PCR for quantification of eleven individual *Fusarium* species in cereals'. *J Microbiol Methods*, **76**, 234–40.

Niessen L. (2007). 'PCR-based diagnosis and quantification of mycotoxin producing fungi'. *Int J Food Microbiol*, **119**, 38–46.

Niessen L. and Vogel R.F. (2010). 'Detection of *Fusarium graminearum* DNA using a loop-mediated isothermal amplification (LAMP) assay'. *Int J Food Microbiol*, **40**(2–3), 183–91.

Niessen L., Schmidt H. and Vogel R.F. (2004). 'The use of *TRI5* gene sequences for PCR detection and taxonomy of trichothecene-producing species in the *Fusarium* section *Sporotrichiella*'. *Int J Food Microbiol*, **95**, 305–19.

Niessen L., Schmidt H., Mühlencoer T.E., Färber P., Karolewiez A. and Geisen R. (2005). 'Advances in the molecular diagnosis of ochratoxin A-producing fungi'. *Food Addit Contam*, **22**, 324–34.

Nilsson J. and Nissen P. (2005). 'Elongation factors on the ribosome'. *Curr Opin Struct Biol*, **15**, 349–54.

Nitschke E., Nihlgard M. and Varrelemann M. (2009). 'Differentiation of eleven *Fusarium* spp. isolated from sugar beet, using restriction fragment analysis of a polymerase chain reaction – amplified translation elongation factor la gene fragment'. *Phytopathology*, **99**, 921–9.

Noonim P., W. Mahakarnchanakul, K.F. Nielsen, J.C. Frisvad and R.A. Samson (2009). 'Fumonisin B$_2$ production by *Aspergillus niger* in Thai coffee beans. Food Additives and Contaminants Part a-Chemistry Analysis Control Exposure & Risk Assessment, 26, 94–100.

Notomi T., Okayama H., Masubuchi H., Yonekawa T., Watanabe K., Amino N. and Hase T. (2000). 'Loop-mediated isothermal amplification of DNA'. *Nucl Acids Res*, **28**, e63i–vii.

O'Callaghan J., Stapleton P.C. and Dobson A.D.W. (2006). 'Ochratoxin A biosynthetic genes in *Aspergillus ochraceus* are differentially regulated by pH and nutritional stimuli'. *Fungal Genetics Biol*, **43**, 213–21.

Ohori A., Endo S., Sano A., Yokoyama K., Yarita K., Yamaguchi M., Kamei K, Miyaji M. and Nishimura K. (2006). 'Rapid identification of *Ochroconis gallopava* by a loop-mediated isothermal amplification (LAMP) method'. *Veterinary Microbiol*, **114**, 359–65.

Parry D.W. and Nicholson P. (1996). 'Development of a PCR assay to detect *Fusarium poae* in wheat'. *Plant Pathol*, **45**, 383–91.

Paterson R.R.M (2006a). 'Primers from the isoepoxydon dehydrogenase gene of the patulin biosynthetic pathway to indicate critical control points for patulin contamination of apples'. *Food Control*, **17**, 741–4.

Paterson R.R.M (2006b). 'Identification and quantification of mycotoxigenic fungi by PCR'. *Process Biochem*, **41**, 1467–74.

Paterson R.R.M., Kozakiewicz Z., Locke T., Brayford D. and Jones S.C.B. (2003). 'Novel use of the isoepoxydon dehydrogenase gene probe of the patulin metabolic pathway and chromatography to test penicillia isolated from apple production systems for the potential to contaminate apple juice with patulin'. *Food Microbiol*, **20**, 359–64.

Patiño B., Mirete S, González-Jaén MT, Mulé G., Rodríguez MT and Vázquez C. (2004). 'PCR detection assay of fumonisin-producing *Fusarium verticillioides* strains'. *J Food Prot*, **67**, 1278–83.

Patiño B., González-Salgado A., González-Jaén M.T. and Vázquez C. (2005). 'PCR detection assays for the ochratoxin-producing *Aspergillus carbonarius* and *Aspergillus ochraceus* species'. *Int J Food Microbiol*, **104**, 207–14.

Pel H.J., *et al.* (2007). 'Genome sequencing and analysis of the versatile cell factory *Aspergillus niger* CBS 513.88'. *Nature Biotechnol*, **25**, 221–31.

Pelegrinelli-Fungaro M.H., Vissotto P.C., Sartori D., Vilas-Boas L.A., Furlaneto M.C. and Taniwaki M.H. (2004). 'A molecular method for detection of *Aspergillus carbonarius* in coffee beans'. *Current Microbiol*, **49**, 123–7.

Perrone G., Susca A., Stea G. and Mulè G. (2004). 'PCR assay for identification of *Aspergillus carbonarius* and *Aspergillus japonicus*'. *Eur J Plant Pathol*, **110**, 641–69.

Proctor R.H., Desjardins A.E., Plattner R.D. and Hohn T.M. (1999). 'A polyketide synthase gene required for biosynthesis of fumonisin mycotoxins in *Gibberella fujikuroi* mating population A'. *Fungal Genetics Biol*, **27**, 100–12.

Proctor R.H., Brown D.W., Plattner R.D. and Desjardins A.E. (2003). 'Co-expression of 15 contiguous genes delineates a fumonisin biosynthetic gene cluster in *Gibberella moniliformis*'. *Fungal Genetics Biol*, **38**, 237–49.

Proctor R.H., Plattner R.D., Brown D.W., Seo J.A. and Lee Y.W. (2004). 'Discontinuous distribution of fumonisin biosynthetic genes in the *Gibberella fujikuroi* species complex'. *Mycological Res*, **108**, 815–22.

Proctor R.H., Busman M., Seo J.A., Lee Y.W. and Plattner R.D. (2008). 'A fumonisin biosynthetic gene cluster in *Fusarium oxysporum* strain O-1890 and the genetic basis for B versus C fumonisin production'. *Fungal Genetics Biol*, **45**, 1016–26.

Proctor R.H., McCormick S.P., Alexander N.J. and Desjardins A.E. (2009). 'Evidence that a secondary metabolic biosynthetic gene cluster has grown by gene relocation during evolution of the filamentous fungus *Fusarium*'. *Molec Microbiol*, **74**, 1128–42.

Puel O., Tadrist S., Delaforge M., Oswald I.P. and Lebrihi A. (2007). 'The inability of *Byssochlamys fulva* to produce patulin is related to absence of 6-methylsalicylic acid synthase and isoepoxydon dehydrogenase genes'. *Int J Food Microbiol*, **115**, 131–9.

Quarta A., Mita G., Haidukowski M., Logrieco A., Mulè G. and Visconti A. (2006). 'Multiplex PCR assay for the identification of nivalenol, 3- and 15-acetyl-deoxynivalenol chemotypes in *Fusarium*'. *FEMS Microbiol Lett*, **259**, 7–13.

Raff E.C. (1984). 'Genetics of microtubule systems'. *The Journal of Cell Biology*, **99**, 1–10.

Rath P.-M. and Ansorg R. (2000). 'Identification of medically important *Aspergillus* species by single strand conformational polymorphism (SSCP) of the PCR-amplified intergenic spacer region. *Mycoses*, **43**, 381–6.

Reischer G.H., Lemmens M., Farnleitner A., Adler A. and Mach R.L. (2004). 'Quantification of *Fusarium graminearum* in infected wheat by species specific real-time PCR applying a TaqMan Probe'. *J Microbiol Methods*, **59**, 141–6.

Richard G.-F., Kerrest A. and Dujon B. (2008). 'Comparative genomics and molecular dynamics of DNA repeats in eukaryotes'. *Microbiol Mol Biol Rev*, **72**, 686–727.

Sánchez-Rangel D., SanJuan-Badillo A. and Plasencia J. (2005). 'Fumonisin production by *Fusarium verticillioides* strains isolated from maize in Mexico and development of a polymerase chain reaction to detect potential toxigenic strains in grains'. *J Agric Food Chem*, **53**, 8565–71.

Sano A. and Itano E.N. (2010). 'Applications of loop-mediated isothermal amplificaton methods (LAMP) for identification and diagnosis of mycotic diseases: Paracoccidioidomycosis and *Ochroconis gallopava* infection', in *Molecular Identification of Fungi*. Gherbawy Y. and Voigt K. (eds). Springer Berlin Heidelberg, 417–37.

Sartori D., Furlaneto M.C., Martins M.K., Ferreira de Paula M.R., Pizzirani-Kleiner A.A., Taniwaki M.H. and Fungaro M.H.P. (2006). 'PCR method for the detection of potential ochratoxin-producing *Aspergillus* species in coffee beans'. *Res Microbiol*, **157**, 350–4.

Sartori D., Taniwaki M.H., Iamanaka B. and Fungaro M.H.P. (2010). *Molecular Diagnosis of Ochratoxigenic Fungi*, in *Molecular Identification of Fungi*. Gherbawy Y. and Voigt H. (eds). Springer Berlin Heidelberg, 195–212.

Scherm B., Palomba M, Serra D., Marcello A. and Migheli Q. (2005). 'Detection of transcripts of the aflatoxin genes *aflD*, *aflO*, and *aflP* by reverse transcription-polymerase chain reaction allows differentiation of aflatoxin-producing and non-producing isolates of *Aspergillus flavus* and *Aspergillus parasiticus*'. *Int J Food Microbiol*, **98**, 201–10.

Schilling A.G., Moller E.M. and Geiger H.H. (1996). 'Polymerase chain reaction-based assays for species-specific detection of *Fusarium culmorum*, *F. graminearum*, and *F. avenaceum*'. *Phytopathology*, **86**, 515–22.

Schmidt-Heydt M. and Geisen R. (2007). 'A microarray for monitoring the production of mycotoxins in food'. *Int J Food Microbiol*, **117**, 131–40.

Schmidt H., Ehrmann M., Vogel R.F., Taniwaki M.H. and Niessen L. (2003). 'Molecular typing of *Aspergillus ochraceus* and construction of species specific SCAR-Primers based on AFLP'. *Systematic Appl Microbiol*, **26**, 138–46.

Schmidt H., Bannier M., Vogel R.F. and Niessen L. (2004a). 'Detection and quantification of *Aspergillus ochraceus* in green coffee by PCR'. *Lett Appl Microbiol*, **38**, 464–9.

Schmidt H., Taniwaki M.H., Vogel R.F. and Niessen L. (2004b). 'Utilization of AFLP markers for PCR-based identification of *Aspergillus carbonarius* and indication of its presence in green coffee samples'. *J Appl Microbiol*, **97**, 899–909.

Seifert K.A., Samson R.A., de Waard J.R., Houbraken J., Lévesque C.A., Moncalvo J.-M., Louis-Seize G. and Hebert P.D.N. (2007). 'Prospects for fungus identification using CO1 DNA barcodes, with *Penicillium* as a test case'. *Proc Natl Acad Sci USA*, **104**, 3901–6.

Selma M.V., Martinez-Culebras P.V. and Aznar R. (2008). 'Real-time PCR based procedures for detection and quantification of *Aspergillus carbonarius* in wine grapes'. *Int J Food Microbiol*, **122**, 126–34.

Shapira R., Paster N., Eyal O., Menasherov M., Mett A. and Salomon R. (1996). 'Detection of aflatoxigenic molds in grains by PCR'. *Appl Environ Microbiol*, **62**, 3270–3.

Shinzato N., Namihira T., Tamaki Y., Tsukahara M. and Matsui T. (2009). 'Application of random amplified polymorphic DNA (RAPD) analysis coupled with microchip electrophoresis for high-resolution identification of *Monascus* strains'. *Appli Microbiol Biotechnol*, **82**, 1187–93.

Sreenivasa M.Y., Dass R., Charith Raj A.P. and Janardhana G.R. (2008a). 'PCR method for the detection of genus *Fusarium* and fumonisin-producing isolates from freshly harvested sorghum grains grown in Karnataka, India'. *J Food Safety*, **28**, 236–47.

Sreenivasa M.Y., González Jaen M.T., Sharmila Dass R., Charith Raj A.P. and Janardhana G.R. (2008b). 'A PCR-based assay for the detection and differentiation of potential fumonisin-producing *Fusarium verticillioides* isolated from Indian maize kernels'. *Food Biotechnol*, **22**, 160–70.

Suanthie Y., Cousin M.A. and Woloshuk C.P. (2009). 'Multiplex real-time PCR for detection and quantification of mycotoxigenic *Aspergillus*, *Penicillium* and *Fusarium*'. *J Stored Products Res*, **45**, 139–45.

Susca A., Stea G. and Perrone G. (2007a). 'Rapid polymerase chain reaction (PCR)-single-stranded conformational polymorphism (SSCP) screening method for the identification of *Aspergillus* section *Nigri* species by the detection of calmodulin nucleotide variations'. *Food Addit Contam*, **24**, 1148–53.

Susca A., Stea G., Mulè G. and Perrone G. (2007b). 'Polymerase chain reaction (PCR) identification of *Aspergillus niger* and *Aspergillus tubingensis* based on the calmodulin gene'. *Food Addit Contam*, **24**, 1154–60.

Suzuki F., Koba A. and Nakajima T. (2010). 'Simultaneous identification of species and trichothecene chemotypes of *Fusarium asiaticum* and *F. graminearum* sensu stricto by multiplex PCR'. *J General Plant Pathol*, **76**, 31–6.

Sweeney M.J., Pamies P. and Dobson A.D.W. (2000). 'The use of reverse transcription-polymerase chain reaction (RT-PCR) for monitoring aflatoxin production in *Aspergillus parasiticus* 439'. *Int J Food Microbiol*, **56**, 97–103.

Tenzer I., degli Ivanissevich S., Morgante M. and Gessler C. (1999). 'Identification of microsatellite markers and their application to population genetics of *Venturia inaequalis*'.

Phytopathology, **98**, 748–53.

Tomlinson J.A., Barker I. and Boonham N. (2007). 'Faster, simpler, more-specific methods for improved molecular detection of *Phytophthora ramorum* in the field'. *Appl Environ Microbiol*, **73**, 4040–7.

Turner A.S., Lees A.K., Rezanoor H.N. and Nicholson P. (1998). 'Refinement of PCR-detection of *Fusarium avenaceum* and evidence from DNA marker studies for phenetic relatedness to *Fusarium tricinctum*'. *Plant Pathol*, **47**, 278–88.

Van Hove F., Waalwijk C., González-Jaén M.T., Mulè G., Moretti A. and Munaut F. (2008). 'A *Fusarium* population from banana missing the fumonisin gene cluster: a new species closely related to *F. verticillioides*?' *J Plant Pathol*, **90**(S3), 22.

Van Hove F., Waalwijk C., Logrieco A., Munaut F. and Moretti A. (2011). '*Gibberella musae* (*Fusarium musae*) sp. nov.: a new species from banana closely related to *F. verticillioides*'. *Mycologia*, in press.

Visentin I., Tamietti G., Valentino D., Portis E., Karlovsky P., Moretti A. and Cardinale F. (2009). 'The ITS region as a taxonomic discriminator between *Fusarium verticillioides* and *Fusarium proliferatum*'. *Mycological Res*, **113**, 1137–45.

Vos P., Hogers R., Bleeker M., Reijans M., van der Lee T., Hornes M., Friters A., Pot J., Paleman J., Kuiper M. and Zabeau M. (1995). 'AFLP: a new technique for DNA finger-printing'. *Nucl Acids Res*, **23**, 4407–14.

Waalwijk C., van der Lee T., de Vries I., Hesselink T., Arts J. and Kema G.H.J. (2004a). 'Synteny in toxigenic *Fusarium* species: the fumonisin gene cluster and the mating type region as examples'. *Eur J Plant Pathol*, **110**, 533–44.

Waalwijk C., van der Heide R., de Vries I., van der Lee T., Schoen C., Costrel-de Corainville G., Häuser-Hahn I., Kastelein P., Köhl J., Lonnet P., Demarquet T. and Kema G. (2004b). 'Quantitative detection of *Fusarium* species in wheat using TaqMan'. *Eur J Plant Pathol*, **110**, 481–94.

Wang J., Wang X., Zhou Y., Du L. and Wang Q. (2010). 'Fumonisin detection and analysis of potential fumonisin-producing *Fusarium* spp. in asparagus (*Asparagus officinalis* L.) in Zhejiang Province of China'. *J Sci Food Agric*, **90**, 836–42.

White T.M., Bruns T., Lee S. and Taylor T. (1990). 'Amplification and direct sequencing of fungal ribosomal RNA for phylogenetics'. In *PCR Protocols: A Guide to Methods and Applications*, Innis M.A., Gelfand D.H., Sninsky J.J. and White T.J. (eds), Academic Press, San Diego, CA, 315–21.

Williams J.G.K., Kubelik J.R., Livak K.J., Rafalski J.A. and Tingey S.V. (1990). 'DNA polymorphisms amplified by arbitrary primers are useful as genetic markers'. *Nucl Acids Res*, **18**, 6531–5.

Wilson A., Simpson D., Chandler E., Jennings P. and Nicholson P. (2004). 'Development of PCR assays for the detection and differentiation of *Fusarium sporotrichioides* and *Fusarium langsethiae*'. *FEMS Microbiol Letts*, **233**, 69–76.

Yang Z.Y., Shim W.B., Kim J.H., Park S.J., Kang S.J., Nam B.S. and Chung D.H. (2004). 'Detection of aflatoxin-producing molds in Korean fermented foods and grains by multiplex PCR'. *J Food Protect*, **67**, 2622–6.

Yli-Mattila T., Paavanen-Huhtala S., Jestoi M., Parikka P., Hietaniemi V., Gagkaeva T., Sarlin T., Haikara A., Laaksonen S. and Rizzo A. (2008). 'Real-time PCR detection and quantification of *Fusarium poae*, *F. graminearum*, *F. sporotrichioides* and *F. langsethiae* in cereal grains in Finland and Russia'. *Archiv Phytopathol Plant Protect*, **41**, 243–60.

Yu J.J., Cary J.W., Bhatnagar D., Cleveland T.E., Keller N.P. and Chu F.S. (1993). 'Cloning and characterization of a cDNA from *Aspergillus parasiticus* encoding an O-methyltransferase involved in aflatoxin biosynthesis'. *Appl Environ Microbiol*, **59**, 3564–71.

Yu J.J., Bhatnagar D. and Cleveland T.E. (2004). 'Completed sequence of aflatoxin pathway gene cluster in *Aspergillus parasiticus*'. *FEBS Lett*, **564**, 126–30.

Yuan G., Liu C. and Chen C. (1995). 'Differentiation of *Aspergillus parasiticus* from *Aspergillus sojae* by random amplification of polymorphic DNA'. *Appl Environ Microbiol*,

61, 2384–7.

Zambounis A.G., Paplomatas E. and Tsaftaris A.S. (2007). 'Intergenic spacer–RFLP analysis and direct quantification of Australian *Fusarium oxysporum* f. sp. *vasinfectum* isolates from soil and infected cotton tissues'. *Plant Disease*, **91**, 1564–73.

Zanzotto A., Burruano S. and Marciano P. (2006). 'Digestion of DNA regions to discriminate ochratoxigenic and non-ochratoxigenic strains in the *Aspergillus niger* aggregate'. *Int J Food Microbiol*, **110**, 155–9.

Zur G., Shimoni E., Hallerman E. and Kashi Y. (2002). 'Detection of *Alternaria* fungal contamination in cereal grains by a polymerase chain reaction-based assay'. *J Food Protect*, **65**, 1433–40.

12

Identification of genes and gene clusters involved in mycotoxin synthesis

D. W. Brown, R. A. E. Butchko and R. H. Proctor, United States Department of Agriculture, Agricultural Research Service (USDA-ARS), USA

Abstract: Research methods to identify and characterize genes involved in mycotoxin biosynthetic pathways have evolved considerably. Before whole genome sequences were available (e.g. pre-genomics), work focused primarily on chemistry, biosynthetic mutant strains and molecular analysis of single or relatively small numbers of genes. In recent years, reductions in the cost of DNA sequencing technologies have made genomic methods more widely available. The availability of whole genome sequence for multiple mycotoxin-producing fungi has led to important discoveries both within single genomes and between genomes via comparative genomics (post-genomic). The ability to analyze simultaneously the expression of a large number of genes through expressed sequenced tags (ESTs) and microarrays (transcriptomics) has also had a significant impact on gene cluster identification. This chapter discusses the pre-genomics strategies used to identify toxin biosynthetic genes/gene clusters in fungi as well as more recently developed genomic strategies that greatly enhance the efficiency of the identification process. Pre-genomic researchers focused on analysis of a small number of genes, while post-genomic researchers can compare genomes and examine transcription patterns of thousands of genes at a time to help define biosynthetically related genes involved in mycotoxin synthesis. An examination of two *Fusarium* secondary metabolites (SMs), the mycotoxins fumonisins and a perithecial pigment, provide models for similar studies of other fungi. The identification of additional mycotoxin gene clusters in other fungi will proceed significantly faster in the future using genomic technologies.

Key words: *Fusarium verticillioides*, genomic, microarrays, mycotoxins, secondary metabolites.

12.1 Introduction

Food safety is a critical concern and a highly visible public health issue. Increasing international trade in food and frequent media reports of foodborne illnesses in the public are reminders to health officials and scientists of the need to develop knowledge and new technologies to address current and future health issues. Many national and international agencies are tasked with maintaining a clean and healthy food supply and limiting diseases caused by foodborne pathogens or toxins. An essential component assuring a safe food supply is a thorough understanding of the life cycle of microorganisms that cause foodborne diseases, commodity losses or contamination with toxic chemicals. The filamentous fungi *Aspergillus* and *Fusarium* occur worldwide and include multiple mycotoxin-producing species. Species of each genus can have either a broad or a restricted host range (including humans) and may be endophytes or saprophytes (or both). In addition to the diseases and mycotoxin contamination problems they cause, a few of these fungi are used for industrial production of small acids (e.g. citric acid), high value proteins and pharmaceuticals.

Fungal secondary metabolites (SMs), including mycotoxins, are generally defined as chemicals that are not required to sustain daily life or for propagation (Keller *et al.*, 2005). The veritable 'rainbow of colors' produced by fungi has fascinated people for ages and was an early and clear indication of their chemical potential. The discovery of penicillin and its health and economic value led to a boom in discovery of natural products from fungi and other microorganisms. Studies by major pharmaceutical companies, as well as academic and government research groups, have identified thousands of fungal metabolites (Cole *et al.*, 2003). Some of these, particularly mycotoxins, can be detrimental and others, such as pharmaceuticals, can be beneficial to human activities. Examples of mycotoxins include aflatoxin, fumonisins, trichothecenes, zearalenone, ochratoxins, patulin, fusaric acid and fusarins (Fig. 12.1). Some examples of fungal-derived pharmaceuticals are the anti cholesterol agent lovastatin and the antibiotic cephalosporin (Fig. 12.1). Although the function of most SMs is not known, some (e.g. deoxynivalenol and gibberellins produced by different *Fusarium* species) can play an important role in the plant disease process (Proctor *et al.*, 1995). The work of Gloer and Wicklow, which has spanned the last 23 years and has resulted in the discovery of hundreds of novel compounds from a wide variety of fungi, exemplifies the richness of fungi as a source of novel chemicals (Gloer *et al.*, 1988; Wicklow and Poling, 2009).

12.2 Genetic basis for secondary metabolite biosynthesis

The majority of fungal SMs are synthesized from relative simple building blocks (e.g. acetate or amino acids) via the activities of a small group of enzymes including terpene synthases (TSs), polyketide synthases (PKSs), non-ribosomal peptide synthetase (NRPS) and PKS-NRPS hybrids (Keller *et al.*, 2005). Most of

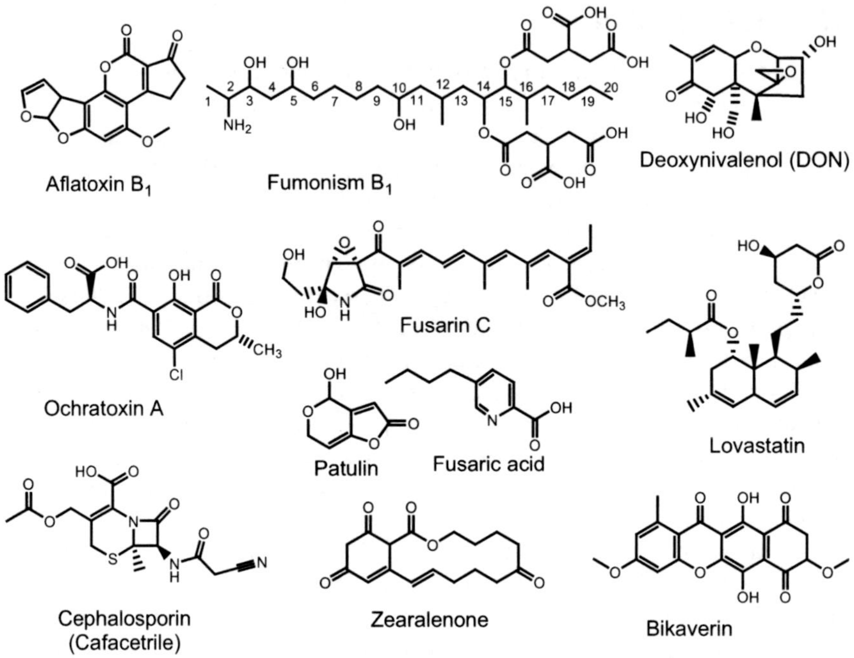

Fig. 12.1 Fungal secondary metabolites. Aflatoxin B_1 of *Aspergillus* species, fumonisin B_1 of *Fusarium verticillioides*, deoxynivalenol (DON) and zearalenone of *F. graminearum*, ochratoxin A and patulin of *Aspergillus* and *Penicillium* species, fusaric acid, bikaverin and fusarin C of *Fusarium* species, lovastatin of *A. terreus* and cephalosporin of *Cephalosporium acremonium*. At present, only fusaric acid synthesis has not yet been linked to a secondary metabolite biosynthetic gene cluster.

the products of these enzyme reactions undergo further enzyme-catalyzed modifications, such as oxygenation, cyclization and isomerization, to complete their respective biochemical pathways. The first intermediate in the aflatoxin, fumonisin, zearalenone and bikaverin biosynthetic pathways are polyketides; the first intermediate in the biosynthesis of the trichothecene deoxynivalenol (DON) is a terpene; and the first intermediate in the fusarin C biosynthetic pathway is a polyketide-peptide hybrid. For most fungal secondary metabolite biosynthetic (SMB) pathways that have been examined to date, genes encoding the synthase enzyme as well as the modifying enzymes are located adjacent to one another in a gene cluster. The discovery that fungal biosynthetic genes were clustered was a paradigm shift in eukaryotic biology (Keller and Hohn, 1997).

Fungal SMB gene clusters have two characteristics in common: (1) they consist of two or more genes located adjacent to each other in a genome and (2) the clustered genes exhibit similar patterns of expression. SMB cluster genes can be divided into distinct categories. All clusters have one to three genes that encode proteins responsible for constructing the carbon skeleton of the terpene, polyketide, peptide or hybrid molecule. The most numerous category of cluster genes are those that encode enzymes that catalyze structural modifications of the carbon skeleton

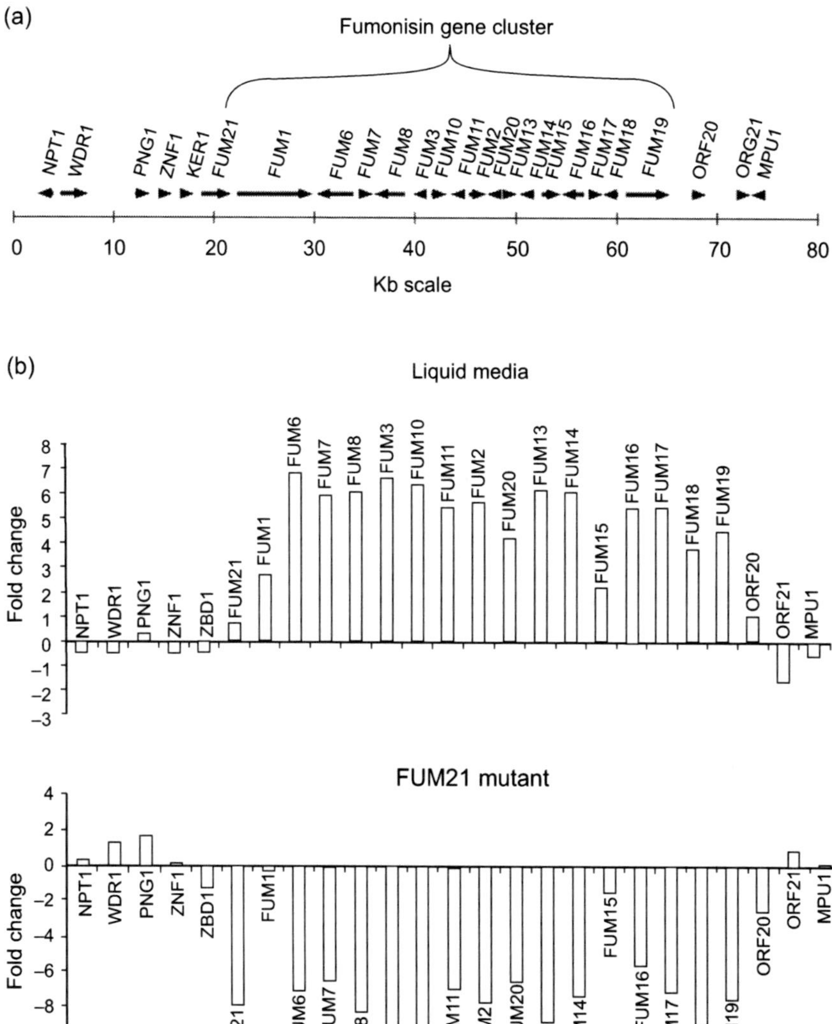

Fig. 12.2 Fumonisin gene cluster. (a) Graphic representation of the fumonisin gene cluster and flanking genes spanning 80 kb. (b) Microarray analysis. The bars in the top graph represent wild-type *F. verticillioides* gene expression fold changes at four days compared to one day growth in liquid fumonisin medium. The bars in the bottom graph represent fold changes in gene expression in a Δ*FUM21* mutant compared to the wild-type *F. verticillioides.*

(e.g. hydroxylases, acyl transferases). Clusters can also include one or more genes or two involved in chemical transport or self-protection (or both). And finally, most clusters include genes involved in either transcriptional or translational regulation. Two archetypical SMB clusters are the 17-gene fumonisin (Fig. 12.2(a)) and 25-gene aflatoxin clusters in *F. verticillioides* and *A. parasiticus*, respectively (Proctor *et al.*, 2003; Yu *et al.*, 2004). Both clusters include genes that encode proteins to synthesize the first intermediate, modify the skeleton, regulate transcription of clustergenes and transport chemical products. Multiple studies have shown co-expression of genes located in each cluster (e.g. see Fig. 12.2(b) for co-expression of fumonisin cluster). In contrast, genes flanking the cluster tend to exhibit different patterns of expression.

Researchers tasked with limiting the impact of a toxic fungal SM on society generally first determine the chemical structure of the metabolite and then examine the biochemistry and genetics of its biosynthetic pathway. The identification and characterization of the genetic basis of biosynthesis of SMs in microorganisms has enjoyed spectacular success and has mirrored an exponential growth in technology in recent years. Historically, interest in SMs focused on biological activity, either as toxins or potential pharmaceuticals. The discovery that production of trichothecene mycotoxins is critical to the ability of *F. graminearum* to cause wheat head blight is the centerpiece of an effort by the agriculture/biotechnology company Syngenta to produce toxin-resistant wheat varieties (Hohn *et al.*, 2002) in order to control the crop disease and resulting mycotoxin contamination problem.

This chapter will focus on the general strategies used to identify genes involved in fungal SM synthesis as well as the new technologies being employed to unravel the potential chemical repertoire of fungal communities associated with food crops and which, therefore, pose a risk to food safety. Today, multiple genomic technologies, including DNA sequencing, microarrays, transcriptomics, comparative genomics, proteomics and metabolomics are utilized to identify and characterize SMB genes. Prior to the availability of genomic sequence data, knowledge of the biochemistry and structure of known metabolites drove studies on the genetic basis of mycotoxin biosynthesis. By contrast, in the genomic era of today, knowledge of existing fungal gene clusters is driving efforts to identify fungal metabolites.

12.3   Gene and gene cluster identification: pre-genomics

Early research on the genetic basis of SM biosynthesis used a variety of techniques to identify key biosynthetic genes. For example, the first trichothecene biosynthetic gene in *F. sporotrichioides* was identified by isolating the terpene synthase that catalyzed synthesis of trichodiene, the first intermediate or parent compound in trichothecene biosynthesis. The purified enzyme was used to generate an antibody, which was in turn used to screen a library of *Fusarium* genomic DNA expressed in *Escherichia coli* (Hohn and Beremand, 1989). A second technique to identify key SM biosynthetic genes utilized mutants that were blocked in production of the mycotoxin. For example, a key aflatoxin biosynthetic gene was identified by complementation of an *A. parasiticus* mutant blocked in aflatoxin production. The mutant was complemented by transformation with cosmid DNA from a wild-type strain (Skory *et al.*, 1992). A third technique used to identify key biosynthetic genes employed conserved sequences within synthase genes coupled with differential gene expression. In this case, the fumonisin PKS gene was first identified in *F. verticillioides* using degenerate PCR primers that were designed based on highly conserved motifs in PKSs from other organisms and a template of DNA generated from mRNA isolated from cultures of the fungus during fumonisin synthesis (Proctor *et al.*, 1999).

Another technique that utilizes differences in gene expression is differential display. This approach was used to identify the bikaverin PKS gene in *F. fujikuroi* (Linnemannstons *et al.*, 2002). In order to identify the gene, mRNA was isolated from the fungus grown under conditions that induced or suppressed bikaverin production. The RNA was then converted to cDNA, which was used in PCR with 27 combinations of random primers. Sequence and subsequent analyses of the resulting amplicons revealed that one, derived from the bikaverin-inducing conditions, was part of the bikaverin PKS gene (Linnemannstons *et al.*, 2002).

The next step all research programs took to identify new SM biosynthetic genes was to determine the nucleotide sequence of the DNA directly flanking both sides of the genes described above, usually on cosmids or by PCR/genome walking. Potential open reading frames (ORFs) were identified and those that encode proteins with predicted functions that are consistent with enzymatic activities thought to be necessary for the formation of the SM(s) in question (e.g. SMB-like genes) and are co-expressed with the original biosynthetic gene, were further analyzed by gene deletion and chemical analysis.

Sequence and transcription analysis of genes flanking the fumonisin PKS (*FUM1*) identified 15 contiguous co-expressed genes with predicted functions consistent with SMB synthesis (Proctor *et al.*, 2003). In such an analysis, the predicted functions of the proteins are based on sequence similarities to gene/proteins with known functions. In some cases, the predicted function of a gene matches very closely with the biochemical data such that the assignment of genes to one pathway step is very clear, given the structure of the metabolite in question. For example, *FUM8* encodes a putative oxoamine synthase, a class of enzymes that catalyzes the condensation of an acyl function with an amino acid.

Precursor feeding studies had previously shown that the amine function of fumonisin B_1 was derived from alanine. Gene deletion and complementation studies provided evidence for this assignment (Alexander *et al.*, 2009). In other cases, multiple genes within a cluster can share the same predicted function. For example, initial characterization of the *FUM* cluster indicated that it included three P450 monooxygenase genes, each of which could have played a role in formation of the C-5 or C-10 hydroxyl groups (Proctor *et al.*, 2003). Gene deletion studies were able to assign two of the three P450s to specific biochemical steps while the function of the third is unclear (Alexander *et al.*, 2009). It is important to note that not all genes in some SMB gene clusters appear to be required for the metabolite synthesis, transport or regulation. Both the fumonisin and aflatoxin gene clusters contain co-expressed genes that do not appear to be required for toxin synthesis (Alexander *et al.*, 2009; Yu *et al.*, 2004). One possibility is that these genes play a role in the biological function of the toxin or the synthesis of minor structural variants.

The discovery of gene clusters in one organism has been used to identify similar clusters, involved in the synthesis of identical or structurally related SMs, in other fungi. An early example of this approach involved the identification of the first gene in *A. nidulans* required for sterigmatocystin synthesis based on its sequence similarity to a gene required for aflatoxin biosynthesis in *A. parasiticus* (Keller *et al.*, 1994). A second example was the identification of the *F. graminearum* core–trichothecene gene cluster involved in DON synthesis based on sequence identity of the *F. sporotrichioides* T-2 toxin synthesis cluster (Brown *et al.*, 2001). More recently, homologous fumonisin clusters have been identified in *F. oxysporum*, *F. proliferatum* and *A. niger* using sequence data from *F. verticillioides* (Baker, 2006; Waalwijk *et al.*, 2004; Proctor *et al.*, 2008).

Although many fungal SM gene clusters have been characterized to date, variations in the single cluster paradigm have been found. For example, the characterization of trichothecene biosynthetic genes (*TRI* genes) in a number of fusaria has revealed that multiple genetic loci are required for toxin synthesis in some while in others, all *TRI* genes appear to be located in a single locus. In *F. sporotrichioides* and *F. graminearum*, *TRI* genes are located at three loci: a 10–12 gene core cluster, a locus with 1 or 2 genes (*TRI1* and *TRI16*) and a third locus with a single gene (*TRI101*) (Brown *et al.*, 2003, 2004). Co-expression of genes at each locus is regulated by *TRI6*, a C_2H_2 DNA binding transcription factor located within the core gene cluster. In contrast the *F. equiseti* core *TRI* cluster includes *TRI1* and *TRI101* (Proctor *et al.*, 2009). Phylogenetic analyses of multiple fusaria support the hypothesis that *TRI1* and *TRI101* moved into the core cluster directly or indirectly from the small *TRI* loci during the evolution of *F. equiseti* and related species (Proctor *et al.*, 2009).

12.4   Gene and gene cluster identification: early-genomics

Expressed sequence tags (ESTs) were the first genomic technology to become available to fungal molecular biologists studying toxin synthesis. An EST is a short nucleotide sequence (generally less than 800 nts) generated from a single sequencing run using a DNA template derived from mRNA. An EST library is created by isolat-ing total RNA from an organism, converting the RNA to DNA (cDNA) and then determining the sequence for ~400 to 800 nucleotides at the 5' and/or 3' ends of thousands to tens of thousands of different cDNA clones. Sequences derived from the same cDNA represent an individual transcript that was present at the point in the life cycle of the fungus when the RNA was isolated. The presence of an EST and its relative abundance in an EST library is a powerful tool (e.g. transcriptomics) to ident-ify genes involved in the same or related biological process. Together, the thousands of sequences in an EST library can represent a large proportion of the total nucleotide sequence of an organism's genome. For example, the ~87,000 EST sequences in the *F. verticillioides* EST library represent at least partial sequences for 78% of the ~14 200 genes estimated to be in the genome of this fungus (Brown *et al.*, 2008).

Fusarium EST collections have had a significant impact on discovery and analysis of mycotoxin/SMB gene clusters. For example, a *F. sporotrichioides* EST library generated from a mutant strain that overexpressed genes involved in trichothecene biosynthesis was used to identify new trichothecene gene(s) (www.genome.ou.edu/fsporo.html) (Brown *et al.*, 2003; Peplow *et al.*, 2003b). In this case, a set of highly expressed ESTs with a similarity to a P450 monooxygenase led to the identification of the *TRI1/TRI16* gene cluster involved in T-2 toxin synthesis (Brown *et al.*, 2003). It is interesting to note that the *TRI16* enzyme catalyzes acylation of the hydroxyl group on carbon atom 8 (C-8) of the core trichothecene molecule and that *TRI16* is a pseudogene in *Fusarium* species (e.g. *F. graminearum*) that produce trichothecenes lacking the C-8 acyl function (Brown *et al.*, 2003; Meek *et al.*, 2003; Peplow *et al.*, 2003a).

In a second example, analysis of ESTs from nine different *F. verticillioides* libraries led to the discovery of two genes in the fumonisin biosynthetic gene cluster that were not identified in the initial characterization of the cluster. The first gene (*FUM21*) encodes a Zn(II)2Cys6 protein that is a transcriptional regulator of the entire fumonisin gene cluster and the second gene (*FUM20*) is of unknown function (Brown *et al.*, 2007). Finally, analysis of ESTs from 13 different *F. graminearum* libraries identified a cluster of co-expressed genes, of which at least one is required for synthesis of the mycotoxin butenolide (Harris *et al.*, 2007).

The biggest impact on SMB gene discovery in fungi has been the development and public availability of whole genomic sequence databases. Today, such databases are available for dozens of filamentous fungi (fungalgenomes.org/wiki/ Fungal_Genome_Links) and additional genome databases for other fungi are being developed. The sizes of genomes of filamentous fungi range, on average, from 20 to 60 million base pairs and code for 8000 to 17 000 genes. One of the first lessons learned from initial annotation efforts was that fungi contain more 'SM biosynthetic potential' than expected. Based on collections of published fungal natural products (Cole *et al.*, 2003), it was expected that genomic sequence data would reveal the presence of a few SMB gene clusters per genome. The minimal and unique combination of chemicals synthesized by a fungus has served as a phenotypic character to identify fungal species (referred to as chemotaxonomy) (Frisvad *et al.*, 2007). However, sequence and annotation of genomes of four species of *Aspergillus* revealed that each contains hundreds of SMB associated genes, a finding that greatly surpassed expectations (Galagan *et al.*, 2005; Machida *et al.*, 2005; Nierman *et al.*, 2005). Despite years of culturing under an array of conditions, production of only a few SMs has been detected in each of these species. This could be due to a variety of factors, including production of metabolites below detection limits or lack of SMB gene expression under the culture conditions employed.

Another lesson learned from early analyses of fungal genomes is that closely related fungi can have significant differences in SMB genes. For example, *A. flavus* and *A. oryzae* share 98% identity at the DNA level. However, each species has numerous SMB-associated genes that are absent in the other. This finding suggests that SMB gene diversity is an important attribute of fungi and may play a significant role in their ability to grow under or adapt to different environmental conditions (Payne *et al.*, 2006; Yu *et al.*, 2008).

Today, a number of research programs are pursuing the production and identification of new SMs by genetically manipulating fungi to induce expression of SMB gene clusters that exhibit little or no expression under typical laboratory conditions (Chiang *et al.*, 2009). One approach is to identify a transcriptional regulatory gene within a predicted SMB cluster and then alter the promoter sequence of the gene so that its expression can be induced experimentally. If the transcriptional regulator controls expression of the other genes in the cluster, inducing its expression should induce expression of the other cluster genes and, therefore, induce production of the SM(s) for which the cluster is responsible.

Concomitant with the increasing availability of affordable sequence data was the development of new bioinformatic tools to identify genes of interest from among millions of base pairs of sequence data. The first step after acquiring a newly aligned genomic sequence is to define all possible genes or ORFs using a variety of annotation programs (Ma *et al.*, 2010; Martinez *et al.*, 2009). In general, between 50 and 60% of the predicted ORFs are assigned a role or function based on sequence identity to genes/proteins whose functions have been previously determined.

One popular collection of known protein families is the *Pfam* database (Finn *et al.*, 2008). To facilitate the identification of fungal SMB gene clusters, an automated prediction program (SMURF) was recently developed based on the proximity of *Pfam* domains associated with secondary metabolism of PKS, TS and NRPS genes (Ma *et al.*, 2010). SMURF, the secondary metabolite unique regions finder, is a web-based tool that can be found at www.jcvi.org/smurf/index.php. Using SMURF, a total of 46 SMB gene clusters were found in whole genome sequences of three *Fusarium* species centered around either a PKS or a TS (Ma *et al.*, 2010). Each cluster contains eight genes on average, with a minimum of four and a maximum of 17 genes (Table 12.1).

The uniqueness of each PKS gene set within *Fusarium* was evident in a phylogenetic analysis of conserved portions of the PKSs: only three PKSs were conserved among four *Fusarium* species (Ma *et al.*, 2010). Two of the three conserved PKSs lie within predicted gene clusters. One of the PKS genes, *PGL1* (formerly *GmPKS3*), lies within the non-reducing PKS clade and is involved in

Table 12.1   SMURF secondary metabolite biosynthetic (SMB) gene clusters in *Fusarium*

Fungus	Total # clusters	R-PKS[1]	NR-PKS	TS
F. verticillioides	16	2	13	1
F. oxysporum	12	2	9	1
F. graminearum	18	5	8	5

[1]Core cluster biosynthetic enzyme: R-PKS refers to reducing type polyketide synthase, NR-PKS refers to non-reducing PKS and TS refers to terpene synthase. Adapted from Ma *et al.*, 2010.

perithecial pigment synthesis, while the other, *GmPKS7*, lies within the reducing PKS clade 1 and encodes a protein of unknown function. It is important to note that none of the SMURF clusters described included an NRPS (Ma *et al.*, 2010). As mentioned previously, NRPSs do play an important role in the synthesis of fungal SMs (e.g. beauvercin and enniatins) (reviewed in Desjardins and Proctor, 2007). The recent identification of 20 NRPS genes in *F. graminearum* (Bushley and Turgeon, 2010) indicates the significant potential role that this gene family may play in SM synthesis in this fungus as well as other *Fusarium* species.

12.5   Gene and gene cluster identification: post-genome genomics

12.5.1 Transcriptomics via microarrays

The post-genome genomic tool that has had the most significant impact to date on our understanding of gene clusters is microarray analysis. Microarrays are designed to examine the level of transcript expression of all genes in a genome at the same time. A microarray consists of tens of thousands of small pieces (e.g. 25 to 60-mer oligonucleotide) of DNA (or probes) that can collectively correspond to all known genes in the genome of the target organism. The probes are covalently attached to the surface of a glass slide; they can be synthesized beforehand and subsequently attached to discrete positions on the slide or synthesized directly on the slide. In microarray analysis, the probes on the slide are hybridized to mRNA-derived cDNA, and hybridization is detected and quantified via fluorescent or chemiluminescent markers that are attached to the cDNA. There are two basic types of information generated by microarrays: 1) What genes are on or off at one time (or experimental condition) and 2) Differences in expression. In the first case, the microarray data provides information about the relative gene expression levels of all of the genes represented on the array at a specific time/experimental condition. In the second case, the expression level observed at one condition is compared to the expression level observed at another condition. Common types of microarray experiments can compare gene expression in an organism grown in different media, for different lengths of time in the same medium, or in response to different stimuli (e.g. light, chemical or temperature). Microarray experiments can also be used to compare gene expression in different strains of the same organism (e.g. mutant versus wild-type progenitor).

Analysis of the fumonisin cluster genes by microarray provides an example of the power of microarray technology to identify co-regulated genes (Fig. 12.2(b)). The microarrays used to examine transcription of fumonisin biosynthetic genes were designed in collaboration with JCV (The J. Craig Venter Institute) and were generated by NimbleGen Systems (Madison, WI). The microarrays consist of approximately 336 000 sixty-nucleotide probes, with each gene in the *F. verticillioides* genome represented by a set of up to 12 probes. Nucleotide labeling, hybridization and data acquisition were done by NimbleGen Systems. Data were normalized for each chip (replica of the microarray) and between chips using NimbleScan Bioanalyzer software, utilizing Robust Multichip Analysis (Irizarry *et al.*, 2003) and analyzed using Acuity 4.0 from Molecular Devices Corporation (Sunnyvale, CA). Total RNA from two biological replications were extracted using Trizol reagent (Invitrogen™) following established protocols (Brown *et al.*, 2008). In one microarray experiment, gene expression was compared in mycelia of wild-type *F. verticillioides* and a *FUM21* deletion mutant (Δ*FUM21*) at 24, 48, 72, 96 and 120 hours growth in liquid fumonisin production medium. The analysis detected transcripts for 16 of the 17 fumonisin cluster genes in the wild type after 48 hours and confirmed that the genes were differentially expressed over time relative to genes flanking the cluster (Fig. 12.2(b)). In contrast, expression of the 16 fumonisin biosynthetic genes was not detected in the Δ*FUM21* mutant. These results are consistent with previous work and confirm that *FUM21* is a pathway-specific, positive-acting transcriptional activator of other fumonisin biosynthetic genes (Brown *et al.*, 2007).

Keller and co-workers have elegantly exploited microarray analysis along with LaeA, a global regulator of secondary metabolism in *Aspergillus*, to identify novel SM gene clusters. Microarray analysis revealed that deletion of the LaeA gene (*laeA*) blocks the expression of multiple gene clusters while overexpression of *laeA* enhances expression of other clusters (Bok and Keller, 2004). This approach was used successfully to identify and characterize a gene cluster involved in the synthesis of the antitumor compound terrequinone A that had not previously been described from *A. nidulans* (Bok *et al.*, 2006). More recently, Keller and co-workers have identified a second global regulator of gene expression called CclA that is involved in the chromatin-level regulation of SMB gene cluster expression. Deletion of the CclA gene (*cclA*) resulted in expression of previously silent gene clusters and the production of metabolites also not previously observed in *A. nidulans* cultures (Bok *et al.*, 2009). LaeA homologs in other filamentous fungi have also proven to be global regulators of SMBs. In *F. fujikuroi* and *F. verticillioides*, the production of different SMs were differentially affected in *laeA* deletion mutants (Wiemann *et al.*, 2010 and Butchko, unpublished observations). The *F. fujikuroi laeA* homolog, *FflaeA*, can simultaneously act as a positive (gibberellins and fumonisins) and negative (bikaverin) regulator of secondary metabolism (Wiemann *et al.*, 2010). The identification and exploitation of additional global regulators of secondary metabolism will continue to have a significant impact on the characterization of fungal gene clusters in the future.

12.5.2 Transcriptomics and comparative genomics

Analysis of orthologues of the PKS gene *PGL1* and flanking regions in *F. graminearum*, *F. verticillioides* and *N. haematococca* (anamorph *F. solani* f. sp. *pisi*) provides an example of another strategy that can aid in identification of genes and gene clusters involved in SM biosynthesis. Initial analysis of *F. graminearum* genomic sequence identified 15 putative PKS genes and deletion analysis demonstrated that one, *PGL1*, was required for production of the dark violet pigment in

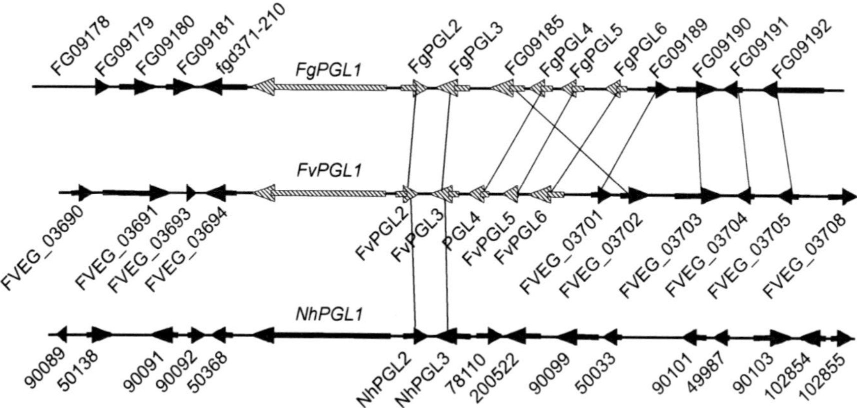

Fig. 12.3 Comparative genomic and microarray analysis of *pgl1* gene cluster of *F. graminearum* (top), *F. verticillioides* (middle) and *N. haematococca* (bottom). Gene designations correspond to the gene models assigned by the Broad and Joint Genome Institute (JGI) sequencing centers, respectively. Lines drawn between genes from *F. graminearum* and *F. verticillioides* indicated orthologues. Shaded arrows indicate genes that appear to be co-regulated.

the walls of sexual fruiting bodies (perithecia) (Gaffoor *et al.*, 2005). Although the involvement of *PGL1*-flanking genes in synthesis of the pigment was not assessed, the predicted proteins of some of these genes share similarity to SMB genes. Thus, it is possible that some of the flanking genes are involved in synthesis of the pigment.

Further evidence for a biosynthetic gene cluster for the perithecial pigment was obtained from analysis of genomic sequence from *F. verticillioides* and *N. haematococca*. The genomes of these two species each include a PKS gene sequence that shared between 70% and 82% identity with the *F. graminearum PGL1* (*FgPGL1*) (Proctor *et al.*, 2007). Gene deletion analysis confirmed that the *F. verticillioides* orthologue (*FvPGL1*) is also required for perithecial pigmentation (Proctor *et al.*, 2007). In contrast, *N. haematococca* perithecial pigments are red and require a different PKS (Graziani *et al.*, 2004; Proctor *et al.*, 2007). Therefore, Proctor *et al.* (2007) reasoned that genes flanking *PGL1* that are also required for synthesis of the perithecial pigment would be conserved in *F. graminearum* and *F. verticillioides*. Analysis of ~12 and 25 kb of DNA flanking the *PGL1* orthologs in the three fungi revealed that within the 3'-flanking region of the three species, none of the genes were orthologous. In contrast, within the 5'-flanking region, the two genes immediately upstream of *PGL1* were orthologous in all three species, and the eight genes immediately upstream were orthologous in *F. graminearum* and *F. verticillioides*, although the order and orientation of one gene differed in the two species (Fig. 12.3) (Proctor *et al.*, 2007).

In another study, analysis of Pfam domains flanking *PGL1* in *F. verticillioides* and *F. graminearum* identified six and seven likely SM genes respectively (Ma *et al.*, 2010). In *F. graminearum*, the location of FG09185, encoding a protein of

Table 12.2   Differential expression of the perithecial pigment gene cluster

Fg (0 to 96 hr)[a]	Fold change	Fv (24 to 96 hr)[b]	Fold change
Unknown A[c] (Fgd371-210_at)	4.0	Oxidase (FVEG_03694)	−3.5
PKS (FgPGL1 or Fg12125_at)	32	**PKS** (FVEG_03695)	3.1
Methyltransferase (FgFGL2 or Fgd371-230_at)	64	**Methyltransferase** (FVEG_03696)	5.8
Monooxygenase (FgPGL3 or Fgd371-240_at)	~700	**Monooxygenase** (FVEG_03697)	6
Unknown B (FG09185 or Fgd371-250_at)	-0.6		
Oxidoreductase (FgPGL4 or Fgd371-260_at)	4	**Oxidoreductase** (FVEG_03698)	3.8
Dehydrogenase (FgPGL5 or Fg09187_s_at)	3.5	**Dehydrogenase** (FVEG_03699)	6.2
Zn2Cys6 (FgPGL6 or Fgd371-280_at)	90	**Zn2Cys6** (FVEG_03700)	3.2
Unknown C (FG09189 or Fgd371-290_at)	2	Unknown C (FVEG_03701)	0.8
		Unknown B (FVEG_03702)	0.6
Transporter (FG09190 or Fgd371-300_at)	1.8	Transporter (FVEG_03703)	−0.8

[a]The *F. graminearum* microarray data was obtained from www.plexdb.org/plex.php?database=Fusarium; Experiment FG5. In this experiment, total *F. graminearum* RNA was isolated in 24-hr increments after sexual development was induced. The genes shown in bold are believed to be part of the *PGL1* gene cluster. Fold change is the difference between 0 hr and 96 hr after induction.
[b]RNA was extracted from cultures of *F. verticillioides* grown on liquid GYAM for 24, 48, 72, 96 and 120 hr. Fold change is the difference between 0 hr and 96 hr after inoculation.
[c]Fg (*F. graminearum*) and Fv (*F. verticillioides*) genes designated with the same names are putative homologs. The gene name in parenthesis are designations included at the Broad Institute web site (www.broadinstitute.org/annotation/genome/fusarium_verticillioides/MultiHome.html).

unknown function, near *PGL1* and predicted SMB genes suggests that it may also be involved in perithecial pigment synthesis. In contrast, the observation that the *F. verticillioides* homolog (FVEG_03702) is not at the same relative genomic location suggests that FG09185 may not be involved. In addition, microarray data from both *F. verticillioides* and *F. graminearum* indicate that the pattern of expression of FVEG_03702/FG09185 is quite different from the expression of *PGL1* and six other SMB-like genes flanking *PGL1* (Table 12.2). Determination of whether *PGL1*-flanking genes are involved in perithecial pigment biosynthesis, and therefore whether they and *PGL1* constitute a SMB gene cluster, will require functional analysis of the genes.

12.6   Future trends

Fungal secondary metabolites, mycotoxins and food safety will continue to be of critical interest to a variety of researchers for years to come. The impact of genomic technologies on the discovery of novel metabolites has been phenomenal and is reflected by the tremendous increase in and quality of publications of studies on fungal secondary metabolism. Today, one critical bottleneck in such studies is methods to determine gene function. Technical difficulties related to low fungal transformation efficiencies coupled with even lower frequency of homologous integration events needed to create deletion mutants, as well as the need to create individual disruption vectors for each target gene, continue to hinder our ability to assess SMB gene function.

The development of new technologies to increase the efficiencies of methods used for analysis of gene function is critical for future progress. For example, RNAi technology involving the transcriptional inhibition of a target gene after heterologous integration of transforming DNA, is having a limited but growing impact on the field. The availability of a better characterized model system will also improve understanding of the genetic basis of fungal secondary metabolism. The NIH funded *Neurospora crassa* arrayed knockout project targeting up to 10 000 genes will have a transformative effect on our understanding of both primary and secondary metabolic activities in filamentous fungi (www.dartmouth. edu/~neurosporagenome/).

Finally, new advances in transcriptomics, proteomics and metabolomics will continue to advance understanding of fungal secondary metabolism. The most exciting advance in transcriptomics is the development of next-generation sequencing technology (Tan *et al.*, 2009). Roche, ABI and Illumina machines can generate sequences covering 5 to 10 times the average fungal genome in two weeks for ~1% of the cost of just a few years ago. Direct transcript profiling is now possible and provides greater sensitivity than microarray analysis as well as less bias because it does not require *a priori* knowledge of gene sequences. The major advantage of direct transcript profiling is also a drawback; this technology generates an enormous quantity of data and will require development of new bioinformatic tools before it can be more widely used. As more fungal genomes are sequenced and more gene expression studies are published, our understanding of the chemical biosynthetic potential of fungi will improve. This improved understanding will provide insight into how to reduce mycotoxin contamination of crop plants and the food/feed derived thereof.

12.7   Acknowledgements

The authors thank Chris McGovern, Crystal Probyn and Marcie Moore for technical assistance. Mention of trade names or commercial products in this article is solely for the purpose of providing specific information and does not imply recommendation or endorsement by the US Department of Agriculture.

12.8   References

Alexander, N.J., Proctor, R.H. and McCormick, S.P. (2009). 'Genes, gene clusters, and biosynthesis of trichothecenes and fumonisins in *Fusarium*'. *Toxin Rev.*, **24**, 198–215.

Baker, S.E. (2006). '*Aspergillus niger* genomics: past, present and into the future'. *Med. Mycol.*, **44** Suppl 1, S17–21.

Bok, J.W. and Keller, N.P. (2004). 'LaeA, a regulator of secondary metabolism in *Aspergillus spp*'. *Eukaryot. Cell.*, **3**, 527–35.

Bok, J.W., Hoffmeister, D., Maggio-Hall, L.A., Murillo, R., Glasner, J.D. and Keller, N.P. (2006). 'Genomic mining for *Aspergillus* natural products'. *Chem. Biol.*, **13**, 31–7.

Bok, J.W., Chiang, Y.M., Szewczyk, E., Reyes-Dominguez, Y., Davidson, A.D., Sanchez, J.F., Lo, H.C., Watanabe, K., Strauss, J., Oakley, B.R., Wang, C.C. and Keller, N.P. (2009). 'Chromatin-level regulation of biosynthetic gene clusters'. *Nat. Chem. Biol.*, **5**, 462–4.

Brown, D.W., McCormick, S.P., Alexander, N.J., Proctor, R.H. and Desjardins, A.E. (2001). 'A genetic and biochemical approach to study trichothecene diversity in *Fusarium sporotrichioides* and *Fusarium graminearum*'. *Fungal Genet. Biol.*, **32**, 121–33.

Brown, D.W., Proctor, R.H., Dyer, R.B. and Plattner, R.D. (2003). 'Characterization of a *Fusarium* 2-gene cluster involved in trichothecene C-8 modification'. *J. Agric. Food Chem.*, **51**, 7936–44.

Brown, D.W., Dyer, R.B., McCormick, S.P., Kendra, D.F. and Plattner, R.D. (2004). 'Functional demarcation of the *Fusarium* core trichothecene gene cluster'. *Fungal Genet. Biol.*, **41**, 454–62.

Brown, D.W., Butchko, R.A., Busman, M. and Proctor, R.H. (2007). 'The *Fusarium verticillioides FUM* gene cluster encodes a Zn(II)2Cys6 protein that affects *FUM* gene expression and fumonisin production'. *Eukaryot. Cell*, **6**, 1210–8.

Brown, D.W., Butchko, R.A. and Proctor, R.H. (2008). 'Genomic analysis of *Fusarium verticillioides*'. *Food Addit. Contam.*, **25**, 1158–65.

Bushley, K.E. and Turgeon, B.G. (2010). 'Phylogenomics reveals subfamilies of fungal nonribosomal peptide synthetases and their evolutionary relationships'. *BMC Evol. Biol.*, **10**, 26.

Chiang, Y.M., Lee, K.H., Sanchez, J.F., Keller, N.P. and Wang, C.C. (2009). 'Unlocking fungal cryptic natural products'. *Nat. Prod. Commun.*, **4**, 1505–10.

Cole, R.J., Jarvis, B.B. and Schweikert, M.A. (2003). *Handbook of Secondary Fungal Metabolites*. Academic Press, San Diego.

Desjardins, A.E. and Proctor, R.H. (2007). 'Molecular biology of *Fusarium* mycotoxins'. *Int. J. Food Microbiol.*, **119**, 47–50.

Finn, R.D., Tate, J., Mistry, J., Coggill, P.C., Sammut, S.J., Hotz, H.R., Ceric, G., Forslund, K., Eddy, S.R., Sonnhammer, E.L. and Bateman, A. (2008). 'The Pfam protein families database'. *Nucleic Acids Res.*, **36**, D281–8.

Frisvad, J.C., Larsen, T.O., de Vries, R., Meijer, M., Houbraken, J., Cabanes, F.J., Ehrlich, K. and Samson, R.A. (2007). 'Secondary metabolite profiling, growth profiles and other tools for species recognition and important *Aspergillus* mycotoxins'. *Stud. Mycol.*, **59**, 31–7.

Gaffoor, I., Brown, D.W., Plattner, R., Proctor, R.H., Qi, W., and Trail, F. (2005). 'Functional analysis of the polyketide synthase genes in the filamentous fungus *Gibberella zeae* (anamorph *Fusarium graminearum*)'. *Eukaryot. Cell*, **4**, 1926–33.

Galagan, J.E. *et al.* (2005). 'Sequencing of *Aspergillus nidulans* and comparative analysis with *A. fumigatus* and *A. oryzae*'. *Nature*, **438**, 1105–15.

Gloer, J.B., TePaske, M.R., Sima, J.S., Wicklow, D.T. and Dowd, P.F. (1988). 'Antiinsectan aflavinine derivatives from the sclerotia of *Aspergillus flavus*'. *J. Org. Chem.*, **53**, 5457–60.

Graziani, S., Vasnier, C. and Daboussi, M.J. (2004). 'Novel polyketide synthase from *Nectria haematococca*'. *Appl. Environ. Microbiol.*, **70**, 2984–8.

Harris, L.J., Alexander, N.J., Saparno, A., Blackwell, B., McCormick, S.P., Desjardins, A.E., Robert, L.S., Tinker, N., Hattori, J., Piche, C., Schernthaner, J.P., Watson, R. and Ouellet, T. (2007). 'A novel gene cluster in *Fusarium graminearum* contains a gene that contributes to butenolide synthesis'. *Fungal Genet. Biol.*, **44**, 293–306.

Hohn, T.M. and Beremand, M.N. (1989). 'Isolation and nucleotide sequence of a sesquiterpene cyclase gene from the trichothecene-producing fungus *Fusarium sporotrichioides*'. *Gene*, **79**, 131–8.

Hohn, T.M., Peters, C. and Salmeron, J. (2002). '*Trichothecene-resistant Transgenic Plants*. Syngenta Participations AG. US Patent 6,346,655.

Irizarry, R.A., Hobbs, B., Collin, F., Beazer-Barclay, Y.D., Antonellis, K.J., Scherf, U. and Speed, T.P. (2003). 'Exploration, normalization, and summaries of high density oligonucleotide array probe level data'. *Biostatistics*, **4**, 249–64.

Keller, N.P. and Hohn, T.M. (1997). 'Metabolic pathway gene clusters in filamentous fungi'. *Fungal Genet. Biol.*, **21**, 17–29.

Keller, N.P., Kantz, N.J. and Adams, T.H. (1994). '*Aspergillus nidulans verA* is required for production of the mycotoxin sterigmatocystin'. *Appl. Environ. Microbiol.*, **60**, 1444–50.

Keller, N.P., Turner, G. and Bennett, J.W. (2005). 'Fungal secondary metabolism – from biochemistry to genomics'. *Nat. Rev. Microbiol.*, **3**, 937–47.

Linnemannstons, P., Schulte, J., del Mar Prado, M., Proctor, R.H., Avalos, J. and Tudzynski, B. (2002). 'The polyketide synthase gene *pks4* from *Gibberella fujikuroi* encodes a key enzyme in the biosynthesis of the red pigment bikaverin'. *Fungal Genet. Biol.*, **37**, 134–48.

Ma, L.-J. *et al.* (2010). 'Comparative genomics reveals mobile pathogenicity chromosomes in *Fusarium oxysporum*'. *Nature*, **464**, 367–73.

Machida, M. *et al.* (2005). 'Genome sequencing and analysis of *Aspergillus oryzae*'. *Nature*, **438**, 1157–61.

Martinez, D. *et al.* (2009). 'Genome, transcriptome, and secretome analysis of wood decay fungus *Postia placenta* supports unique mechanisms of lignocellulose conversion'. *Proc. Natl. Acad. Sci.*, **106**, 1954–9.

Meek, I.B., Peplow, A.W., Ake Jr, C., Jr., Phillips, T.D. and Beremand, M.N. (2003). '*Tri1* encodes the cytochrome P450 monooxygenase for C-8 hydroxylation during trichothecene biosynthesis in *Fusarium sporotrichioides* and resides upstream of another new *Tri* gene. Appl. Environ. Microbiol., **69**, 1607–1613.

Nierman, W.C. *et al.* (2005). 'Genomic sequence of the pathogenic and allergenic filamentous fungus *Aspergillus fumigatus*'. *Nature*, **438**, 1151–6.

Payne, G.A., Nierman, W.C., Wortman, J.R., Pritchard, B.L., Brown, D., Dean, R.A., Bhatnagar, D., Cleveland, T.E., Machida, M. and Yu, J. (2006). 'Whole genome comparison of *Aspergillus flavus* and *A. oryzae*'. *Med. Mycol.*, **44**, 9–11.

Peplow, A.W., Meek, I.B., Wiles, M.C., Phillips, T.D. and Beremand, M.N. (2003a). '*Tri16* Is required for esterification of position C-8 during trichothecene mycotoxin production by *Fusarium sporotrichioides*'. *Appl. Environ. Microbiol*, **69**, 5935–40.

Peplow, A.W., Tag, A.G., Garifullina, G.F. and Beremand, M.N. (2003b). 'Identification of new genes positively regulated by *Tri10* and a regulatory network for trichothecene mycotoxin production'. *Appl. Environ. Microbiol.*, **69**, 2731–6.

Proctor, R.H., Hohn, T.M. and McCormick, S.P. (1995). 'Reduced virulence of *Gibberella zeae* caused by disruption of a trichothecene toxin biosynthetic gene'. *Mol. Plant Microbe Interact.*, **8**, 593–601.

Proctor, R.H., Desjardins, A.E., Plattner, R.D. and Hohn, T.M. (1999). 'A polyketide synthase gene required for biosynthesis of fumonisin mycotoxins in *Gibberella fujikuroi* mating population A'. *Fungal Genet. Biol.*, **27**, 100–12.

Proctor, R.H., Brown, D.W., Plattner, R.D. and Desjardins, A.E. (2003). 'Co-expression of 15 contiguous genes delineates a fumonisin biosynthetic gene cluster in *Gibberella moniliformis*'. *Fungal Genet. Biol.*, **38**, 237–49.

Proctor, R.H., Butchko, R.A., Brown, D.W. and Moretti, A. (2007). 'Functional characterization, sequence comparisons and distribution of a polyketide synthase gene required for perithecial pigmentation in some *Fusarium* species'. *Food Addit. Contam.*, **24**, 1076–87.

Proctor, R.H., Busman, M., Seo, J.A., Lee, Y.W. and Plattner, R.D. (2008). 'A fumonisin biosynthetic gene cluster in *Fusarium oxysporum* strain O-1890 and the genetic basis for B versus C fumonisin production'. *Fungal Genet. Biol.*, **45**, 1016–26.

Proctor, R.H., McCormick, S.P., Alexander, N.J. and Desjardins, A.E. (2009). 'Evidence that a secondary metabolic biosynthetic gene cluster has grown by gene relocation during evolution of the filamentous fungus *Fusarium*'. *Mol. Microbiol.*, **74**, 1128–42.

Skory, C.D., Chang, P., Cary, J. and Linz, J.E. (1992). 'Isolation and characterization of a gene from *Aspergillus parasiticus* associated with the conversion of versicolorin A to sterigmatocystin in aflatoxin biosynthesis'. *Appl. Environ. Microbiol.*, **58**, 3527–37.

Tan, K.C., Ipcho, S.V., Trengove, R.D., Oliver, R.P. and Solomon, P.S. (2009). 'Assessing the impact of transcriptomics, proteomics and metabolomics on fungal phytopathology'. *Mol. Plant. Pathol.*, **10**, 703–15.

Waalwijk, C., van der Lee, T., de Vries, I., Hesselink, T., Arts, J. and Kema, G.H.J. (2004). 'Synteny in toxigenic *Fusarium* species: The fumosin gene cluster and the mating type region as examples.' *Eur. J. Plant. Pathol.*, **110**, 533–34.

Wicklow, D.T. and Poling, S.M. (2009). 'Antimicrobial activity of pyrrocidines from *Acremonium zeae* against endophytes and pathogens of maize'. *Phytopathology*, **99**, 109–15.

Wiemann, P., Brown, D.W., Kleigrewe, K., Bok, J.W., Keller, N.P., Humpf, H.U. and Tudzynski, B. (2010). 'FfVel1 and FfLae1, components of a velvet-like complex in *Fusarium fujikuroi*, affect differentiation, secondary metabolism and virulence'. *Mol. Microbiol.*, **77**(4), 972–94.

Yu, J., Chang, P.K., Ehrlich, K.C., Cary, J.W., Bhatnagar, D., Cleveland, T.E., Payne, G.A., Linz, J.E., Woloshuk, C.P. and Bennett, J.W. (2004). 'Clustered pathway genes in aflatoxin biosynthesis'. *Appl. Environ. Microbiol.*, **70**, 1253–62.

Yu, J., Payne, G.A., Nierman, W.C., Machida, M., Bennett, J.W., Campbell, B.C., Robens, J.F., Bhatnagar, D., Dean, R.A. and Cleveland, T.E. (2008). '*Aspergillus flavus* genomics as a tool for studying the mechanism of aflatoxin formation'. *Food Addit. Contam.*, **25**, 1152–7.

13

DNA barcoding of toxigenic fungi: a perspective

M. Santamaria, National Research Council (CNR), Italy

Abstract: The 'DNA barcode' is a recent initiative aiming to provide a rapid and cost-effective protocol to identify unequivocally species in a particular taxonomic group on the basis of a short DNA fragment taken from a standardized genome portion. To date, this approach has been tested most widely in the animal kingdom where an approximately 650-pb long region belonging to the mitochondrial cytochrome c oxidase 1 (CO1) gene seems to have a high resolution power at the species level. The possibility of applying DNA barcoding to the identification of fungal species has been investigated. In particular, it is possible to imagine how much this type of standardized and rapid species recognition approach could improve the diagnosis of toxigenic fungi in food. Lately several nuclear DNA sequences have been widely used in order to classify taxonomical relationships in fungi but poor attention has been paid to standardization. Instead the DNA barcode introduces the important innovation of a highly standardized protocol, which is so rapid and practical as to be easily used both by researchers involved in fungi species definition studies and by non-experts for practical uses. For example, an effective fungi barcode system would have a great research and commercial value if it was applied to *Fusarium,* which is often the principal object of interest of plant pathologists and regulators guarding food safety. The possibility of adopting definite mitochondrial regions as fungi barcode markers, for example the CO1 gene, has been considered. Unfortunately, the presence of mobile introns in almost all fungal mitochondrial genes makes PCR and bioinformatic surveys of these markers seriously difficult. Currently, the most attractive alternative choice is represented by the nuclear ribosomal internal transcribed spacer (ITS), which has already been widely used as a species marker for fungi. Indeed, it has been proposed to the 'Consortium for the Barcode of Life' (CBOL) as a standard barcode for this kingdom.

Key words: DNA barcode, fungal species, species molecular marker, standardization.

13.1   DNA barcode: a new opportunity to discriminate fungi species

DNA has been used for a long time to classify taxonomical relationships, especially within the fungi where it is often difficult to discriminate species on the basis of their morphological traits, but recently a new initiative called 'DNA Barcode of Life' has updated this concept, supported by the advent of efficient DNA amplification and sequencing methods, combined with advances in computing and information technology. The integration of the classical species identification methods with a rapid taxonomic discrimination system based on DNA has been the main aim of the creation of the Consortium for the Barcode of Life (CBOL) (Barcode of Life website http://www.barcodinglife.org/) which aspires to associate a DNA barcode to every species.

Indeed, the general aim of DNA barcoding is unequivocally to identify a species in a particular domain of life, on the basis of a short DNA fragment taken from a standardized portion of the genome. An essential requirement for the efficacy of a barcode marker is that it should exhibit sufficient variability to allow discrimination between species and low variability within individuals belonging to the same species (Hebert *et al.*, 2003). Moreover, the barcode marker capacity of discriminating between closely related taxa should be complemented by the possibility of designing DNA primers for polymerase chain reaction (PCR) amplification in wide taxonomic ranges.

The main novelty of DNA barcoding is the great attention paid to standardization: the same PCR primers are adopted to amplify a fixed small set of molecular markers able to discriminate species in a definite huge taxonomic range, the same protocol is performed for species-markers sequence data production and analysis and common reference databases inclusive of all barcoded species associated to the same set of annotations are created. The high degree of standardization emphasizes the practical implications of this new method which could be used also by non-specialists to identify species quickly and cheaply.

To date, this approach has been tested most widely in the animal kingdom where an approximately 650-pb long region belonging to the mitochondrial cytochrome c oxidase 1 (CO1) gene seems to have a high resolution power at a species level. Indeed, the results of these analyses suggest that this protocol can prove to be very effective and provide a high-throughput in the assignment of unknown specimens to known species in several large groups of animals, such as fish (Ward *et al.*, 2005), birds (Hebert *et al.*, 2004a), spiders (Barret and Hebert, 2005), cowries (Meyer and Paulay, 2005) and lepidoptera (Hebert *et al.*, 2004b). Currently a DNA barcoding protocol has being created also for plants (Kress *et al.*, 2005; Lahaye *et al*, 2008), macroalgae (Saunders, 2005), protists (Scicluna *et al.*, 2006) and fungi (Summerbell *et al.*, 2005). However, despite its broad effectiveness in species discrimination and identification, it seems that the DNA barcode is not very helpful in resolving higher taxonomic groups and inappropriate for reconstructing phylogenies (Greenstone *et al.*, 2005). Nonetheless it may provide a useful tool for taxonomists when they must identify species in a sample.

Alternative genome regions have been suggested for barcode species identification when CO1 gene sequences cannot be produced robustly or when there are too many differences within the same species (Vences *et al.*, 2005). For example 18S rDNA has been used as a species molecular marker in soil nematodes (Blaxter, 2004). A particular case is represented by plants where the mitochondrial DNA sequences, including CO1, evolve too slowly to be proposed for discriminating at the species level. Researchers are evaluating alternative species markers for this group of organisms (Kress *et al.*, 2005).

In conclusion DNA barcoding, based on one or a few universal species markers, could have good potential for much wider use in application to all kingdoms of life: it is based on DNA, whose language and techniques are universally shared. In particular, it is possible to imagine how much such a standardized and rapid species recognition approach could support the 'Fungi species definition challenge' and improve the diagnosis of toxigenic fungi.

13.2   Future trends in DNA barcoding of fungi

The possibility of using a DNA barcoding protocol, which is highly standardized, rapid and sufficiently practical to be easily used both by researchers involved in species definition studies and by non-experts, has been explored in fungi. For example, an effective fungi barcode system would have great research and commercial value if it was applied to species involved in food production, safety and quality, such as those belonging to *Fusarium* genus, which is often the principal object of interest of plant pathologists and regulators guarding food security.

The mitochondrial sequences are very attractive as barcode markers in Fungi, owing to their particular features including simple genetic structure, high copy number, fast rates of evolution compared with the nuclear DNA and rarity of repetitive regions, which is particularly important when orthologous sequences pairs are compared.

The mycologists participating to the Canadian Barcode of Life Network, which began in 2003, investigated the potential use of mitochondrial CO1 gene as species marker in fungi but obtained poor results. Follow-up studies resulted in contradictory conclusions. Seifert *et al.* (2007) demonstrated that the CO1 gene could be easily amplified and aligned, owing to the lack of indels in the subgenus *Penicillium*, providing a large barcode dataset which approximately discriminated two-thirds of the species considered. On the other hand, relatively high similarity of the CO1 sequences between the species belonging to the *Penicillium* genus was observed by Chen *et al.* (2009). Indeed, they observed a certain degree of cross-reaction during their CO1 barcode oligonucleotide array experiments.

There are three main reasons which would exclude the commonly used animal species molecular marker as a fungi barcode: the first is that such marker is not 'universal' enough and therefore will be fruitful only for some fungal taxa; the second is the presence of mobile introns in almost all the fungal mitochondrial genes which potentially interfere with PCR amplification of the barcode region; and finally, fungal universal primer pairs are not available for the CO1 gene. Recent papers show that, even if, in general, the mitochondrial genes are characterized by a high degree of genetic divergence, suggesting good resolution power for low-level taxonomic relationships, they also show a high pervasiveness of non-coding regions in *Penicillium, Basidiomycota* and *Ascomycota* (Seifert *et al.*, 2007; Santamaria *et al.*, 2009; Vialle, 2009). The introns pervasiveness and the variable resolution power of the marker which is observed among the different fungal taxa could seriously compromise the standardization of a barcode protocol. Finally another obstacle to the use of mitochondrial genes as barcodes, not only in fungi, is represented by the possibility of meeting nuclear copies of mitochondrial DNA (NUMTs) which could produce unaccountable results in comparative analyses.

However, the remarkable work to standardize the protocol continues. The nuclear ribosomal RNA (rRNA) cluster has been extensively used for the fungal phylogeny reconstruction. Usually, the 18S and the 28S sequences allow the discrimination of higher taxonomic groups, such as families or genera, while the internal transcribed spacer regions (ITS) are more effective in the differentiation of species. Because of the broad research on the ITS as a molecular taxonomic marker in fungi (Druzhinina *et al.*, 2005), the barcode scientists have focused their attention on this gene which has been proposed to the CBOL for adoption as fungal barcode (Seifert, 2009). Indeed the Barcode of Life Data System (www.barcoding life.org; Ratnasingham and Hebert, 2007) was the first database to include ITS sequences as barcodes (Seifert, 2008).

ITS can be easily amplified by means of PCR using a definite set of primers (Gardes and Bruns, 1993). Moreover the inter-species genetic variability of this region is often larger than the intra-species one. This feature has allowed effective use of this marker for fungal taxa identification (Katsu *et al.*, 2004; Vialle, 2009).

However, several tests to evaluate the potential of the ITS as fungal barcode marker have shown that important limits exist for some taxonomic groups. Indeed, sometimes a very low genetic variability has been observed in this region when closely related species are analysed. In particular ITS appeared to be ineffective in discriminating well-characterized species in *Penicillium, Fusarium, Heterobasidion* and *Armillaria* (Chillali *et al.*, 1998; Skouboe *et al.*, 1999; Bruns, 2001; Seifert *et al.*, 2007) and in revealing cryptic species in fungal species complexes. Furthermore the ITS sequences of some fungi groups show a certain degree of intra-species and intra-individual variability (Lim *et al.*, 2008) which could confuse the molecular demarcation of species boundary. Moreover, the ITS seems sometimes to be too short to be considered as an optimal barcode marker.

The matter has not yet been solved because several recent studies demonstrated that the reduction of the barcode marker length strongly affects the accuracy of the resulting phylogenetic topology, although surprisingly the species discrimination is still effective (Min and Hickey, 2007; Meusnier *et al.*, 2008). In several *Ascomycota* genera, including *Fusarium* (O'Donnel and Cigelnik, 1997) and *Penicillium* (Skouboe *et al.*, 1999) the short length of the PCR amplicons belonging to ITS causes a too small degree of variability to discriminate species. Moreover, in subgenus *Penicillium* both ITS and other rDNA regions show a low species resolution power (Skouboe *et al.*, 1999; Peterson, 2000). However, β-tubulin gene seems to provide a more resolved phylogeny (Samson *et al.*, 2004).

Finally, the occurrence of multiple and heterogeneous copies of ITS in the same fungal isolate could present a remarkable problem when the PCR products are sequenced and compared between different isolates with the aim of discriminating them at the species level (Aanen *et al.*, 2001).

In conclusion, further studies are needed to confirm that ITS could be a universal barcode in fungi. Probably one or more additional markers will be associated with ITS in order to reach the most effective discrimination power (at species level) in a broad range of fungal taxonomic groups. These additional molecular markers could be identified among the nuclear DNA markers currently used for the identification of species in fungi, such as the 'nuclear large ribosomal subunit' (LSU rDNA) (Kurtzman and Robnett, 1998), the 'nuclear small ribosomal subunit' (SSU rDNA) (Baayen *et al.*, 2001), 'β-tubulin (BenA)' (Geiser *et al.*, 2007), 'elongation factor 1-α (EF-1-α)' (O'Donnell *et al.*, 2008) and the 'second largest subunit of RNA polymerase II (RPB2)' (Ertz *et al.*, 2009).

Mitochondrial markers can also be considered, when using appropriate large-scale bioinformatics tools (Santamaria *et al.*, 2009) to select those mitochondrial genome regions where introns are missing. Indeed mitochondrial genes have many favourable features, among which are their high copy number which enables easier and cheaper recovery, the paucity of repetitive regions which could produce misleading results from the comparison of non-orthologous sequences pairs and the scarcity of deletions and insertions which could consistently disturb sequence alignment and comparison. Moreover, mitochondrial genes could have appropriate intra- and inter-species variability (Seifert *et al.*, 2007).

In conclusion, an approach based on just one molecular marker is probably not sufficient to discriminate fungal species: a multi-locus barcode approach may be more effective.

13.3 Sources of further information and advice about the DNA barcode

A wide overview of the Barcode of Life initiative as well as of its partners and resources can be explored by visiting the website of CBOL: http://www.barcoding. si.edu/ An online resource supporting the collection, management, analysis and use of DNA barcodes is represented by the Barcode of Life Data Systems (BOLD), which can be visited at http://www.barcodinglife.org/views/login.php.

Other important web resources describing specific barcode activities and projects are the International Barcode of Life (iBOL) website (http://www.dnabar coding.org/), the Canadian Centre for DNA Barcoding webpages (http://www.dna barcoding.ca/) and the Barcode Blog (http://phe.rockefeller.edu/barcode/blog/).

There are several networks and barcode-like resources for the molecular taxonomic identification of fungi. Assembling the Fungal Tree of Life (AFTOL) is a multi-laboratory consortium aiming to build a complete phylogeny of fungi. The Dictionary of the Fungi (Hibbett *et al.*, 2007) is a mycology reference source developed on the basis of the phylogeny by AFTOL. The Fungal Environmental Sampling and Informatics Network (FESIN) aims to develop DNA barcode databases for fungi by focusing on nuclear ribosomal markers. UNITE (Kåljalg *et al.*, 2005) is a research network linked to a root-associated fungi database including ITS sequences from identified voucher specimens.

13.4   References

Aanen DK, Kuyper TW and Hoekstra RF (2001). 'A widely distributed ITS polymorphism within a biological species of the ectomycorrhizal fungus *Hebeloma velutipes*'. *Mycological Res*, **105**, 284–90.

Baayen RP, O'Donnell K, Breeuwsma S, Geiser DM and Waalwijk C (2001). 'Molecular Relationships of Fungi Within the Fusarium redolens–F. hostae clade'. *Phytopathology*, **91**(11), 1037–44.

Barret RDH and Hebert PDN (2005). 'Identifying spiders through DNA barcodes'. *Can J Zool*, **83**, 481–91.

Blaxter ML (2004). 'The promise of a DNA taxonomy'. *Phil Trans R Soc Lond B Biol Sci*, **359**(1444), 669–79.

Bruns TD (2001). 'ITS reality'. *Inoculum*, **52**, 2.

Chen W, Seifert KA and Lévesque CA (2009). 'A high density COX1 barcode oligonucleotide array for identification and detection of species of *Penicillium* subgenus *Penicillium*'. *Molec Ecol Resources*, **9**(Suppl. 1), 114–29.

Chillali M, Idder-Ighili H, Guillaumin JJ, Mohammed C, Escarmant BL and Botton B (1998). 'Variation in the ITS and IGS regions of ribosomal DNA among the biological species of European Armillaria'. *Mycological Res*, **102**, 533–40.

Druzhinina IS, Kopchinskiy AG, Komoñ M, Bissett J, Szakacs G and Kubicek CP (2005). 'An oligonucleotide barcode for species identification in *Trichoderma* and *Hypocrea*'. *Fungal Genet Biol*, **42**(10), 813–28.

Ertz D, Miadlikowska J, Lutzoni F, Dessein S, Raspé O, Vigneron N, Hofstetter V and Diederich P (2008). 'Towards a new classification of the *Arthoniales* (*Ascomycota*) based on a three-gene phylogeny focussing on the genus Opegrapha'. *Mycological Res*, **113**(1), 141–52.

Gardes M and Bruns TD (1993). 'ITS primers with enhanced specificity for basidiomycetes – application to the identification of mycorrhizae and rusts'. *Molec Ecol*, **2**(2), 113–8.

Geiser DM, Klich MA, Frisvad JC, Peterson SW, Varga J and Samson RA (2007). 'The current status of species recognition and identification in *Aspergillus*'. *Studies Mycology*, **59**, 1–10.

Greenstone MH, Rowley DL, Heimbach U, Lundgren JG, Pfannenstiel RS and Rehner SA (2005). 'Barcoding generalist predators by polymerase chain reaction: carabids and spiders'. *Molec Ecol*, **14**(10), 3247–66.

Hebert PDN, Ratnasingham S and deWaard JR (2003). 'Barcoding animal life: cytochrome c oxidase subunit 1 divergences among closely related species'. *Proc R Soc Lond B*, **270**(Suppl 03BL0066), 1–4.

Hebert PD, Stoeckle MY, Zemlak TS and Francis CM (2004a). 'Identification of birds through DNA barcodes'. *PLoS Biol*, **2**(10), e312.

Hebert PD, Penton EH, Burns JM, Janzen DH and Hallwachs W (2004b). 'Ten species in one: DNA barcoding reveals cryptic species in the neotropical skipper butterfly *Astraptes fulgerator*'. *Proc Natl Acad Sci USA*, **101**(41), 14812–7.

Hibbett DS *et al.* (2007). 'A higher-level phylogenetic classification of the Fungi'. *Mycological Res*, **111**(5), 509–47 (Review).

Katsu M, Kidd S, Ando A, Moretti-Branchini ML, Mikami Y, Nishimura K and Meyer W (2004). 'The internal transcribed spacers and 5.8S rRNA gene show extensive diversity among isolates of the *Cryptococcus neoformans* species complex'. *FEMS Yeast Res*, **4**(4–5), 377–88.

Kõljalg U, Larsson KH, Abarenkov K, Nilsson RH, Alexander IJ, Eberhardt U, Erland S, Høiland K, Kjøller R, Larsson E, Pennanen T, Sen R, Taylor AF, Tedersoo L, Vrålstad T and Ursing BM (2005). 'UNITE: a database providing web-based methods for the molecular identification of ectomycorrhizal fungi'. *New Phytol*, **166**(3), 1063–8.

Kress WJ, Wurdack KJ, Zimmer EA, Weigt LA and Janzen DH (2005). 'Use of DNA barcodes to identify flowering plants'. *Proc Natl Acad Sci USA*, **102**(23), 8369–74.

Kurtzman CP and Robnett CJ (1998). 'Identification and phylogeny of ascomycetous yeasts from analysis of nuclear large subunit (26S) ribosomal DNA partial sequences'. Antonie Van Leeuwenhoek, **73**(4), 331–71.

Lahaye R, van der Bank M, Bogarin D, Warner J, Pupulin F, Gigot G, Maurin O, Duthoit S, Barraclough TG and Savolainen V (2008). 'DNA barcoding the floras of biodiversity hotspots'. *Proc Natl Acad Sci USA*, **105**(8), 2923–8.

Lim YW , Sturrock R, Leal I, Pellow K, Yamaguchi T and Breuil C (2008). 'Distinguishing homokaryons and heterokaryons in *Phellinus sulphurascens* using pairing tests and ITS polymorphisms'. *Antonie van Leeuwenhoek Int J General Molecular Microbiol*, **93**, 99–110.

Meusnier I, Singer GA, Landry JF, Hickey DA, Hebert PD and Hajibabaei M (2008). 'A universal DNA mini-barcode for biodiversity analysis'. *BMC Genomics*, **9**, 214.

Meyer CP and Paulay G (2005). 'DNA barcoding: error rates based on comprehensive sampling'. *PLoS Biol*, **3**(12), e422.

Min XJ and Hickey DA (2007). 'Assessing the effect of varying sequence length on DNA barcoding of fungi'. *Molec Ecol Notes*, **7**(3), 365–73.

O'Donnell K and Cigelnik E (1997). 'Two divergent intragenomic rDNA ITS2 types within a monophyletic lineage of the fungus *Fusarium* are nonorthologous'. *Mol Phylogenet Evol*, **7**(1),103–16.

O'Donnell K, Sutton DA, Fothergill A, McCarthy D, Rinaldi MG, Brandt ME, Zhang N and Geiser DM (2008) 'Molecular phylogenetic diversity, multilocus haplotype nomenclature, and in vitro antifungal resistance within the *Fusarium solani* species complex'. *J Clin Microbiol*, **46**(8), 2477–90.

Peterson SW (2000). 'Phylogenetic analysis of *Penicillium* species based on ITS and LSU-rDNA nucleotide sequences'. In *Integration of Modern Taxonomic Methods for Penicillium and Aspergillus*, Samson RA and Pitt JI (eds), Harwood Academic Publishers, Amsterdam, The Netherlands, 1163–78.

Ratnasingham S and Hebert PD (2007). 'Bold: The Barcode of Life Data System (http://www.barcodinglife.org)'. *Molec Ecol Notes*, **7**(3), 355–64.

Samson RA, Seifert KA, Kuijpers AFA, Houbraken JAMP and Frisvad JC (2004). 'Phylogenetic analysis of *Penicillium* subgenus *Penicillum* using partial β-tubulin sequences'. *Studies Mycology*, **49**, 175–200.

Santamaria M, Vicario S, Pappadà G, Scioscia G, Scazzocchio C and Saccone C (2009). 'Towards barcode markers in Fungi: an intron map of *Ascomycota mitochondria*'. *BMC Bioinformatics*, **10**(Suppl 6), S15.

Saunders GW (2005). 'Applying DNA barcoding to red macroalgae: a preliminary appraisal holds promise for future applications'. *Phil Trans R Soc Lond B Biol Sci*, **360**(1462), 1879–88.

Scicluna SM, Tawari B and Clark CG (2006). 'DNA barcoding of blastocystis. *Protist*, **157**(1),77–85.

Seifert KA (2008). 'Integrating DNA barcoding into the mycological sciences'. *Persoonia*, **21**, 162–6.

Seifert KA (2009). 'Progress towards DNA barcoding of Fungi'. *Molec Ecol Resources*, **9**(Suppl1), 83–9.

Seifert KA, Samson RA, Dewaard JR, Houbraken J, Lévesque CA, Moncalvo JM, Louis-Seize G and Hebert PD (2007). 'Prospects for fungus identification using CO1 DNA barcodes, with *Penicillium* as a test case'. *Proc Natl Acad Sci USA*, **104**(10), 3901–6.

Skouboe P, Frisvad JC, Taylor JW, Lauritsen D, Boysen M and Rossen L (1999). 'Phylogenetic analysis of nucleotide sequences from the ITS region of terverticillate *Penicillium* species'. *Mycological Res*, **103**, 873–81.

Summerbell RC, Lévesque CA, Seifert KA, Bovers M, Fell JW, Diaz MR, Boekhout T, de Hoog GS, Stalpers J and Crous PW (2005). 'Microcoding: the second step in DNA barcoding'. *Phil Trans R Soc Lond B Biol Sci*, **360**(1462),1897–903 (Review).

Vences M, Thomas M, Bonett RM and Vieites DR (2005). 'Deciphering amphibian diversity through DNA barcoding: chances and challenges'. *Phil Trans R Soc Lond B Biol Sci*, **360**(1462),1859–68.

Vialle A (2009). 'Evaluation of mitochondrial genes as DNA barcode for Basidiomycota'. *Molec Ecol Resources*, **9**(Suppl1), 99–113.

Ward RD, Zemlak TS, Innes BH, Last PR and Hebert PD (2005). 'DNA barcoding Australia's fish species'. *Phil Trans R Soc Lond B Biol Sci*, **360**(1462), 1847–57.

Part V

Emerging methods for mycotoxin analysis in food and feed

14

Emerging bio-sensing methods for mycotoxin analysis

I. E. Tothill, Cranfield University, UK

Abstract: Mycotoxins analysis is an important area in food quality and safety to eliminate and control the risk of eating contaminated foods. Therefore, the ability to detect and monitor the toxins at the required legislative limit is a worldwide priority. Bio-sensing methods are able to fulfil these requirements and also compete with traditional techniques. To date, novel bio-sensing methods are developing at a rapid rate with vast advances taking place in all aspects of the technology. The devices have the ability to provide rapid, sensitive, specific, robust and cost effective quantitative and qualitative analysis for mycotoxins in food samples. Hence, a wide range of biosensor devices have been developed and also reported in the literature for this specific application, exploiting a range of transducers as the sensor platforms and also biological and synthetic receptors as the sensing layers. With the advent of nanotechnology and its impact on developing ultrasensitive devices, mycotoxins analysis is benefiting greatly from the advances taking place in the development of micro/nano-array devices and also nanomaterials applications. This chapter covers all aspects of the bio-sensing technology and gives examples of devices developed for mycotoxins analysis.

Key words: analysis, biosensors, molecular receptors, mycotoxins, nanotechnology, transducers.

14.1 Introduction

Analysis of food and feed to date for mycotoxins contamination is an important process for ensuring quality control and managing any risk of contaminants entering the food chain. Most analyses are conducted using off-site techniques and require the samples to be transferred to an accredited laboratory for testing and confirmation of risks. Current methods of analysis of mycotoxins are primarily based on techniques such as thin layer chromatography (TLC), high performance liquid chromatography (HPLC), gas liquid chromatography (GC) and hyphenated techniques with mass-spectrometry (MS) such as HPLC–MS and GC–MS. These techniques are coupled with extraction and sample clean up using solid phase extraction (SPE) columns such as C18 or immunoaffinity separation to improve mycotoxins recovery.

Most of these techniques require extensively trained operators and in some cases very expensive equipment. Easy-to-use and less expensive lateral flow devices and enzyme linked immunosorbant assays (ELISA) have also been developed for a range of mycotoxins and these have become more popular in recent years. To date there are emerging and innovative technologies being developed for mycotoxins analysis and these can range from invasive, minimally invasive to non-invasive technologies. These include biosensors and affinity sensors as rapid diagnostic devices, new artificial receptors used as sensing layers and a diverse range of chemical sensors that can be applied in finger profile analysis such as the e-nose techniques. These sensing methods, depending on the technique used can allow accurate quantification, high recovery rates and low detection limits. Using robust, low cost, portable and rapid technologies will allow the determination of target toxin at the sampling site. This will also overcome the problems associated with the need to send the samples for laboratory analysis as is the case for analysis using the traditional techniques. Such screening methods can save valuable time and resources especially in food analysis and processing. To insure optimum quality is delivered to the consumer, rapid assessment using cost-effective measurements is required in the food industry. Food analysis methods have to be sensitive and accurate to comply with the legislation. An overview of the emerging bio-sensing technologies available to date and those that are being developed for mycotoxins analysis will be covered in the following sections.

14.2   Biosensors as diagnostics tools in mycotoxin analysis

A biosensor is defined as a bioanalytical device incorporating a molecular recognition element associated or integrated with a physicochemical transducer (Tothill and Turner, 2003). To date there are five types of transducers used in biosensor devices and these include electrochemical, optical, mass sensitive, calorimetric and magnetic devices. The official IUPAC definition states 'A biosensor is a self-contained integrated device, which is capable of providing specific quantitative or semi-quantitative analytical information using a biological recognition element (biochemical receptor) which is retained in direct spatial contact with a transduction element' (Thévenot et al., 1999). Biosensors can bring the capability of analysing food and feed samples on-site such as in the farm or at the food factory. This is because biosensors have the advantage over traditional methods of being simple, rapid, cost effective and portable devices that are sensitive and specific to the target mycotoxin. These devices can also be designed for single or multi-analyte testing, disposable for single use, reusable for several analyses or for continuous monitoring.

There are over 190 moulds identified that are able to produce mycotoxins and also several toxins can be produced by the same mould (Bhatnagar *et al.*, 2002; Gilbert and Anklam, 2002). To date, a range of biosensors have been developed and reported in the literature for mycotoxins analysis some of which have the capability of performing multi-array mycotoxins diagnosis (multiplexing capabilities). Sample preparation can usually be incorporated as part of the sensor procedure system. Some of the extraction and clean-up procedures may be similar to the procedures used for HPLC or GC analysis especially if the samples are solid foods. To date sensors can contain microfluidics and a membrane separation system as part of their design for extracted food samples handling. Examples in the literature which review the methods used for mycotoxin analysis include Patel (2004), Magan and Olsen (2004), Zheng *et al.*, (2006), Pohanka *et al.*, (2007) and Wang and Wang (2008).

14.2.1 Molecular recognition element

In order to recognise specifically the mycotoxin of interest, optimal recognition materials (sensing receptor) need be integrated on the surface of the sensor. This is especially important for small molecular weight toxins with diverse structural similarities as the sensitivity and specificity of the sensing molecules will play an important part in the success of the sensor device. A range of molecular recognition receptors that can be used as sensing layers have been applied and developed for mycotoxins recognition and detection. The range of the different types of sensing receptors that can be used as sensing layers in biosensors has been reviewed recently (Zourob, 2010). The most widely used is the antibody molecule, which provides the specificity and sensitivity required for low level of toxins detection. More recently, synthetic (artificial) molecular recognition elements have been designed and fabricated as affinity materials and applied to analyte detection and analysis. This type of material can include nanomaterials and membrane structures and can comprise molecular imprinted polymers (MIPs), aptamers, phage display peptides, binding proteins and synthetic peptides as well as metal oxides materials (Tothill, 2001, Tothill *et al.*, 2001, Shaikh *et al.*, 2005, Collett *et al.*, 2005). A recent review of the development of novel materials for mycotoxin analysis has been reported by Maragos (2009).

Antibodies (monoclonal and polyclonal) or antibody fragments have been used in the development of immunosensors some of which are targeting mycotoxins analysis. Polyclonal antibodies can be raised against any mycotoxin as long as the molecular mass and the structure of the mycotoxin is able to induce specific high affinity antibody production. With the introduction of high throughput techniques, applying these molecules in sensors has been successful (Chambers and Johnston, 2003). The use of monoclonal antibodies however, results in more specific tests. A drawback is that monoclonal antibodies are more difficult to maintain and can be more expensive than polyclonal antibodies.

A range of antibodies are now commercially available for most mycotoxins with the exception of patulin. Efforts have been made worldwide to produce a specific antibody for patulin, but these have had very low success owing to the very small mass structure of this toxin and the similarity of its chemical structure to compounds present in fruits and other food matrices. Many ELISA tests, dipsticks and assays have been developed and are commercially available for a range of mycotoxins using antibodies (for reviews please refer to Chapter 5 on ELISA development in this book).

Antibody fragments and molecular engineered antibodies have been developed for biosensor applications (Zourob, 2010). The use of direct and combinatorial mutagenesis has been applied to enhance the affinity and selectivity of recombinant antibodies. However, owing to lack of stability of the antibody molecule, this can make it unsuitable for long-term biosensor device storage and can hinder wider application and commercialization in some instances. Replacing natural biomolecules with artificial receptors or biomimics has therefore become an attractive area of research in recent years (Tothill, 2003). The advantages of using these molecules is that they are robust, more stable and cheap to produce in some cases, since they can be made using wet chemistry and can be modified easily to aid immobilization on the sensor surface and to add labels as the marker for detection. They also obviate the use of animals in the production of affinity receptors as is the case in antibodies production.

The use of phage display technology to screen libraries for specific receptor molecules is being rapidly applied in the diagnostics area. This is to identify specific peptide sequences with affinity towards target analytes (Oyama *et al.*, 2003; McGuire *et al.*, 2004 a,b; Chang *et al.*, 2005; Shukla and Krag, 2005). However, a large amount of literature in this area focuses mainly on medical applications. Combinatorial libraries with peptides synthesized on beads are mainly used for drug discovery but more recently they have been applied to receptor ligands selection by screening them against specific analyte targets (Brown, 2003, Chen *et al.*, 1998). There are many examples in the literature where peptides have been specifically synthesized to target specific molecules such as oestradiol (Giraudi *et al.*, 2003), oestrogen (Tozzi *et al.*, 2002) and aflatoxins (Tozzi *et al.*, 2003). Recent work at Cranfield University (Parker, 2008; Heurich, 2008), used a computational approach for the design of peptide receptors for aflatoxin M_1 and ochratoxin A, respectively (Fig. 14. 1).

Using molecular modelling software, virtual libraries of functional monomers, in this case amino acids, were screened against the target toxins, which were employed as a virtual template for affinity study analysis. The amino acids giving the highest binding energy were selected and used in a molecular dynamics process to investigate their interaction with the toxins. Peptide sequences with the highest affinity to the toxins were generated and screened against the toxins of interest and redesigned and screened to achieve the optimal sequence with the highest affinity towards the toxin. Selected peptides with a high affinity score observed from these studies were then synthesized using peptide synthesizers and applied in affinity screening using the Biacore ™ optical biosensor and ELISA assay techniques. The affinity of the peptides was confirmed during these analyses; however further

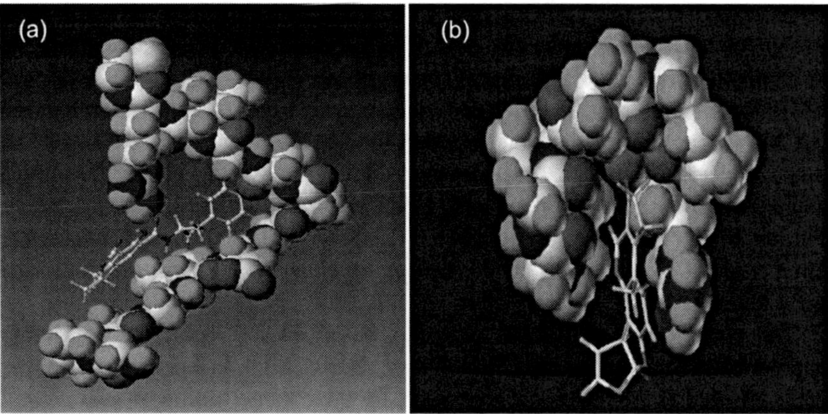

Fig. 14.1 The *de novo* designed peptide receptors showing interaction with the mycotoxin (a) peptide designed for ochratoxin A interacting with the toxin. (b) Peptide designed for aflatoxin M_1, showing interaction with the toxin (Parker, 2008; Heurich, 2008).

studies were required to increase the stability of the toxin peptide receptor complex.

Peptides can be chemically synthesized in large quantities once the optimal peptide with the required affinity has been identified. However, peptides have relatively lower affinity ($K_d \sim 10^{-6}$ to 10^{-7} M) when compared to antibodies and therefore, in designing the assay it is important to develop a successful sensing method when incorporating them in sensors. Research into the development of higher affinity peptides as sensing layers needs further investment. Peptides as receptors for biosensors application have been recently reviewed (Tothill, 2010b).

Molecular imprinted polymers (MIP) have been used to replace antibodies as sensing receptors. The process uses organic polymers to imprint the target molecule specifically to produce synthetic materials that can be used for affinity recognition of that target molecule after removing the target compound from the polymer (Mosbach, 1994, Mosbach and Ramstrom, 1996). Molecular imprinting polymerization has been the focus of intense research interest in the past 15 years and is being considered in a wide range of application areas, for example in the preparation of selective separation materials and as sensing layers in sensor devices (Batra and Shea, 2003; Piletsky *et al.*, 2006). MIPs are usually produced by forming a highly cross-linked organic polymer around the molecule of interest (analyte target/template) and therefore any impurities will produce less specific affinity media (Tozzi *et al.*, 2003).

For mycotoxins applications, MIPs have mainly been used as clean-up media for sample extraction and pretreatment. Maier *et al.* (2004) developed MIPs for ochratoxin A extraction from red wine. However, the recovery was reported to be < 66%. Urraca *et al.* (2006) developed a MIP for zearalenone which showed low recognition of the toxin (Urraca *et al.*, 2006; Weiss, 2003). Yu and Lai (2005) developed a MIP membrane on a gold surface plasmon resonance (SPR)-based sensor chip for the detection of ochratoxin A achieving a detection limit in the range of 0.05–0.5 mg l⁻¹. At Cranfield University diverse research has taken place to develop MIPs using molecular modelling software for mycotoxin analysis. As an example Turner and co-workers (Turner, 2004; Turner *et al.*, 2004) developed a MIP for ochratoxin A extraction and analysis using a quartz crystal microbalance (QCM) sensor which showed a detection limit in the range of 50–100 µg l⁻¹. A range of other MIPs has also been developed for ochratoxin A and used for sample extraction and clean-up from red wine (Jodlbauer *et al.*, 2002; Maier *et al.*, 2004). Further, MIPs developed for fumonisin B (De Smet *et al.*, 2009) and T-2 toxin (De Smet *et al.*, 2010) proved to be alternatives for clean up and preconcentration of these toxins in food matrices.

The construction of peptide receptors based on the molecular imprinting approach has been carried out (Giraudi et *al.*, 2000, 2003). In this approach amino acids are used as the functional monomers for the polymerization process around the template which is the molecular analyte target. The use of molecular modelling has been applied for the selection of the highest affinity monomers to be used in the construction of MIPs. The construction of a virtual library of functional monomers with regard to a target molecule followed by the selection of the highest binding score monomers as the building blocks has been undertaken in many studies (Chianella *et al.*, 2002, 2003; Lotierzo *et al.*, 2004). The software facilitates calculations using different dielectric constants to reflect the polarity of the environment (solvents) in which the polymers have to be prepared or used. MIPs are conceptually attractive owing to their ease of preparation, high thermal and chemical stability and long shelf-life at ambient temperature and humidity. However, the major disadvantage is that MIPs require a template (pure form of the analyte) to produce the synthetic polymer which can in some cases have a significant cost. Their applications as sensing layers in sensor devices have also been limited owing to their low specificity and sensitivity when compared to the antibody molecule.

Aptamers (derived from the latin *aptus*, meaning 'to fit') are synthetic oligonucleotide ligands (DNA or RNA). These can be selected from combinatorial libraries of synthetic nucleic acids that have specific affinity for a target molecule. Using the synthetic evolution of ligands (SELEX) process, aptamers can be selected from random pools and then be provided chemically (Ulrich *et al.*, 2001; Cerchia *et al.*, 2002). These can be selected with regard to drugs, proteins, toxins and other analytes and used as receptors in diagnostics and sensor devices (Brody and Gold, 2000; Mascini *et al.*, 2001; O'Sullivan, 2002; Clark and Remcho, 2002). Interest in the use of aptamers has increased in recent years especially for use in the area of diagnosis as sensing molecules (Luzi *et al.*, 2003; Tombelli *et al.*, 2005). A range of aptamers has been synthesized with affinity towards target analytes and these have been tested for implementation in future devices. An aptamer that is specific for ochratoxin A has been reported recently by Cruz-Aguado and Penner (2008a). The authors used fluorescence polarization immunoassay where the tracer molecules were fluorescently labelled oligonucleotides and the assay was reported to achieve a 2 ng ml⁻¹ detection limit (Cruz-Aguado and Penner, 2008b). The same aptamer was also implemented in an SPE colum for sample extraction, before analysis. Aptamers have been selected that have high affinity ($K_d \sim 10^{-8}$ to 10^{-9} M) for target molecules. However, their specificity in most cases is lower than that of the antibodies. Detection of aptamer–protein interactions using a label-free electrochemical detection system based on impedance spectroscopic transduction has been reported (Rodriguez *et al.*, 2005).

The use of a bilayer lipid membrane as the sensing layer for mycotoxin detection has been reported by Andreous and Nikolelis (1998), Siontorou *et al.* (2000) and Gilbert and Vargas (2003). These layers suffer from low stability and are more difficult to implement in sensor devices when compared to other receptors.

14.2.2 Transducers

A transducer is a device that converts the recognition signal events into a digital signal which can be electrochemical (amperometry, potentiometry, conductimetry/ impedimetry), optical (colorimetry, fluorescence, luminescence, interferometry, spectroscopy of optical waveguides and SPR), calorimetric (thermistor), mass sensitive (piezoelectric/acoustic wave) or magnetic (Tothill and Turner, 2003). Electrochemical transducers are the most widely used in sensor technology; however, recently optical and piezoelectric sensors have gained popularity and their use in biosensor development is expanding. The main reason for this is that most electrochemical systems still require the use of a labelled molecule such as an enzyme where a substrate mixture needs to be added for the assay results to be monitored; this is with the exception of conductimetry/impedimetry transducers. In an optical system such as SPR and a waveguide system, the binding event between the receptor used on the sensor surface and the analyte can be detected directly. In a piezoelectric/acoustic system the change in mass caused by the binding between the analyte and the receptor on the sensor can also be monitored directly. Figure 14.2, shows the construction of a biosensor device. A range of sensors has been developed for mycotoxins analysis, an example of which is listed in Table 14.1.

Electrochemical transducers

Electrochemical biosensors are used for mycotoxins analysis since they are portable, simple, easy to use, cost effective, highly sensitive and in most cases can be disposable. The electrochemical instruments used with the biosensors have been miniaturized to small pocket-size devices which makes them applicable for on-site use in the farm and the factory and at the point of source. A range of electrochemical transducers is used and these include amperometry, potentiometry and impedimetry/conductivity transducers. A large number of biosensors use amperometry as the transducer with the oxygen electrode developed by Clark in 1973 as the first example (Tothill and Turner, 2003). This was followed by the glucose biosensor which is the most widely used example of an electrochemical biosensor based on a screen-printed amperometric disposable electrode. This type of biosensor has been used widely throughout the world for glucose testing at home, bringing diagnosis to on-site analysis. Another development in this area is

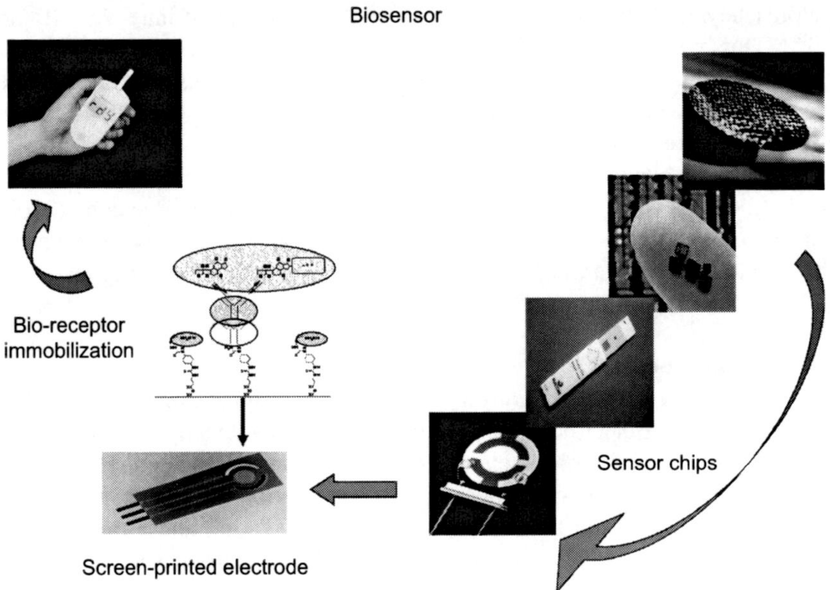

Fig. 14.2   Biosensor construction.

the hand-held i.STAT clinical analyser which combines several electrochemical biosensors on a single chip and is used for multiple electrolytes and metabolites in clinical samples. Amperometric and potentiometric transducers are the most commonly used, but attention in recent years has also been devoted to impedance and conductivity-based transducers since they are classified as label-free detection sensors.

For mycotoxins analysis, an electrochemical transducer will be the most useful specifically because of the cost and portability of these devices. Two types of electrochemical sensor principles have been reported for mycotoxins analysis and these include enzyme inhibition and affinity-based sensors. Very few biosensors based on enzyme inhibition for mycotoxin analysis have been reported in the literature. These types of sensor are based on the inhibition of the enzyme immobilized on the sensor surface by the mycotoxin in the sample. Therefore, these are mainly toxicity-based sensors and will not be able to identify the toxin in the sample, but it gives a general indication of total toxicity analysis (Tothill, 2010a). Moressi *et al.* (1999) used the enzyme tyrosinase on a carbon electrode to detect the mycotoxins produced by *Alternaria* spp. Detecting mycotoxins directly on a carbon electrode using amperometry has shown minimum success since the sensitivity of these devices is not adequate for the requirements of mycotoxins analysis. The sensors also lack the specificity in this type of systems.

Electrochemical affinity sensors offer great selectivity and sensitivity for rapid and sensitive mycotoxins analysis. ELISA-based assay conducted on the electrode surface is the most frequently used technique for different mycotoxins identifica-

Table 14.1   Examples of biosensors developed for mycotoxins analysis

Mycotoxin	Matrix	Biosensor principle	Limit of detection	Reference
Aflatoxin B$_1$	Buffer	Optical	3 µg l^{-1}	Daly *et al.* (2000)
	Buffer		1 µg l^{-1}	Carlson *et al.* (2000)
	Maize, sorgum, nuts		5 µg kg^{-1}	Nasir and Jolley (2002)
	Nuts, oat		0.2 µg kg^{-1}	Gaag *et al.* (2003)
	Buffer		0.5 µg l^{-1}	Adanyi *et al.* (2007)
	Buffer		0.5 µg l^{-1}	Sapsford *et al.* (2006)
	Barley	Electro-chemical	0.03 µg kg^{-1}	Ammida *et al.* (2004)
	Buffer		0.15 µg l^{-1}	Pemberton *et al.* (2006)
	Rice		0.06 µg l^{-1}	Tan *et al.* (2009)
	Human serum, Grape		0.05 µg l^{-1}	Sun *et al.* (2008)
	Corn		0.03 µg kg^{-1}	Piermarini *et al.* (2007)
	Buffer	QCM	0.01 µg l^{-1}	Jin *et al.* (2009)
Aflatoxin M$_1$	Liver	Optical	1 µg kg^{-1}	Chiavaro *et al.* (2005)
	Milk	Electro-chemical	0.02 µg l^{-1}	Badea *et al.* (2004)
	Milk		0.05 µg l^{-1}	Micheli *et al.* (2005)
	Milk		0.039 µg l^{-1}	Parker and Tothill (2009)
	Milk		0.008 µg l^{-1}	Parker *et al.* (2009)
Fumonisin B$_1$	Corn	Optical	50 µg kg^{-1}	Mullett *et al.* (1998)
	Corn		10 µg l^{-1}	Thompson and Maragos (1996)
	Buffer		50 µg l^{-1}	Gaag *et al.* (2003)
	Buffer		0.5 µg l^{-1}	Sapsford *et al.* (2006)
	Corn	Electro-chemical	5 µg l^{-1}	Abdul Kadir and Tothill (2010)
Ochratoxin A	Cereals	Optical	3.8 to 100 µg kg^{-1}	Ngundi *et al.*, (2005)
	Coffee		7 µg kg^{-1}	
	Wine		38 µg kg^{-1}	
	Wheat	Electro-chemical	0.06 µg l^{-1}	Alarcon *et al.* (2006)
	Wine		0.7 µg l^{-1}	Prieto-Simon *et al.* (2007)
	Buffer		10 µg l^{-1}	Khan and Dhayal (2008)
	Wine		0.05 µg l^{-1}	Heurich *et al.* (2010)
	Buffer	QCM	16.1 µg l^{-1}	Tsai and Hsieh (2007)
Deoxynivalenol	Wheat	Optical	2.5 µg l^{-1}	Schnerr *et al.*, (2002)
	Wheat		2.5 µg l^{-1}	Tudos *et al.* (2003)
	Buffer		0.2 µg l^{-1}	Ngundi *et al.* (2006)
	Oats		50 µg kg^{-1}	

tion and quantification. These sensors are mainly mass produced using screen-printing technology which is an easy method of fabricating the sensors and is also cost effective. Sensors comprise a three-electrode configuration with a working, reference and counter (auxiliary) electrode. In this system the antibody (or antigen) is labelled with an enzyme such as horseradish peroxidase (HRP), or alkaline phosphatase (AP) and these will catalyse an added substrate and produce an electroactive species which is then detected on the sensor surface.

Currently, immunosensors based on an electrochemical transducer are widely reported in the literature for mycotoxins analysis; most of these are for one-shot analysis. For example Ammida *et al.* (2004) used indirect competitive ELISA assay on a disposable carbon screen-printed electrode with differential pulse voltammetry (DPV), achieving low detection limit for aflatoxin B_1 analysis. Micheli *et al.* (2005) developed a screen-printed electrode using a carbon working electrode for aflatoxin M_1 analysis in milk samples using a direct competitive assay with HRP as the enzyme label. In other literature, for example Piermarini *et al.* (2006) and Pemberton *et al.* (2006) also developed screen-printed carbon-based electrodes for aflatoxin B_1 analysis. Parker and Tothill (2009), fabricated a disposable screen-printed carbon electrode for aflatoxin M_1 analysis to be used directly with milk samples. This sensor gave a low detection limit for aflatoxin M_1 (limit of detection 0.039 µg l^{-1} in milk samples). The use of gold as the working electrode has been expanding in recent years since it is more stable against oxidation and easy to functionalize for covalent immobilization. It is also less prone to non-specific binding and provides better conductivity compared to carbon electrodes. Heurich *et al.* (2010) and Abdul Kadir and Tothill (2010) used a gold screen-printed electrode for ochratoxin A analysis in wine and fumonisins in corn, respectively. Change in the electrochemical properties (conductivity or capacitance) on the sensor surface caused by the antibody antigen interaction can also be used as a detection principle. These devices are very attractive as label-free techniques; however, they can be less sensitive for small molecules such as mycotoxins.

Table 14.1 lists examples of electrochemical biosensors which have been developed for mycotoxins analysis. However, many of the devices are still at the prototype stage and, so far, no commercially available product exists yet on the market.

Optical transducers
Optical transducers used in biosensors include fluorescence, interferometry and spectroscopy of optical waveguides and surface plasmon resonance (SPR). For further information on the principle of using optical transducers in biosensors the reader is referred to recent literature (Ligler and Taitt, 2008; Borisov and Wolfbeis (2008). Many commercially available platforms use fluorescence labels as the detection system. This labelling can impose extra time and cost on the detection system used. Fluorescent labels can in some cases interfere with the molecular interaction by occluding a binding site leading to false negative results and assays can also suffer from high background binding leading to false positives (Cooper, 2002). The instruments used for signal readout of fluorescent labels are usually expensive and are more suitable in a laboratory setting. A hand-held fibre-optics device for the detection of aflatoxin B_1 based on evanescent waves has been developed by Maragos and Thompson (1999). An optical sensing technique for aflatoxins based on fluorescence polarization was also reported by Nasir and Jolley (2002), which can be portable and powered by a laptop computer.

Other optical biosensor platforms that exploit grating coupler, waveguides, resonant mirror and surface plasmon have been widely used for analyte diagnosis and biomolecular interaction events studies. These are classified as label-free and real-time affinity detection systems since they allow the determination of the affinity and kinetics of molecular interactions without the need for a label or a tag.

The use of SPR and related techniques which exploit evanescent waves as a method of interrogation of affinity interactions was introduced in the 1980s (Liedberg *et al.*, 1983; Flanagan, and Pantell (1984). There are many papers in the literature describing these techniques in general (Kretschmann, 1971; Kurihara and Suzuki; 2002; Sarkar and Somasundaran, 2002). Their application for biomolecular interactions has been increasingly described in scientific publications that cover most disciplines (Leatherbarrow and Edwards, 1999; Myszka, 1999; Weimar, 2000; Cooper, 2002). These techniques allow the user to study the interaction between an analyte and its receptor in real time without the use of labelling compounds. The receptors are usually immobilized to the biosensor surface and the analyte is in solution above the sensor. Binding the analyte molecules to the immobilized receptors alters the refractive index near the sensor surface which can be monitored accurately. This can then be interpreted to provide information about the analyte concentrations and the strength of the affinity reaction. The BiaCore™ biosensor instrument (GE Health Care, USA) is the most widely used SPR-based optical transducer system as a biosensor platform. This instrument is fully automated with a computer controlled system. The sensor chip contains four channels which allow four interactions to be analysed at the same time, one of which can be a control. The sensor is fabricated using a bare gold chip with a range of different chemical pathways available commercially that allow a range of molecules to be immobilized on the chip surface. Figure 14.3 is a diagram of the SPR optical sensor system (Cooper, 2002). Other companies have also produced SPR sensors such as Texas Instruments (SPREETA range), Nippon Laser, IBIS Technologies B.V (IBIS-iSPR) Electronics Laboratories and Sierra Sensors (Germany). Other manufacturers of optical biosensors based on waveguide technology (such as Farfield Sensors, Affinity Sensing Instruments) and grating-coupler (Aviv Instruments, HTS Biosystems, Luna Analytics) produce specific products for specific applications.

The use of SPR for mycotoxins detection has been reported for several types of toxin (Mullett *et al.*, 1998; Daly *et al.*, 2000; Gaag *et al.*, 2003; Tudos *et al.*, 2003). The assay design is usually based on an inhibition assay (competitive ELISA assay) on the sensor surface since all mycotoxins are of low molecular weight. Either the antibody or the mycotoxin can be immobilized on the sensor surface using the range of different chemical pathways available for surface immobilization. The interaction between the antibody and the mycotoxin is reversible 'non-covalent' and therefore it is easy to regenerate the sensor for multi-use applications. Gaag *et al.* (2003), used a Biacore 2000 instrument to detect multiple mycotoxins including aflatoxin B$_1$, fumonisin, deoxynivalenol and zearalenone in the same sample using inhibition assay simultaneously in relevant concentrations after sample extraction and clean up.

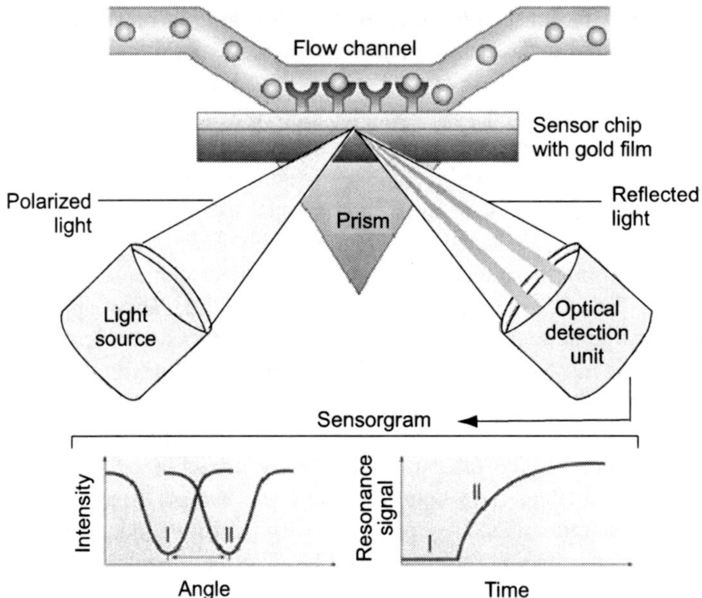

Fig. 14.3   Schematic illustration of surface plasmon resonance (Cooper, 2002).

Recent advances in optical biosensors instrumentation and experimental design have led to an increase in applying these instruments in a wide range of applications. There are also an increasing number of commercially available instruments based on an optical transducer system, with novel sensor platforms which allow most analytes to be screened using different immobilization and surface modification chemistry. However, optical biosensors have the disadvantage of being more expensive than other biosensor technologies such as electrochemical sensors (Tothill and Turner, 2003). Different optical-based biosensors have been developed for mycotoxins detection based on the above optical systems, some of which are reported in Table 14.1.

Mass sensitive transducers
Piezoelectric/acoustic sensors such as quartz crystal microbalance (QCM) and surface acoustic devices (SAW) are also classified as being label-free technology. Piezoelectric sensors comprise a quartz crystal coated with gold electrode and are used as a microbalance sensitive to changes in the mass on the sensor surface (O'Sullivan and Guilbault, 1999; Cooper and Singleton, 2007). QCM devices have been used for a wide range of applications in the medical field (Uludag and Tothill, 2010). However, QCM immunosensors are potentially more suitable for detection of pathogens and large molecular weight proteins since the sensing principle is mass deposition dependent. Owing to the low molecular mass of mycotoxins it is difficult to produce a sensitive sensor based on this principle. The assay design has to be in a competitive format with an amplification system using, for example, nanoparticles rather than a direct capture format to achieve an

acceptable detection. Piezoelectric immunosensors are usually designed to detect analytes where the specific antibody is immobilized on the sensor chip.

Ochratoxin A was measured in liquid food using a QCM sensor (Hauck *et al.*, 1998). The sensor gave a linear range of 2–100 µg l^{-1} using a resonance frequency of 20 MHz. Tsai and Hsieh (2007) reported the development of a QCM sensor using a 16-mercaptohadecanoic acid (16-MHDA) modified electrode to immobilize the anti-ochratoxin A antibody and competitively capture the toxin against a BSA–OTA conjugate. The sensor exhibited a detection limit of 16.1 ng ml^{-1} and a working range of 50–1000 ng ml^{-1} in buffer. Jin *et al.* (2009) developed a QCM-based sensor for aflatoxin B$_1$ detection using a bio-catalyzed deposition amplification system on the sensor chip. The quartz crystal surface was coated with 3-mercaptopropionic acid (MPA) for covalent immobilization of the BSA–AFB$_1$ conjugate which competes with free AFB$_1$ in the sample for the anti-AFB$_1$ sites.

To achieve the required sensitivity an amplification system was used based on the addition of an anti-IgG-horseradish peroxidase enzyme conjugate, which will attach to the anti-AFB$_1$ antibody. The enzyme (HRP) will then catalyse the oxidation of 4-chloro-1-naphthol and H$_2$O$_2$ to yield an insoluble product on the sensor surface, resulting in an increase in mass which will amplify the signal. This is a complicated method with several steps, but was able to achieve a detection range of 0.01–10.0 µg l^{-1}. The use of piezoelectric sensor combined with a flow system using a flow cell has been used to provide real-time data on binding events between the analyte and its receptor. This sensor was used for microcystin-LR toxins analysis using molecular imprinted polymer as the receptor (Chianella *et al.*, 2003). Turner (2004) used the same set up to develop a sensor for ochratoxin A analysis with a MIP design against ochratoxin A.

More advanced QCM devices are now available on the market, for example the QCM-1 Sensor instrument (Sierra Sensors GmbH, Hamburg, Germany) which is a fully automated instrument. This is being used at Cranfield University (England, UK) to develop a range of affinity-based sensors for a range of analytes. The assay is based on the use of gold chip with an antibody immobilized on a self-assembled monolayer (SAM) surface.

Cantilever-based sensors have also been introduced for analyte detection. The affinity interaction between the antibody on the surface of the cantilever and the analyte is detected by the amount of bending of the sensor caused by a change in mass which is detected as a change in the resonant frequency (Baller *et al.*, 2000; Hansen and Thundat, 2005). Cantilever technology was initially used in atomic force microscopy (AFM), but more recently microfabricated cantilever has been produced as a promising technology in bio-sensing applications (Waggoner and Craighead, 2007). Cantilevers can operate either in static mode measuring cantilever bending caused by analyte binding, or in dynamic mode, measuring frequency changes as the analyte binds to the surface. These techniques have been applied mainly for pathogen detection and fungal pathogens such as *Aspergillus niger*, achieving high sensitivity (10^3 cfu ml^{-1}) (Nugaeva *et al.*, 2007). For mycotoxins analysis, so far no work has been reported in the literature.

14.3   Lab-on-a-chip for multiplex detection

Multi-mycotoxins detection has also been reported in the literature using different types of sensor platforms. Many are based on the use of multi ELISA assays on the same chip to detect several analytes at the same time or for the detection of the same analyte in several samples at the same time. Therefore multi-toxins can be detected on a single microelectrode array chip with a multi-array working electrode, where a different antibody is immobilized to detect a specific mycotoxin. The use of lab-on-a-chip is expanding in all areas of analysis owing to the advantages of using small samples to analyse several markers/toxins, in other words that offer high throughput analysis.

These devices can be based on macro types of transducers such as receptor spots on a glass slide or can be easily fabricated using screen-printing technology with a multi-working electrode array. More recently a range of sensor chips have been fabricated using semiconductor fabrication techniques with micro/nano-array patterns. In order to develop practical devices for commercial development problems of binding between heterogeneous antigens and antibodies used in the sensor assays and the high background signals need to be eliminated and controlled. These types of device will be attractive for mycotoxin analysis since several toxins may exist in the same food or feed sample. Examples are the electrochemical assay developed by Piermarini *et al.* (2007) using a 96-well screen printed microplate to detect aflatoxin B_1 in corn samples. Detection was carried out using alkaline phosphatase as the label enzyme with the array used to detect the toxins in several samples simultaneously. Other examples of multi-mycotoxins analysis sensors include those reported by Ngundi *et al.* (2005, 2006) using a fluorescence-labelled antibody with a sensor arrays to detect ochratoxin A and deoxynivalenol. Gaag *et al.* (2003) developed an SPR biosensor array to monitor aflatoxin B_1, and deoxynivalenol. However, no real samples were analysed using this sensor. Ligler *et al.* (2003) reported the use of a biosensor consisting of capture antibodies array immobilized on a planar waveguide. A fluorescent assay is then performed and the spots are captured using a CCD camera. Several authors reported the use of competitive immunofluorescent assays on a biosensor array for the simultaneous detection of several mycotoxins such as aflatoxin B_1, fumonisin, ochratoxin A and deoxynivalenol (Ligler *et al.*, 2003; Sapsford *et al.*, 2006).

Parker *et al.* (2009), developed an electrochemical microarray with 35 arrays and which was used for the detection of aflatoxin M_1 (Fig. 14.4). The whole microarray was used to immobilize the monoclonal antibody for aflatoxin M_1 analysis with very sensitive detection limit achieved (8 ng l^{-1} in milk) using amperometry as the detection system. This has now also been used for aflatoxin B_1 and fumonisins detection (unpublished data). The use of microarrays for mycotoxin analysis is still progressing and as shown above it can produce very highly sensitive sensors for mycotoxins analysis.

In order to develop a hand-held biosensor device that can be used outside the laboratory a microfluidic system needs to be incorporated as part of the sensor device. Integrated microfluidics will enable samples and fluids to move in a

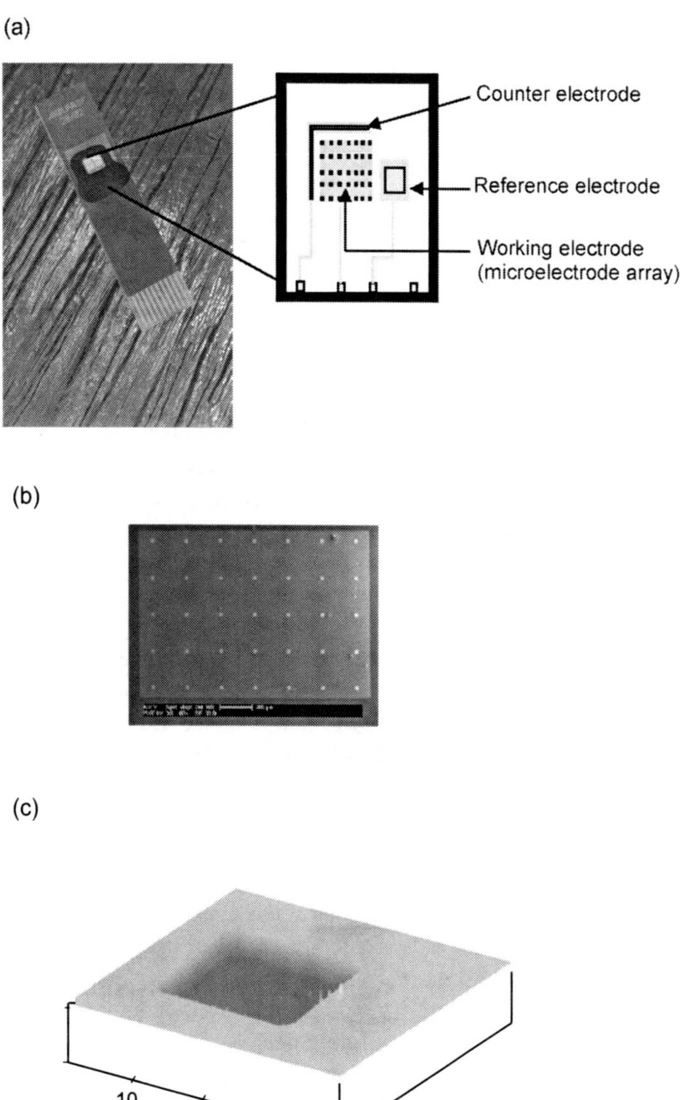

Fig. 14.4 (a) The 3-electrode chip fabricated with one working electrode area (35 electrodes in the array), a counter electrode and a reference electrode area. (b) Scanning electron microscopy (SEM) image of the whole working microelectrode of the untreated surface at 80× magnification. (c) Atomic force microscopy (AFM) image of a single element of the array for the untreated working microelectrode (image 40 µm × 40 µm) (Parker *et al.*, 2009).

lab-on-a-chip device so that the reaction takes place efficiently on the sensor surface (Abrantes *et al.*, 2001; Wang *et al.*, 2002). Advances in the development of microfluidics is increasing and different moulding and fabrication technology have been applied, such as photolithography and using quartz, silica, glass or composite materials and chips. This will then produce a fully integrated multiplex sensor system that allows multi-analyte analysis.

14.4   Nanomaterials and their use in biosensors for mycotoxin analysis

The use of nanomaterials and structures such as semiconductors and conducting polymer nanowires for biosensor applications is expanding rapidly and many comprehensive review articles have been published (Willner *et al.*, 2007; Katz and Willner, 2005; Katz *et al.*, 2004). The application of nanotechnology in biosensors can range from the transducer device to the recognition ligand, the label and the running systems. Their application in sensor development has been due to the excellent advantages offered by these materials in miniaturization of the devices, signal enhancement and amplification of signal by the use of nanoparticles as labels. These can increase the sensitivity of the final devices and also allow the fabrication of multiplex sensor systems such as high density protein arrays (Jain, 2004). The high surface to volume ratio offered by nanomaterials makes these devices very sensitive and can allow a single molecule detection which is very attractive in contaminant monitoring of toxins. The use of nanowire transducers can also offer greater sensitivity in affinity sensors (Wang *et al.*, 2005; Woolley *et al.*, 2000).

The use of luminescent nanocrystals (quantum dots) as molecular labels to replace fluorophores has created new applications for nanomaterials in labelling and visualization. These nanocrystals can be attached as labels to antibodies and other molecules to detect different analytes at the same time (multiplex sensing). Quantum dots show distinct advantages over other markers owing to their spectroscopic properties and narrow emission peaks and therefore their use in multiplexed analysis is increasing. Their high emission quantum yield results in an improved signal/noise ratio and therefore a decrease in false readings (negative and positive). The use of quantum dots in sensing applications has been expanding across a diverse range of analytes (Goldman *et al.*, 2004; Medintz *et al.*, 2005; Jorge *et al.*, 2007). However, the stability of quantum dots needs improvement, reduced aggregation in use conditions and reduced cost as they are still expensive.

The use of stripping voltammetry to detect metal nanoparticles has been applied where these metals have been used as marker tags. Gold and silver nanoparticles can be used in these methods including different inorganic nanocrystals (e.g. ZnS, PbS, CdS) for analyte detection. The unique physical and chemical properties of nanoparticles such as colloidal gold can provide excellent applications in a wide range of bio-sensing techniques (Rosi and Mirkin, 2005). Several products are available on the market such as Oxanica (UK) Quantum dots and MultiPlxBeads™

from Crystalplex Corp., USA. Nanoparticles can also be exploited in conductivity-based sensors where they can induce a change in the signal upon the attachment of the nanoparticles antibody tagged with the captured antigen on the sensor surface. The application of gold nanoparticles for mycotoxin analysis has been recently reported by Liu *et al.* (2006) where they used the particles as tags in the assay design. Gold nanoparticles are easy to functionalize and they are used for antibody immobilization and then applied in the ELISA test on the surface of the electrode. At Cranfield, gold nanoparticles are being used for mycotoxin analysis to enhance the enzyme signal achieved on the surface of the electrochemical transducer.

The development of micro/nano-sensor devices for toxins analysis is increasing owing to their extremely attractive characteristics for this application. Their novel electron transport properties make them highly sensitive for low level detection (Wang, 2005; Logrieco *et al.*, 2005). The multiplex analysis capability is also very attractive for multi-biomarker analysis.

14.5   Electronic nose method for mycotoxin analysis

Rapid development in the area of the 'artificial nose' took place during the 1990s and their implementation for fungal growth and mycotoxin detection has been investigated since then. Electronic noses have been applied for medical, environmental and food sensing (Gardner and Bartlett, 1999; Tothill *et al.*, 2001; Pearce *et al.*, 2002, Turner and Magan 2004).

Generic electronic noses (e-nose) or volatile mapping systems have been defined as 'a volatile compound, which is presented to an active material of a sensor which converts a chemical input into an electrical signal'. Based on this concept, a great deal of research in the development and commercialization of the electronic nose has taken place with applications ranging from microbial detection to toxic volatile compounds analysis. The sensors used mimic the olfactory system in the nose and the instrument usually consists of an array of gas sensors with different selectivity patterns. The circuitry represents the conversion of the chemical reactions on the sensor to electrical signals and data-analysis software which analyses the signal by pattern recognition methods (Tothill and Turner, 2003). These include principal component analysis (PCA), discriminate function analysis (DFA), cluster analysis and artificial neural networks (ANN). Tothill and Magan (2003) list the commercial instruments available on the market that can be used for mycotoxins analysis.

The results achieved for the e-nose are usually a comparative rather than a quantitative 'fingerprint'. Although largely qualitative or semi-quantitative, this approach can be ideal for rapid risk assessment of toxic volatile compounds. A diverse range of sensors are used such as metal oxides and conducting polymers sensors. Other types of devices have been developed such as metal oxide silicon field-effect sensors (sensitive to organic compounds), piezoelectric crystals, optical sensors and electrochemical sensors. The polymers can be highly sensitive but not specific and can respond to volatile compounds with a molecular weight ranging from 30 to 300 (Reineccius, 1996). Molecules such as alcohols, ketones, fatty acids and esters give a strong response, while fully oxidized species, such as CO_2, NO_2 and H_2O have a lower response. The sensor array can also recognize molecules containing sulphur and amine groups.

Application of the e-nose for mycotoxin detection has been focused on detecting the fungi which produce the toxins rather than detecting the toxins themselves since it is difficult to distinguish the toxins using the e-nose system. Therefore, there has been particular interest in early and rapid detection of mycotoxigenic moulds using the e-nose. Significant interest has also been focused on discriminating between strains of species with the ability to produce specific mycotoxins (Aldred *et al.*, 2004). The e-nose has been successful in discriminating between mouldy from good grain (Magan and Evans, 2000) and studies in this area also indicated that the e-nose can discriminate between mycotoxin-producing fungi and non-producers (Olssen *et al.*, 2002; Keshri and Magan, 2000). There is a considerable amount of literature reviewing this in detail, showing that the e-nose is good for this particular application (Sahgal *et al.*, 2007; Cabanes *et al.*, 2009).

E-nose technology is still in development and so far has had a limited commercial application in the diagnostic area since it is more used to identify variations rather than quantification of polluting and toxin molecules. The instrument needs to be trained and optimized for the sensing application it is required for and to be able to distinguish between samples before it can be used. However, e-nose systems have a recognized advantage in use in the areas of safety and quality monitoring.

14.6   Future trends

The field of biosensors is moving forward at a very rapid pace with developments and innovation taking place at all levels including the sensing receptor, the transducer and the accompanying electronics and software. As we progress from single analyte testing to multi-analyte analysis, miniaturization and nanotechnology are playing a large part in producing highly sensitive and cost-effective devices. This chapter has shown that there are very attractive technologies being developed for mycotoxins analysis which can be applied at all levels whether it is in the farm or the factory and can be operated for on-site analysis by unskilled personnel. Trends towards further developing and producing chip-based micro/nano-arrays for multi-mycotoxins analysis will continue and this will have a significant impact on risk assessment testing. The introduction of a diverse array of nanomaterials such as gold and silver nanoparticles and other metal oxides such as quantum dots for diagnostic applications will enhance the capability of the biosensor technology. Advances in silicon fabrication technologies is producing more defined and reproducible array devices and this will add further improvement to the final sensing devices. This however, needs to be combined with developments in sampling acquisition and sample handling procedures.

14.7   Conclusions

Biosensors have the potential to provide rapid and specific sensing for food quality assurance. Analysing mycotoxins at the required legislative limit requires very highly sensitive devices that allow rapid analysis. It is advantageous for these techniques to be portable since a large number of mycotoxin analyses could benefit from on-site testing for risk assessment and management. Therefore, a need exists for simple and sensitive diagnostics methods that can detect multiple mycotoxins which occur at low concentrations in different food and feed matrices. Biosensors can fulfil these requirements. However, biosensor devices need to be further developed to face these new challenges such as multiplex analysis of several toxins where arrays of sensors need to be developed at the same chip. Innovation in biosensor technology to include analysis software and microfluidics that can handle directly the food sample can make these devices of great potential in this application area. Applying nanomaterials in the development of sensors for mycotoxins will make these devices highly sensitive and more applicable for lab-on-a chip diagnosis. Based on all the investigations reported in this chapter, it is apparent that innovation in this area is taking place at a rapid pace. Early and sensitive detection will aid in eliminating these toxins from entering the food chain and preventing illness and protecting life. Therefore these rapid technologies need to be developed further using appropriate funding to move the technology from research to commercial products.

14.8   Acknowledgements

The author acknowledges the European Commission through Project FP6-IST1–508774–IP 'GOODFOOD: Food safety and quality with microsystems technology' and the Malaysian Agricultural Research and Development Institute (MARDI) for supporting PhD research in this area. Acknowledgement also goes to Charlie Parker, Meike Heurich and Mohamad Kamal Abdul Kadir: my PhD students whose results were presented in this chapter.

14.9   References

Abdul Kadir, M.K. and Tothill, I.E. (2010). 'Development of an electrochemical immunosensor for fumonisins detection in foods'. *Toxins*, **2**, 382–98.

Abrantes, M., Magone, M.T., Boyd, L. F and Schuck, P. (2001). 'Adaptation of a surface plasmon resonance biosensor with microfluidics for use with small sample volumes and long contact times'. *Anal. Chem.*, **73**, 2828–35.

Adanyi, N., Levkovet, I,A., Rodriguez-Gil, S., Ronald, A., Varadi, M. and Sdenro, I. (2007). 'Development of immunosensor based on OWSL technique for detemining of aflatoxin B$_1$ and ochratoxin A'. *Biosensor and Bioelectronic*, **22**, 797–806.

Alarcon, S.H., Palleschi, G., Campahnone, D., Pascale, M., Visconti, A. and Barna-Vetro, I. (2006). 'Monoclonal Antibody Based Electrochemical Immunosensor for the Determination of Ochratoxin a in Wheat'. *Talanta*, **69**(4), 1031–7.

Aldred D., Olsen M. and Magan N. (2004). 'HACCP and mycotoxin control in the food chain'. In *Mycotoxin in Food: Detection and Control*. Magan N. andOlsen M. (eds), Woodhead, Cambridge, 139–73.

Ammida, N.H.S., Micheli, L. and Palleschi, G. (2004). 'Electrochemical immunosensor for determination of aflatoxin B_1 in barley'. *Anal. Chim. Acta*, **520**, 159–64.

Andreous, V.G. and Nikolelis, D.P. (1998). 'Flow injection monitoring of aflatoxin M_1 in milk and milk preparations using filter supported bilayer lipid membranes'. *Anal. Chem.*, **70**, 2366–71.

Badea, M., Micheli, L., Messia, M.C., Candigliota, T., Marconi, E., Mottram, T., Velasco-Garcia, M., Moscone, D. and Palleschi. G. (2004). 'Alfatoxin M_1 determination in raw milk using a flow-injection immunoassay system'. *Anal. Chim. Acta*, **520**, 141–8.

Baller, M.K., Lang, H.P., Fritz, J., Gerber, C., Gimzewsk, J.K., Drechsler, U., Rothuizen, H., Despont, M., Vettiger, P., Battiston, F.M., Ramseyer, J.P., Fornaro, P., Meyer, E. and Guntherodt, H.J. (2000). 'A cantilever array-based artifcial nose'. *Ultramicroscopy*, **82**, 1–9.

Batra, D. and Shea, K.J. (2003). 'Combinatorial methods in molecular imprinting'. *Current Opinions Chem. Biol.*, **7**, 434–42.

Bhatnagar, D., Yu, J. and Ehrlich, K.C. (2002). 'Toxins of filamenous fungi'. In *Fungal Allergy and Pathogenicity*, Breitenbach, M., Cremeri, R. and Lehrer, S.B. (eds). Chemical Immunology, Karger, Basel, Volume 81, 167–206.

Borisov S.M. and Wolfbeis O.S. (2008). 'Optical Biosensors', *Chem. Rev.*, **108**(2),423–61.

Brody E.N. and Gold L. (2000). 'Aptamers as therapeutic and diagnostic agents'. *J. Biotechnol.*, **74**(1), 5–13.

Brown, K.C. (2003). 'New approaches for cell-specific targeting: identification of cell-selective peptides from combinatorial libraries'. *Curr. Opin. Chem. Biol.*, **4**, 16–21.

Cabanes, F.J., Sahgal, N., Bragulat, R. M. and Magan, N. (2009). 'Early discrimination of fungal species responsible of ochratoxin A contamination of wine and other grape products using an electronic nose'. *Mycotoxin Res.*, **25**, 187–92.

Carlson, M.A., Bargeron, C.B., Benson, R.C., Fraser A.B., Philips, T.E., Velky J.T., Groopman, J.D., Strickland, P.T. and Ko, H.W. (2000). 'An automated, handheld biosensor for aflatoxin'. *Biosensors and Bioelectronics*, **14**, 841–8.

Cerchia, L., Hamm, J., Libri, D., Tavitian, B. and de Franciscis, V. (2002). 'Nucleic acid aptamers in cancer medicine'. *FEBS*, **528**, 12–6.

Chambers, R.S. and Johnston, S.A. (2003). 'High-level generation of polyclonal antibodies by genetic immunization'. *Nature Biotechnol.*, **21**, 1088–92.

Chang, CY., Abdo, J., Hartney, T. and McDonnell, DP. (2005). 'Development of peptide antagonists for the androgen receptor using combinatorial peptide phage display'. *Mol. Endocrinol.*, **19**, 2478–90.

Chen, B., Bestetti, G. and Turner, A.P.F. (1998). 'The synthesis and screening of a combinatorial library for affinity ligands for glycosylated haemoglobin', *Biosensors and Bioelectronics*, **13**, 779–85.

Chianella, I., Lotierzo, M., Piletsky, S., Tothill, IE., Chen, B. and Turner, APF. (2002). 'Rational design of a polymer specific for microcystin-LR using a computational approach'. *Anal. Chem.*, **74**, 1288–93.

Chianella, I, Piletsky, S.A., Tothill, I.E., Chen, B. and Turner, A.P.F. (2003). 'Combination of solid phase extraction cartridges and MIP-based sensor for detection of microcystin-LR'. *Biosensors and Bioelectronics*, **18**, 119–27.

Chiavaro, E., Caccgiolo, C., Berni, E. and Spotto, E. (2005). 'Immunoaffinity cleanup and direct fluorescence measurement of aflatoxins B_1 and M_1 in pig liver: Comparison with high-performance liquid chromatography determination'. *Food Addit. Contam.*, **22**, 1154–61.

Clark S.L. and Remcho V.T. (2002). 'Aptamers as analytical reagents'. *Electrophoresis*, **23**(9), 1335–40.

Collett, J.R., Cho, E.J., Lee, J.F., Levy, M., Hood, A.J., Wan, C. and Ellington, A.D. (2005).

'Functional RNA microarrays for high-throughput screening of antiprotein aptamers'. *Anal. Biochem.*, **338**, 113–23.

Cooper, M A. (2002). 'Optical biosensors in drug discovery', *Nature Reviews Drug Discovery*, **1**, 515–28.

Cooper, M.A. and Singleton, V.T. (2007). 'A survey of the 2001 to 2005 quartz crystal microbalance biosensor literature: applications of acoustic physics to the analysis of biomolecular interactions'. *J. Molec. Recognit.* **20**, 154–84.

Cruz-Aguado, J.A. and Penner, G. (2008a). 'Determination of ochratoxin A with a DNA aptamer'. *J. Agric. Food Chem.*, **56**, 10456–61.

Cruz-Aguado, J.A. and Penner, G. (2008b). 'Fluorescence polarization based displacement assay for the determination of small molecules with aptamers'. *Anal. Chem.*, **80**, 8853–5.

Daly, S.J., Keating, G.J., Dillon, P.P., Manning, B.M., O'Kennedy, R., Lee, H A. and Morgan, R.A. (2000). 'Development of surface plasmon resonance-based immunoassay for aflatoxin B(1)'. *J. Agric. Food Chem.*, **48**, 5097–104.

De Smet D, Dubruel P, Van Peteghem C, Schacht E and De Saeger S. (2009). 'Molecularly imprinted solid-phase extraction of fumonisin B analogues in bell pepper, rice and corn flakes'. *Food Addit. Contam. Part A – Chem. Anal. Control Exposure and Risk Assessment*, **26**(6), 874–84.

De Smet D, Monbaliu S., Dubruel P., Van Peteghem C., Schacht E. and De Saeger S. (2010). 'Synthesis and application of a T-2 toxin imprinted polymer'. *J. Chromatogr. A*, **1217**, 2879–86.

Flanagan, M.T. and Pantell, R.H. (1984). 'Surface plasm3on resonance and immunosensors'. *Electron Lett.*, **20**, 968–70.

Gaag, van der., B., Spath, S., Dietrich, H., Stigter, E., Boonzaaijer, G., Osenbruggen, T. and Koopal, K. (2003). 'Biosensors and multiple mycotoxin analysis'. *Food Control*, **14**, 251–4.

Gardner, J.W and Bartlett, P.N. (1999). *Electronic Noses: Principle and Applications*, Oxford University Press, Oxford.

Gilbert, J. and Anklam, E. (2002). 'Validation of analytical methods for determining mycotoxins in foodstuffs', *Trends Anal. Chem.*, **21**, 468–86.

Gilbert, J. and Vargas, E.A. (2003). 'Advances in sampling and analysis for aflatoxins in food and animal feed'. *J. Toxicol. – Toxin Rev.*, **22**, 381–422.

Giraudi, G., Giovannoli, C., Tozzi, C., Baggiani, C. and Anfossi, L. (2000). 'Estradiol binding synthetic polypeptides'. *Chem. Comm.*, **13**, 1135–6.

Giraudi, G., Giovannoli, C., Tozzi, C., Baggiani, C. and Anfossi, L. (2003) 'Molecular recognition properties of peptide mixtures obtained by polymerisation of amino acids in the presence of estradiol'. *Anal. Chim. Acta*, **481**, 41–53.

Goldman, E.R., Clapp, A.R., Anderson, G.P., Uyeda, H.T., Mauro, J.M., Medintz, I.L. and Mattoussi, H. (2004). 'Multiplexed Toxin Analysis Using Four Colors of Quantum Dot Fluororeagents'. *Anal. Chem.*, **76**, 684–8.

Hansen, K.M. and Thundat, T. (2005). 'Microcantilever biosensors'. *Methods*, **37**, 57–64.

Hauck, S., Kosslinger, C., Drost S. and Wolf H. (1998). 'Biosensor system to determine ochratoxin A'. *Lebensmittelchemie*, **52**, 158.

Heurich, M. (2008). *Development of an Affinity Sensors for Ochratoxin A.* PhD Thesis, Cranfield Health, Cranfield University, Bedford, UK.

Heurich, M., Abdul Kadir, M.K and Tothill, I.E. (2010). 'Development of an electrochemical immunosensor-based on CM-dextran modified gold electrode for the detection of ochratoxin A in wine samples', (submitted).

Jain, K.K. (2004). 'Applications of biochips: from diagnostics to personalized medicine'. *Curr. Opin. Drug Discov. Dev.*, **7**, 285–9.

Jin, X. , Jin, X., Liu, X., Chen, L., Jiang, J., Shen, G. and Yu, R. (2009). 'Biocatalyzed deposition amplification for detection of aflatoxin B$_1$ based on quartz crystal microbalance', *Anal. Chim. Acta*, **645**, 92–7.

Jodlbauer, J., Maier, N. M. and Lindner, W. (2002). 'Towards ochratoxin A selective molecularly
imprinted polymers for solid phase extraction.' *J. Chromatogr. A*, **945**, 45–63.

Jorge, P., Martins, M.A., Trindade , T., Santos, J. L. and Farahi, F. (2007) 'Optical fiber sensing using quantum dots'. *Sensors*, **7**, 3489–534.

Katz, E. and Willner, I. (2005). 'Biomolecular-functionalized carbon nanotubes: applications in nanobioelectronics'. *Chem. Phys. Chem.*, **5**, 1085–91.

Katz, E., Willner, I. and Wang, J. (2004). 'Electroanalytical and bioelectroanalytical systems based on metal and semiconductive nanoparticles', *Electroanalysis*, **16**, 19–44.

Keshri, G. and Magan, N. (2000). 'Detection and differentiation between mycotoxigenic and non-mycotoxigenic strains of *Fusarium* spp. Using volatile production profiles and hydrolytic enzymes'. J. *Appl. Microbiol.*, **89**, 825–33.

Khan, R. and Dhayal, M. (2008). 'Nanocrystalline-bioactive TiO2–chitosan impedimetric immunosensor for ochratoxin A'. *Electrochem. Commun.*, **10**, 492–5.

Kretschmann, E. (1971). 'Determination of optical constants of metal by excitation of surface plasmons'. *Zeits. für Physik*, **241**, 313–24.

Kurihara, K. and Suzuki, K. (2002). 'Theoretical understanding of an absorption based surface plasmon resonance sensor based on Kretchmann's theory'. *Anal. Chemi.*, **74**, 696–701.

Leatherbarrow, R.J. and Edwards, P.R. (1999). 'Analysis of molecular recognition using optical biosensors'. *Curr. Opin. Chem. Biol.*, **3**, 544–7.

Liedberg, B., Nylander, C. and Lundstroem, I. (1983). 'Surface plasmon resonance for gas detection and biosensing'. *Lab. Sensors Actuators*, **4**, 299–304.

Ligler, F.S. and Taitt, C.R. (2008). *Optical Biosensors: Today and Tomorrow*. 2nd edition, Elsevier BV, Amsterdam, The Netherlands, 712.

Ligler, F.S., Taitt, C.R., Shriver-Lake, L.C., Sapsford, K.E., Shubin, Y. and Golden, J.P. (2003). 'Array biosensor for detection of toxins'. *Anal. Bioanal. Chem.*, **377**, 469–77.

Liu, Y., Qin, Z., Wu, X. and Jiang, H. (2006). 'Immune-biosensor for aflatoxin B_1 based bio-electrocatalytic reaction on micro-comb electrode'. *Biochem. Eng. J.*, **32**, 211–7.

Logrieco, A., Arrigan, D. W. M. , Brengel-Pesce, K., Siciliano P. and Tothill, I.E. (2005). 'Microsystens technology solutions for rapid detection of toxigenic fungi and mycotoxins'. *J. Food Addit. Contam.*, **22**(4), 335–44.

Lotierzo, M., Henry, O.Y.F., Piletsky, S., Tothill, I., Cullen, D., Kania, M., Hock B. and Turner, A.P.F. (2004). 'Surface plasmon resonance sensor for domoic acid based on grafted imprinted polymer'. *Biosensors and Bioelectronics*, **20**, 145–52.

Luzi, E., Minunni, M., Tombelli, S. and Mascini, M. (2003). 'New trends in affinity sensing: aptamers for ligand binding'. *Trends Anal. Chem.*, **22**, 810–8.

Magan, N. and Evans, P. (2000). 'Volatiles in grain as an indicator of fungal spoilage, order descriptors for classifying spoiled grain and the potential for early detection using electronic nose technology: A review'. *J. Stored Product Protect.*, **36**, 319–40.

Magan N. and Olsen M. (eds) (2004). *Mycotoxins in Food: Detection and Control*. Woodhead, Cambridge.

Maier, N.M., Buttinger, G., Welhartiziki, S., Gaviolo, E. and Lindner, W. (2004). 'Molecularly imprinted polymer-assisted sample clean-up of ochratoxin A from red wine: merits and limitations.' *J. Chromatogr. B*, **804**(1), 103–11.

Maragos, C. M. (2009). 'Recent advances in the development of novel materials for mycotoxin analysis'. *Anal. Bioanal. Chem.*, **395**, 1205–13.

Maragos, C.M. and Thompson, V.S. (1999). 'Fiber-optic immunosensor for mycotoxins'. *Natural Toxins*, **7**, 371–6.

Mascini, M., Palchetti, I. and Marraza, G. (2001). 'DNA electrochemical biosensors'. *Fresenius J. Anal. Chem.*, **369**, 15–20.

McGuire, M.J., Samli, K.N., Johnston, S.A. and Brown, K.C. (2004a). '*In vitro* selection of a peptide with high selectivity for cardiomyocytes *in vivo*'. *J. Mol. Biol.*, **342**, 171–82.

McGuire, M.J., Sykes, K.F., Samli, K.N., Timares, L., Barry, M.A., Stemke- Hale, K.,

Tagliaferri, F., Logan, M., Jansa, K., Takashima, A., Brown, K.C. and Johnston, S.A. (2004b). 'A library-selected, Langerhans cell-targeting peptide enhances an immune response'. *DNA Cell Biol.*, **23**, 742–52.

Medintz , I.L., Uyeda, H. T., Goldman, E. R. and Mattoussi, H. (2005). 'Quantum dot bioconjugates for imaging, labelling and sensing'. *Nature Materials* **4**, 435 – 46.

Micheli, L., Grecco, R., Badea, M., Moscone, D. and Palleschi, G. (2005) 'An electrochemical immunosensor for aflatoxin M_1 determination in milk using screen printed electrodes'. *Biosensors and Bioelectronics*, **21**, 588–96.

Moressi, M.B., Zon, A., Fernández, H., Rivas, G. and Solis, V. (1999). 'Amperometric quantification of alternaria mycotoxins with a mushroom tyrosinase modified carbon paste electrode'. *Electrochem. Commun.*, **1**, 472–6.

Mosbach, K. (1994). 'Molecular imprinting'. *Trends Biochem. Sci.*, **19**, 9–14.

Mosbach, K. and Ramstrom, O. (1996). 'The emerging technique of molecular imprinting and its future impact on biotechnology'. *Biotechnology*, **14**, 165–70.

Mullett, W., Lai, E.P.C. and Yeung, J.M. (1998). 'Immunoassay of fumonisins by a surface plasmon resonance biosensor'. *Anal. Biochem.*, **258**, 161–7.

Myszka, D.G. (1999). 'Survey of the 1998 optical biosensor literature'. *J. Mol. Recogn.*, **12**, 390–408.

Nasir, M.S. and Jolley, M.E., (2002). 'Development of a fluorescence polarization assay for the determination of aflatoxins in grains'. *J. Agric. Food Chem.*, **50**, 3116–21.

Ngundi, M.M., Shriver-Lake, L.C., Moore, M H., Lassman, M.E., Ligler, F.S., and Taitt, C R. (2005). 'Array biosensor for detection of ochratoxin A in cereals and beverages'. *Anal. Chem.*, **77**, 148–54.

Ngundi, M.M., Shriver-Lake, L.C., Moore, M.H. Ligler, F.S. and Taitt, C.R. (2006). 'Multiplexed detection of mycotoxins in foods with a regenerable array'. *J. Food Prot.*, **69**, 3047–51.

Nugaeva N., Gfeller, K.Y., Backmann, N., Duggelin, M., Lang, H.P., Guntherodt H.J. and Hegner M. (2007). 'An antibody-sensitized microfabricated cantilever for the growth detection of *Aspergillus niger* spores'. *Microscopy Microanal.*, **13**(1), 13–7.

Olssen, J., Borjesson, T., Lundstedt, T. and Schnuere R. (2002). 'Detection and quantification of ochratoxin A and deoxynivalenol in barley by GC–MS and electronic nose'. *Int. J. Food Microbiol.*, **72**, 203–14.

O'Sullivan, C.K. (2002). 'Aptasensors – the future of biosensors?' *Anal. Bioanal. Chem.*, **372**, 44–8.

O'Sullivan, C.K. and Guilbault, G.G. (1999). 'Commercial quartz crystal microbalances – theory and applications'. *Biosensors and Bioelectronics*, **14**, 663–70.

Oyama, T., Sykes, K.F., Samli, K.N., Minna, J.D., Johnston, S.A. and Brown, K.C. (2003). 'Isolation of lung tumor specific peptides from a random peptide library: generation of diagnostic and cell-targeting reagents'. *Cancer Lett.*, **202**, 219–30.

Parker, C. (2008) *Development of an Affinity Sensor for the Detection of Aflatoxin M_1 in Milk*. PhD Thesis, Cranfield Health, Cranfield University, Bedford, UK.

Parker, C. and Tothill, I.E. (2009). 'Development of an electrochemical immunosensor for aflatoxin M_1 in milk with focus on matrix interference'. *Biosensors and Bioelectronics*, **24**, 2452–7.

Parker, C., Lanyon, Y., Manning, M. Arrigan, D. W. M. and Tothill I.E (2009). 'Development of microsensor for Aflatoxin M_1'. *Anal. Chem.*, **81**, 5291–98.

Patel, P. (2004). 'Mycotoxin analysis: current and emerging technologies'. In *Mycotoxins in Food, Detection and Contol*, Magan N. and Olsen M. (eds). Woodhead Publishing Limited, Cambridge, England, 88–106.

Pearce, T.C., Schiffman, S.S., Nagle, H.T. and Gardner, J.W. (eds) (2002). *Handbook of Machine Olfaction: Electronic Nose Technology*, Wiley, Chichester, UK.

Pemberton, M.R., Pittson, R., Biddle, N., Drago, G.A. and Hart, J.P. (2006). 'Studies towards the development of a screen printed carbon electrochemical immunosensor array for mycotoxin: A sensor for aflatoxin B_1'. *Anal. Lett.*, **39**, 1573–86.

Pettersson H. and Aberg L.(2003). 'Near infrared spectroscopy for determination of mycotoxins in cereals'. *Food Control*, **14**, 229–32.

Piermarini, S., Micheli, L., Ammida, N.H.S., Palleschi, G. and Moscone, D. (2006). 'Electrochemical immunosensor array using a 96-well screen-printed microplate for aflatoxin B_1 detection'. *Biosensors and Bioelectronics*, **22**, 1434–40.

Piletsky S., Piletska E. V., Sergeyeva T. A., Nicholls I., Weston D. and Turner A. (2006). 'Synthesis of biologically active molecules by imprinting polymerization'. *Biopolymers and Cell*, **22**, 63–8.

Pohanka, M., Jun, D. and Kuca, K. (2007). 'Mycotoxin Assays Using Biosensor Technology: A Review'. *Drug Chem. Toxicol.*, **30**, 253–61.

Prieto-Simón, B., Noguer, T. and Campàs, M. (2007). 'Emerging biotools for assessment of mycotoxins in the past decade'. *TrAC Trends Anal. Chem.*, **26**(7), 689–702.

Reineccius, G. (1996). 'Instrumental means of monitoring the flavour quality of foods'. *ACS. Symp. Series*, **631**, 241–52.

Rodriguez, M., Kawde N. and Wang, J. (2005). 'Aptamer biosensor for label-free impedance spectroscopy detection of proteins based on recognition-induced switching of surface change'. *Chem. Commun.*, 4267–9.

Rosi N.L. and Mirkin C.A. (2005). 'Nanostructure and biodiagnostics'. *Chem. Rev.*, **105**, 1547–62.

Sahgal, N., Needham, R., Cabanes, F.J. and Magan, N. (2007). 'Potential for detection and discrimination between mycotoxigenic and non-toxigenic spoilage moulds using volatile production patterns: A review'. *Food Addit. Contam.*, **24**, 1161–8.

Sapsford, K.E., Ngundi, M.M., Moore, M.H., Lassman, M.E., Shriver-Lake, L.C., Taitt, C.R. and Ligler, F.S. (2006). 'Rapid detection of foodbome contaminants using an array biosensor'. *Sensors and Actuators B Chemical*, **113**, 599–607.

Sarkar, D. and Somasundaran, P. (2002) 'Overcoming contamination in surface plasmon resonance spectroscopy'. *Langmuir*, **18**, 8271–7.

Schnerr, H., Vogel, R.F. and Niessen, L. (2002). 'A biosensor-based immunoassay for rapid screening of deoxynivalenol contamination in wheat'. *Food Agric. Immunol.*, **14**, 313–21.

Shaikh, K.A., Ryu, K.S., Goluch, E.D., Nam, J.M., Liu, J., Thaxton, C.S., Chiesl, T.N., Barron, A.E., Lu, Y., Mirkin, C.A. and Liu, C. (2005). 'A modular microfluidic architecture for integrated biochemical analysis'. *Proc. Natl Acad. Sci. USA*, **102**, 9745–50.

Shukla G.S. and Krag D.N. (2005). 'Selection of tumor-targeting agents on freshly excised human breast tumors using a phage display library'. *Oncol. Rep.*, **13**, 757–64.

Siontorou, C.G., Andreou, V.G., Nikolelis, D.P. and Krull, U.J. (2000). 'Flow injection monitoring of aflatoxin M_1 in cheese using filter supported bilayer lipid membranes with incorporated DNA'. *Electroanalysis*, **12**, 747–51.

Sun, A-L., Qi, Q-A., Dong, Z-L. and Liang, K Z. (2008). 'An electrochemical enzyme immunoassay for aflatoxin B_1 based on bio-electrocatalytic reaction with room-temperature ionic liquid and nanoparticle-modified electrodes'. *Sensing Instrumentation Food Qual.*, **2**, 43–50.

Tan,Y., Chu, Z., Shen, G-L.and Yu, R-Q (2009). 'A signal-amplified electrochemical immunosensor for aflatoxin B_1 determination in rice'. *Anal. Biochem.*, **387**, 82–6.

Thévenot, D.R., Toth, K., Durst, R.A. and Wilson, G.S. (1999). 'Electrochemical biosensors: recommended definitions and classification'. *Pure Appl. Chem.*, **71**, 2333–48.

Thompson, V.S. and Maragos, C.M. (1996). 'Fiber-optic immunosensor for the detection of fumonisin B_1'. *J. Agric. Food Chem.*, **44**, 1041–6.

Tombelli, S. Minunni, M. and Mascini, M. (2005). 'Analytical applications of aptamers (Review)'. *Biosensors and Bioelectronics*, **20**, 2424–34.

Tothill, I.E (2001). 'Biosensors Developments and potential applications in the agricultural diagnosis sector'. *Computers and Electronics in Agriculture*, **30**, 205–18.

Tothill, I.E. (2003). 'On-line immunochemical assays for contaminants analysis'. In *Rapid and On-line Instrumentation for Food Quality Assurance*, I.E. Tothill (ed.). Woodhead Publishing Limited, England, 14–39.

Tothill, I.E. (2010a) 'Advances in the development of toxicity sensors'. In *Sensors for Chemical and Biological Applications*, M.K. Ram and V.R. Bhethanabotla (eds.), Taylor and Francis, CRC Press, USA, 384.

Tothill I.E. (2010b), 'Peptides as molecular receptors'. In *Recognition Receptors in Biosensors*, M. Zourob, Editor (ed.), Springer, New York, 249–74.

Tothill, I.E. and Magan, N. (2003). 'Rapid detection methods of microbial contamination'. In *Rapid and On-line Instrumentation for Food Quality Assurance*, I.E. Tothill (ed.). Woodhead Publishing Limited, England, 136–60.

Tothill, I.E and Turner, A.P.F. (2003). 'Biosensors'. In *Encyclopaedia of Food Sciences and Nutrition* (2nd edn), B. Caballero, L. Trugo and P. Finglas (eds), Academic Press, 489–499.

Tothill, I.E., Piletsky, S., Magan, N. and Turner, A.P.F. (2001). 'New Biosensors'. In *Instrumentation and Sensors for the Food Industry*, 2nd edn, E. Kress-Rogers and C.J.B. Brimelow (eds).Woodhead Publishing Limited, Cambridge, England, 760–76.

Tozzi, C., Anfossi, L., Giraudi, G., Giovannoli, C., Baggiani, C. and Vanni, A. (2002) 'Chromatographic characterisation of an estrogen-binding affinity column containing tetrapeptides selected by a combinatorial-binding approach'. *J. Chromatogr. A*, **966**, 71–9.

Tozzi, C., Anfossi, L., Baggiani, C., Giovannoli, C. and Giraudi, G. (2003). 'A combinatorial approach to obtain affinity media with biding properties towards the aflatoxins'. *Anal. Bioanal. Chem.*, **375**, 994–9.

Tsai, W.-C. and Hsieh, C.-K. (2007). 'QCM-Based Immunosensor for the Determination of Ochratoxin A'. *Anal. Lett.*, **40**, 1979–91.

Tudos, A.J., Lucas van den Bos, E.R. and Stigter, E.C.A. (2003). 'Rapid surface plasmon resonance based inhibition assay of deoxynivalenol'. *J. Agric. Food Chem.*, **51**, 5843–58.

Turner, N.S. (2004). *Molecular Imprinted Polymers for Ochratoxin*. PhD Thesis, Institute of BioScience and Technology, Cranfield University, UK.

Turner, A.P.F. and Magan, N (2004). 'Electronic noses and disease diagnostics'. *Nature Rev*, **2**, 1–6.

Turner, N.W., Piletska, E.V., Karim, K., Whitcombe, M.J., Malecha, M.M., Magan, N., Baggiani, C. and Piletsky, S.A. (2004). 'Effect of the solvent on recognition properties of molecularly imprinted polymer specific for ochratoxin A'. *Biosensors and Bioelectronics*, **20**: 1060–7.

Ulrich H., Alves M.J.M. and Colli W. (2001). 'RNA and DNA aptamers as potential tools to prevent cell adhesion in disease'. *Brazil J. Med. Biol. Res.*, **34**, 295–300.

Uludag, Y. and Tothill, I.E. (2010). 'Development of a sensitive detection method for cancer biomarkers in human serum (75%) using a quartz crystal microbalance sensor and nanoparticles amplification system', *Talanta*, **82**, 277–82.

Urraca, J.L., Marazuela, M.D., Merino, E.R., Orellana, G. and Moreno-Bondi, M.C. (2006) 'Molecularly imprinted polymers with a streamlined mimic for zearalenone analysis'. *J. Chromatogr. A*, **1116**, 127–34.

Wang, J. (2005). 'Nanomaterial-based electrochemical biosensors'. *Analyst*, **130**, 421–6.

Wang, P.C., Gao, J. and Lee, C.S. (2002). 'High-resolution chiral separation using microfluidics-based membrane chromatography'. *J. Chromatogr. A*, **942**, 115–22.

Wang W.U., Chen C., Lin K.H, Fang Y. and Lieber C.M. (2005). 'Label-free detection of small molecule–protein interactions by using nanowire nanosensors'. *Proc Natl Acad Sci USA*, **102**, 3208–12.

Wang, X-H. and Wang S. (2008). 'Sensors and biosensors for the determination of small molecule biological toxins'. *Sensors*, **8**, 6045–54.

Waggoner, P.S. and Craighead, H.G. (2007). 'Micro- and nanomechanical sensors for environmental, chemical, and biological detection'. *Lab on a Chip*, **7**, 1238–55.

Weimar, T. (2000). 'Recent trends in the application of evanescent wave biosensors'. *Angew. Chem. Int. Edn Engl.*, **39**, 1219–21.

Weiss, R., Freudenschuss, M., Krska, R. and Mizaikoff, B. (2003) 'Improving methods of analysis for mycotoxins: molecularly imprinted polymers for deoxynivalenol and zearalenone'. *Food Addit. Contam.*, **20**, 386–95.

Willner, I., Baron R. and Willner B. (2007). 'Integrated nanoparticle – biomolecule systems for biosensing and bioelectronics'. *Biosensors and Bioelectronics*, **22**, 1841–52.

Woolley, A.T., Guillemette, C., Cheung, C.L., Housman, D.E. and Lieber, C.M. (2000). 'Direct haplotyping of kilobase-size DNA using carbon nanotube probes'. *Nature Biotechnol.*, **18**, 760–3.

Yu, J.C.C. and Lai, E.P.C. (2005). 'Interaction of ochratoxin A with molecularly imprinted polypyrrole film on surface plasmon resonance sensor'. *Reactive and Functional Polymers*, **63**, 171–6.

Zheng, M., Richard, J. and Binder, J.A. (2006).' Review of rapid methods for the analysis of mycotoxins'. *Mycopathologia*, **161**(5), 261–73.

Zourob, M. (ed.) (2010). *Recognition Receptors in Biosensors*. Springer, New York.

15

Masked mycotoxins in food and feed: challenges and analytical approaches

J. Diana Di Mavungu and S. De Saeger, Ghent University, Belgium

Abstract: Conjugated mycotoxins, in which the toxin is usually bound to a more polar substance like glucose, are referred to as masked mycotoxins, as they escape commonly used methods of analysis but can release their toxic precursors after hydrolysis. Nowadays, there is a growing interest related to the determination of these hidden forms of mycotoxins in food and feed commodities, as they represent an additional potential risk for the consumer. This chapter reviews the current state of knowledge regarding masked mycotoxins and the current analytical strategies for their detection and quantification in food and feed. The main problems related to the extraction of these highly polar derivatives as well as the different alternatives employed for their simultaneous determination together with the precursor toxins are discussed.

Key words: conjugated mycotoxins, food processing conjugates, liquid chromatography–mass spectrometry (LC–MS), masked mycotoxins, plant metabolism.

15.1 Introduction

In the last few years it has become more and more clear that in mycotoxin-contaminated commodities, many structurally related compounds generated by plant metabolism, fungi or food processing can coexist together with the native toxins. These mycotoxin derivatives may have a very different chemical behaviour, compared to the parent molecules and therefore, they can easily escape commonly used analytical methods. However, although derivatives can be less toxic than the precursor toxins (Poppenberger *et al.*, 2003; Wu *et al.*, 2007), the possibility that they can be converted back into the original toxic form by hydrolysis during food/feed processing or in the digestive tract of humans and animals cannot be excluded (Gareis *et al.*, 1990).

Transformation of mycotoxins by plants has mainly been described for *Fusarium* toxins (Böswald *et al.*, 1995; Gareis *et al.*, 1990) but has also been shown for ochratoxin A (Ruhland *et al.*, 1996a,b; Ruhland *et al.*, 1997). As a defence mechanism, plants can convert the relatively apolar mycotoxins into more polar derivatives via conjugation with sugars, amino acids or sulfate groups and compartmentalize them in vacuoles (Engelhardt *et al.*, 1999). These conjugated mycotoxins (reviewed by Berthiller *et al.*, 2006a, 2007, 2009a), are often referred to as masked mycotoxins as to emphasize the fact that they are usually not detected by routine analysis of food and feed, but contribute to the total mycotoxin content. Besides plant metabolism, the technological process also has an important role in the masking mechanism, in particular in cereal-derived products. Indeed, mechanical or thermal energy during the transformation process may cause significant modification, for instance the induction of reactions with macromolecular components such as polysaccharides, proteins or lipids, or the release of the native toxins through decomposition of the masked derivatives. In the case of fumonisins, the phenomenon has been described as 'the fumonisin paradox' because apparently low contaminated commodities have also been found to induce toxic effects, highlighting the problem of 'bound' or 'hidden' fumonisins, which may be released upon alkaline hydrolysis (Dall'Asta *et al.*, 2009a; Galaverna *et al.*, 2009).

In consequence, masked mycotoxins have been incorporated into the analytical methods in order to have a better insight into the presence of mycotoxins in different matrices. This chapter focuses on the study of the significant mycotoxin conjugates found in food and feed samples, indicating the main analytical techniques used for the extraction and determination of these types of compounds and highlighting advances in instrumentation that may facilitate the structural elucidation and identification of novel conjugates. A distinction is made by some authors between soluble conjugates referred to as masked mycotoxins and insoluble conjugates referred to as bound mycotoxins. In this chapter we shall refer to both as masked mycotoxins.

15.2   Occurrence of masked mycotoxins in food and feed

15.2.1 Plant conjugates

Plants, like most other organisms, have a remarkable capability to defend themselves against the potentially toxic effects of xenobiotics. This defence reaction typically involves oxidation, reduction, hydrolysis and conjugation with glutathione, sugars, amino acids or sulfates. Eventually, xenobiotics or xenobiotic conjugates are converted into insoluble bound residues, which are deposited in specific organelles (e.g. vacuoles, chloroplasts) or in extracytoplasmic space (e.g. cell wall) of plant cells (Conn, 1985; Engelhardt *et al.*, 1999; Sandermann, 1992).

Transformation of mycotoxins by plants has long been suspected from studies intended to assess deoxynivalenol (DON) accumulation in corn. Miller *et al.* (1983) observed that DON concentration of field maize inoculated with *F. graminearum* decreased during the growing season. The authors speculated that this decrease might be due to a chemical transformation of the toxin by plant enzymes. Shortly thereafter, it was reported that the DON content of yeast fermented food products was higher than that of the contaminated wheat flour used for their production (Young *et al.*, 1984). This increase was attributed to the enzymatic conversion of a DON conjugate, probably generated by the metabolization of the toxin by wheat. Savard (1991) chemically synthesized glucoside and fatty acid of DON, providing analytical standards for further study. A year later Sewald *et al.* (1992) showed that DON-3-glucoside (DON-3-G) could be formed from DON in maize cell suspension. In the model plant *Arabidopsis thaliana*, a gene encoding a uridine diphosphate glycosyltransferase (UDP-glycosyltransferase) that is able to convert DON into DON-3-G was isolated and characterized (Poppenberger *et al.*, 2003). However, it remained unclear whether this DON conjugate could also occur in cereals until 2005, when the natural occurrence of DON-3-G was reported for the first time in artificially (Berthiller *et al.*, 2005a; Dall'Asta *et al.*, 2005) and in naturally (Berthiller *et al.*, 2005a) infected cereals. More recently, the same authors investigated the frequency of occurrence as well as the relative and absolute concentrations of DON-3-G in naturally infected cereals (Berthiller *et al.*, 2009b). This minisurvey showed that DON-3-G can exceed 1000 µg kg^{-1} in naturally contaminated wheat and can reach over 70% of the DON concentration in maize. Sasanya *et al.* (2008) even found DON-3-G at levels exceeding the DON concentration in some hard red spring wheat samples. The occurrence of DON-3-G in naturally contaminated barley has also been reported (Lancova *et al.*, 2008).

In 1988, zearalenone-4-glucoside (ZON-4-G) had already been shown to be a metabolite of plants, after ZON was transformed by maize cell suspension cultures into its glucoside (Engelhardt *et al.*, 1988; Zill *et al.*, 1990a). However, it was in 2002 that ZON-4-G was reported for the first time in naturally contaminated wheat (Schneweis *et al.*, 2002). This study is indeed of great importance in research on masked mycotoxins, since it demonstrated for the very first time the occurrence of a mycotoxin conjugate in naturally contaminated cereals. The ZON-4-G standard used in this study was produced according to a previously described procedure (Zill *et al.*, 1990a,b). Recently, a fermentation procedure to produce ZON-4-G from ZON using an engineered *Saccharomyces cerevisiae* strain, expressing the *Arabidopsis thaliana* UDP-glucosyltransferase UGT73C6 was reported (Berthiller *et al.*, 2009c). Unambiguous structural assignment using two-dimensional nuclear magnetic resonance (2D NMR) confirmed that the glucose moiety is attached at the C-4 position of the precursor molecule.

Ochratoxin A (OTA) has been shown to be converted in suspension cultures of various plant cells into ochratoxin α, ochratoxin A methylester, (4*r*)- and (4*s*)-4-hydroxyochratoxin A and the glucosides and methyl esters of both isomers of hydroxyochratoxin A (Ruhland *et al.*, 1996a,b). OTA transformation also occurred in artificially contaminated vegetables and cereals, leading to the same metabolites as those reported in cell-suspension cultures (Ruhland *et al.*, 1997). However, to the best of our knowledge, the natural occurrence of these metabolites in food commodities has not been reported yet.

Recently, it has been reported that bound fumonisins can occur not only in thermally treated maize-based products but also in mild processed or even raw products (Dall'Asta *et al.*, 2008), indicating the possibility of chemical transformation of fumonisins *in planta*. However, the exact chemical nature of these bound fumonisins is still unknown.

15.2.2   Food processing conjugates

Besides the chemical transformation *in planta*, processing of food commodities is another source of mycotoxin conjugate formation. The effect of thermal food processing on the chemical structure and the toxicity of fumonisins has been extensively studied and a relevant review of the topic has been published (Humpf and Voss, 2004). A major reaction occurring in heat-treated food involves fumonisin B_1 (FB_1) and reducing sugars to form N-(carboxymethyl)-FB_1 (Howard *et al.*, 1998) and N-(1-deoxy-D-fructos-1-yl)-FB_1 (Poling *et al.*, 2002). The occurrence of these fumonisin derivatives in corn products has been reported (Seefelder *et al.*, 2001). On the other hand, low recoveries of FB_1 have been observed in different matrices such as rice flour, cornstarch, cornmeal and glucose after thermal treatment (Kim *et al.*, 2002). It was postulated that FB_1 interacted with food macroconstituents such as protein or starch. In order to yield evidence that FB_1 might bind to matrix components in thermally treated food, model experiments have been performed (Kim *et al.*, 2003; Park *et al.*, 2004; Seefelder *et al.*, 2003). The results indicated that fumonisins can bind to polysaccharides and proteins via their two tricarballylic acid groups and that binding to starch occurs to a greater extent than in proteins. The occurrence of protein-bound fumonisins in commercial corn flakes has also been shown (Dall'Asta *et al.*, 2008; Kim *et al.*, 2003; Park *et al.*, 2004).

With regard to trichothecenes, the occurrence of food processing conjugates has mainly been shown with DON. Indeed, a significant increase in DON-3-G levels was observed during the malting and brewing processes, using barley naturally and artificially infected with *Fusarium* (Kostelanska *et al.*, 2009; Lancova *et al.*, 2008). The DON-3-G in malt even increased over the sum of DON and DON-3-G levels measured in the grain used. The nature of this increase could not be completely explained. The authors hypothesized that under the malting conditions, fungus present in the infected barley could produce additional mycotoxin that is conjugated efficiently by the metabolically highly active germling. On the other hand, enzymes produced during the mashing of malt grist could degrade cell walls, membrane-bound proteins and starch depots in kernels, thus releasing DON-3-G from insoluble forms.

15.3   Analysis of masked mycotoxins in food and feed

15.3.1   Extraction of masked mycotoxins and clean up of sample extracts

Analytical methods for the determination of mycotoxins in food and feed usually require solvent extraction to liberate the toxin from the sample matrix and subsequent clean-up of the extract to reduce matrix interferences. This sample treatment can be the most tedious step in the analytical procedure, as it can take over 60% of the analyst's time (Picó *et al.*, 2006). As a consequence, there is currently a trend towards the simplification of this stage, trying to find simple methods that allow the determination of a wide range of compounds, using a single extraction step. However, a problem related to the determination of masked mycotoxins is that these conjugated derivatives show a relatively higher polarity than their precursors, which can make the simultaneous determination of both types of compounds difficult. Therefore, commonly used extraction and clean-up procedures for free mycotoxins may need to be adapted.

In literature, there are few sample pretreatment protocols that have been applied to methods for determination of masked mycotoxins in food. Schneweis *et al.* (2002) used the solvent mixture acetonitrile/water (21:4) to simultaneously extract ZON-4-G and ZON from wheat samples. For the sample clean up, solid phase extraction (SPE) was applied, using Florisil as sorbent. Apparent recovery rates of 69% were achieved for ZON-4-G. The solvent mixture acetonitrile/water (21:4) was also used by Berthiller *et al.* (2005a; 2005b) to extract DON-3-G and other trichothecenes simultaneously in artificially and naturally contaminated wheat samples. Sample extracts were purified using a MycoSep® 230 column. The average apparent recovery for DON-3-G was 59%. Recently, the above mentioned protocol has also been applied to a set of matrices occurring during malting and brewing processes (Lancova *et al.*, 2008; Kostelanska *et al.*, 2009). Apparent recoveries for DON-3-G ranged from 34–121% depending on the matrix.

Immunoaffinity columns (IACs) for the clean up of different mycotoxins are commercially available. Generally, antibodies with high specificity are used, allowing the removal of cross-reactive and structurally analogous compounds during clean-up. However, in some instances, antibodies that can recognize a group of structurally similar toxins (e.g. aflatoxins B_1, B_2, G_1 and G_2) have been reported (Senyuva and Gilbert, 2009). Similarly, existing IACs can be evaluated for their ability to bind mycotoxin conjugates as well. In this regard, the use of IAC in a method devoted to the analysis of DON and DON-3-G in malt and beer has been reported (Zachariasova *et al.*, 2009). On the other hand, single clean-up strategy was not successful in a liquid chromatography tandem mass spectrometry (LC–MS/MS) method intended to determine DON, ZON and their metabolites simultaneously in cereals and cereal-based food (Vendl *et al.*, 2009). The authors evaluated different clean-up strategies including C18-SPE, primary and secondary amines (PSA), MycoSep push-through column, and IACs. None of the procedures investigated proved to be applicable, owing to the wide polarity range of the set of compounds included. Therefore, in the final method, no sample clean up was applied. Indeed, with highly sensitive LC–MS instruments, it has become feasible simply to dilute sample extracts to minimize possible matrix effects (dilute and shoot approach) (Marazuela and Bogialli, 2009).

Considering the difficulties of differentiating between masked mycotoxins and precursor toxins in a single analysis, in an alternative approach, a special pretreatment (enzymatic, acidic or basic hydrolysation) is applied to release the toxin from the conjugate binding prior to extraction and subsequent clean-up of the extract. In this case, the masked mycotoxin is determined by comparing the results obtained with and without hydrolysis. The advantages of these protocols are that measurement standards of conjugated forms are not mandatory for quantification and that unknown mycotoxin conjugates might also be included in the final result.

A major drawback is that without calibrants, the efficiency of the hydrolysis process cannot be verified, which additionally makes the estimation of the uncertainty of the results obtained very difficult (Berthiller *et al.*, 2009a). Another drawback is that this indirect approach does not take into account the possible difference in toxicities of individual substances. Furthermore, sample preparations are usually tedious. Nevertheless, a variety of protocols using this approach have been described in the literature. For instance, β-glucosidase was used in great molar excess to cleave ZON-4-G into ZON for 18 hours at 37 °C in cereal samples (Gareis *et al.*, 1990). However, this enzyme failed to cleave DON-3-G into DON (Sewald *et al.*, 1992; Berthiller *et al.*, 2009a). Other enzymes have been used to treat *Fusarium* head blight (FHB)-infected barley samples and their effect on the liberation of DON has been investigated (Zhou *et al.*, 2008). This included either amylolytic (α-amylase and amyloglucosidase), proteolytic (papain) or cell wall degrading (cellulase and xylanase) enzymes. Papain treatment resulted in significant increases (up to 28%) in the amount of DON detected in five of the seven samples tested compared with the untreated samples. Treatment with cellulase/ xylanase resulted in increased DON detection in three of the seven samples, whereas amylase/amyloglucosidase resulted in increased DON detection in only one sample. The results strongly indicated that FHB-infected barley samples can contain bound DON that might not be detected in routine quantification but can be released by proteolytic or cell wall-degrading activity.

Acidic and thermal treatments of cereals to liberate conjugated forms of DON have also been reported. In this regard, a solvolysis procedure for the determination of DON conjugates in *Fusarium*-infected wheat was developed (Liu *et al.*, 2005). A trichloroacetic acid treatment at 140 °C was applied for 40 min, after samples were homogenized with acetonitrile/water. During method development, 15-acetyl-deoxynivalenol (15-ADON) was chosen as a model compound to set free DON in an acid-catalyzed de-esterification reaction. The result obtained indicated that 3–63% of the total DON consisted of non-extractable DON. More recently, a Doehlert matrix design was performed to determine the optimal conditions for determination of non-extractable DON in barley grain, using above-mentioned protocol with a trifluoroacetic acid (TFA) treatment (Zhou *et al.*, 2007). In the optimized conditions, samples were treated with 1.25 N TFA in 86:14 acetontrile/water for 54 min at 133 °C. This treatment resulted in an increase of 9–88% in DON concentrations in a set of verification samples. However, it remains unclear whether or not the relatively more acid stable DON-3-G might also be converted to DON with these procedures.

Alkaline hydrolysis of masked mycotoxins has been mainly described for fumonisin conjugates (Kim *et al.*, 2003; Dall'Asta *et al.*, 2009a). For instance, Kim *et al.* (2003) found a 2.6-fold average increase of FB_1 concentration in corn flake samples after treatment with 1% sodium dodecyl sulfate (SDS) solution and hydrolysis with 2 M potassium hydroxide (KOH), and subsequent extraction of FB_1 (as hydroxy-fumonisin B_1, HFB_1) with ethyl acetate. Later, this method was improved by Park *et al.* (2004) and applied to 30 retail samples of heat-processed corn foods. About 1.3 times higher content of FB_1 was found compared to that determined by traditional analysis. More recently, this increase of fumonisin concentration was confirmed using optimized alkaline hydrolysis conditions for thermally treated and untreated maize-based foods (Dall'Asta *et al.*, 2008).

Considering the drawbacks already mentioned of the indirect approach for determination of masked mycotoxins and provided that suitable reference standards are available, one may look for a versatile sorbent for the extraction of a wide range of mycotoxins and their conjugates. Since Oasis HLB SPE cartridges allow a more generic extraction (Diana Di Mavungu *et al.*, 2009; Lattanzio *et al.*, 2009) because of their hydrophilic and lipophilic characteristics, it is worth checking their suitability for the simultaneous determination of masked and free mycotoxins.

15.3.2 Determination of masked mycotoxins

The most important and widely used separation technique for masked mycotoxin determination in food and feed is liquid chromatography (LC). While the first quantitative method for a mycotoxin conjugate employed LC with fluorescence detection (FLD) (Zill *et al.*, 1990b), LC coupled with MS has become the method of choice for the determination of masked mycotoxins. Indeed, with this hyphenated technique, mycotoxin conjugates can be identified by their mass and their collision-induced dissociation (CID) fragmentation behaviour. Food and feed commodities can be screened for unknown conjugates with different MS/MS techniques. If standards are available, analytical methods can be developed and known conjugates can be quantified. However, a major constraint for the determination of conjugated mycotoxins is the limited availability of measurement standards, which are required for accurate quantification. Currently, only DON-3-G is commercially available. For this reason, in several methods, the precursor toxin is quantified before and after enzymatic, acidic or basic hydrolyzation of the conjugate (see Section 15.3.1). Using this approach, measurement standards are only needed for the parent molecules, which are usually available. Schneweis *et al.* (2002) developed an LC–MS electrospray ionization (LC–MS-ESI) method allowing direct determination of ZON-4-G in wheat. They synthesized the ZON-4-G to be used as the analytical standard. ZON was found at levels above the quantification limit in 22 out of 24 samples analysed. Of the 22 samples positive for ZON, 10 samples (42%) were also positive for ZON-4-G. Approximately 10–20% of total ZON content of these samples was detected as ZON-4-G.

An LC–MS atmospheric pressure chemical ionization (LC–MS-APCI) method for the direct determination of DON-3-G in cereals has been described with respect to DON conjugates (Berthiller *et al.*, 2005a,b). The method was applied to the analysis of naturally contaminated wheat and maize samples. DON-3 G was identified and characterized in these samples by comparing the fragmentation pattern with a chemically synthesized DON-3-G standard. The results showed that DON-3-G is the major form of masked DON, constituting up to 12% of total DON found in *Fusarium*-infected wheat and maize. In addition, minor levels of a substance with a MS signal and presumed molecular fragmentation behaviour corresponding to 15-acetyl-DON-3-glucoside were detected. LC–MS-APCI was also used to monitor directly the occurrence of DON-3-G in beer production (Lancova *et al.*, 2008; Kostelanska *et al.*, 2009). A significant increase of DON-3-G levels during malting and brewing processes was observed, compared to the original barley. Sasanya *et al.* (2008) reported an LC–MS–ESI method that allows the direct determination DON-3-G. The author quantified DON and DON-3-G in hard red spring wheat. Purposively and randomly selected samples were investigated. In some of the purposively selected samples DON-3-G concentrations were even found to exceed the DON concentration. Recently, an LC–MS/MS method with ESI/APCI switching, that allows the simultaneous determination of DON, ZON and eight of their masked metabolites has been developed and validated for four cereal-based food matrices, namely corn flour, porridge, beer and pasta (Vendl *et al.*, 2009). The authors reported that the application of this method to approximately 100 different food samples will be published soon. LC–MS has also been used for the indirect determination of DON-3-G after enzymatic (Zhou *et al.*, 2008) or acid (Liu *et al.*, 2005) treatment. On the other hand, it is worth noting that DON-3-G and ZON-4-G have also been included in LC–MS/MS methods for multi-mycotoxin determination in cereals (Sulyok *et al.*, 2006; Sulyok *et al.*, 2007a,b).

Determination of bound fumonisins in food commodities has been performed after alkaline treatment, using LC-FLD (Kim *et al.*, 2003) or LC–MS (Park *et al.*, 2004; Dall'Asta *et al.*, 2008) methods. In the latter paper, bound fumonisins were found to be present, not only in thermally treated maize-based products, but also in mild processed or even raw products (pasta, bread, cakes, crisps, flour) and they were always present in almost similar or even higher amounts than the free forms. Recently, Dall'Asta *et al.* (2009b) compared the results obtained by five different methods, based on LC-FLD and LC–MS, for the quantification of FB_1, fumonisin B_2 (FB_2) and fumonisin B_3 (FB_3) in raw maize. Although each method was validated and common calibrants were used, a poor agreement about fumonisin contamination levels was observed. Investigation of the occurrence of fumonisin derivatives indicated that significant amounts of hidden fumonisins were present in the analysed samples. Furthermore, the application of an *in vitro* digestion protocol to raw maize allowed a higher recovery of native fumonisins. Depending on the analytical method as well as the maize sample, only 37–68% of the total fumonisin concentrations were found to be extractable from the samples. These results highlight the actual difficulties in ascertaining the trueness of a method for fumonisin determination and, further, in assessing the risk for consumers.

With respect to immunochemical methods, their use in masked mycotoxin analysis is conceivable. However, as mycotoxin conjugates normally co-occur with the parent toxins, an independent measurement of each of the varieties is possible only if highly selective antibodies are used. Enzyme-linked immunosorbent assays (ELISAs) for instance are often employed for the control of DON in different food commodities. A study conducted to assess the accuracy of DON determination by immunoassays in cereal-based matrices revealed that DON-3-G was strongly cross-reacting in all examined commercial ELISA kits tested (Zachariasova *et al.*, 2008). This might, besides matrix effects, explain the overestimation of mycotoxin concentrations determined by ELISA compared with LC–MS (Avantaggiato *et al.*, 2008; Gallo *et al.*, 2008; Thongrussamee *et al.*, 2008). One approach would definitely be the use of currently available ELISA kits to assess mycotoxin conjugates indirectly, after conversion to the parent toxins.

15.3.3 Analytical perspectives

In the reviewed literature, reversed-phase LC (RP-LC) has been commonly applied to the simultaneous analysis of parent mycotoxins and their masked metabolites, although the use of normal-phase LC (NP-LC) has also been reported (Zill *et al.*, 1990b). In order to separate the free and the conjugated mycotoxins, which exhibit very different polarity, use has been made of RP-LC columns with increased retention of highly polar compounds, namely Synergi Polar-RP (Vendl *et al.*, 2009) and Synergi Hydro-RP (Lancova *et al.*, 2008; Kostelanska *et al.*, 2009). An alternative for the determination of extremely polar compounds is the application of hydrophilic interaction chromatography (HILIC). This technique has been used as an alternative to ion-pair LC for the separation of very polar compounds (e.g. aminoglycoside antibiotics) that are not retained in typical reversed-phase columns (Zhu *et al.*, 2008; Van Holthoon *et al.*, 2009). This separation modality is similar to NP-LC, but it uses polar mobile phases that are compatible with MS detection. The use of HILIC has not been reported yet for masked mycotoxins. However, since mycotoxin conjugates are generally polar compounds and some of them are highly polar, this technique could be an interesting alternative to common LC separation. On the other hand, the utilization of ultra performance liquid chromatography (UPLC) may also be considered. The main advantages of UPLC are well known (e.g. shorter run time or narrower peaks compared to conventional LC) and the utilization of this technology is rapidly increasing. Apart from typical LC or UPLC, other LC modes such as multidimensional LC (LC × LC) might also be considered for the separation of masked mycotoxins.

With regard to the detection technique, most of the work reviewed has been carried out using ultraviolet, fluorescence or MS detectors. MS offers high sensitivity and selectivity and no derivatization is needed when coupled to LC. In addition, identification and confirmation can be carried out in a single step. For MS analysis of masked mycotoxins, several strategies can be used: targeted and non-targeted compound analysis. In targeted compound analysis, masked mycotoxins are known and included within a defined MS method and they can be monitored in routine analysis. This approach has been applied to the analysis of masked mycotoxins, using low-resolution MS (LRMS) instruments such as quadrupole (Schneweis *et al.*, 2002; Berthiller *et al.*, 2005a), ion-trap (Lancova *et al.*, 2008; Kostelanska *et al.*, 2009) or hybrid triple quadrupole/linear ion-trap (Vendl *et al.*, 2009) mass analysers, mainly operating in tandem MS (MS/MS).

The triple quadrupole analyser is the most utilized and it allows the application of the four existing MS/MS modes: product ion scan, precursor ion scan, neutral loss scan and selected reaction monitoring (SRM). When triple quadrupole analyser is operated in SRM mode, it shows several advantages and interesting characteristics for target analysis such as an increase of selectivity, reduction of number of interferences and high sensitivity. Another important point is the possibility of diminishing the analysis time, including extraction and instrumental determination, owing to the increase in selectivity and sensitivity. Indeed, these properties can reduce the need for exhaustive sample pretreatment and the high speed acquisition in SRM allows the simultaneous monitoring of a great number of transitions. In relation to quantification in targeted compound analysis, the matrix effect can be an important source of errors and different strategies can be used. The use of isotope labelled internal standards (IS) is the best choice to compensate for matrix effects, but the lack of standards for most of masked mycotoxins and, probably, their high cost make necessary the use of alternative procedures such as matrix-matched standard calibration.

The utilization of a triple quadrupole analyser requires the previous selection of the appropriate SRM transitions of the mycotoxin conjugates prior to analysis by using the corresponding standards. However, information related to non-target and/or unknown conjugates not included in the study will be missed. For this purpose, MS analysis can be carried out using time-of-flight (TOF) analysers, which are high-resolution MS (HRMS) instruments allowing the performance of accurate mass measurements. An important feature of TOF instruments is that they record full scan spectra. As a consequence, it is possible to review and to look for additional compounds (known or unknown) not included in the initial analysis. However, one of the main problems associated with TOF instruments is that it is not possible to isolate a precursor-ion to obtain clean MS/MS spectra and fragmentation can only be enhanced if higher fragmentator voltages are used (Soler *et al.*, 2007). Therefore, hybrid instruments such as the quadrupole TOF (Q-TOF) analysers have been introduced, allowing a more accurate identification. These hybrid instruments may facilitate the elucidation of masked mycotoxins. On the other hand, new analysers, such as the Orbitrap, allow accurate mass measurement in multiple MS stages (MS^n) experiments and, therefore, additional information can be obtained (Picó and Barceló, 2008). In this regard, the suitability of Orbitrap for the analysis of free and masked mycotoxins in malt and beer has been shown (Zachariasova *et al.*, 2009).

In the same way, an ion-trap analyser in MS^n experiments can be used for the determination of masked mycotoxins in food and feed, providing qualitative information that could be used to ascertain whether these mycotoxin derivatives or any other substances were present or not in the sample. Furthermore, new analysers such as quadrupole linear ion-trap (Q-LIT) and ion-trap TOF (IT-TOF) can be used in these studies. In this regard, Berthiller *et al.* (2006b) used a Q-LIT in the SRM mode to monitor assumed ZON metabolites in the model plant *Arabidopsis thaliana*. In the same way as for the IT-TOF, this hybrid analyser combines the possibility of performing MS^n (as the Orbitrap) and that of obtaining mass accuracy measurements for different ions obtained by MS^n, increasing the structural data available.

Although the use of hybrid TOF instruments or an Orbitrap mass analyser can overcome some shortcomings related to the elucidation of masked mycotoxins, some problems can still remain. For instance, it might not be possible to obtain an unequivocal elucidation of the structure of a given masked mycotoxin and owing to the lack of commercial standards, the combination of this technique with others such as nuclear magnetic resonance (NMR) may provide important structural information. Taking into account the fact that no single instrument is able to provide all the information required in the analysis of novel mycotoxin conjugates, the utilization of several instruments providing complementary information seems to be necessary.

15.4   Conclusions

In the last few years it has become more and more clear that in mycotoxin-contaminated commodities many structurally related compounds generated by plant metabolism or by food processing can coexist together with the native toxins, underlining the need to include these compounds in analytical methods of mycotoxin analysis. A few mycotoxin conjugates have already been identified and their occurrence in food commodities has been shown. However, the unavailability of measurement standards limits the development of suitable analytical methods. Nevertheless, some known mycotoxin conjugates have been isolated by research groups and therefore it is expected that they will be commercially available in the near future. In this regard, LC–MS/MS using triple quadrupole will provide unambiguous identification and quantification of masked mycotoxins in complex matrix samples, because of its high specificity and sensitivity in the SRM mode. Besides the known mycotoxin conjugates, other unknown mycotoxin derivatives can be present in food commodities at relevant concentrations. Unequivocal structure elucidation can only be achieved by combining several analytical techniques (e.g. LC–TOF–MS, MS^n and NMR).

Regardless of the analytical technique, the availability of measurement standards for masked mycotoxins is a key factor, since the determination of such compounds in food and feed is only possible whenever commercial standards are in the market. Otherwise, the study of masked mycotoxins is limited to a qualitative stage. Moreover, mycotoxin legislation is focused on well-known compounds and this knowledge is always preceded by a corresponding quantitative study in a variety of matrices. In this sense, considering the reviewed literature, further research should be performed in order to achieve a more comprehensive overall view of the occurrence of masked mycotoxins in food and feed by developing multi-mycotoxin methods for monitoring a variety of parent mycotoxins and their conjugates. Taking into account all these aspects, more efforts must be applied to the analysis of masked mycotoxins in different food commodities, paying attention to the structural elucidation of these compounds and complete validation of the proposed methods. On the other hand, the toxicity and bioavailability of mycotoxin conjugates are not well known. Preliminary research conducted by Berthiller *et al.* (2009a) has indicated that DON-3-G is not hydrolyzed *in vitro* under the acidic conditions of the mammal stomach and by some enzymes, while this DON conjugate is converted back into the parent toxin by several bacteria that are found in the intestinal tract of humans. Further a DON-3-G feeding experiment with rat performed by the same authors indicates that conversion of DON-3-G into DON also occurs *in vivo*. More research is needed to establish clearly the fate of masked mycotoxins after digestion. Nonetheless, these mycotoxin conjugates should be regarded as potentially hazardous for human and animal health.

15.5   References

Avantaggiato G, Quaranta F, Aureli G, Melloni S, D'Egidio M G and Visconti A (2008), 'ELISA and HPLC analyses of deoxynivalenol in durum wheat varieties grown in organic farming in Italy'. A paper presented at the *International Durum Wheat Symposium 2008*, Bologna, Italy.

Berthiller F, Dall'Asta C, Schuhmacher R, Lemmens M, Adam G and Krska R (2005a), 'Masked mycotoxins: determination of a deoxynivalenol glucoside in artificially and naturally contaminated wheat by liquid chromatography-tandem mass spectrometry'. *Journal of Agricultural and Food Chemistry*, **53**, 3421–5.

Berthiller F, Krska R, Dall'Asta C, Lemmens M, Adam G and Schuhmacher R (2005b), 'Determination of DON-3-glucoside in artificially and naturally contaminated wheat with LC–MS/MS'. *Mycotoxin Research*, **21**, 205–8.

Berthiller F, Schuhmacher R, Poppenberger B, Lucyshyn D, Lemmens M, Adam G and Krska R (2006a), 'Determination of masked mycotoxins using HPLC-tandem mass spectrometry,' In *Mycotoxins and Phycotoxins: Advances in determination, toxicology and exposure management*, Njapeu H, Van Egmond H and Park D (eds), Wageningen Academic Publishers, Wageningen, The Netherlands, 125–32.

Berthiller F, Werner U, Sulyok M, Krska R, Hauser M T and Schuhmacher R (2006b), 'Liquid chromatography coupled to tandem mass spectrometry (LC–MS/MS) determination of phase II metabolites of the mycotoxin zearalenone in the model plant *Arabidopsis thaliana*'. *Food Additives and Contaminants*, **23**, 1194–200.

Berthiller F, Lemmens M, Werner U, Krska R, Hauser M, Adam G and Schuhmacher R (2007), 'Short review: Metabolism of the *Fusarium* mycotoxins deoxynivalenol and zearalenone in plants'. *Mycotoxin Research*, **23**, 68–72.

Berthiller F, Schuhmacher R, Adam G and Krska R (2009a), 'Formation, determination and significance of masked and other conjugated mycotoxins'. *Analytical and Bioanalytical Chemistry*, **395**, 1243–52.

Berthiller F, Dall'Asta C, Corradini R, Marchelli R, Sulyok M, Krska R, Adam G and Schuhmacher R (2009b), 'Occurrence of deoxynivalenol and its 3-beta-D-glucoside in wheat and maize'. *Food Additives and Contaminants*, **26**, 507–11.

Berthiller F, Hametner C, Krenn P, Schweiger W, Ludwig R, Adam G, Krska R and Schuhmacher R (2009c), 'Preparation and characterization of the conjugated *Fusarium* mycotoxins zearalenone-4*O*-beta-D-glucopyranoside, alpha-zearalenol-4*O*-beta-D-glucopyranoside and beta-zearalenol-4*O*-beta-D-glucopyranoside by MS/MS and two-dimensional NMR'. *Food Additives and Contaminants*, **26**, 207–13.

Böswald C, Engelhardt G, Vogel H and Wallnöfer P R (1995), 'Metabolism of the *Fusarium* mycotoxins zearalenone and deoxynivalenol by yeast strains of technological relevance'. *Natural Toxins*, **3**, 138–44.

Conn, E (1985), 'Chemical conjugation and compartmentalization: plant adaptations to toxic natural products', In *Cellular and Molecular Biology of Plant Stress*, Key J L and Kosuge T (eds), Centre for Agricultural Publishing, New York, 351–65.

Dall'Asta C, Berthiller F, Schuhmacher G, Adam G, Lemmens M and Krska R (2005), 'DON-glycosides: Characterisation of synthesis products and screening for their occurrence in DON-treated wheat samples'. *Mycotoxin Research*, **21**, 123–7.

Dall'Asta C, Galaverna G, Aureli G, Dossena A and Marchelli R (2008), 'A LC/MS/MS method for the simultaneous quantification of free and masked fumonisins in maize and maize-based products'. *World Mycotoxin Journal*, **1**, 237–46.

Dall'Asta C, Galaverna G, Mangia M, Sforza S, Dossena A and Marchelli R (2009a), 'Free and bound fumonisins in gluten-free food products'. *Molecular Nutrition and Food Research*, **53**, 492–9.

Dall'Asta C, Mangia M, Berthiller F, Molinelli A, Sulyok M, Schuhmacher R, Krska R, Galaverna G, Dossena A and Marchelli R (2009b), 'Difficulties in fumonisin determination: the issue of hidden fumonisins'. *Analytical and Bioanalytical Chemistry*, **395**, 1335–45.

Diana Di Mavungu J, Monbaliu S, Scippo M L, Maghuin-Rogister G, Schneider Y J, Larondelle Y, Callebaut A, Robbens J, Van Peteghem C and De Saeger S (2009), 'LC–MS/MS multi-analyte method for mycotoxin determination in food supplements'. *Food Additives and Contaminants*, **26**, 885–95.

Engelhardt G, Zill G, Wohner B and Wallnofer P R (1988), 'Transformation of the *Fusarium* mycotoxin zearalenone in maize cell suspension cultures'. *Naturwissenschaften*, **75**, 309–10.

Engelhardt G, Ruhland M and Wallnöfer P (1999), 'Metabolism of mycotoxins in plants'. *Advances in Food Sciences*, **21**, 71–8.

Galaverna G, Dall'Asta C, Mangia M A, Dossena A and Marchelli R (2009), 'Masked Mycotoxins: an Emerging Issue for Food Safety'. *Czech Journal of Food Sciences*, **27**, S89–S92.

Gallo G, Lo B M, Bognanni R and Saimbene G (2008), 'Mycotoxins in durum wheat grain: hygienic-health quality of sicilian production'. *Journal of Food Science*, **73**, T42–T47.

Gareis M, Bauer J, Thiem J, Plank G, Grabley S and Gedek B (1990), 'Cleavage of zearalenone-glycoside, a "masked" mycotoxin, during digestion in swine'. *Zentralblatt für Veterinarmedizin B*, **37**, 236–40.

Howard P C, Churchwell M I, Couch L H, Marques M M and Doerge D R (1998), 'Formation of *N*-(carboxymethyl)fumonisin B_1, following the reaction of fumonisin B_1 with reducing sugars'. *Journal of Agricultural and Food Chemistry*, **46**, 3546–57.

Humpf H U and Voss K A (2004), 'Effects of thermal food processing on the chemical structure and toxicity of fumonisin mycotoxins'. *Molecular Nutrition and Food Research*, **48**, 255–69.

Kim E K, Scott P M, Lau B P Y and Lewis D A (2002), 'Extraction of fumonisins B_1 and B_2 from white rice flour and their stability in white rice flour, cornstarch, cornmeal, and glucose'. *Journal of Agricultural and Food Chemistry*, **50**, 3614–20.

Kim E K, Scott P M and Lau B P (2003), 'Hidden fumonisin in corn flakes. *Food Additives and Contaminants*, **20**, 161–9.

Kostelanska M, Hajslova J, Zachariasova M, Malachova A, Kalachova K, Poustka J, Fiala J, Scott P M, Berthiller F and Krska R (2009), 'Occurrence of deoxynivalenol and its major conjugate, deoxynivalenol-3-glucoside, in beer and some brewing intermediates'. *Journal of Agricultural and Food Chemistry*, **57**, 3187–94.

Lancova K, Hajslova J, Poustka J, Krplova A, Zachariasova M, Dostalek P and Sachambula L (2008), 'Transfer of *Fusarium* mycotoxins and 'masked' deoxynivalenol (deoxynivalenol-3-glucoside) from field barley through malt to beer'. *Food Additives and Contaminants*, **25**, 732–44.

Lattanzio V M T, Pascale M and Visconti A (2009), 'Current analytical methods for trichothecene mycotoxins in cereals'. *Trends in Analytical Chemistry*, **28**, 758–68.

Liu Y, Walker F, Hoeglinger B and Buchenauer H (2005), 'Solvolysis procedures for the determination of bound residues of the mycotoxin deoxynivalenol in fusarium species infected grain of two winter wheat cultivars preinfected with barley yellow dwarf virus'. *Agricultural and Food Chemistry*, **53**, 6864–9.

Marazuela M D and Bogialli S (2009), 'A review of novel strategies of sample preparation for the determination of antibacterial residues in foodstuffs using liquid chromatography-based analytical methods'. *Analytica Chimica Acta*, **645**, 5–17.

Miller J D, Young J C and Trenholm H L (1983), '*Fusarium* toxins in field corn.I. Time course of fungal growth and production of deoxynivalenol and other mycotoxins'. *Canadian Journal of Botany*, **61**, 3080–7.

Park J W, Scott P M, Lau B P Y and Lewis D A (2004), 'Analysis of heat-processed corn foods for fumonisins and bound fumonisins'. *Food Additives and Contaminants*, **21**, 1168–78.

Picó Y and Barceló D (2008), 'The expanding role of LC–MS in analyzing metabolites and degradation products of food contaminants'. *Trends in Analytical Chemistry*, **27**, 821–35.

Picó Y, Font G, Ruiz M J and Fernandez M (2006), 'Control of pesticide residues by liquid chromatography-mass spectrometry to ensure food safety'. *Mass Spectrometry Reviews*, **25**, 917–60.

Poling S M, Plattner R D and Weisleder D (2002), 'N-(1-deoxy-D-fructos-1-yl) fumonisin B-1, the initial reaction product of fumonisin B-1 and D-glucose. *Journal of Agricultural and Food Chemistry*, **50**, 1318–1324.

Poppenberger B, Berthiller F, Lucyshyn D, Sieberer T, Schuhmacher R, Krska R, Kuchler K, Glossl J, Luschnig C and Adam G (2003), 'Detoxification of the *Fusarium* mycotoxin deoxynivalenol by a UDP-glucosyltransferase from *Arabidopsis thaliana*'. *Journal of Biological Chemistry*, **278**, 47905–14.

Ruhland M, Engelhardt G, Schafer W and Wallnofer P R (1996a), 'Transformation of the mycotoxin ochratoxin A in plants: 1. Isolation and identification of metabolites formed in cell suspension cultures of wheat and maize'. *Natural Toxins*, **4**, 254–60.

Ruhland M, Engelhardt G and Wallnofer P R (1996b), 'Transformation of the mycotoxin ochratoxin A in plants: 2. Time course and rates of degradation and metabolite production in cell-suspension cultures of different crop plants'. *Mycopathologia*, **134**, 97–102.

Ruhland M, Engelhardt G and Wallnöfer P R (1997), 'Transformation of the mycotoxin ochratoxin A in artificially contaminated vegetables and cereals'. *Mycotoxin Research*, **13**, 54–60.

Sandermann J (1992), 'Plant metabolism of xenobiotics'. *Trends in Biochemical Sciences*, **17**, 82–4.

Sasanya J J, Hall C and Wolf-Hall C (2008), 'Analysis of deoxynivalenol, masked deoxynivalenol, and *Fusarium graminearum* pigment in wheat samples, using liquid chromatography–UV-mass spectrometry'. *Journal of Food Protection*, **71**, 1205–13.

Savard M E (1991), 'Deoxynivalenol fatty-acid and glucoside conjugates'. *Journal of Agricultural and Food Chemistry*, **39**, 570–4.

Schneweis I, Meyer K, Engelhardt G and Bauer J (2002), 'Occurrence of zearalenone-4-beta-D-glucopyranoside in wheat'. *Journal of Agricultural and Food Chemistry*, **50**, 1736–8.

Seefelder W, Hartl M and Humpf H U (2001), 'Determination of N-(carboxymethyl)fumonisin B-1 in corn products by liquid chromatography/electrospray ionization-mass spectrometry. *Journal of Agricultural and Food Chemistry*, 49, 2146–2151.

Seefelder W, Knecht A and Humpf H U (2003), 'Bound fumonisin B-1: Analysis of fumonisin-B-1 glyco and amino acid conjugates by liquid chromatography–electrospray ionization–tandem mass spectrometry. *Journal of Agricultural and Food Chemistry*, **51**, 5567–73.

Senyuva H Z and Gilbert J (2009), 'Immunoaffinity column clean-up techniques in food analysis: A review'. *Journal of Chromatography B*, **878**, 115–32.

Sewald N, von Gleissenthall J L, Schuster M, Muller G and Aplin R T (1992), 'Structure elucidation of a plant metabolite of 4-desoxynivalenol'. *Tetrahedron*, **3**, 953–60.

Soler C, Hamilton B, Furey A, James K J, Manes J and Pico Y (2007), 'Liquid chromatography quadrupole time-of-flight mass spectrometry analysis of carbosulfan, carbofuran, 3-hydroxycarbofuran, and other metabolites in food'. *Analytical Chemistry*, **79**, 1492–501.

Sulyok M, Berthiller F, Krska R and Schuhmacher R (2006), 'Development and validation of a liquid chromatography/tandem mass spectrometric method for the determination of 39 mycotoxins in wheat and maize'. *Rapid Communications in Mass Spectrometry*, **20**, 2649–59.

Sulyok M, Krska R and Schuhmacher R (2007a), 'A liquid chromatography/tandem mass spectrometric multi-mycotoxin method for the quantification of 87 analytes and its application to semi-quantitative screening of moldy food samples'. *Analytical and Bioanalytical Chemistry*, **389**, 1505–23.

Sulyok M, Krska R and Schuhmacher R (2007b), 'Application of a liquid chromatography-tandem mass spectrometric method to multi-mycotoxin determination in raw cereals and evaluation of matrix effects. *Food Additives and Contaminants*, **24**, 1184–95.

Thongrussamee T, Kuzmina N S, Shim W B, Jiratpong T, Eremin S A, Intrasook J and Chung D H (2008), 'Monoclonal-based enzyme-linked immunosorbent assay for the detection of zearalenone in cereals'. *Food Additives and Contaminants*, **25**, 997–1006.

Van Holthoon F L, Essers M L, Mulder P J, Stead S L, Caldow M, Ashwin H M and Sharman M (2009), 'A generic method for the quantitative analysis of aminoglycosides (and spectinomycin) in animal tissue using methylated internal standards and liquid chromatography tandem mass spectrometry'. *Analytica Chimica Acta*, **637**, 135–43.

Vendl O, Berthiller F, Crews C and Krska R (2009), 'Simultaneous determination of deoxynivalenol, zearalenone, and their major masked metabolites in cereal-based food by LC–MS-MS'. *Analytical and Bioanalytical Chemistry*, **395**, 1347–54.

Wu X, Murphy P, Cunnick J and Hendrich S (2007), 'Synthesis and characterization of deoxynivalenol glucuronide: its comparative immunotoxicity with deoxynivalenol'. *Food and Chemical Toxicology,* **45**, 1846–55.

Young J C, Fulcher R G, Hayhoe J H, Scott P M and Dexter J E (1984), 'Effect of milling and baking on deoxynivalenol (vomitoxin) content of eastern Canadian wheats'. *Journal of Agricultural and Food Chemistry*, **32**, 659–64.

Zachariasova M, Hajslova J, Kostelanska M, Poustka J, Krplova A, Cuhra P and Hochel I (2008), 'Deoxynivalenol and its conjugates in beer: A critical assessment of data obtained by enzyme-linked immunosorbent assay and liquid chromatography coupled to tandem mass spectrometry'. *Analytica Chimica Acta*, **625**, 77–86.

Zachariasova M, Cajka T, Godula M, Kostelanska M, Malachova A, Poustka J and Hajslova Jana (2009), 'The benefits of Orbitrap high resolution mass spectrometry for analysis of free and masked mycotoxins in malt and beer'. A poster presented at the *4th International Symposium on Recent Advances in Food Analysis,* Prague, Czech Republic.

Zhou B, Li Y, Gillespie J, He G Q, Horsley R and Schwarz P (2007), 'Doehlert matrix design for optimization of the determination of bound deoxynivalenol in barley grain with trifluoroacetic acid (TFA)'. *Journal of Agricultural and Food Chemistry*, **55**, 10141–9.

Zhou B, Schwarz P, He G Q, Gillespie J and Horsley R (2008), 'Effect of enzyme pretreatments on the determination of deoxynivalenol in barley'. *Journal of the American Society of Brewing Chemists*, **66**, 103–8.

Zhu W X, Yang J Z, Wei W, Liu Y F and Zhang S S (2008), 'Simultaneous determination of 13 aminoglycoside residues in foods of animal origin by liquid chromatography–

electrospray ionization tandem mass spectrometry with two consecutive solid-phase extraction steps'. *Journal of Chromatography A*, **1207**, 29–37.

Zill G, Engelhardt G, Wohner B and Wallnoefer P R (1990a), 'The fate of the *Fusarium* mycotoxin zearalenone in maize cell suspension cultures'. *Mycotoxin Research*, **6**, 31–40.

Zill G, Ziegler W, Engelhardt G and Wallnöfer P R (1990b), 'Chemically and biologically synthesized zearalenone-4-β-D-glucopyranoside: Comparison and convenient determination by gradient HPLC'. *Chemosphere*, **21**, 435–42.

16

Spectroscopic techniques for fungi and mycotoxins detection

C. B. Singh and D. S. Jayas, University of Manitoba, Canada

Abstract: Spectroscopic techniques have become a well-established tool for routine compositional and functional as well as structural analyses of several agri-food materials. This chapter describes various types of spectroscopic techniques such as infrared (IR), Raman, near infrared (NIR), nuclear magnetic resonance (NMR) and ultraviolet (UV) and their basic working principles, advantages, disadvantages and limitations. The chapter discusses various applications of these techniques for fungi and mycotoxins detection.

Key words: food safety, FTIR spectroscopy, fungi, mycotoxins, NIR spectroscopy, NMR spectroscopy, Raman spectroscopy, ultraviolet spectroscopy.

16.1 Introduction

Pathogens are a serious cause of foodborne illnesses in humans and pose a big challenge to food safety. The World Health Organization reported that around 30% of the population in the industrialized world suffers from foodborne illnesses every year and in USA alone, there are approximately 76 million estimated cases of foodborne diseases resulting in 325 000 hospitalizations and 5000 deaths annually (WHO, 2007). This figure will be much higher for developing countries and cause an adverse social and health impact globally. Fungi which produce mycotoxins are a serious cause of food poisoning in humans or animals, if present in the consumed food or feed. There are more than 400 mycotoxins mostly produced by *Penicillium, Aspergillus* and *Fusarium* species, the most common being aflatoxins, ochratoxin A, fumonisins, trichothecenes and zearalenone. Owing to an increase in the number of foodborne outbreaks and increasing consumer demand for safer and high-quality food, there is a need to inspect the food in the production line more accurately and rapidly for the presence of mycotoxins and toxigenic fungi. Current methods used for fungal detection are chemical, immunological, selective and differential media, fungal volatiles, most probable number, Howard mold count, polymerase chain reaction, microbial culture and microscopy (Narvankar *et al.*, 2009). Methods involving wet analysis are destructive and time consuming and although the results are reliable, by the time results/infections are known, the product might already have reached the distribution systems or in some cases the consumers. Effective recall of the product is challenging and results in huge losses to food/feed processors and suppliers. Food processing industries are always looking for an alternative rapid, accurate and non-destructive method for real-time inspection of their products. In a recent review, Ghosh and Jayas (2009) have explored the usefulness of using spectral data from various spectroscopic techniques for automation in the food processing industry. Spectroscopic techniques have high potential to become an alternative or supplemental tool for automated inspection of agri-food products for their quality and safety. In this chapter, major spectroscopic techniques and their application to detect fungi, mycotoxins or both are discussed.

16.2   Spectroscopic techniques

Spectroscopic techniques are used to disperse the light (electromagnetic radiation) on a material to form a spectrum. The spectra arise as a result of interaction of electromagnetic energy with the molecules which may occur in the form of emission (nuclear magnetic resonance spectroscopy), absorption (near-infrared, infrared and ultraviolet spectroscopy), or scattering (Raman spectroscopy). Spectra are a form of graph that records the amount of energy emitted, absorbed, or scattered by the samples over a wide range of electromagnetic radiation. The full electromagnetic spectrum spans, in descending order of energy, gamma rays, X-rays, ultraviolet (UV), visible (VIS), near-infrared (NIR), infrared (IR), microwaves to radiowaves (Fig. 16.1). Based on the region of electromagnetic radiation, different types of spectroscopic techniques are available. The characteristics of the spectrum depend on the specific range of electromagnetic radiation and type of instrument used for the analysis.

A molecule in space can have several forms of energy such as vibrational, rotational, or electronic. The energy associated with molecular transition can be described by vibrational, electronic and rotational energy. Vibrational energy is the result of periodic displacement of atoms from their equilibrium position and rotational energy is the result of rotation about a center of gravity. The energy required for the rotational state is low compared to the energy required for change in the vibrational state (NIR-mid IR). Rotational energy is absorbed in the far IR and microwave regions and except for the gases, the spectra in this region are very

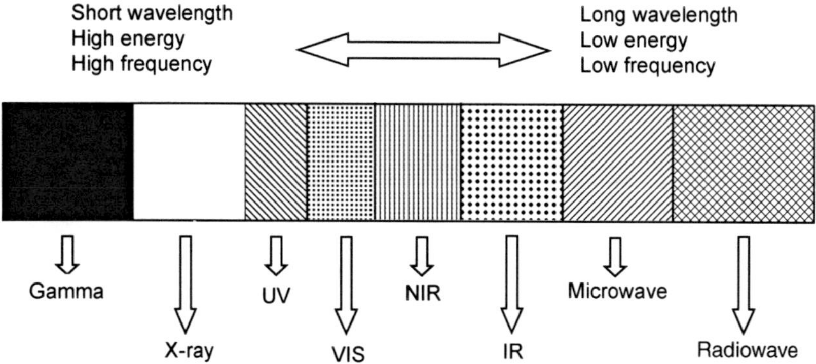

Fig. 16.1   The electromagnetic spectrum.

broad. In general, only vibrational motion is considered in the spectral analysis of solids and liquids (Osborne *et al.*, 1993). Electronic energy is the result of electron excitation (ionization) in the UV–VIS region.

In this chapter, different spectroscopic techniques, namely IR, Raman, NIR, nuclear magnetic resonance (NMR) and UV spectroscopy, covering wide range of electromagnetic radiation and various modes of energy transitions, are discussed. Since each of the following techniques is a major field of study in itself, only basic conceptual information is provided in this chapter. Readers are referred to specific books (Osborne, 1993; Williams and Norris, 2001; Cooper, 1980; Mossoba, 1998; Pavia *et al.*, 1979) on these techniques for an in-depth study. At the end of the chapter various applications of these techniques for fungal and mycotoxins identification are presented.

16.2.1 Vibrational spectroscopy

There are three types of spectroscopic techniques which are part of vibrational techniques namely, IR, Raman, and NIR spectroscopy.

Infrared (IR) spectroscopy

Molecular vibration in the IR region can be considered to be a simple harmonic model (spring model) that has a harmonic oscillatory path with nearly zero potential energy at equilibrium which follows the Hook's law (Fig. 16.2). This mechanical model is useful in understanding the concept of vibrational energy and in explaining many spectral observations. A covalent bond between two atoms in a molecule follows the same principle. If a covalent bond is stretched from the equilibrium position, the electromagnetic force between the electrons in the covalent bond becomes weaker and breaks when the potential energy reaches the bond energy. Transition of energy in the quantum mechanical model can only occur between adjacent energy states, that is the allowable change in the vibra-tional quantum number is $v = \pm 1$ (Pasquini, 2003). If the photon energy of the electromagnetic radiation is equal to the bond energy of a diatomic molecule then

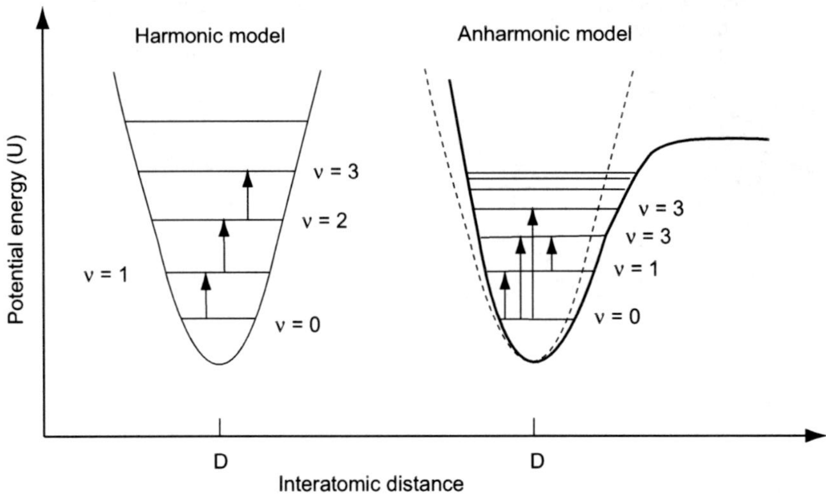

Fig. 16.2   Simple harmonic and anharmonic models with their respective potential energies.

the photon is absorbed and the molecule jumps from the ground state to an excited state, instantly causing vibration. The molecule releases the absorbed energy as heat and returns to the ground state. Photons in the IR region have lower energies that correspond to covalent bond stretch and bend vibrations which can cause only a fundamental transition from the ground state to the first excited state (single band).

An IR absorption spectrum can be produced by collecting the radiation after its interaction with the material. Fundamental vibrations, which occur in the mid-IR region (4000–650 cm^{-1}), are commonly studied in spectral IR analysis. IR vibration has strong water absorption band but the effect can be overcome by subtraction. The IR spectroscopic analysis has several advantages over NIR in terms of sensitivity, spectral assignment, resolution and quantification. The IR spectrum has well-resolved peaks whereas broad overlapping absorption bands in the NIR spectrum make structural selectivity and band assignment to a specific chemical compound very difficult (Wilson and Goodfellow, 1994). IR spectra are very useful for both structural and qualitative analysis. IR spectroscopy is capable of analyzing solid, liquid and gas samples. A disadvantage of the IR technique is the sample preparation requirement which limits rapid analysis.

There are several IR techniques such as dispersion IR (using grating devices) and Fourier transform infrared (FTIR). Data collection using a dispersive grating takes several minutes whereas FTIR instruments take only a few seconds to record the spectra. In FTIR, an interferogram is recorded by the detector and then transformed into a spectrum by applying Fourier transformation. There are several advantages of using FTIR spectroscopy such as high sensitivity, increased brightness, high resolution and very low background noise as only the interference signal is used in Fourier transformation.

A typical FTIR instrument consists of a collimated IR radiation light source, a fixed mirror, a moving mirror, a beam splitter and a detector. The beam splitter splits the incoming radiation into two beams and directs them to the fixed and moving mirrors. These two beams are then reflected back to the beam splitter from the mirror giving rise to interference which is sensed by the detector after passing through the sample. If the light returning back to the beam splitter directed from both mirrors has the same phase as if the mirrors are at an equal distance, constructive interference occurs and maximum energy is passed by the beam splitter to the detector. If one mirror moves away, a path difference is created and constructive interference will not occur because of the destruction caused by the phase difference. Thus the detector output is a fluctuating cosine wave produced by movement of one mirror. This cosine wave (function of optical retardation) is called an interferogram. Interferogram has all the information that is contained in a spectrum; however, it is much more complex and difficult to analyze so Fourier transformation is used to transform this data into wavelength and intensity.

Raman spectroscopy
Raman spectroscopy is also a vibrational spectroscopic technique; however, the vibrations causing Raman spectra are symmetric and the main criterion is the polarizability of the compound. When the radiation (photon) strikes a molecule, the incoming radiation is scattered. In the scattering process, some of the incident light might scatter at the same frequency as that of the incident light if the collision between the striking photon and the molecule is elastic. This type of scattering is called Rayleigh scattering. However, if the collision is inelastic, the scattered light has a different frequency from the striking photons owing to energy exchange between photons and molecules. This type of scattering is called Raman scattering. Not all the vibrating molecules in a sample are present in the ground energy state; some occur in the excited energy state. When the incoming radiation (photon) collides with a molecule of low energy in the ground energy state, the molecule gains energy resulting in the scattered photon having low energy or frequency. This phenomenon is called Stokes Raman scattering. If the photon strikes a molecule in the higher energy state, the photon gains energy from the molecule, resulting in a scattered photon with high energy and frequency. This type of energy transition gives rise to anti-Stokes scattering. Since Stokes scattering lines are stronger than anti-Stokes lines owing to the high energy of the scattered photon and the presence of a majority of the vibrating molecules in the ground energy state, researchers are more interested in analyzing Stokes lines. A Raman spectrometer consists of a scattering radiation source and usually a charge-coupled device (CCD) detector.

Both IR and Raman spectra offer complimentary information to the analyst for structural exposition. Peaks in IR spectra arise from molecular vibration associated with a change in the dipole moment of the molecule, whereas peaks in Raman spectra arise from vibrations associated with a change in the electronic polarizabilty, in other words polar bonds (C–H, O–H, C=O, C=O, C–O, C–F, Si–O, and N–H) give strong IR absorption peaks but homonuclear apolar symmetric bonds (C=C,

C=C, C–C, S–S) are poor absorber of IR radiation. However, these symmetric bonds in symmetric vibrations give strong Raman Stokes lines. Raman spectra have the advantage that water contributes significantly less to them. The disadvantages of Raman spectroscopy are: small sample volume, interference by fluorescence, low efficiency, low intensity of desired Stokes scattering and high intensity of the undesired Rayleigh scattering.

Near infrared (NIR) spectroscopy

In the NIR region several absorption bands are found which cannot be explained by the harmonic model, owing to the assumption of equally spaced discrete energy levels. The harmonic model is only able to explain the vibrations occurring in the IR region due to fundamental modes but cannot explain the overtones of the NIR region. An anharmonic model (Fig. 16.2) can explain this, based on the phenomenon that two nuclei cause a rapid increase in potential energy compared to a harmonic model when these are stretched. This asymmetric distribution results in quantized but unequally spaced energy levels, therefore, absorption bands are not an example multiple of fundamental absorption bands (Osborne *et al.*, 1993). This phenomenon can be described using the Morse function (Pasquini, 2003).

The anharmonic model can predict the transition energy for transition from $\nu = 0$ to the higher transition levels $\nu = 2$ and $\nu = 3$, which correspond to the first and second overtones, respectively. In the anharmonic model, the vibrations of the molecules depend on each other and can interact with them. Combination bands occur as a result of a combination of overtones and fundamental vibration in various combinations. Overtones and combination vibrations occur in the NIR region (700–2500 nm or 14 300–4000 cm^{-1}). Overtones and combination bands are weaker than fundamental absorptions. This is an advantage that NIR radiation has over IR radiation and samples with thicknesses of several mm or cm, such as intact whole seeds can be analyzed without any sample preparation. A typical NIR instrument consists of a radiation source (tungsten-halogen lamps, light emitting diodes), a wavelength filtering device (grating devices, interference filters or advanced electronically tunable filters) and a diode array detector, for example Silicon (Si), lead sulphide (PbS) and indium gallium arsenide (InGaAs). Spectral data can be analyzed using various advanced chemometric tools (Naes *et al.*, 2002).

16.2.2 Nuclear magnetic resonance (NMR) spectroscopy

The nuclear magnetic resonance (NMR) phenomena arises from the fact that an atomic nucleus with an odd atomic mass and/or atomic number has a spinning charged particle that results in a magnetic moment. Nuclei with an even atomic mass and/or atomic number do not have any magnetic moment associated with them due to zero spin. In nuclei with spins, the number of allowed spin states for each nucleus is determined by the nuclear spin number. In the absence of an externally applied magnetic field, the magnetic moment associated with the spinning charged particles of nuclei is pointed randomly in any direction as all the spin states have the same energy. However, once the external magnetic field is applied, the magnetic moments align themselves either with the field or opposite to the field. The spin state with lower energy aligns in the direction of the applied magnetic field whereas the spin state with higher energy aligns in the opposite direction of the applied magnetic field. If the appropriate amount of radiofrequency energy is supplied, the energy is absorbed by the nuclei, resulting in a change in energy state from lower to higher energy or a change in spin. This absorption phenomenon is also known as resonance which results in NMR spectra. Since every nucleus in a molecule will have a resonance or absorption at a significantly different energy level or frequency, NMR spectra can be used to identify different types and numbers of atoms in a molecule.

A number of NMR techniques such as ^{1}H-NMR, ^{13}C-NMR, ^{29}Si-NMR and ^{31}P-NMR spectroscopy are used for analysis of hydrogen, carbon, silicon and phosphorus, respectively. Hydrogen NMR spectroscopy (^{1}H-NMR) is the simplest and most commonly used NMR spectroscopic technique. A general NMR spectrometer consists of a large magnet, a radiofrequency generator (energy source), a coil inserted between the poles of magnet and attached to the generator, and a radiofrequency detector. Resonance can be induced either by changing the frequency of the generator or by changing the magnetic field strength of the magnet (by changing the current flow into it). An NMR spectrum is recorded by the detector.

The NMR spectrum does not have a peak at a specific wavelength as in other absorption spectra (NIR, IR and UV) owing to dependency of the NMR spectrum on both the radiofrequency and the externally applied electric field. Therefore, NMR spectra are characterized, and the location of a specific NMR signal is assigned, using the chemical shift. The chemical shift is defined as the ratio of the shift in resonance frequency to the basic operating frequency of the spectrometer and is independent of the applied field and proton resonance frequency (Pavia *et al.*, 1979). The frequency shift in an NMR signal is measured relative to a reference material such as tetramethylsilane or TMC and quantified in ppm (parts per million). Resonance peaks are recorded against chemical shift (δ) or a converted scale called tau (τ) (Cooper, 1980). Samples are first dissolved into a solvent which has no resonance (peaks) in the region of interest. The common solvents used in NMR spectroscopic analysis are CCl_4, $CDCl_3$, and D_2O. The fully dissolved samples in NMR tubes are placed between the poles of magnet. Since the samples have to be dissolved and this requires sample preparation, its applicability for in-line rapid process monitoring may be limited.

16.2.3 Ultraviolet (UV) spectroscopy

In UV absorption spectroscopy, the absorption spectra arise as a result of electron excitation at higher energy levels, sometimes referred to as electron spectroscopy. The UV spectrometer consists of a detector and a light source, such as a hydrogen discharge tube or incandescent bulbs, which emits light in the 100–400 nm wavelength range. A wavelength region of less than 200 nm is not regularly used in UV spectral analysis owing to the restriction of the instrumentation for handling

UV radiation. The detector output is recorded giving a spectrum which is generally in the form of a plot of wavelength versus transmittance. The detector and light source are housed in a chamber to prevent stray light reaching the detector. In UV spectroscopic analysis, the obtained structural information is less detailed compared to IR and NMR analysis because broad and overlapping UV spectra arise from similar functional groups in the molecule (Hurst, 1994).

Samples can be solid, liquid or gases; however, liquid is the most common sampling method. Solid samples must be dissolved in a solvent before the analysis. The solvent should be selected based on its absorption characteristics and it should not absorb UV radiation in the region of interest of the solute. Ethanol and hexane are the two most common solvents. Ultraviolet spectroscopy can be used both for qualitative and quantitative analyses.

16.3 Applications

A summary of the applications of various spectroscopic techniques for fungal and mycotoxins analysis is given in Table 16.1 (vibrational spectroscopy) and Table 16.2 (NMR and UV spectroscopy) and discussed in this section.

Table 16.1   Fungi and mycotoxin detection by vibrational spectroscopy

Technique	Analysis	Sample	Reference
FTIR	Fungi	Corn	Greene *et al.* (1992)
FTIR	Fungi	Culture	Fischer *et al.* (2006)
FTIR	Fungi	Corn	Kos *et al.* (2004)
FTIR	Ochratoxin A	Vine fruit	Galvis-Sanchez *et al.* (2008)
Raman	Bacteria	Culture	Chu *et al.* (2008)
FT-Raman	Fungi	Culture	Edwards *et al.* (1995)
NIR	Aflatoxin	Corn	Pearson *et al.* (2001)
NIR	Fumonisin	Corn	Dowell *et al.* (2002)
NIR	Deoxynivalenol	Wheat	Petterson and Aberg (2003)
NIR	Fungi	Corn	Wang *et al.* (2004)
FT-NIR	Aflatoxin	Chilli	Tripathi and Mishra (2009)
NIR	Fungi	Tomato	Hahn (2002)

Table 16.2   Fungi and mycotoxin detection by NMR and UV spectroscopy

Spectroscopic technique	Analysis	Sample	Reference
NMR	Degree of acetylation	*Aspergillus niger*	Heux *et al.* (2000)
NMR	Secondary metabolites	Trichothecenes, altertoxins, fumonisins	Mazzola (1998)
NMR	New product	Fungal extract	Schroeder *et al.* (2007)
UV–VIS	Secondary metabolites	*Penicillium*	Svendsen and Frisvad (1994)

16.3.1 Applications of infrared (IR)/Fourier transform infrared (FTIR) spectroscopy

Greene *et al.* (1992) investigated the potential of the FTIR technique coupled with photoacoustic spectroscopy (PAS) and diffuse reflectance spectroscopy (DRS) to detect corn kernels infected with fungi *Fusarium moniliforme* and *Aspergillus flavus*. The spectra of the infected corn kernels differed significantly from those of healthy kernels. Several peaks related to amide I and amide II absorption bands (1650 cm^{-1} and 1550 cm^{-1}), ester carbonyl (1740 cm^{-1}), methylene doublet (2855 cm^{-1} and 2925 cm^{-1}), and broad OH and NH stretch absorption regions (3000–3600 cm^{-1}) were found to be promising for fungal detection and their quantification. They observed that a PAS-based technique was more sensitive to fungal detection; however, only a single kernel could be placed in the sample holder and analyzed at a time. The advantage of this FTIR–PAS technique is that whole intact corn kernels can be analyzed, whereas in case of FTIR–DRS the samples need to be pulverized.

In a recent review, Santos *et al.* (2010) described the FTIR technique as a powerful tool for rapid identification, characterization and authentication of several filamentous fungi and yeast strains. They outlined the potential use of the FTIR technique for identification and quantification of bio-compounds produced by filamentous fungi and yeasts as well as for detection of biomarkers. Fischer *et al.* (2006) used FTIR spectroscopy for rapid identification and intra-species characterization of airborne filamentous fungi. They differentiated fungal species of *Aspergillus* and *Penicillium* on the generic, species basis and strain level. Peaks relating to several chemical structures such as the C–H region dominated by fatty acids (3050–2800 cm^{-1}), amide I and amide II regions (1600–1700 cm^{-1}), the N–H absorption region (1500–1600 cm^{-1}), the spectral region between 1500 cm^{-1} and 1200 cm^{-1}, the polysaccharides (1 200–900 cm^{-1}) and the region from 900 cm^{-1} to 700 cm^{-1} (also known as the fingerprint region) were observed and used in window selection for species identification and strain characterization of microfungi by cluster analysis. Their study gave strong evidence that species identification and strain characterization are possible in microfungi by using FTIR spectroscopic analysis.

Kos *et al.* (2004) did a comparative study between two IR spectroscopic techniques, namely diffuse reflectance (DR) and attenuated total reflectance (ATR) for the classification and quantification of fungal infected corn kernels. They used a spectral region of only 1800–800 cm^{-1} as the region from 1800–3300 cm^{-1} was not found to be suitable for analysis owing to the large intensity variation in OH-stretching vibration (3300 cm^{-1}), ubiquitous C–H stretching bands (2925 cm^{-1} and 2855 cm^{-1}), and absorption from an internal reflection element (1800–2340 cm^{-1}). The performance of the ATR-based technique was better in both classification and prediction of deoxynivalenol (DON) concentration rather than the DR technique. Apart from a better multivariate model for classification and prediction, ATR has the additional advantage in terms of sample preparation and instrumentation.

Galvis-Sanchez *et al.* (2008) explored the potential of the FTIR-ATR technique for determining ochratoxin A (OTA) in dried vine fruits in the concentration range

2–50 µg kg^{-1} of OTA within the permissible European Union limit (maximum of 10 µg kg^{-1} of OTA). In a partial least square (PLS) analysis of spectral data, significance peaks associated with OTA were observed in the regions 1737–1669 cm^{-1} (carbonyl), 1650 cm^{-1} (aromatic breathing) and 1538 cm^{-1} (amide II). A good correlation was found between IR spectral data and the OTA reference concentration ($R^2 = 0.85$) in the PLS analysis.

16.3.2 Application of Raman spectroscopy

Chu *et al.* (2008) used surface-enhanced Raman scattering (SERS) to detect pathogenic bacteria, namely *Escherichia coli, E. coli* O157:H7, *E. coli* DH 5α, *Staphylococcus aureus, S. epidermidis*, *Salmonella typhimurium* and their mixtures. A silver nanorod was developed for use as substrate. Principal component analysis was applied to analyse the Raman spectra and to classify bacteria according to Gram types, different species and strains. Edwards *et al.* (1995) used Fourier transform (FT) Raman spectroscopy to study fungi and were the first to report FT Raman spectra of three species, namely *Agaricus bisporus*, *Mortierella* and *Mucor*. Their results demonstrated the potential of using FT Raman spectroscopy to discriminate different mixed species of fungi in a culture media.

16.3.3 Application of near infrared (NIR) spectroscopy

Pearson *et al.* (2001) studied the potential of applying the NIR spectroscopic technique to detect aflatoxin contaminated corn kernels. Reflectance and transmittance spectra were transformed to absorbance units and absorbance ratios were computed at 15-nm intervals. The best ratio features were selected and classification models were developed using PLS and discriminant analyses. The classification model correctly classified 95% of kernels as containing either high (>100 ppb) or low (<10 ppb) levels of aflatoxin. Model accuracy was lower for kernels in between 10 to 100 ppb aflatoxin concentration level but these kernels were not significant as they would not affect the total sample concentration.

Dowell *et al.* (2002) detected fumonisin content in corn kernels using NIR reflectance and transmittance spectroscopy. The beta coefficients from PLS analysis of transmittance data found peaks at 650, 710, 935 and 990 nm contributing to the classification. In reflectance mode, wavelengths of 590, 995, 1200, and 1410 nm were identified as those contributing to PLS classification models. Their study demonstrated the potential of NIR spectroscopic technique for online detection of fumonisin in corn kernels.

Petterson and Aberg (2003) used NIR spectroscopy to determine DON in wheat kernels by analyzing NIR transmittance spectral data using principal component analysis (PCA) and PLS and developing a regression model. Their study demonstrated the capability of the NIR technique to measure the mycotoxin content in cereals rapidly.

Wang *et al.* (2004) classified healthy and fungal-damaged soybean seeds and differentiated between various types of fungal damage using NIR spectroscopy.

They used PLS and an artificial neural network (ANN) to develop calibration models. A two-class PLS model correctly classified more than 99% of damaged seeds. However, in five-class models, ANN gave better classification results and correctly classified healthy seeds, Phomopsis, *Cercospora kikuchii* Matsumoto and Tomoy, soybean mosaic virus (SMV) and downy mildew damaged seeds with 100, 99, 84, 94 and 96% accuracy, respectively.

Tripathi and Mishra (2009) used FT-NIR spectroscopy to determine the aflatoxin B_1 content in red chilli powder. Fourier transform-based spectroscopy uses an interferometer which gives better spectral resolution and reproducibility using dispersive wavelength filtering devices compared to the NIR instruments. They developed a partial least square model and achieved high prediction accuracy for the concentration range 15–500 µg kg^{-1}.

Hahn (2002) used NIR spectroscopy to detect fungal spores on tomatoes. Spectral data (500–1000 nm) were collected before and after inoculation of tomatoes with *Fusarium oxysporum* and *Rhizopus stolonifer* conidia. Discriminant analysis of fast Fourier transformed (FFT) spectral data gave very high detection rate.

16.3.4 Application of nuclear magnetic resonance (NMR) spectroscopy

Nuclear magnetic resonance spectroscopy has been successfully used for structural elucidation and characterization of several naturally occurring food contaminants. The secondary metabolites (mycotoxins) examined/identified by NMR spectroscopy include several mycotoxins from the trichothecenes, altertoxins, and fumonisins families (Mazzola, 1998). Heux *et al.* (2000) used NMR spectroscopy to determine the degree of acetylation (DA) of chitin and chitosan in *Aspergillus niger*. The degree of acetylation is important as polymers with higher DAs are soluble in only a few solvents. The DA was evaluated for a wide range using ^{1}H liquid-state NMR and ^{13}C and ^{15}N cross polarization (CP)–magic angle spinning (MAS) solid-state NMR. Solid-state NMR gave good results with a detection threshold limit of 5%. ^{15}N CP-MAS NMR is considered a powerful tool for the evaluation of acetyl content in chitin and other polysaccharides in complex mixtures. Schroeder *et al.* (2007) used NMR spectroscopy to detect new products from the fungal extract. They analyzed the NMR spectral data by simple two-dimensional differential analysis. In this work, the authors focused more on extracting structural information rather than quantifying it. The NMR spectral analysis provided a useful tool for non-discriminatory characterization of small-molecule mixtures and demonstrated several potential applications in metabolomics and natural product chemistry.

16.3.5 Application of Ultraviolet (UV) spectroscopy

UV spectroscopy can be used for mycotoxin detection in food products. The protocols used for calibrating UV spectroscopic instruments for identification and quantification of mycotoxins can be found in Nesheim and Stack (2001). The UV spectroscopy has been used to examine the secondary metabolites of fungi. Svendsen and Frisvad (1994) analyzed the secondary metabolites of *Penicillium* with HPLC and diode array detection (DAD). The UV-VIS spectra were recorded and metabolites were grouped into different families using a chromophore (functional group absorbing light) and applying average linkage clustering multivariate statistical analysis. Members of the same biosynthetic family tend to have very similar chromophores and UV-VIS spectra. A total of 219 chromophore families of secondary metabolites were observed. Profiles of chromophore families demonstrated clear separation of *Penicillium arenicola*, *P. lanosum*, *P. nalgiovense* and *P. oxalicum* and terverticillate taxa and two new taxa were indentified. However, UV spectroscopy cannot be used for detection of some mycotoxins, for example fumonisins and deoxynivalenol, owing to the absence of a suitable chromophore in these molecules (Nesheim and Stack, 2001).

16.4   Summary

Various spectroscopic techniques have been discussed. The selection of a particular type of technique depends on the intended application. For on-line screening for fungi and mycotoxins in food and feed materials, the technique should be non-destructive, rapid and require no or minimal sample preparation. Near-infrared (NIR) spectroscopy meets these criteria and has shown potential for real-time applications. The limitation of this technique is that it cannot provide structural information. Raman spectroscopy also offers similar advantages but has not yet been fully explored owing to small sampling and instrumentation issues for real-time monitoring. Fourier transform infrared (FTIR) spectroscopy which is also known as the fingerprint technique can be used for both quantification and structural analysis. NMR spectroscopy has been used for structural analysis of fungi and mycotoxins in agri-food products. UV spectroscopy can potentially be used for on-line process monitoring and for both quantitative and qualitative analyses.

16.5   References

Chu H, Huang Y and Zhao Y (2008) 'Silver nanorod arrays as a surface-enhanced Raman scattering substrate for foodborne pathogenic bacteria detection', *Appl Spectrosc*, **62**(8), 922–31.

Cooper J W (1980), *Spectroscopic Techniques for Organic Chemists*, John Wiley and Sons, New York.

Dowell F E, Pearson T C, Elizabeth B, Maghirang F X and Wicklow D T (2002) 'Reflectance and transmittance spectroscopy applied to detecting fumonisin in single corn kernels infected with fusarium verticillioides', *Cereal Chem*, **79**(2), 222–6.

Edwards H G M, Russell N C, Weinstein R and Wynn-Williams D D (1995) 'Fourier transform Raman spectroscopic study of fungi', *J Raman Spectrosc*, **26**, 911–6.

Fischer G, Braun S, Thissen R and Dott W (2006) 'FT-IR spectroscopy as a tool for rapid identification and intra-species characterization of airborne filamentous fungi', *J Microbiol Methods*, **64**(1), 63–77.

Galvis-Sanchez A C, Barros A S and Delgadillo I (2008) 'Method for analysis dried vine fruits contaminated with ochratoxin A', *Anal Chim Acta*, **617**, 59–63.

Ghosh P K and Jayas D S (2009) 'Use of spectroscopic data for automation in food processing industry', *Sensing Instrum Food Qual Safety*, **3**, 3–11.

Greene R V, Gordon S H, Jackson M A and Bennett G A (1992) 'Detection of fungal contamination in corn: potential of FTIR-PAS and -DRS', *J Agric Food Chem*, **40**, 1144–9.

Hahn F (2002) 'Fungal spore detection on tomatoes using spectral Fourier signatures', *Biosys Eng*, **81**(3), 249–59.

Heux L, Brugnerotto J, Desbrieres, J, Versali M F and Rinaudo M (2000) 'Solid state NMR for determination of degree of acetylation of chitin and chitosan', *Biomacromolecules*, **1**, 746–51.

Hurst, W J (1994), 'Ultraviolet/visible methods', in *Spectroscopic Techniques for Food Analysis*, Wilson R H (ed.). VCH Publishers, New York.

Kos G, Krska R, Lohninger H and Griffiths P R (2004) 'A comparative study of mid-infrared diffuse reflection (DR) and attenuated total reflection (ATR) spectroscopy for the detection of fungal infection on RWA2-corn', *Anal Bioanal Chem*, **378**, 159–66.

Mazzola E P (1998) 'Applications of nuclear magnetic resonance spectroscopy to food additive and contaminant problems', in *Spectral Methods in Food Analysis*, Mossoba M M (ed.). Marcel Dekker, New York, 89–123.

Mossoba M M (1998), *Spectral Methods in Food Analysis*, Marcel Dekker, New York.

Naes T, Isaksson T, Fearn T and Davies T (2002) *A User-friendly Guide to Multivariate Calibration and Classification*, NIR Publications, Chichester, UK.

Narvankar D S, Singh C B, Jayas D S and White N D G (2009) 'Assessment of soft X-ray imaging for detection of fungal infection in wheat', *Biosys Eng*, **103**, 49–56.

Nesheim S and Stack M E (2001), 'Preparation of mycotoxin standards', in *Mycotoxin Protocols*, Trucksess M W and Pohland A E (eds). Springer, New York, 31–6.

Osborne B G, Fearn T and Hindle P H (1993), *Near Infrared Spectroscopy in Food Analysis*, Longman Singapore, Singapore.

Pasquini C (2003) 'Near infrared spectroscopy: fundamentals, practical aspects and analytical applications', *J Brazilian Chem Soc*, **14**(2), 198–219.

Pavia D L, Lampman G M and Kriz G S (1979), *Introduction to Spectroscopy: A Guide for Students of Organic Chemistry*, W B Saunders Company, Toronto, ON.

Pearson T C, Wicklow D T, Maghirang E B, Xie F and Dowell F E (2001) 'Detecting aflatoxin in single corn kernels by using transmittance and reflectance spectroscopy', *Trans ASAE*, **44**, 1247–54.

Petterson H and Aberg L (2003) 'Near infrared spectroscopy for determination of mycotoxins in cereals', *Food Control*, **14**, 229–32.

Santos C, Fraga M E, Kozakiewicz Z and Lima N (2010) 'Fourier transform infrared as a powerful technique for the identification and characterization of filamentous fungi and yeasts', *Res Microbiol*, **161**(2), 168–75.

Schroeder F C, Gibson D M, Churchill A C L, Sojikul P, Wursthorn E J, Krasnoff S B and Clardy J (2007) 'Differential analysis of 2D NMR spectra: New natural products from a pilot-scale fungal extract library', *Angew Chem Int Ed*, **46**, 901–904.

Svendsen A and Frisvad J C (1994) 'A chemotaxonomic study of the terverticillate penicillia based on high performance liquid chromatography of secondary metabolities', *Mycological Res*, **98**(11), 1317–28.

Tripathi S and Mishra H N (2009) 'A rapid FT-NIR method for estimation of aflatoxin B_1 in red chili powder', *Food Control*, **20**, 840–6.

Wang D, Dowell F E, Ram M S and Schapaugh W T (2004) 'Classification of fungal-damaged soybean seeds using near-infrared spectroscopy', *Int J Food Prop*, **7**, 75–82.

WHO (2007), *Food Safety and Foodborne Illness*, Geneva, World Health Organization. Available from: http://www.who.int/mediacentre/factsheets/fs237/en/ (Accessed 25 May 25, 2010).

Williams P and Norris K (2001), *Near-infrared Technology in the Agriculture and Food Industries*, American Association of Cereal Chemists, St. Paul, Minnesota.

Wilson R H (1994), *Spectroscopic Techniques for Food Analysis*, VCH, New York.

Wilson R H, and Goodfellow B J (1994), 'Mid-infrared spectroscopy', in *Spectroscopic Techniques for Food Analysis*, Wilson R H (ed.). VCH, New York.

Index

Lightning Source UK Ltd.
Milton Keynes UK
UKOW051700210313

207983UK00001B/77/P